# A Primer of
# Multivariate Statistics

# A Primer of Multivariate Statistics

*RICHARD J. HARRIS*
*University of New Mexico*

ACADEMIC PRESS    New York   San Francisco   London

A Subsidiary of Harcourt Brace Jovanovich, Publishers

ACADEMIC PRESS, INC.
111 Fifth Avenue, New York, New York 10003

*United Kingdom Edition published by*
ACADEMIC PRESS, INC. (LONDON) LTD.
24/28 Oval Road, London NW1

Library of Congress Cataloging in Publication Data

Harris, Richard J
    A primer of multivariate statistics.

    Bibliography: p.
    1.    Multivariate analysis.    I.    Title.
QA278.H35          519.5'3          74-1636
ISBN 0−12−327250−5

# CONTENTS

## Chapter 3   HOTELLING'S $T^2$
### Tests on One or Two Mean Vectors

## Chapter 4   MULTIVARIATE ANALYSIS OF VARIANCE
### Differences among Several Groups on Several Measures

## *Chapter 8*   THE FOREST REVISITED

## *Digression 1*   FINDING MAXIMA AND MINIMA OF POLYNOMIALS

## *Digression 2*   MATRIX ALGEBRA

## *Digression 3*   SOLUTION OF CUBIC EQUATIONS

# PREFACE

The purpose of this text is to provide an introduction to the more common multivariate techniques for students interested in research in sociology, psychology, anthropology, and other social and behavioral sciences who have had at least one introductory course in statistics. Several excellent comprehensive textbooks on multivariate statistical techniques (for example, Anderson, 1958; Dempster, 1969; Morrison, 1967) were already available when preparation of the Primer was first considered. But the author's experience—as instructor of a course in multivariate statistics—has been that none of these otherwise excellent texts could be understood by mathematically unsophisticated students, despite the surprising simplicity of the basic goals of and the approaches taken to achieve these goals in multivariate statistics.

Each of the commonly employed techniques in multivariate statistics is a very straightforward generalization of some univariate statistical tool, with the single variable being replaced by a linear combination of several original variables. All the high-powered matrix algebra and calculus that makes multivariate statistics seem so inaccessible is really just a substitute for trial-and-error selection of weights for the different variables entering into this linear combination so as to achieve the "best" possible set of weights according to some predetermined criterion of optimality. It thus seemed that by concentrating on "heuristic" descriptions of what all the math was designed to accomplish, for example, by emphasizing optimality criteria rather than the mathematical tools used in satisfying them, it should be possible to make multivariate statistical techniques available to a much broader audience than has been the case. The Primer is my personal test of this hypothesis.

No prior knowledge of matrix algebra or calculus is assumed, though background in these areas should speed the reader's progress through the book. These and other mathematical tools are introduced, as is the use of canned computer programs, as the student encounters the need for them. Moreover, the text provides verbal explanations of what each application of one of these tools accomplishes, so that the mathematical details may be skipped—at the usual expense of an increase in the level of faith required from the reader—without loss of continuity.

The approach will often seem verbose and inelegant to the mathematically sophisticated reader, but it is hoped that even he will find something of value in the examples used to illustrate and/or challenge common interpretations of the mathematical results on which multivariate statistics is based. For instance, in what sense do principal components "explain" the intercorrelations among a set of variables? What grounds are there for the assertion that inserting communalities in the main diagonal before factoring a correlation matrix "removes" error variance from the resulting factors? Is canonical correlation equivalent, as Harman (1967) claims, to computing a Pearson product–moment correlation between the first principal factors of the two sets of variables?

The material in the Primer is probably too much for a one-semester course, though I myself present the material in one semester by giving a very cursory treatment of factor analysis. If two semesters are available, the appropriate dividing line would probably fall between Chapters 5 and 6, since the first five chapters deal with relationships between two or more sets of variables, while Chapters 6 and 7 concentrate on relationships within a single set of variables. Whether one semester or two, however, Chapters 1, 2, and 8 should be included—Chapters 1 and 8 for obvious reasons, and Chapter 2 because multiple regression is used as the vehicle (hopefully an already relatively familiar one) for demonstrating the usefulness of and introducing the reader to various fundamental tools such as matrix algebra, differential calculus, and the use of "canned" computer programs.

Readers of this text who have also read Morrison's (1967) work will notice a similarity in range and organization of topics which is far from accidental. Morrison's text provided me with my introduction to multivariate statistics. I hope the Primer has succeeded in making the underlying simplicity of multivariate approaches as apparent to nonmathematical readers as Morrison's text does to those who feel comfortable with matrix algebra and mathematical derivations.

# ACKNOWLEDGMENTS

It is questionable whether I or Professor John Cotton of the Graduate School of Education, University of California at Santa Barbara, devoted more total hours to this book. Professor Cotton was one of the two initial reviewers of the manuscript in its incomplete, dittoed form, and he continued in that capacity through all subsequent stages, carefully pointing out grammatical, orthographical, mathematical, statistical, stylistic, and logical errors in such profusion as to demonstrate once again that an author is a poor choice to proofread his own manuscript. Professor William L. Sawrey of the Psychology Department, California State College at Hayward, also participated in the initial review.

Providing an independent and very valuable final review of the Primer was Professor Robert Pruzek of the Department of Educational Psychology and Statistics, State University of New York at Albany. Professor Pruzek had the courage to use a prepublication draft of the Primer in his own multivariate statistics class. A number of his students provided me with written comments and corrections, those of Charles G. Ackerson and J. R. Tibbetts being especially helpful. Another of Professor Pruzek's students, Ronald S. Burt, provided such a cogent review (together with his own monograph on path analysis, which monograph, due to time pressures, has had less impact on this Primer than it probably should) that I mistook him for a colleague and fellow faculty member in preparing the initial version of this preface.

My interactions with students in my own multivariate statistics class these past five spring semesters provided both the impetus and the "experimental material" for this book. The criticisms (and, bless them, encouragement) provided by the students in the past three classes while suffering patiently as guinea pigs through three successive drafts of the manuscript have been invaluable. Special mention should be made of the conscientious, independent checking of problem answers, derivations, etc., by Ralph Liguori, Nancy Lee, and Renée Silleroy.

I have in addition had the benefit of numerous face-to-face discussions of statistical and general methodological issues with Professor Thomas P. Friden, a colleague in the Psychology Department here at UNM, as well as feedback from colleagues in other departments. Professor Stoughton Bell, Director of the Computing Center, and Mr. William H. McMahan, Manager of Applications Programming, have been very generous in providing time for and advice on computer use, as well as commenting on Digression 4.

I have no doubt whatever that the Primer is a sounder text because of the advice given me by those mentioned above, and I certainly could not reject the hypothesis that it would have been even better had I adopted a larger proportion of their suggestions.

My very meager secretarial talents have been greatly supplemented by a number of people. The major burdens, however, have fallen on Mrs. Eleanor Orth, Mrs. Elna Parks, and Mrs. Doris Basile, who have suffered patiently my (literally) last-minute need for copies of the section of this Primer which my multivariate class was to discuss that morning, as well as shepherding the manuscript through successively refined drafts.

Finally, my admiration and gratitude go to Dr. Mary B. Harris of the Department of Educational Foundations, College of Education, University of New Mexico, whose example of research productivity despite less than optimal reinforcement contingencies and whose gentle nagging were instrumental in spurring me finally to complete this book. And thanks to Christopher and Jennifer Harris for providing excuses for deadlines I would have missed anyway.

# 1 | INTRODUCTION
## The Forest before the Trees

## 1.0  Why Statistics?

This text and its author subscribe to the importance of sensitivity to data and of the wedding of humanitarian impulse to scientific rigor. Therefore, it seems appropriate to discuss the present author's conception of the role of statistics in the overall research process. This section assumes familiarity with the general principles of research methodology. It also assumes some acquaintance with the use of statistics, especially significance tests, in research. If this latter is a poor assumption, the reader is urged to delay reading this section until after having read Section 1.2.

Statistics is a form of social control over the professional behavior of researchers. The ultimate justification for any statistical procedure lies in the kinds of research behavior it encourages or discourages.

In their descriptive applications, statistical procedures provide a set of tools for summarizing efficiently the researcher's empirical findings in a form which is more readily assimilated by his audience than would be a simple listing of the raw data. The availability and apparent utility of these procedures generate pressure on researchers to employ them in reporting their results, rather than relying on a more discursive approach. On the other hand, most statistics summarize only certain aspects of the data, so that automatic (for example, computerized) computation of standard (cookbook?) statistics without the intermediate step of "living with" the data in all its concrete detail may lead to overlooking important features of these data. A number of authors (cf. especially Anscombe, 1973; Daniel and Wood, 1971; and Mosteller and Tukey, 1968) have recently offered suggestions (such as increased emphasis on graphs of individual data points and plotting these individual points as functions of supposedly irrelevant factors such as order of collection) for preliminary screening of the data so as to ensure that the summary statistics finally selected are truly relevant to the data at hand.

The inferential applications of statistics provide protection against the universal tendency to confuse aspects of the data which are unique to the particular sample of subjects, stimuli, and conditions involved in a study with general properties of the populations from which these subjects, stimuli, and conditions were sampled. For instance, it often proves difficult to convince a subject who has just been through a binary prediction experiment involving, say, predicting which of two lights will be turned on in each of several trials that the experimenter had used a random number table in selecting the sequence of events. Among researchers this tendency expresses itself as a tendency to generate complex *post hoc* explanations of their results which must be constantly revised since they are based in part on aspects of the data which are highly unstable from one replication of the study to the next. Social control is obtained over this tendency and the "garbage rate" for published studies reduced by requiring of each experimenter that he first demonstrate that the null hypothesis of no true relationship in the population between his independent and dependent variables is a clearly *im*plausible explanation of his obtained results before he (or she) is permitted to foist upon his colleagues more complex explanations. The scientific community generally accepts this control over their behavior because:

(a)   Bitter experience with reliance on investigators' informal assessment of the generalizability of their results has shown that some formal system of "screening" data is needed.

(b)   The particular procedure just (crudely) described, which we may label the *null-hypothesis-testing procedure*, has the backing of a highly developed mathematical model which, if certain plausible assumptions are met, provides rather good quantitative estimates of the relative frequency with which we will falsely reject (Type I error) or falsely fail to reject (Type II error) the null hypothesis, together with clear rules as to how to adjust our criteria for rejection and/or the conditions of our experiment (such as number of subjects) so as to set these two "error rates" at prespecified levels.

(c)   The null-hypothesis-testing procedure is usually not a particularly irksome one, thanks to the ready availability, for a broad class of research situations, of formulas, tables, and computer programs to aid in carrying out the testing procedure.

However, acceptance is not uniform. Bayesian statisticians, for instance, point out that the mathematical model underlying the null-hypothesis-testing procedure fits the behavior and beliefs of researchers quite poorly. No one, for example, seriously entertains the null hypothesis, since almost any treatment or background variable will have *some* systematic (though possibly miniscule) effect. Similarly, no scientist accepts or rejects a conceptual hypothesis on the basis of a single study, but instead withholds final judgment until a given phenomenon has been replicated on a variety of studies. Bayesian approaches to statistics thus picture the researcher as beginning each study with some degree of confidence in a particular hypothesis and then revising his confidence in (the subjective probability of) that hypothesis up or down, depending on the outcomes of the study. This is almost certainly a more realistic description of research behavior than that provided by the null-hypothesis-testing model. However, the superiority of the Bayesian approach as a descriptive theory of research behavior does not necessarily make it a better prescriptive (normative) theory than the null-hypothesis-testing model. Bayesian approaches are not nearly so well developed as are null-hypothesis-testing procedures, and they demand more from the user in terms of mathematical sophistication. They also demand more in terms of ability to specify the nature of the researcher's subjective beliefs about the hypotheses about which his study is designed to provide evidence. Further, this dependence of the results of Bayesian analyses on the investigator's subjective beliefs means that Bayesian "conclusions" may differ among different investigators examining precisely the same data, and that the mathematical and computational effort expended by the researcher in performing a Bayesian analysis may be relatively useless to those of his readers who hold different prior subjective beliefs about the phenomenon. (The "may's" in the preceding sentence derive from the fact that many Bayesian procedures are robust across a wide range of prior beliefs.) For these reasons, Bayesian approaches are not employed in the present Primer. Press (1972) has incorporated Bayesian approaches wherever possible.

An increasingly "popular" objection to null hypothesis testing centers around the contention that these procedures have become *too* readily available, thereby seducing researchers and journal editors into allowing the tail (inferential aspects of statistics) to wag the dog (the research process considered as a whole). Many statisticians have recently appealed for one or more of the following reforms in the null-hypothesis-testing procedure:

(a) Heavier emphasis should be placed on the *descriptive* aspects of statistics, including as a minimum careful examination of the individual data points before, after, during, or possibly instead of application of "cookbook" statistical procedures to them.

(b) The research question should dictate the appropriate statistical analysis, rather than letting the ready availability of a statistical technique generate a search for research paradigms which fit the assumptions of that technique.

(c) Statistical procedures should be developed which are less dependent on distributional and sampling assumptions, such as randomization tests (which compute the probability that a completely random reassignment of observations to groups would produce as large an apparent discrepancy from the null hypothesis as does sorting scores on the basis of the treatment or classification actually received by the subject) or jackknifing tests (which are based on the stability of the results under random deletion of portions of the data). These procedures have only recently become viable as high-speed computers have become readily available.

(d) ·Our training of behavioral scientists (and our own practice) should place more emphasis on the hypothesis-*generating* phase of research, including use of *post hoc* examination of the data gathered while testing one hypothesis as a stimulus to theory revision or origination.

Kendall (1968), Mosteller and Tukey (1968), Anscombe (1973), and McGuire (1973) can serve to introduce the reader to this growing "protest literature."

The ultimate answer to all of these problems with traditional statistics is probably what Skellum (1969) refers to as the "mathematization of science" as opposed to the (cosmetic?) application of mathematics to science in the form of very broad statistical models. Mathematization of the behavioral sciences involves the development of mathematically stated theories leading to quantitative predictions of behavior and to derivation from the axioms of the theory of a multitude of empirically testable predictions. An excellent discussion of the advantages of this approach to theory construction *vis à vis* the more typical verbal–intuitive approach is provided by Estes

(1957), and illustrations of its fruitfulness are provided by Atkinson, Bower, and Crothers (1965), Cohen (1963), and Rosenberg (1968). The impact on statistics of adoption of mathematical approaches to theory construction is at least twofold:

(a)   Since an adequate mathematical model must account for variability as well as regularities in behavior, the appropriate statistical model is often implied by the axioms of the model itself, rather than being an *ad hoc* addition to what in a verbal model are usually overtly deterministic predictions.

(b)   Because of the necessity of amassing large amounts of data in order to test the quantitative details of the model's predictions, ability to reject the overall null hypothesis is almost never in doubt, and attention turns instead to measures of goodness of fit relative to other models and to less formal criteria such as the range of phenomena handled by the model and its ability to generate counterintuitive (but subsequently confirmed) predictions.

As an example of the way in which testing global null hypotheses becomes an exercise in belaboring the obvious when a math model is being tested, Harris' (1969) study of the relationship between rating scale responses and pairwise preference probabilities in personality impression formation found, for the most adequate model, a correlation of .9994 between predicted and observed preference frequency for the pooled "psychometric function" (which was a plot of the probability of stating a preference for the higher rated of two stimuli as a function of the absolute value of the difference in their mean ratings). On the other hand, the null hypothesis of a perfect relationship between predicted and observed choice frequencies could be rejected at beyond the $10^{-30}$ level of significance, thanks primarily to the fact that the pooled psychometric function was based on 3500 judgments from each of 19 subjects.

Nevertheless, the behavioral sciences are too young and researchers in these sciences are as yet too unschooled in mathematical approaches to hold out much hope of mathematizing all research in these sciences; nor would such complete conversion to formal, mathematically stated theories be desirable. Applying math models to and collecting massive amounts of data on some phenomenon can be very uneconomical if done prematurely—that is, before a number of at least partly exploratory studies inspired by imperfectly formulated theories have begun to narrow somewhat the range of plausible theories and to point to the most profitable research paradigms in the area. Skellum (1969), who was cited earlier as favoring mathematization of science, also argues for a very broad view of what constitute acceptable models in the early stages of theory construction and testing. Null hypothesis testing can be expected to continue for some decades as a quite serviceable and quite necessary method of social control for most research efforts in the behavioral sciences. Null hypotheses may well be merely convenient fictions, but no more opprobrium need be attached to their fictional status than to the ancient logical technique of *reductio ad absurdum* which null hypothesis testing extends to probabilistic, inductive reasoning.

As will become obvious in the remaining sections of this chapter, the present Primer attempts in part to plug a "loophole" in the current social control exercised over researchers' tendency to read too much from their data. (It also attempts to add a collection of rather powerful techniques to the descriptive tools available to behavioral researchers. Van de Geer (1971) has in fact written a textbook on multivariate statistics which deliberately omits any mention of their inferential applications.) The above discussion has hopefully convinced the reader—including those who, like the author, favor the eventual mathematization of the behavioral sciences—that this will lead to an increment in the quality of research, rather than merely prolonging un-necessarily the professional lives of researchers who, also like the author, find it necessary to carry out exploratory research on verbally stated theories with quanti-ties of data small enough to make the null hypothesis an embarrassingly plausible explanation of our results.

## 1.1   Why Multivariate Statistics?

As the name implies, *multivariate statistics* refers to an assortment of descriptive and inferential techniques that have been developed to handle situations where sets of variables are involved either as predictors or as measures of performance. If researchers were sufficiently narrowminded, or theories and measurement techniques so well developed or nature so simple as to dictate a single independent variable and a single outcome measure as appropriate in each study, there would be no need for multivariate statistical techniques. In the classic scientific experiment involving a single manipulated variable (all other variables being eliminated as possible causal factors either through explicit experimental control or through the statistical control provided by randomization) and a single outcome measure, questions of *patterns* or *optimal combinations* of variables scarcely arise; nor do the problems of *multiple comparisons* becloud the interpretations of any *t*-test or correlation coefficient used to assess the relation between the independent (or predictor) variable and the depen-dent (or outcome) variable. However, researchers in all of the sciences—behavioral, biological, or physical—have long since abandoned sole reliance on the classic, univariate design for very excellent reasons. It has become abundantly clear that a given experimental manipulation (for example, positively reinforcing a class of re-sponses on each of *N* trials) will affect many somewhat different but partially cor-related aspects (for example, speed, strength, consistency, "correctness") of the organism's behavior. Similarly, many different pieces of information about an appli-cant (for example, the socioeconomic status of his parents; his high school grades in math, English, and journalism; his attitude toward authority) may be of value in predicting his grade point average in college, and it is necessary to consider how to combine all of these pieces of information into a single "best" prediction of college performance. (It is widely known—and will be demonstrated in our discussion of multiple regression—that the predictors having the highest correlations with the criterion variable when considered singly might contribute very little to that combin-ation of the predictor variables which correlates most highly with the criterion.)

As is implicit within the discussion of the preceding paragraph, multivariate statistical techniques accomplish two general kinds of things for us, these two func-tions corresponding roughly to the distinction between *descriptive* and *inferential* statistics. On the descriptive side, they provide rules for combining the variables in an optimal way. What is meant by "optimal" varies from one technique to the next, as will be made explicit in the next section. On the inferential side, they provide a solution to the *multiple comparison* problem. Almost any situation in which multivariate techniques are applied could be analyzed through a series of univariate significance tests (for example, *t*-tests), one such univariate test for each possible combination of one of the predictor variables with one of the outcome variables. However, since *each* of the univariate tests is designed to produce a significant result $\alpha \times 100\%$ of the time (where $\alpha$ is the "significance level" of the test) when the null hypothesis is correct, the probability of having at least one of the tests produce a significant result when in fact nothing but chance variation is going on, increases rapidly as the number of tests increases. It is thus highly desirable to have a means of explicitly controlling the experimentwise error rate. Multivariate statistical techniques provide this control.[1] They also, in many cases, provide for *post hoc* comparisons which explore the statistical significance of various possible explanations of the overall statistical significance of the relationship between the predictor and outcome variables.

The descriptive and inferential functions of multivariate statistics are by no means independent. Indeed, one whole class of multivariate techniques (known as *least-squares* or *union–intersection tests*) base their tests of significance on the sampling distribution of the "combined variable" which results when the original variables are combined according to the criterion of optimality employed by that particular multivariate technique. The present notes will concentrate on this approach whenever possible because of its high interpretability. When we achieve statistical significance, we also automatically know which combination of the original set of variables provided the strongest evidence against the overall null hypothesis, and we can compare its efficacy with that of various *a priori* or *post hoc* combining rules. However, in a few situations only likelihood ratio tests (which treat

---

[1] A possible counterexample is provided by the application of analyses of variance (whether univariate or multivariate) to studies employing factorial designs. In the univariate case, for instance, a $k$-way design ($k$ experimental manipulations combined factorially) produces a summary table involving $2^{k-1}$ terms—$k$ main effects and $\binom{k}{2}$ two-way interactions, etc.—each of which typically yields a test having a Type I error rate of .05. The experimentwise error rate is thus $1 - (.95)^{2^{k-1}}$. The usual multivariate extension of this analysis suffers from exactly the same degree of compounding of Type I error rate, since a separate test of the null hypothesis of no difference among the groups on any of the outcome measures (dependent variables) is conducted for each of the components of the among-group variation corresponding to a term in the univariate summary table. In the author's view, this simply reflects the fact that the usual univariate analysis is appropriate only if each of the terms in the summary table represents a truly *a priori* (and, perhaps more important, theoretically relevant) comparison among the groups, so that treating each comparison independently makes sense. Otherwise, the analysis should be treated (statistically at least) as a one-way Anova followed by specific contrasts employing Scheffé's *post hoc* significance criterion (Winer, 1971, p. 198), which holds the experimentwise error rate (that is, the probability that any one or more of the potentially infinite number of comparisons among the means yields rejection of $H_0$ when it is in fact true) to at most $\alpha$. Scheffé's procedure is in fact a multivariate technique designed to take into account a multiplicity of *independent* variables. Further, Anova can be shown to be a special case of multiple regression, with the overall $F$ testing the hypothesis of no true differences among any of the population means corresponding to the test of the significance of the multiple regression coefficient. This overall test is not customarily included in Anova applied to factorial designs, which simply illustrates that the existence of a situation for which multivariate techniques are appropriate does not guarantee that the researcher will apply them.

the assessment of the probability of the data, given that the null hypothesis holds, as separate from the problem of finding optimal combining rules) have been developed.

An important difference between descriptive and inferential uses of multivariate techniques is that the former require no assumptions whatever about the distributions from which the observations are sampled, while most of the commonly used multivariate significance tests, like their univariate counterparts, are based on homogeneity and normality assumptions. It is well known that the common univariate tests are generally quite robust, that is, insensitive to any but gross departures from normality and homogeneity assumptions. Less is known about the robustness of the corresponding multivariate techniques, but the evidence so far suggests a similar degree of robustness.

## 1.2 A Heuristic Survey of Statistical Techniques

Most statistical formulas come in two versions: a *heuristic* version which is suggestive of the rationale underlying that procedure, and an algebraically equivalent *computational* version which is more useful for rapid and accurate numerical calculations. For instance, the formula for the variance of a sample of observations can be written either as

$$\sum (X - \bar{X})^2/(N - 1),$$

which makes clear the status of the variance as approximately the mean of the squared deviations of the observations about the sample mean, or as

$$[N \sum X^2 - (\sum X)^2]/[N(N - 1)],$$

which produces the same numerical value as the heuristic formula but avoids having to deal with negative numbers and decimal fractions. The present survey of statistical techniques concentrates—in the same way as heuristic versions of statistical formulas—on what it is that each technique is designed to accomplish, rather than on the computational tools used to reach that goal.

In general in any situation we will have $m$ predictor variables, which may be either manipulated by the experimenter or simply measured by him, and $p$ outcome variables. The distinctions among various statistical techniques are based primarily on the size of these two sets of variables, and are summarized in Table 1.1.

The techniques listed in the right-hand column of Table 1.1 have not been explicitly labeled as univariate or multivariate. Here, as is so often the case, common useage and logic clash. If we define multivariate techniques (as the present author prefers) as those applied when two or more variables are employed either as independent (predictor) or dependent (outcome) variables, then higher-order Anova must certainly be included, and possibly one-way Anova as well. However, Anova and Ancova involving only one control variable and thus only bivariate regression has traditionally not been included in treatments of multivariate statistics. (Note that "bivariate regression" is a univariate statistical technique.) If we define multivariate techniques as those applied to situations involving multiple *dependent* measures, then

*Table 1.1*

*Statistical Techniques[a]*

| Predictor variables | Outcome variables | Criterion for combining variables | Name of technique(s) |
| --- | --- | --- | --- |
| 1 manipulated, 2 levels | 1 | — | *t*-test |
| 1 manipulated, > 2 levels | 1 | — | One-way analysis of variance (Anova) |
| ≥ 2 manipulated | 1 | — | Higher-order Anova |
| 1 measured | 1 | — | Pearson *r*, bivariate regression |
| ≥ 2 measured | 1 | Maximize correlation of combined variable with outcome variable | Multiple correlation, multiple regression analysis (MRA) |
| Mixture of manipulated, measured | 1 | Maximize correlation of measured variables with outcome variable within levels of the manipulated variable(s) | Analysis of covariance (Ancova) |
| 1 manipulated, 2 levels | ≥ 2 | *t*-ratio on combined variable as large as possible | Hotelling's $T^2$, discriminant analysis |
| 1 manipulated, > 2 levels | ≥ 2 | Maximize one-way $F$ on combined variable | One-way multivariate analysis of variance (Manova) |
| ≥ 2 manipulated | ≥ 2 | $F$ on combined variable maximized for each effect | Higher-order Manova |
| Mixture of manipulated, measured | ≥ 2 | Maximize correlation of measured variables with outcome variables within levels of the manipulated variable(s) | Multivariate Ancova (Mancova) |
| ≥ 2 measured | ≥ 2 | Maximize correlation of combined predictor variable with combined outcome variable | Canonical correlation, canonical analysis (Canona) |
| ≥ 2 measured | | Maximize variance of combined variable | Principal component analysis (PCA) |
| ≥ 2 measured | | Reproduce intercorrelation matrix of original variables as accurately as possible | Factor analysis (FA) |

[a] In every technique except factor analysis, the combined variable is simply a weighted sum of the original variables. The predictor–outcome distinction is irrelevant in principal component analysis and in factor analysis, where only one set of variables is involved. In Manova, Ancova, Canona, PCA, and FA, several combined variables are obtained, each successive combined variable being selected in accordance with the same criterion used in selecting the first combined variable, but only after the preceding variables have been selected, and subject to the additional constraints that it be uncorrelated with the preceding combined variables.

multiple regression is excluded, which again runs counter to common practice. To some extent the usual practice of including multiple regression but excluding Anova (despite the fact that the latter can be viewed as a special case of the former) from "multivariate statistics" reflects the historical accident that multivariate techniques have come to be associated with correlational research while Anova is used primarily by "hard-nosed" researchers who conduct laboratory experiments. It is hoped that the present text will weaken this association. However, the issues involved in higher-order Anova have become so involved and so specialized, and so many excellent texts are available in this area of statistics (for example, Winer, 1971) that detailed treatment of Anova is omitted from the present Primer on practical, rather than logical, grounds.

※[2] At a practical level, what distinguishes techniques considered univariate from those considered multivariate is whether matrix manipulations (cf. Sections 1.3 and 2.2.3, and Digression 2) are required. As we shall see in subsequent chapters, any of the traditional approaches to statistics can be defined in terms of matrix operations. Univariate statistical techniques are simply those in which the matrix operations "degenerate" into scalar operations, that is, matrix operations on $1 \times 1$ matrices. (If you have had no previous experience with matrix operations, you will probably find that this definition makes more sense to you after reading Chapter 2.) This "operational" definition of multivariate statistics resolves the paradox mentioned in the preceding paragraph that Anova is generally considered a univariate technique, despite being a special case of multiple regression, which is commonly viewed as a multivariate technique. When the matrix formulas for a multiple regression analysis (MRA) are applied to a situation in which the predictor variables consist of $k - 1$ dichotomous group-membership variables ($X_i = 1$ if the subject is a member of group $i$; 0 otherwise; for $i = 1, 2, \ldots, k - 1$), and when these group-membership variables are uncorrelated (as they are when the same number of observations is obtained for each possible combination of levels of the independent variables), the matrix operations lead to relatively simple single-symbol expressions which are the familiar Anova computational formulas, so that matrix operations and the connection with MRA need not arise at all in carrying out the Anova. However, unequal cell sizes in a higher-order Anova lead to correlated group-membership variables, and force us to employ the matrix manipulations of standard MRA in order to obtain uncorrelated, least-squares estimates of the population parameters corresponding to the effects of our treatments. Thus higher-order Anova with equal cell sizes is a univariate technique, while higher-order Anova with unequal cell sizes is a multivariate technique.

If the union–intersection approach (versus the likelihood-ratio approach) to multivariate significance tests is adopted, the operational definition just offered translates into the statement that multivariate techniques are those in which one or more combined variables—linear combinations of the original variables—are derived on the basis of some criterion of optimality.

---

[2] ※ Paragraphs or whole sections preceded by the symbol ※ may be skipped without loss of continuity.

### 1.2.1  Student's *t*-Test

The most widely known of the experimental researcher's statistical tools is Student's *t*-test. This is used in two situations. The first exists when we have a single sample of observations whose mean is to be compared with an *a priori* value. Two kinds of evidence bear on the null hypothesis that the observed value of $\bar{X}$ arose through random sampling from a population having a mean of $\mu_0$:

(a)  the difference between $\bar{X}$ and $\mu_0$, and
(b)  the amount of fluctuation we observe among the different numbers in the sample.

The latter provides an estimate of $\sigma$, the standard deviation of the population from which the sample was presumably drawn, and thus—via the well-known relationship between the variance of a population and the variance of the sampling distribution of means of samples drawn from that population—provides an estimate of $\sigma_{\bar{X}}$. The ratio between these two figures is known to follow Student's *t*-distribution with $N - 1$ degrees of freedom when the parent population has a normal distribution or when the size of the sample is sufficiently large for the central limit theorem—which says, among other things, that means of large samples are normally distributed irrespective of the shape of the parent population (provided only that this parent population have finite variance)—to apply. The hypothesis that $\mu = \mu_0$ is therefore rejected at the $\alpha$ significance level if and only if the absolute value of

$$t = (\bar{X} - \mu)/\sqrt{\sum (X - \bar{X})^2/[N(N - 1)]}$$

is greater than the $100(1 - \alpha/2)$ percentile of Student's *t*-distribution with $N - 1$ degrees of freedom.

The second, and more common situation in which the *t*-test is used arises when the significance of the difference between two independent sample means is to be evaluated. The basic rationale is the same, except that now a pooled estimate of $\sigma^2$, namely

$$s_c^2 = [\sum (X_1 - \bar{X}_1)^2 + \sum (X_2 - \bar{X}_2)^2]/(N_1 + N_2 - 2),$$

is obtained. This pooled estimate takes into consideration only variability within each group and is therefore independent of any differences among the population means. Thus large absolute values of

$$t = (\bar{X}_1 - \bar{X}_2)/\sqrt{s_c^2(1/N_1 + 1/N_2)}$$

lead to rejection of the null hypothesis that $\mu_1 = \mu_2$.

### 1.2.2  One-Way Analysis of Variance

When more than two levels of the independent variable (and thus more than two groups of *S*s) are involved, the null hypothesis that $\mu_1 = \mu_2 = \cdots = \mu_k$ is tested by comparing a *direct measure* of the variance of the *k* sample means, namely

$$\sum (\bar{X}_j - \bar{X})^2/(k - 1),$$

with an *indirect estimate* of how much these $k$ means would be expected to vary if in fact they simply represented $k$ random samples from the same population: $s_w^2/n$, where $n$ is the size of each group and $s_w^2$ is computed in the same way $s_c^2$ was for the *t*-test, except that $k$ instead of only 2 sums of squared deviations must be combined. Large values of $F$, the ratio between the direct estimate and the indirect estimate (or, in Anova terminology, between the between-group mean square and the within-group mean square), provide evidence against the null hypothesis, with a table of the $F$-distribution providing the precise value of $F$ needed to reject the overall null hypothesis at, say, the .05 level.

More formally, for those readers who benefit from formality, the between-group (better, among-group) mean square,

$$\text{MS}_a = n \sum (\bar{X}_j - \bar{\bar{X}})^2/(k-1),$$

has expected value

$$\sigma^2 + n \sum (\mu_j - \mu)^2/(k-1) = \sigma^2 + n\sigma_\tau^2,$$

where $n$ is the number of observations in each of the $k$ groups, and $\sigma_\tau^2$ represents the variance of the $k$ population means. (A more complicated constant replaces $n$ in the preceding expressions if sample sizes are unequal.) The within-group mean square, $\text{MS}_w = s_w^2$, has expected value $\sigma^2$, the variance common to the $k$ populations from which the samples were drawn. Under the null hypothesis, $\sigma_\tau^2 = 0$, so that both the numerator and denominator of $F = \text{MS}_a/\text{MS}_w$ have the same expected value. (The expected value of their ratio $E(F)$ is nevertheless not equal to unity but to $\text{df}_w/(\text{df}_w - 2)$, where $\text{df}_w$ is the number of degrees of freedom associated with $\text{MS}_w$, namely $N - k$, where $N$ is the total number of observations in all $k$ groups.)

If the assumptions of independently and randomly sampled observations, of normality of the distributions from which the observations were drawn, and of homogeneity of the variances of these underlying distributions are met, then $\text{MS}_a/\text{MS}_w$ is distributed as $F$. Numerous empirical sampling studies (for example, D. W. Norton's (1952) study) have shown this test to be very robust, with true Type I error rates differing very little from those obtained from the $F$-distribution even when population variances differ by as much as a ratio of 9 and even when the populations are as grossly nonnormal as the rectangular distribution. Much more detailed discussion of these and other matters involved in one-way Anova is available in the book by Winer (1971) or in his somewhat more readable 1962 edition.

Since nothing in this discussion hinges on the precise value of $k$, we might expect that the *t*-test for the difference between *two* means would prove to be a special case of one-way Anova. This expectation is correct, for when $k = 2$ the $F$-ratio obtained from Anova computations precisely equals the square of the *t*-ratio computed on the same data, and the critical values listed in the $\text{df} = 1$ column of a table of the $F$-distribution are identically equal to the squares of the corresponding critical values listed in a *t*-table. However, we can *not* generalize in the other direction by, say, testing every pair of sample means by an $\alpha$-level *t*-test, since the probability that at least one of these $k(k-1)/2$ pairwise tests would produce a significant result by chance is considerably greater than $\alpha$ when $k$ is greater than 2. The overall $F$-ratio thus provides protection against the inflation of the overall ("experimentwise") probability of a Type I error which running all possible $t$-tests (or simply conducting a $t$-test on the largest difference between any pair of sample means) would produce.

Once it has been concluded that the variation among the various group means represents something other than just chance fluctuation, there still remains the problem of specifying the way in which the various treatment effects contribute to the significant overall $F$-ratio: Is the difference between group 2 and all other groups the only difference of any magnitude, or are the differences between "adjacent" treatment means all about the same?

How best to conduct such specific comparisons among the various means, and especially how to assess the statistical significance of such comparisons, are the center of a lively debate among statisticians. The approach favored by the present author is Scheffé's contrast method. Scheffé's approach permits testing the significance of the difference between any two linear combinations of the sample means, for example,

$$\tfrac{1}{2}(\bar{X}_1 + \bar{X}_2) - \tfrac{1}{3}(\bar{X}_3 + \bar{X}_5 + \bar{X}_6)$$

or

$$(5\bar{X}_1 + 3\bar{X}_2 + \bar{X}_3) - (\bar{X}_4 + 3\bar{X}_5 + 5\bar{X}_6).$$

The only restriction is that the weights employed within each of the two linear combinations sum to the same number or, equivalently, that the sum of the weights (some of which may be zero) assigned to all variables sum to zero.

Moreover, the significance tests for these contrasts can be adjusted, simply through multiplication by an appropriate constant, to fix at $\alpha$ either the probability that a *particular* preplanned contrast will lead to a false rejection of the null hypothesis, or the probability that *any* one or more of the infinite number of comparisons that might be performed among the means will lead to a Type I error. Finally, associated with each contrast is a "sum of squares for contrast" which can be compared with the sum of the squared deviations of the group means about the grand mean of all observations to yield a precise statement of the percentage of the variation among the group means which is attributable to that particular contrast.

This "percentage of variance" interpretation is made possible by the fact that the sums of squares associated with any $k - 1$ *independent* contrasts among the means add to a sum identical to $SS_a$, the sum of squares associated with the total variation among the $k$ means. Independence, which simply requires that the two contrasts be uncorrelated, and thus each be providing unique information, is easily checked since two contrasts are independent if and only if the sum of the products of the weights they assign corresponding variables $\sum c_1 c_2$ is zero. (A slightly more complex check is necessary when sample sizes are unequal.)

The Scheffé contrast procedure may, however, be much too broad for the researcher who is interested only in certain kinds of comparisons among the means, such as a prespecified set of contrasts dictated by the hypotheses of the study, or all possible pairwise comparisons, or all possible comparisons of the unweighted means of any two distinct subsets of the means. For situations like these where the researcher can specify in advance of examining the data the size of the set of comparisons that are candidates for examination, the use of Bonferroni critical values (cf. Kirk, 1968) is appropriate and will often yield more powerful tests than the Scheffé approach. The Bonferroni approach simply adjusts the critical values of whatever test statistic is employed in such a way that $\sum_{i=1}^{b} \alpha_i = \alpha$, where $\alpha_i$ is the $\alpha$-level (Type I error rate, that is, probability of falsely rejecting the null hypothesis) for the $i$th of $b$ null hypotheses corresponding to the set of $b$ comparisons which include all comparisons the researcher might conceivably wish to make. This procedure sets at less than or equal to $\alpha$ the overall probability that any one or more of the $b$ null hypotheses will be rejected when in fact all $b$ of them are true. An advantage of this approach, in addition to the fact that it often yields more powerful tests than the Scheffé procedure, is that it permits the researcher to use less stringent $\alpha_i$s for (and thus provide relatively more powerful tests of) the contrasts of greatest interest and/or importance to him.

### 1.2.3 Hotelling's $T^2$

It is somewhat ironic that "hard-nosed" experimentalists, who are quick to see the need for one-way analysis of variance when the number of experimental manipulations (and thus of groups) is large, are often blithely unaware of the inflated error rates that result if $t$-test procedures are applied to more than one outcome measure. Any background variable (age, sex, socioeconomic status, etc.) which we employ as a basis for sorting subjects into two different groups or for distinguishing two sub-categories of some population almost inevitably has associated with it many different other variables, so that the two groups (subpopulations) so defined differ on a wide variety of dimensions. Similarly, only very rarely are the effects of an experimental manipulation (high versus low anxiety, low versus high dosage level, etc.) confined to a single behavioral manifestation. Even where our theoretical orientation tells us that the effect of the manipulation should be on a single conceptual dependent variable, there are usually several imperfectly correlated operational definitions of (means of measuring) that conceptual variable. For instance, learning can be measured by percentage correct responses, latency of response, trials to criterion, etc.; and hunger can be measured by subjective report, strength of stomach contractions, hours since last consumption of food, etc. Thus all but the most trivial studies of the difference between two groups do in fact produce several different measures on each of the two groups, unless the researcher has deliberately chosen to ignore all but one measure in order to simplify his data analysis task. Hotelling's $T^2$ provides a means of testing the overall null hypothesis that the two populations from which the two groups were sampled do not differ in their means on *any* of the $p$ measures. Heuristically, the method by which this is done is quite simple. The $p$ outcome measures from each subject are combined into a single number by the simple process of multiplying the subject's score on each of the original variables by a weight associated with that variable and then adding these products together. (The same weights are, of course, used for every subject.) Somewhat more formally, the combined variable for each subject is defined by

$$W_i = w_1 X_{i,1} + w_2 X_{i,2} + \cdots + w_p X_{i,p}.$$

A univariate *t*-ratio based on the difference between the two groups in their mean values of *W* is computed. A second set of weights which may produce a larger *t*-ratio is tried out, and this process is continued until that set of weights which makes the univariate *t*-ratio on the combined variable as large as it can possibly be has been found. (Computationally, of course, the mathematical tools of calculus and matrix algebra are used to provide an analytic solution for the weights, but this simply represents a short cut to the search procedure described here.) This set of optimal weights—optimal in the sense that it maximally discriminates between the two groups—is called the *discriminant function*, and has considerable interpretative value in its own right. The square of the *t*-ratio which results from the maximizing procedure is known as $T^2$, and its sampling distribution has been found to be identical in shape with a member of the family of *F*-distributions. Thus it provides a test of the overall null hypothesis of identical "profiles" in the populations from which the two samples were drawn.

### 1.2.4  One-Way Multivariate Analysis of Variance

Just as Hotelling's $T^2$ provides a generalization of the univariate *t*-test, so does one-way multivariate analysis of variance provide a generalization of one-way (univariate) analysis of variance. One-way Manova is applicable whenever there are several groups of *S*s, with more than one measure being obtained on each *S*. Just as in the case of $T^2$, the test of the overall null hypothesis is accomplished by reducing the set of *p* measures on each *S* (his *response profile* or *outcome vector*) to a single number by applying a linear combining rule

$$W_i = \sum_j w_j X_{i,j}$$

to his scores on the original outcome variables. A univariate *F*-ratio is then computed on the combined variable, and new sets of weights are selected until that set of weights which makes the *F*-ratio as large as it possibly can be has been found. This set of weights is the (multiple) discriminant function, and the largest possible *F*-value is the basis for the significance test of the overall null hypothesis. The distribution of such maximum *F*-statistics (known, because of the mathematical tools used to find the maximum, as "greatest characteristic root" statistics) is complex, and deciding whether to accept or reject the null hypothesis requires the use of a series of tables based on the work of Heck (1960), Pillai (1965, 1967), and Venables (1973), and are found in Appendix A. Note that the most widely used computer program for Manova (BMDX69) employs an entirely different criterion (the likelihood ratio, defined as the ratio of the determinants of the between- and within-group covariance matrices) in assessing statistical significance.

### 1.2.5  Higher-Order Analysis of Variance

The procedures of one-way analysis of variance can be applied no matter how large the number of groups. However, most designs involving a large number of groups of $Ss$ arise because the researcher wishes to assess the effects of two or more independent (manipulated) variables by administering all possible combinations of a particular level of one variable with particular levels of the other variables. An example of such a factorial design would be a study in which various groups of $Ss$ perform tasks of high, low, or medium difficulty after having been led to be slightly, moderately, or highly anxious, thus leading to a total of nine groups (lo–slight, lo–mod, lo–hi, med–slight, med–mod, med–hi, hi–slight, hi–mod, and hi–hi). The investigator will almost always wish to assess the *main effect* of each of his independent variables (that is, the amount of variation among the means for the various levels of that variable, where the mean for each level involves averaging across all groups receiving that level), and will in addition wish to assess the *interactions* among the different independent variables (the extent to which the relative spacing of the means for the levels of one variable differs, depending on which level of a second variable the subjects received). For this reason, inclusion of contrasts corresponding to these main effects and interactions as preplanned comparisons has become almost automatic, and computational procedures have been formalized under the headings of two-way (two independent variables), three-way (three independent variables), etc., analysis of variance. In addition, researchers and statisticians have become aware of the need to distinguish between two kinds of independent variables: *fixed* variables (for example, sex of $S$), identified as those variables whose levels are selected on *a priori* grounds and for which $E$ would therefore select precisely the same levels in any subsequent replication of the study, and *random* variables (for example, litter from which $S$ is sampled), whose levels in any particular study are a random sample from a large population of possible levels and for which $E$ would probably select a new set of levels were he to replicate the study. Our estimate of the amount of variation among treatment means we would expect if the null hypothesis of no true variation among the population means were correct will depend on whether we are dealing with a fixed or a random independent variable. This selection of an appropriate "error term" for assessing the statistical significance of each main effect and each interaction in a study which involves both fixed and random independent variables can be a matter of some complexity. Excellent treatments of this and other problems involved in higher-order analysis of variance are available in several texts (cf. especially Winer, 1971 and Hays and Winkler, 1971), and will not be a focus of this text.

### 1.2.6   Higher-Order Manova

What *is* important to realize is that a multivariate counterpart exists for every univariate analysis of variance design, with the multivariate analysis always involving (heuristically) a search for that linear combination of the various outcome variables which makes the univariate $F$-ratio (computed on that single, combined variable) for a particular main effect or interaction as large as possible. Note that the linear combination of outcome variables used in evaluating the effects of one independent variable may not assign the same weights to the different outcome variables as the linear combination used in assessing the effects of some other independent variable in the same analysis. This may or may not be desirable from the viewpoint of interpretation of results. Methods are, however, available for ensuring that the same linear combination is used for all tests in a given Manova.

A multivariate extension of Scheffé's contrast procedure is available which permits the researcher to make as many comparisons among linear combinations of independent and/or outcome variables as he wishes with the assurance that his probability of falsely identifying any one or more of these comparisons as representing true underlying population differences is lower than some prespecified value.

Finally, it should be pointed out that there exists a class of situations in which univariate analysis of variance and Manova techniques are in "competition" with each other. These are situations in which each subject receives more than one experimental treatment, or is subjected to the same experimental treatment on a number of different occasions. Clearly the most straightforward approach to such *within-subject* and *repeated measures* designs is to consider the set of $S$'s responses to the different treatments administered to him, or his responses on the successive trials of the experiment, as a single outcome vector for that $S$, and then apply Manova techniques to the $N$ outcome vectors produced by the $N$ subjects. However, if certain rather stringent conditions are met (these conditions involving primarily the uniformity of the correlations between $Ss$' responses to the various possible pairs of treatments or trials), it is possible to use the computational formulas of univariate Anova in conducting the significance tests on the within-subjects variables. This has the advantage of unifying somewhat the terminology employed in describing the between- and within-subjects effects, though it is probable that the popularity of the univariate approach to these designs rests more on its avoidance of matrix algebra.

### 1.2.7  Pearson *r* and Bivariate Regression

There are many situations where we wish to assess the relationship between some outcome variable (for example, attitude toward authority) and a predictor variable (for example, age) which the researcher either cannot or does not choose to manipulate. The researcher's data will consist of a number of pairs of measurements (one score on the predictor variable and one on the outcome variable), each pair having been obtained from one of a (hopefully random) sample of $Ss$. Note that bivariate regression is a univariate statistical technique. The predictor variable (traditionally labeled as $X$) will be of value in predicting scores on the outcome variable (traditionally labeled as $Y$) if, *in the population*, it is true either that high values of $X$ are consistently paired with high values of $Y$, and low $X$ values with low $Y$ values; or that $Ss$ having high scores on $X$ are consistently found to have low scores on $Y$ and vice versa. Since the unit and origin of our scales of measurement are usually quite arbitrary, "high" and "low" are defined in terms of the number of standard deviations above or below the mean of all observations a given observation falls. In other words, the question of the relationship between $X$ and $Y$ is converted to a question as to how closely $z_x = (X - \mu_x)/\sigma_x$ matches $z_y = (Y - \mu_y)/\sigma_y$, on the average. The classic measure of the degree of relationship between the two variables in the population is

$$\rho = \sum z_x z_y / N,$$

where the summation includes all $N$ members of the population. $S$s for whom $X$ and $Y$ lie on the same side of the mean contribute positive cross-product $(z_x z_y)$ terms to the summation, while $S$s whose $X$ and $Y$ scores lie on opposite sides of their respective means contribute negative terms, with $\rho$ (the Pearson product–moment coefficient of correlation) taking on its maximum possible value of $+1$ when $z_x = z_y$ for all $S$s and its minimum possible value of $-1$ when $z_x = -z_y$ for all $S$s. The researcher obtains an estimate of $\rho$ through the simple expedient of replacing $\mu$s and $\sigma$s with sample means and standard deviations, whence

$$r = \sum z_x z_y/(N - 1) = \sum (X - \overline{X})(Y - \overline{Y})/\sqrt{\sum (X - \overline{X})^2 \cdot \sum (Y - \overline{Y})^2}.$$

The Pearson $r$ (or $\rho$) is very closely related to the problem of selecting the best linear equation for predicting $Y$ scores from $X$ scores. If we try out different values of $b_0$ and $b_1$ in the equation, $\hat{z}_y = b_0 + b_1 z_y$ (where $\hat{z}_y$ is a predicted value of $z_y$), we will eventually discover (as we can do somewhat more directly by employing a bit of calculus) that the best possible choices for $b_0$ and $b_1$ (in the sense that they make the sum of the squared differences between $\hat{z}_y$ and $z_y$ as small as possible) are $b_0 = 0$ and $b_1 = r$. (The formula for predicting raw scores on $Y$ from raw scores on $X$ follows directly from the $z$-score version of the *regression equation* by substituting $(X - \overline{X})/s_x$ and $(Y - \overline{Y})/s_y$ for $z_x$ and $z_y$, respectively, and then isolating $\hat{Y}$, the predicted raw score on $Y$, on the left-hand side of the equation.) Furthermore, the total variation among the $Y$ scores $s_y^2$ can be partitioned into a component attributable to $Y$'s relationship to $X$, $s_{\hat{y}}^2$, and a second component representing the mean squared error of prediction $s_{y \cdot x}^2$. (The square root of this latter term is called the "standard error of estimate.") It turns out that the ratio between $s_{\hat{y}}^2$ and $s_y^2$—that is, the percentage of the variance in $Y$ which is "accounted for" by knowledge of $S$'s score on $X$—is identically equal to $r^2$, the square of the Pearson product–moment correlation coefficient computed on this sample of data. (This suggests, incidentally, that $r^2$ is a more directly meaningful measure of the relationship between the two variables than is $r$. Tradition and the fact that $r^2$ is always smaller and less impressive than $r$ have, however, ensconced $r$ as by far the more commonly reported measure.) This close relationship between the correlation coefficient and the linear regression equation (linear because no squares or higher powers of $z_x$ are employed) should alert us to the fact that $r$ is a measure of the degree of *linear* relationship between the two variables, and may be extremely misleading if we forget that the relationship may involve large nonlinear components. The problem of possible curvilinearity is a persistent (though often unrecognized) one in all applications of correlation and regression.

Quite large sample values of $r$ can of course arise as a consequence of random fluctuation even if $\rho$ in the population from which the $S$s were sampled truly equals zero. The null hypothesis that $\rho = 0$ is tested (and hopefully rejected) via a statistic having a $t$-distribution, namely

$$t = r/\sqrt{(1 - r^2)/(N - 2)}.$$

The researcher seldom has to employ this formula, however, as tables of critical values of $r$ are readily available.

The Pearson $r$ was developed specifically for normally distributed variables. A number of alternative measures of correlation have been developed for situations where one or both measures are dichotomous or rank–order variables, but with few exceptions (such as are provided by Kendall's tau and the Goodman–Kruskal measures of relationship between nominal variables), these alternative measures are either numerically identical to Pearson $r$ blindly applied to the decidedly nonnormal ranks or 0–1 measures (Spearman's coefficient of rank–order correlation, phi coefficient, point biserial coefficient) or are exercises in wishful thinking whose numerical value is equal to the value of Pearson $r$ which "would have" been obtained had the normally distributed measures presumably underlying the imperfect data at hand been available for analysis (tetrachoric coefficient, biserial $r$).[3] None of the descriptive aspects of Pearson $r$ are altered by applying it to "imperfect" data, and the critical values for testing the hypothesis of no relationship that have been derived for the alternative measures are often nearly identical to the corresponding critical values for the Pearson $r$ applied to impeccable data. The present text therefore will not use the more specialized name (phi coefficient, for example) when Pearson $r$s are computed on dichotomous or "merely" ordinal data.

One particularly revealing application of Pearson $r$ to dichotomous data arises (or should arise) whenever a $t$-test for the difference between two independent means is appropriate. If the dependent measure is labeled as $Y$ and each subject is additionally assigned an $X$ score of one or zero, depending on whether he was a member of the first or the second group, then the $t$-test of the significance of the difference between the $r$ computed on these pairs of scores and zero will be numerically identical to the usual $t$-test for the significance of the difference between $\bar{X}_1$ and $\bar{X}_2$. This is intuitively reasonable, since in testing the hypothesis that $\mu_1 = \mu_2$ we are essentially asking whether there is any relationship between our independent variable (whatever empirical operation differentiates group 1 from group 2) and our outcome measure. More importantly, $r^2 = t^2/[t^2 + (N-2)]$ provides a measure of the percentage of the total variation among the $S$s in their scores on the dependent measure which is attributable to (predictable from a knowledge of) the group to which they were assigned. Many an experimentalist who sneeers at the low percentage of variance

[3] The nature of the situations in which the Pearson $r$, as compared with alternative measures of correlation such as Spearman's rho, is appropriate is sometimes stated in terms of the strength of measurement required for the operations involved in computing $r$ (for example, addition, subtraction, multiplication of deviation scores) to be meaningful. Stevens (1951, 1968) and others (for example, Siegel, 1956; Senders, 1958) would restrict "parametric" statistics such as the Pearson $r$ and the $t$-test to situations where the data have been shown to have been obtained through a measurement process having at least interval scale properties (versus "weaker" scales such as ordinal measurement). A number of other authors (for example, Anderson, 1961, and an especially entertaining article by Lord, 1953) have pointed out that statistical conclusions are valid whenever the distributions of numbers from which the data are sampled meet the assumptions (typically, normality and homogeneity of variance) used to derive the particular techniques being applied, irrespective of the measurement process which generated those numbers. Moreover, the validity of parametric statistics (for example, the correspondence between actual and theoretically derived Type I error rates) is often affected very little by even relatively gross departures from these assumptions. The author agrees with this latter position, though he (along with most of Stevens's critics) agrees that level of measurement may be very relevant to the researcher's efforts to fit his statistical conclusions into a theoretical framework. These points are discussed further in Chapter 8.

accounted for by some paper-and-pencil measure would be appalled at the very low values of $r^2$ underlying the highly significant (statistically) values of $t$ produced by his experimental manipulations.


### 1.2.8 Multiple Correlation and Regression

We often have available several measures from which to predict scores on some criterion variable. In order to determine the best way of using these measures, we collect a sample of Ss for each of whom we have scores on each of the predictor variables as well as a score on the criterion variable. We wish to have a measure of the overall degree of relationship between the set of predictor variables and the criterion (outcome) measure. Mystical, complex, and Gestalt-like as such an overall measure sounds, the coefficient of multiple correlation (multiple $R$) is really nothing more than the old, familiar Pearson $r$ between $Y_i$ (our outcome measure on each subject $i$) and $W_i = \sum_j w_j X_{i,j}$, a linear combination of subject $i$'s scores on the predictor variables. The particular weights employed are simply those we discover (by trial and error or somewhat more efficiently by calculus and matrix algebra) produce the largest possible value of $R$. These weights turn out to be identical to those values of the $b_j$ in the *multiple regression* equation

$$\hat{Y} = b_0 + b_1 X_1 + b_2 X_2 + \cdots + b_m X_m$$

which make $\sum_i (\hat{Y}_i - Y_i)^2$ as small as possible. The null hypothesis that the population value of multiple $R$ is truly zero is tested by comparing the amount of variance in $Y$ accounted for by knowledge of scores on the $X$s with the amount left unexplained. When $Y$ and the $X$s have a multivariate normal distribution, the ratio

$$F = (N - k)(R^2)/[(k - 1)(1 - R^2)]$$

has an $F$-distribution with $k - 1$ and $N - k$ degrees of freedom. Specific comparisons among particular $b$-coefficients are provided by procedures analogous to Scheffé's contrast methods for Anova. Indeed, as the last few sentences might suggest, Anova is just a special case of multiple regression where the predictor variables are all dichotomous, zero–one group membership variables, with $X_{i,j} = 1$ if and only if subject $i$ is a member of group $j$.

The interaction terms of higher-order Anova are derived from multiple regression either through contrasts computed on the $b_j$s or by including as separate predictor variables cross-products, $X_i X_j$, of the group-membership variables. This approach to taking into account possible nonadditive relationships among the predictor variables is of course available (though seldom used) when the predictor variables are all measured, rather than manipulated. Similarly, curvilinear relationships between any of the predictor variables and $Y$ can be provided for (and tested for significance) by including as additional predictors the squares, cubes, etc., of the original predictors. Unfortunately, adding these variables cuts into the degrees of freedom available for significance tests. Even if the population value of multiple $R$ is truly zero, the expected value of the *sample R* is equal to $\sqrt{k/(N - 1)}$, where $k$ is the number of predictor variables and $N$ is the number of $S$s. Thus as the number of predictor variables approaches the number of subjects, it becomes increasingly difficult to discriminate between chance fluctuation and true relationships.

### 1.2.9  Canonical Correlation

More often than the statistical techniques used in the literature would suggest, we have several outcome measures as well as several predictor variables. An overall measure of the relationship between the two sets of variables is provided by canonical $R$, which is simply a Pearson $r$ calculated on two numbers for each subject: $W_i = \sum w_j X_{ij}$ and $V_i = \sum v_i Y_{ij}$, where the $X$s are predictor variables and the $Y$s are outcome measures. Heuristically, the $w_j$ and the $v_j$, which are called *canonical coefficients* for the predictor and outcome measures, respectively, are obtained by trying out different sets of weights until that pair of sets of weights which produces the maximum possible value of canonical $R$ $(R_c)$ has been obtained.

Actually, when each set of variables contains two or more variables the analysis of their interrelationships need not stop with the computation of canonical $R$. A second pair of sets of weights can be sought which will produce the maximum possible Pearson $r$ between the two combined variables, subject to the constraint that these two new combined variables be uncorrelated with the first two combined variables. A total of $\min(p, m) = p$ or $m$ (whichever is smaller) pairs of sets of weights, each with a corresponding coefficient of canonical correlation, each uncorrelated with any of the preceding sets of weights, and each accounting for successively less of the variation shared by the two sets of variables, can be derived.

Just as univariate analysis of variance can be considered a special case of multiple regression, so multivariate analysis of variance can be considered a special case of canonical correlation—the special case that arises when one of the two sets of variables consists entirely of dichotomous group-membership variables.

It is possible to generalize canonical correlation—at least in its significance-testing aspects—to situations where more than two sets of variables are involved. For instance, we might have three paper-and-pencil measures of authoritarianism, ratings by four different psychiatrists of the degree of authoritarianism shown by our $S$s in an interview, and observations based on a formal recording system such as Bales' inter-action process analysis of the behavior of each $S$ in a newly formed committee. We could then test the overall null hypothesis that these three sets of measures of authoritarianism has nothing to do with each other. Unfortunately the generalization is based on the likelihood-ratio approach to significance testing and therefore tells us nothing about the magnitude of any relationship which does exist, or about ways to combine the variables within each set so as to maximize some overall measure of the degree of interrelationship among the sets.

### 1.2.10  Analysis of Covariance

The analyses we have considered so far have included only two kinds of variables: predictor variables and outcome variables. (This predictor–outcome distinction is blurred in many situations where we are interested only in a measure of the overall relationship between the two sets, such as the Pearson $r$ or canonical $R$.) All of the variables in either set are of interest for the contribution they make to strengthening the interrelations between the two sets. In many situations, however, a third kind of variable is included in the analysis. We might, rather uncharitably, call these variables "nuisance" or "distractor" variables. A more neutral term, and the one we shall adopt, is "covariates." A covariate is a variable which is related to (covaries with) the predictor and/or the outcome variables, and whose effects we wish to control for statistically as a substitute for experimental control. For instance, we may be interested in studying the relative efficacy of various diet supplements on muscular development, and therefore administer each supplement to a different randomly selected group of adult males. Then, after each group had used the supplement for, say, six months, we would measure, for example, the number of pounds of pressure each could generate on an ergograph (a hand-grip type of measuring device). We could then run a one-way analysis of variance on the ergograph scores. However, we know that there are large individual differences in muscular strength, and while the procedure of randomly assigning groups will prevent these individual differences from exerting any systematic bias on our comparisons among the different supplements, the high variability among our $S$s in how much pressure they can exert *with or without* the diet supplement will provide a very noisy "background" (high error variability) against which only extremely large differences in the effectiveness of the supplements can be detected. If this variability in postsupplement scores due to individual differences in the genetic heritage and/or past history of our subjects could somehow be removed from our estimate of error variance, we would have a more precise experiment which would be capable of reliably detecting relatively small differences among the effects of the various diet supplements. This is what analysis of covariance (Ancova) does for us.

Ancova is essentially just a two-step analysis of variance consisting of a multiple regression analysis of the relationship between the covariates and the outcome variable, followed by an Ancova on the difference between each $S$'s score and that score predicted for him on the basis of his score (or scores) on the covariate(s). In the first step, that regression equation which best (in the least-squares sense) predicts $S$s' scores on the outcome variable is constructed as described in the section on multiple regression. Since any true differences among the various groups would obscure differences due to the covariates (just as the effects of the covariates becloud our assessment of the effects of the experimental manipulations), the regression analysis is actually performed on the difference between $S$'s score on the outcome variable and the mean of his group. A *corrected score* for each $S$ is obtained by subtracting the outcome score predicted for him on the basis of his scores on the predictor variable (that is, his predicted deviation from the mean of his group plus the mean of his group) from his actual outcome score. The corrected scores (one such score per $S$) are then subjected to the usual analysis of variance procedures.

The regression analysis phase of Ancova uses up some of the degrees of freedom in our data. To see this, we need only consider the case where the number of covariates is only one less than the number of $S$s, so that a multiple $R$ of 1.0 is obtained, that is, each $S$'s score on the outcome variable is "predicted" perfectly from knowledge of his scores on the covariates, regardless of whether there is any true relationship in the population between the covariates and the outcome measure. Each $S$'s corrected score would then be identically equal to the mean for his group, and we would have no way of estimating the true population variance of such corrected scores. In order, therefore, for Ancova to produce the greater sensitivity to between-group differences which is its major purpose, the true relationship between the covariates and the outcome variable must be sufficiently large to compensate for the loss of degrees of freedom entailed in introducing these covariates into the analysis.

Multivariate analysis of covariance (Mancova) is a straightforward extension of (univariate) Ancova, and consists of a multivariate analysis of variance performed on a *set* of corrected scores, each corrected score having resulted from a separate regression analysis of the relationship between the covariates and that particular outcome measure.

There would be no need for a separate discussion of covariance analyses were the relationship between Anova and multiple regression and the relationship between Manova and canonical correlation better known. Precisely the same analysis results whether the methods discussed in the previous few paragraphs are employed or, instead, the covariates and appropriate group-membership variables are simply included in the set of predictor variables and a multiple regression or canonical correlation analysis is conducted. One of the infrequently cited properties of regression analysis is that the expected value of any particular $b$-coefficient derived in the analysis is a function only of the population value of the regression coefficient for that particular variable and not of the population parameters representing the effects of any other variable. As a special case of this property, when a group-membership variable and a covariate are included in the same regression analysis, the resulting estimate of the effect of membership in that group is independent (in the sense described above) of, has had removed, is free of, the effects of the covariate on the $S$'s performance. This independence property is also of potentially great value in factorial design experiments where the various groups are of unequal size. One consequence of the unequal group sizes is that the means for the various levels of one independent variable are no longer independent of the means for the various levels of other manipulated variables. Since the estimates of the main effects of the various independent variables are, in analysis of variance, based solely on these means, the estimates of the main effects are interdependent. A high mean for $S$s receiving the highest dosage level, as compared with other dosages may, for instance, be attributable solely to the fact that more $S$s who had been made highly anxious (as opposed to slightly or moderately anxious $S$s) received the high dosage than any other dosage, and *anxiety* is the important variable. This interdependence among estimates of treatment effects is completely absent when a multiple regression including both covariates and group membership variables is performed on the data.

One common situation arises in which Ancova, Anova, and Manova are all in "competition." This occurs when, as in the ergograph example used to introduce Ancova, each $S$ is measured both before and after administration of some set of treatments (different programs of instruction, various drugs, assorted persuasive communications, etc.). Probably the most common approach is to conduct an Anova on *change scores*, that is, on the difference between the pretreatment and posttreatment score for each *S*. Alternately, the two scores for each *S* could be considered as an outcome *vector* and a Manova run on these outcome vectors. Finally, Ancova could be employed, treating the baseline score as the covariate and the posttest score as the single outcome measure. The major disadvantage of Manova in this situation is that it is sensitive to all differences among the groups on either measure, whereas *S*s will usually have been assigned to groups at random so that we know that any between-group differences in baseline scores are the result of random fluctuation. As between the change score approach and Ancova, Ancova is usually the superior choice on purely statistical grounds. It makes use of the correction formula which removes as much as possible of the error variance, while the correction formula used in the change score analysis is an *a priori* one which by definition will effect less of a reduction in error variance. Furthermore, change scores have a built-in tendency, regardless of the nature of the manipulated variable or the outcome measure, to be negatively correlated with baseline scores, with *S*s (or groups) scoring high on the premeasure tending to produce low change scores. The corrected scores derived from Ancova are uncorrelated with baseline performance. The change score is, however, a very "natural" measure, and has the big advantage of ready interpretability. [However, see the work of Harris (1963) for discussions of the often paradoxical problems of measuring change.]

### 1.2.11 Principal Component Analysis

The statistical techniques we have discussed so far all involve relationships *between* sets of variables. In principal component analysis and in factor analysis, however, we concentrate on relationships *within* a single set of variables. Both of these techniques—although authors differ in whether they consider principal component analysis to be a type of factor analysis or a distinct technique—can be used to reduce the *dimensionality* of the set of variables, that is, to describe the *S*s in terms of their scores on a much smaller number of variables with as little loss of information as possible. If this effort is successful, then the new variables (components, factors) can be considered as providing a description of the "structure" of the original set of variables.

The "new variables" derived from the original ones by principal component analysis (PCA) are simply linear combinations of the original variables. The *first principal component* is that linear combination of the original variables which maximally discriminates among the subjects in our sample, that is, whose sample variance is as large as possible. Heuristically we find this first PC by trying out different values of $w_j$ in the formula $W_i = \sum w_j X_{i,j}$ and computing the variance of the $S$s' $W_i$ scores until we have uncovered that set of weights which makes $s_{W_i}^2$ as large as it can possibly be. Actually, we have to put some restriction on the sets of weights we include in our search (the usual one being that the sum of the squares of the weights in any set must equal unity), since otherwise we could make the variance of $W_i$ arbitrarily large by the simple and uninformative expedient of using infinitely large $w_j$'s. At any rate, once we have found the first principal component we begin a search for the *second principal component*. This is that linear combination of the original variables which has the largest possible sample variance, subject to the two constraints that

(a)   the sum of the squares of the weights employed equal unity, and
(b)   scores on the second PC be uncorrelated with scores on the first PC.

This process is continued until a total of $p$ principal components have been "extracted" from the data, with each successive PC accounting for as much of the variance in the original data as possible subject to the condition that scores on that PC be uncorrelated with scores on any of the preceding PCs. The sum of the variances of $Ss$' scores on the $p$ different PCs will exactly equal the sum of the variances of the original variables. Moreover, as implied by the increasing number of restrictions put on the "permissible" sets of weights included in the search for the PCs, each successive PC will have a lower associated sample variance than its predecessor. If the original variables are highly interrelated, it will turn out that the first few PCs will account for a very high percentage of the variation on the original variables, so that each $S$'s scores on the remaining PCs can be ignored with very little loss of information. This condensed set of variables could then be used in all subsequent statistical analyses, thus greatly simplifying the computational labor involved in, for example, multiple regression or Manova. Since the weights assigned the original variables in each principal component are derived in accordance with purely internal criteria rather than on the basis of their relationship with variables outside this set, the value of multiple $R$ or the significance level obtained in a Manova based on $Ss$' scores on the reduced set of PCs will inevitably be less impressive than if scores on the original variables had been employed. In practice, however, the loss of power is rather minimal. A more important justification than simplified math for preceding statistical analyses on a set of variables by a PCA of that set is the fact that the PCs are uncorrelated, thus eliminating duplication in our interpretations of the way $Ss$' responses on this set of variables are affected by or themselves affect other sets of variables.

The results of a PCA are, unlike the results of any of the other techniques we have discussed, affected by linear transformations of the original variables. It must therefore be decided by the investigator before he begins a PCA whether to retain the original units of the variables, or to standardize the scores on the different variables by converting to $z$-scores (or by using some other standardization procedure). This decision will naturally hinge on the meaningfulness of the original units of measurement.

One of the kinds of information which is produced as an offshoot of PCA is the correlation between each original variable and each of the principal components. As might be anticipated, the correlation of variable $i$ with $PC_j$ is closely related to the weight $b_{ij}$ assigned variable $i$ in the linear combination which constitutes the $j$th principal component, being in fact equal to $b_{ij}(s_{PC_j}/s_i)$, where $s_i$ is the sample standard deviation of original variable $X_i$ and $s_{PC_j}$ is the sample standard deviation of the $j$th principal component. By examining those original variables which correlate most highly with each principal component, labels suggestive of the meaning of each PC can be developed. The PCs, when so interpreted, can be seen as providing a description of the original variables, with their complex intercorrelations, in terms of a set of uncorrelated latent variables (factors) which might have generated them. Just as the weights assigned the original variables in $PC_j$ describe the PCs in terms of the original variables, so these $m$ linear equations could be solved (treating the PCs as known this time, and the original variables as unknowns) to produce, for each original variable, an expression for computing each $S$'s score on that original variable as a linear combination of his scores on the $m$ principal components. This information, together with the variance of each PC, would in turn be sufficient to compute the correlations between any two original variables, even if we did not have access to any $S$'s actual score on any variable or any PC. (This follows from the well-known relationship between the variance of a linear combination of variables and the variances of the individual variables entering into that linear combination.)

## 1.2.12    Factor Analysis

Notice that the "explanation" of the relationships among the original variables provided by PCA is not a particularly parsimonious one, since $m$ PCs are needed to reproduce the intercorrelations among the original $m$ variables. (The fact that the PCs are uncorrelated and ranked in order of percentage of variance accounted for may, however, make it a very useful explanation.) If we eliminate all but the first few PCs, we obtain a more parsimonious description of the original data, but at the expense of possible *systematic* error in our reproduction of the intercorrelations, since there may be one or two variables that are so much more highly related to the "missing" PCs than to those that are included as to make our estimates of the intercorrelations of other variables with these one or two highly dependent on the omitted data. Note also that PCA uses all of the information about every variable, though it is almost certain that some of the variation in $S$s' scores on a given variable is "unique" variance, attributable to influences which have nothing to do with the other variables in the set. We might suspect that we could do a better job of explaining the relationships among the variables if this unique variance could somehow be excluded from the analysis. Note also that the criterion used to find the PCs ensures that each successive PC accounts for less of the variance among the original variables than its predecessors. The investigator may, however, have strong grounds for suspecting that the "true" factors underlying the data are all of about equal importance—or for that matter that these latent variables are not uncorrelated with each other, as are the PCs.

*Factor analysis* (FA) refers to a wide variety of techniques which correct for one or more of these "shortcomings" of PCA. All factor analysis models have in common the explicit separation of unique variance from common variance, and the assumption that the intercorrelations among the $m$ original variables are generated by some smaller number of latent variables. Depending on how explicit the researcher's preconceptions about the nature of these underlying variables are, each original variable's *communality* (the percentage of its variance which is held in common with other variables) may either be produced as an offshoot of the analysis or may have to be specified in advance in order to arrive at a factor analytic solution. A factor analytic solution always includes a table indicating the correlation (loading) of each original variable with (on) each latent variable (factor), this table being referred to as the *factor structure*. The usual choice is between advance specification of the number of factors, with the analysis yielding communalities, and advance specification of communalities, with the analysis yielding the number of factors.

What is gained by employing FA versus PCA is the ability to reproduce the original pattern of intercorrelations from a relatively small number of factors, without the systematic errors produced when components are simply omitted from a PCA. What is lost is the straightforward relationship between $Ss$' scores on the various factors and their scores on the original variables. In fact, estimating $Ss$' scores on a given factor requires conducting a multiple regression analysis of the relationship between that factor and the original variables, with the multiple $R$ of these estimates being a decreasing function of the amount of unique variance in the system. [The indeterminacy in the relation between the factors and the original measures is, however, eliminated if Guttman's (1953, 1955, 1956) image analysis procedure—which consists of assuming the communality of each variable to be equal to its squared multiple correlation with the remaining variables—is employed.]

A second loss in almost all methods of factor analysis is the uniqueness of the solution. A given factor structure simply represents a description of the original intercorrelations in terms of a particular frame of reference. That pattern of intercorrelations can be equally well described by any other frame of reference employing the same number of dimensions (factors). Unless additional constraints besides ability to reproduce the intercorrelations are put on the analysis, any one of an infinite number of interrelated factor structures will be an acceptable solution. *Principal factor analysis* (PFA), which is essentially identical to PCA except for the exclusion of unique variance, assures the uniqueness of the factor structure it produces by requiring that each successive factor account for the maximum possible percentage of the common variance while still remaining uncorrelated with the preceding factors. The *triangular decomposition* method requires that the first factor be a general factor on which each variable has a nonzero loading, that the second factor involve all but one of the original variables, the third factor involve $m - 2$ of the original $m$ variables, etc.

All other commonly used factor methods use some arbitrary mathematical constraint (usually one which simplifies subsequent mathematical steps) to obtain a preliminary factor structure and then *rotate* the frame of reference until a factor structure which comes close to some *a priori* set of criteria for *simple structure*, is found. Probably the most common of all approaches to factor analysis is rotation of a solution provided by PFA, despite the fact that such rotation destroys the variance-maximizing properties of PFA. This is probably due primarily to the mathematical simplicity of PFA compared with other methods of obtaining an initial solution. The *maximum-likelihood method* generates factors in such a way as to maximize the probability that the observed pattern of correlations could have arisen through random sampling from a population in which the correlations are perfectly reproducible from the number of factors specified by the researcher. The *minimum residual* (minres) method searches for factor loadings which produce the smallest possible sum of squared discrepancies between the observed and reproduced correlations. The *multiple-group* method requires that $E$ specify in advance various subsets of the original variables, with the variables within each subset being treated as essentially identical. This is the only commonly used method which yields an initial solution involving correlated factors, the researcher's willingness to employ correlated factors in order to obtain a simpler factor structure usually being expressed in his choice of criteria for rotation of the frame of reference employed in the initial solution. The *centroid* method is a technique for obtaining an approximation to PFA when all computations must be done by hand or on a desk calculator, and is rapidly declining in use as computers become generally available. Finally, there is a growing and promising tendency to specify the characteristics of the factor structure as precisely as possible on purely *a priori* grounds, to use optimization techniques similar to the minres method to "fill in" remaining details, and then to use the final value of the optimization criterion as an indication of the adequacy of the researcher's initial assumptions.

As can be seen from the preceding discussion, factor analysis is a very complex area of multivariate statistics which shows rather low internal organization and minimal relationship to other multivariate statistical techniques. Adequate discussion of this important area requires a textbook to itself, and Harman (1967) and Mulaik (1972), among other authors, have provided such texts. The present text will be confined to a fairly full discussion of PFA and a cursory survey of the other techniques.

### 1.3   Learning to Use Multivariate Statistics

This section might well be headed, "Why the rest of the book?" It must be conceded that a full understanding of the heuristic descriptions of the preceding section would put the student at about the 90th percentile of, say, doctoral candidates in psychology in terms of ability to interpret multivariate statistics and in terms of lowered susceptibility to misleading, grandiose claims about multivariate techniques. This heuristic level of understanding would, further, provide a constant frame of reference to guide the student in actually carrying out multivariate analyses on his own data (which usually means submitting these data to a computer program)—suggesting, for instance, several checks on the results—and would establish a firm base from which to speculate about general properties that a particular technique might be expected to have and the consequences of changing some aspect of that technique. This first chapter is an important part of the book to which the student is urged to return often for review of the general goals and heuristic properties of multivariate statistical procedures. The points covered in this chapter will all be reiterated in later portions of the book, but they will nowhere be found in such compact form, so free of distracting formulas, derivations, etc., as in this first chapter.

Why, then, the rest of the book? In part because of the "catch" provided in the second sentence of this section by the phrase, "full understanding." As you no doubt noticed in your initial introduction to statistics, whether it came through a formal course or through self-instruction with the aid of a good text, the warm, self-satisfied feeling of understanding that sometimes follows the reading of the verbal explanation of some technique—perhaps supplemented by a brief scanning of the associated formulas—evaporates quickly (often producing a decided chill in the process) when faced with a batch of actual data to which the technique must be applied. True understanding of any statistical technique resides at least as much in the fingertips (be they caressing a pencil or poised over desk calculator or card punch keys) as in the cortex. The author has yet to meet anyone who could develop an understanding of any area of statistics without performing many analyses of actual sets of data— preferably data that were truly important to him.

In the area of multivariate statistics, however, computational procedures are so lengthy as to make hand computations impractical for any but very small problems. Handling problems of the sort you will probably encounter with "real" data de- mands the use of computer programs to handle the actual computations. One purpose of the rest of the book is therefore *to introduce you to the use of "canned" computer programs* and to discuss some of the more readily available programs for each technique. One approach to gaining an understanding of multivariate statistics would be a "black box" approach in which the heuristic descriptions of the preced- ing section are followed by a set of descriptions of computer programs and the student is urged to run as many batches of data through these programs as possible, with never a thought for what is going on between the input and output stages of the computer run. Such a procedure has two important shortcomings:

(a)   it puts the researcher completely at the mercy of the computer program- mer(s) who wrote the program and/or adapted it to the local computer system; and
(b)   it renders impossible an analytic approach to exploring the properties of a particular technique.

The first of these two shortcomings subsumes at least two limitations. First, no computer program is ever completely "debugged," and a knowledge of the steps which would have been followed had the researcher conducted the analysis "by hand" can be of great value in detecting problems in the computer's analysis of your data—which will of course contain that "one in a million" combination of condi- tions which makes this program "blow up" despite a long past history of troublefree performance. It is also of use in generating small test problems which can be computed both by hand and by the computer program as a check on the latter. Second, the person who programmed the particular analysis may not have provided explicitly for all of the subsidiary analyses in which you are interested, but a few supplementary hand calculations (possible only if you are familiar with the computa- tional formulas) can often be performed on the output that is provided by the program to give you the desired additional information. This is related to a third limitation, that it is difficult to "pick apart" a canned program and examine the relationships among various intermediate steps of the analysis, since such inter- mediate steps are generally not displayed.

With respect to the second shortcoming, computer programs can only tell us what happens when a particular set of data, with all of its unique features, is sub- jected to a particular analysis. This is a very important function, and when applied to a wide assortment of data can begin to suggest some general properties of that technique. However, we can never be certain that the property will not break down (fail to hold) for the next set of data except by examination of the formulas—in algebraic form—used by the computer. (Actually, the heuristic formulas or the stated goals of the technique or the mathematical derivation which gets us from the latter to the former may be of greater use than the computational formulas in deriving general properties. The main point is that "black box" use of canned programs is never a sufficient base for such generalizations.)

A second "service" provided by the present text is thus *the display of the formulas which would be used by the researcher were he to conduct an analysis by hand.* Actually, these formulas are presented in several different forms. For a few simple cases—those involving only two or three variables, and the case of uncorrelated variables—the computational formulas are presented in "basic" algebraic form (the form with which the reader is probably most familiar), in which each datum is represented by a separate algebraic symbol. Thus, for instance, the variance of the linear combination, $a_1 X_1 + a_2 X_2$, of two variables is written as $a_1^2 s_1^2 + a_2^2 s_2^2 + 2a_1 a_2 s_{12}$ in "basic" algebraic form. However, for cases involving many more variables, it becomes much more convenient to write the formulas in matrix algebraic form, in which each *set* of variables is represented by a separate symbol. Thus, for instance, the variance of a linear combination of variables, whether the number of variables is two or 397, can be written in matrix algebraic form as $\mathbf{a'Sa}$, whereas the corresponding basic algebraic expression would involve 210 terms if there were 20 original variables. The matrix algebraic form is usually easier to remember and intuitively more comprehensible than the basic algebraic form. A subsidiary task of the rest of the text will be *to develop a familiarity with the conventions governing the use of these highly compact matrix formulas and equations.* Finally, both the basic and matrix algebra expressions will usually be presented both in heuristic and in computational form.

For many readers, the three kinds of information mentioned so far—the goals of each technique, the formulas used in accomplishing these goals, and the computer programs which actually carry out the required computations—will provide a sufficient working understanding of multivariate statistics. However, an intrepid few—those who have had prior exposure to math or to statistics and/or who dislike the handwaving, black magic approach to teaching—will want to know something about the methods (primarily mathematical) used to get from the goals of a technique to the specific formulas used. The process of filling in this gap between goals and formulas, of justifying the latter step by step in terms of the former, is known as a proof. There will be several such proofs throughout the present text. It is important to keep in mind that the proofs are *not* essential to (though they may be of help in deepening) a working understanding of multivariate statistics. The reader who is willing to take the author's statements on faith should have no compunctions about skipping these proofs entirely (especially on a first reading). Paragraphs or entire sections which may be skipped without loss of continuity are indicated by the symbol ※ in front of the section number or before the first word of the paragraph. On the other hand, the attempt has been made to make even the proofs understandable to the mathematically naive reader. Where a high-powered mathematical tool such as differential calculus must be used, a "digression" is provided at the back of the book in which an intuitive explanation of and a discussion of the reasonableness of that tool are presented. Considerable emphasis is also put on analyzing the properties of each statistical technique through examination of the "behavior" of the computational formulas as various aspects of the data change.

Each chapter introduces a new technique or set of multivariate techniques. The general plan of each chapter is roughly as follows: First, the heuristic description of the technique is reiterated, with a set of deliberately simplified data serving to illustrate the kinds of questions for which the technique is useful. Next the formulas used in conducting an analysis by hand are derived and are applied to the sample data. The relationship between formulas and heuristic properties is explored, and some questions (both obvious and subtle) about the technique are posed and answered. The available computer programs are explained. The student is then asked to work a demonstration problem, that is, to go through the hand computations required for an analysis of some deliberately simplified data, performing a large number of different manipulations on various aspects of the data to demonstrate various properties of the technique. (Yes, Virginia, a $t$-test computed on the discriminant function really does give the same value as all those matrix manipulations used to find $T^2$.) These hand calculations are then compared with the results of feeding the data to a canned computer program. It has been the author's experience that at least $75\%$ of the real learning accomplished by students comes in the course of working these demonstration problems. Do not skip them.

There has been enough talk about how the text is going to go about teaching you multivariate statistics. To borrow one of Roger Brown's (1965) favorite Shakespearean quotes:

> Shall we clap into't roundly, without
> hawking or spitting or saying we are hoarse . . . ?
>
> *As You Like It* V. iii

# 2 | MULTIPLE REGRESSION
## Predicting One Variable from Many

As pointed out in Chapter 1, *multiple regression* is a technique used to predict scores on a single outcome variable $Y$ on the basis of scores on several predictor variables, the $X_i$s. To keep things as concrete as possible for a while (there will be plenty of time for abstract math later), let us consider the hypothetical data in Data Set 1. Say that the $X_1$, $X_2$, and $X_3$ might be IQ as measured by the Stanford–Binet, age in years, and number of speeding citations issued the subject in the past year, from which pieces of information we wish to predict $Y$, proficiency rating as an assembly line worker in an auto factory. In fact, these are purely hypothetical data. Hopefully you will be able to think of numerous other examples of situations where we might be interested in predicting scores on one variable from scores on 3 (or 4 or 2 or 25) other variables. The basic principles used in making such predictions are not at all dependent on the names given the variables, so we might as well call them $X_1$, $X_2, \ldots, X_m$ and $Y$. The $X$s may be called *independent variables*, or *predictor variables*, or just *predictors*, while $Y$ may be referred to as the *dependent variable* or as the *predicted variable* or as the *outcome measure*. Since in many applications of multiple regression all measures ($X$s *and* $Y$) are obtained at the same time, thus blurring the usual independent–dependent distinction, "predictor–predicted" would seem to be the most generally appropriate terminology. In Data Set 1 we have three predictor variables and one predicted variable.

At this point you might question why anyone would be interested in "predicting" $Y$ at all. Surely it is easier (and more accurate) to look up a person's score on $Y$ than

*Data Set 1*

| | | | | | | Intercorrelation matrix | | | | |
|---|---|---|---|---|---|---|---|---|---|---|
| Subject | $X_1$ | $X_2$ | $X_3$ | $Y$ | | $X_1$ | $X_2$ | $X_3$ | $Y$ | Variance |
| 1 | 102 | 17 | 4 | 96 | $X_1$ | 1.0 | .134 | .748 | .546 | 17.5 |
| 2 | 104 | 19 | 5 | 87 | $X_2$ | | 1.0 | .650 | −.437 | 5.0 |
| 3 | 101 | 21 | 7 | 62 | $X_3$ | | | 1.0 | −.065 | 5.0 |
| 4 | 93 | 18 | 1 | 68 | $Y$ | | | | 1.0 | 190.5 |
| 5 | 100 | 15 | 3 | 77 | | | | | | |
| $\bar{X}$ | 100 | 18 | 4 | 78 | | | | | | |

it is to look up his score on three other measures $(X_1, X_2, \text{and } X_3)$ and then plug these numbers into some sort of formula to generate a predicted score on $Y$. There are at least three answers to this question. First: we may be more interested in the prediction formula itself than in the predictions it generates. The *sine qua non* of scientific research has always been the successive refinement of mathematical formulas relating one variable to one or more other variables—$P = VT/C$, $E = mc^2$, Stevens' power law versus Thurstonian scaling procedures, etc. Second, and probably the most common reason for performing multiple regression: we may wish to develop an equation which can be used to predict values on $Y$ for subjects for whom we do *not* already have this information. Thus, for instance, we might wish to use the IQ, age, and number of speeding citations of a prospective employee to predict his probable performance on the job as an aid in deciding whether to hire him. It seems reasonable to select as our prediction equation for this purpose that formula which does the best job of "predicting" the performance of our present and past employees from these same measures. The classic way of approaching this problem is to seek from our available data the best possible *estimates* of the parameters (free constants not specified on *a priori* grounds) of the *population* prediction equation. Not unexpectedly, this "best" approximation to the population prediction equation is precisely the same as the equation which does the best job of "predicting" $Y$ scores in a random sample from the population to which we wish to generalize. Finally, we may wish to obtain a measure of the overall degree of relationship between $Y$, on the one hand, and the $X$s, on the other. An obviously relevant piece of information on which to base such a measure is just how good (or poor) a job we can do of predicting $Y$ from the $X$s. Indeed, one of the outputs from a multiple regression analysis (MRA) is a measure called the *coefficient of multiple correlation* which is simply the correlation between $Y$ and our predicted scores on $Y$, and which has properties and interpretations which very closely parallel those of Pearson's correlation coefficient for the bivariate case.

(If the mention of Pearson $r$ does not bring an instant feeling of familiarity, you should reread Section 1.2.7. If the feeling of familiarity is still not forthcoming, a return to your introductory statistics text is recommended. This prescription is basically the same for any univariate statistical technique mentioned from this point on with which you are not thoroughly familiar. Try rereading the relevant section of Chapter 1 and, if further jogging of your memory is necessary, reread the relevant chapter(s) of the book you used in your introductory statistics course. If this is no longer available to you or has acquired too many traumatic associations, other good texts such as those by Ferguson (1971), Glass and Stanley (1970), or Hays and Winkler (1971) can be consulted. Consulting an unfamiliar text, however, requires additional time for adjusting to that author's notational system.)

## 2.1   The Model

We return to the problem of developing a prediction equation for Data Set 1. What kind of prediction technique would you recommend? Presumably you are thoroughly familiar with the pros and cons of clinical judgment versus statistical

prediction, and are at least willing to concede the usefulness of confining our attention to numerical formulas which can be readily used by anyone and which do not
require Gestalt judgments or clinical experience. Given these restrictions, you will
probably come up with one of the following suggestions or something closely related:

(a)   Take the mean of the subject's scores on $X_1$, $X_2$, and $X_3$ as his predicted
score on $Y$. (Why is this an especially poor choice in the present situation?)

(b)   Take the mean of the subject's z-scores on the three predictor variables as
his predicted z-score on $Y$.

(c)   Use only that one of $X_1$, $X_2$, and $X_3$ which has the highest correlation with
$Y$, predicting the same z-score on $Y$ as on this most highly correlated predictor
variable.

(d)   Use the averaging procedure suggested in (b), but base the prediction on a
*weighted* average in which each z-score receives a weight proportionate to the absolute value of the correlation of that variable with the predicted variable. [A weighted
average is a weighted sum divided by the sum of the weights, or equivalently a
weighted sum in which the weights sum to 1.0. Thus, for instance, the suggested
weighting procedure would involve taking

$$(.546z_1 + .437z_2 + .065z_3)/1.048 = .520z_1 + .418z_2 + .062z_3$$

as our predicted score.]

You may wish to try out each of these suggestions, plus your own, on Data Set 1
and compare the resulting fit with that provided by the multiple regression equation
we eventually develop. The important point for the present purposes is that you are
unlikely to have suggested a formula involving cubes, tangents, hyperbolic sines, etc.
Instead, in the absence of compelling reasons for assuming otherwise, there is a
strong tendency to suggest some sort of *linear* combination of the $X$s, which is
equivalent to assuming that the true relationship between $Y$ and the $X$s is described
by the equation

$$Y_i = \beta_0 + \beta_1 X_{i,1} + \beta_2 X_{i,2} + \cdots + \beta_m X_{i,m} + \varepsilon_i \qquad (2.1)$$

or

$$Y_i = \hat{Y}_i + \varepsilon_i, \qquad (2.2)$$

where $Y_i$ is subject $i$'s score on the outcome variable; $X_{i,j}$ subject $i$'s score on the
predictor variable $j$; $\beta_j$ regression coefficient $j$, that is, the weight applied to predictor
variable $j$ in the population regression equation for predicting $Y$ from the $X_j$s; and $\varepsilon_i$
subject $i$'s *residual* score, that is, the discrepancy between his actual score on $Y$ and
the score predicted for him $\hat{Y}$ on the basis of the presumed linear relationship
between $Y$ and the $X$s.

Equation (2.1) is referred to as *linear* because none of the $X$s are squared, cubed,
or in general raised to any power other than zero—$\beta_0$ can be thought of as $\beta_0 X_0^0$—or
unity. Thus a plot of $Y$ as a function of any one of the $X_j$s ($j = 1, 2, \ldots, m$) would
produce a straight line, as would a plot of $Y$ as a function of the composite variable
$\hat{Y}$. This is not a serious limitation on the kinds of relationships which can be explored
between $Y$ and the $X_j$s, however, since a wide variety of other relationships can be

reduced to an equation of the same form as (2.1) by including as predictor variables various transformations of the original $X_j$s. Thus, for instance, if an experimenter feels that the true relationship between $Y$, $X_1$, and $X_2$ is of the form $Y = cX_1^u X_2^v$, he can, by taking logarithms of both sides of the equation, transform the relationship to

$$\log Y = \log(c) + u \log X_1 + v \log X_2,$$

that is, to

$$Y^* = \beta_0 + \beta_1 X_1^* + \beta_2 X_2^*,$$

where $Y^*$ and $X_i^*$ are transformed values of $Y$ and $X_i$, namely

$$Y^* = \log Y, \qquad \beta_0 = \log(c), \quad \beta_1 = u;$$

$$X_1^* = \log X_1, \qquad \beta_2 = v;$$

and

$$X_2^* = \log X_2.$$

In other words, by taking as our basic variables the logarithms of the scores on the original, untransformed variables, we have produced a *linear* prediction equation to which the techniques discussed in this chapter are fully applicable.

Note that there is a single set of $\beta$s which is used for all individuals (more generally, sampling units) in the population, while $\varepsilon_i$ is generally a different number for each individual. Note, too, that inclusion of the $\varepsilon_i$ term makes our assumption that Eq. (2.1) accurately describes the population a tautology, since $\varepsilon_i$ will be so selected as to force the equality in that equation. However, it will naturally be hoped that the $\varepsilon_i$ will be independent of each other, will have a mean of zero, and will have a normal distribution—in fact, we will make these assumptions in all significance tests arising in multiple regression analysis. Any evidence of nonrandomness in the $\varepsilon_i$s casts serious doubt on the adequacy of a linear regression equation based on these $m$ predictor variables as a predictor of $Y$. It is therefore always wise to examine subjects' residual scores for such things as a tendency to be positive for moderate values of $\hat{Y}$ and negative for extreme values of $\hat{Y}$ (indicating curvilinearity in the relationship between $Y$ and the $X_j$s, possibly correctable by including squared terms) or a systematic trend related to the number of observations sampled prior to this one (indicating lack of independence in the sampling process). Finally, note that the $\beta$s are population parameters which must be estimated from the data at hand, unless these data include the entire population to which we wish to generalize.

We could discuss Eq. (2.1) in terms of $m$-dimensional geometry, with the hypothesis that $m$ independent variables are sufficient to produce perfect prediction of scores on $Y$ (that is, that all the $\varepsilon_i$s are truly zero) being equivalent to the hypothesis that in an $(m + 1)$-dimensional plot of $Y$ as a function of the $m$ $X_j$s, all the data points would lie on an $m$-dimensional "hyperplane." However, most nonmathematicians find it difficult to visualize plots having more than three dimensions. The author therefore prefers to think of Eq. (2.1) as expressing a simple, two-dimensional relationship between the single variable $Y$ and the single *composite variable* $\hat{Y}$ (which does in fact yield a single number for each subject). Thus, multiple regression accomplishes what sounds like a complex, Gestalt-like exploration of the relationship between $Y$ and the

$m$ predictor variables by the simple expedient of reducing the problem right back to a univariate one by constructing a single composite variable which is simply a weighted sum of the original $m$ $X_j$s. This technique for reducing a multivariate problem to essentially a univariate one will be used over and over again as we learn about other multivariate statistical techniques. The trick, of course, comes in picking the weights used in obtaining the weighted sum of the original variables in an "optimal" way. We now turn to this problem.

## 2.2   Choosing Weights

Even if we had the entire population of interest available to us, it would still be necessary to decide what values of $\beta_0$, $\beta_1$, $\beta_2$, etc. provide the "best" fit between the $X$s and $Y$. As it turns out, two different and highly desirable criteria lead to precisely the same numerical values for $b_1$, $b_2$, etc.—our *estimates* of the $\beta$s. These criteria are:

(1)   that the sum of the squared errors of prediction, $E = \sum e_i^2 = \sum (Y_i - \hat{Y}_i)^2$, be as small as possible, and

(2)   that the Pearson product–moment correlation between $Y$ and $\hat{Y}$ be as high as possible.

We shall examine each of these criteria, and will then show that the values of $b_0$, $b_1$, etc. that these two criteria produce also satisfy a third desirable criterion, namely that each estimate of a given $\beta_j$ be a function only of the parameter it is designed to estimate.

### 2.2.1   Least Squares Criterion

Let us return to the hypothetical data in Data Set 1. Eventually for each subject we will compute a predicted score $\hat{Y}_i$ which is simply a linear combination of that $S$'s scores on the various predictor variables. We want to select the weights used in obtaining that linear combination such that

$$E = \sum (Y_i - \hat{Y}_i)^2$$

is as small as possible. Let us work up to the general procedure gradually.

If we are interested in predicting $Y$ from a single predictor variable, say $X_1$, we know from elementary statistics that our best choice for $b_1$ would be the value $r_{1y}(s_y/s_1)$, and our best choice for $b_0$ would be $b_0 = \overline{Y} - b_1\overline{X}$. It may be instructive to consider how these choices arise. The expression for $E$ can be rewritten as

$$E = \sum (Y_i - b_0 - b_1 X_{i,1})^2.$$

We could simply try out different values of $b_0$ and $b_1$ until we saw no room for further improvement. For instance, taking $b_0 = 0$ and $b_1 = 1$ produces an $E$ of

$$6^2 + 17^2 + 39^2 + 25^2 + 23^2 = 3000,$$

and trying out various values of $b_1$ while keeping $b_0 = 0$ reveals that a $b_1$ of about .78

produces the smallest $E$, namely about 610. (The optimum values of $b_0$ and $b_1$ produce an $E$ of about 535.) We could then try several other values of $b_0$, finding for each that value of $b_1$ which minimizes $E$. Then, drawing a curve connecting these points would probably allow us to guess the best choice of $b_1$ fairly accurately. However, even with this aid, the search will be a laborious one, and it should be clear that as soon as we add a second predictor variable, thus giving us three numbers to find, this kind of trial-and-error search will become prohibitively time consuming.

The alternative, of course, is to try calculus. If you have not been introduced to calculus already, you have three options:

(1)  You may accept the results of any application this text makes of calculus on faith.

(2)  You may turn now to Digression 1 at the back of the text, which presents (at a simplified level appropriate to the limited needs of the moment) all the calculus you will need in this text.

(3)  You may test any result arrived at via calculus either by trying out slightly different numerical values and observing their effect on the optimization criterion or by the slightly more elegant process of inserting in the algebraic expressions produced by calculus an additive constant tacked on to each and then showing that the optimization criterion "deteriorates" unless each of these added values equals zero.

At any rate, all that calculus does is to allow us to obtain an algebraic expression for the *rate* at which $E$ changes as a function of each of the variables—in this case, $b_0$ and $b_1$. Recognizing that at the point where $E$ is at a minimum, its rate of change in all directions (with respect to any variable) must be temporarily zero, we then need only set the algebraic expressions for rate of change equal to zero and solve the resulting two simultaneous equations for $b_0$ and $b_1$. Specifically,

$$E = \sum (Y_i - b_0 - b_1 X_{i,1})^2 = \sum (Y^2 + b_0^2 + b_1^2 X_1^2 - 2b_0 Y - 2b_1 X_1 Y + 2b_0 b_1 X_1)$$
$$= \sum Y^2 + N b_0^2 + b_1^2 \sum X_1^2 - 2b_0 \sum Y - 2b_1 \sum X_1 Y + 2b_0 b_1 \sum X_1,$$

whence

$$dE/db_0 = 2b_0 N - 2 \sum Y + 2b_1 \sum X_1 = 0$$

if and only if (iff)

$$b_0 = (\sum Y - b_1 \sum X)/N = \bar{Y} - b_1 \bar{X};$$

and

$$dE/db_1 = 2b_1 \sum X_1^2 - 2 \sum X_1 Y + 2b_0 \sum X_1$$
$$= 2b_1 (\sum X_1^2 - \bar{X}_1 \sum X_1) - 2 \sum X_1 Y + \bar{Y} \sum X_1$$
$$= 2b_1 \sum x_1^2 - 2 \sum x_1 y = 0$$

iff

$$b_1 = \sum x_1 y / \sum x_1^2 = \sum x_1 y \sqrt{\sum y^2 / \sum x_1^2} / \sqrt{\sum x_1^2 \sum y^2} = r_{1y}(s_y/s_1).$$

[Note that we have changed from raw scores to deviation scores in the last line.] It is assumed that the reader is thoroughly familiar with summation notation and its various simplifying conventions—such as omitting the index of summation and the range of the summation when these are clear from the context. If just what is being summed in the above equations is not perfectly clear, an introductory text such as that of Ferguson (1971) or Glass and Stanley (1970) or the one used in your introductory statistics course should be consulted for a review of summation notation.

If we now use both $X_1$ *and* $X_2$ as predictors, we obtain

$$\sum (Y_i - b_0 - b_1 X_{i,1} - b_2 X_{i,2})^2$$
$$= \sum (Y^2 + b_0^2 + b_1^2 X_1^2 + b_2^2 X_2^2 - 2b_0 Y - 2b_1 X_1 Y$$
$$- 2b_2 X_2 Y + 2b_0 b_1 X_1 + 2b_0 b_2 X_2 + 2b_1 b_2 X_1 X_2);$$
$$dE/db_0 = 2Nb_0 - 2\sum Y + 2b_1 \sum X_1 + 2b_2 \sum X_2 = 0$$

iff

$$b_0 = \overline{Y} - b_1 \overline{X}_1 - b_2 \overline{X}_2;$$
$$dE/db_1 = 2b_1 \sum X_1^2 - 2\sum X_1 Y + 2b_0 \sum X_1 + 2b_2 \sum X_1 X_2$$
$$= 2b_1 \sum X_1^2 - 2\sum X_1 Y + 2(\overline{Y} - b_1 \overline{X}_1 - b_2 \overline{X}_2)\sum X_1 + 2b_2 \sum X_1 X_2$$
$$= 2b_1 \sum x_1^2 - 2\sum x_1 y + 2b_2 \sum x_1 x_2;$$

and

$$dE/db_2 = 2b_2 \sum x_2^2 - 2\sum x_2 y + 2b_1 \sum x_1 x_2 .$$

[Note that the present text follows the convention of letting capital letters stand for "raw" observations and lower-case letters for deviation scores. Thus, $\sum x_1 x_2 = \sum (X_1 - \overline{X}_1)(X_2 - \overline{X}_2)$.] Solving for $b_1$ and $b_2$ requires solving the pair of simultaneous equations,

$$(\sum x_1^2)b_1 + (\sum x_1 x_2)b_2 = \sum x_1 y$$

and

$$(\sum x_1 x_2)b_1 + (\sum x_2^2)b_2 = \sum x_2 y;$$

whence, after some laborious algebra (made less laborious if matrix techniques to be discussed shortly are used),

$$b_1 = (\sum x_1 y \sum x_2^2 - \sum x_2 y \sum x_1 x_2)/[\sum x_1^2 \sum x_2^2 - (\sum x_1 x_2)^2]$$
$$= (s_2^2 s_{1y} - s_{12} s_{2y})/(s_1^2 s_2^2 - s_{12}^2)$$

and

$$b_2 = (s_1^2 s_{2y} - s_{12} s_{1y})/(s_1^2 s_2^2 - s_{12}^2),$$

where

$$s_{ij} = [\sum (X_i - \bar{X}_i)(X_j - \bar{X}_j)]/(n - 1) = r_{ij} s_i s_j$$

and

$$s_{iy} = [\sum (X_i - \bar{X}_i)(Y - \bar{Y})]/(N - 1) = r_{iy} s_i s_y.$$

If we now add $X_3$ to the set of predictors, we find (as the reader will naturally wish to verify for himself) that

$$b_0 = \bar{Y} - b_1 \bar{X}_1 - b_2 \bar{X}_2 - b_3 \bar{X}_3;$$

and

$$(\sum x_1^2)b_1 + (\sum x_1 x_2)b_2 + (\sum x_1 x_3)b_3 = \sum x_1 y;$$
$$(\sum x_1 x_2)b_1 + (\sum x_2^2)b_2 + (\sum x_2 x_3)b_3 = \sum x_2 y;$$
$$(\sum x_1 x_3)b_1 + (\sum x_2 x_3)b_2 + (\sum x_3^2)b_3 = \sum x_3 y;$$

whence

$$b_1 = [s_{1y}(s_2^2 s_3^2 - s_{23}^2) + s_{2y}(s_{13}s_{23} - s_{12}s_3^2) + s_{3y}(s_{12}s_{23} - s_{13}s_2^2)]/D;$$
$$b_2 = [s_{1y}(s_{13}s_{23} - s_{12}s_3^2) + s_{2y}(s_1^2 s_3^2 - s_{13}^2) + s_{3y}(s_{12}s_{13} - s_{23}s_1^2)]/D;$$
$$b_3 = [s_{1y}(s_{12}s_{23} - s_{13}s_2^2) + s_{2y}(s_{12}s_{13} - s_{23}s_1^2) + s_{3y}(s_1^2 s_2^2 - s_{12}^2)]/D;$$

where

$$D = s_1^2(s_2^2 s_3^2 - s_{23}^2) + s_{12}(s_{13}s_{23} - s_{12}s_3^2) + s_{13}(s_{12}s_{23} - s_{13}s_2^2)$$
$$= (s_1 s_2 s_3)^2 - (s_1 s_{23})^2 - (s_2 s_{13})^2 - (s_3 s_{12})^2 + 2s_{12}s_{23}s_{13}.$$

Before we consider what the general formula for $m$ predictors would look like, let us apply the formulas we have developed so far to Data Set 1. First, if we are solely interested in $X_1$ as a predictor, then our best-fitting regression equation is

$$\hat{Y} = r_{1y}(s_y/s_1)X_1 + (\bar{Y} - b_1 \bar{X}_1)$$
$$= .546\sqrt{190.5/17.5}\, X_1 + (78 - b_1 \cdot 100)$$
$$= 1.80X_1 - 102.$$

Application of this equation to each of the subjects' scores on $X_1$ yields predicted values of 81.6, 85.2, 79.8, 65.4, and 78.0 for subjects 1–5, respectively, and a total sum of squared deviations of approximately 535.

In order to apply the formulas for the two-predictor case (and, subsequently, for the three-predictor case) we need to know the covariances of scores on our predictor and outcome variables. We could, of course, compute these directly, and the reader will probably wish to do so as a check on the alternative method, which is to "reconstruct" them from the intercorrelations and variances provided for you in the data table. We need only multiply a given correlation coefficient by the product of the standard deviations involved in that correlation to obtain the covariance of the two variables. (Be sure you understand why this works.) This gives values of 1.25, 31.5, and $-13.5$ for $s_{12}$, $s_{1y}$, and $s_{2y}$, respectively. Plugging into the formulas for $b_0$,

$b_1$, and $b_2$ gives a final regression equation of

$$\hat{Y} = 2.029X_1 - 3.202X_2 - 67.3,$$

which yields predictions of 85.3, 82.9, 70.4, 63.8, and 87.6 for the five subjects, and a total sum of squared deviations of 333.1.

Moving now to the three-predictor case, we find, after extensive calculations, that our regression equation is

$$\hat{Y} = 5.18X_1 + 1.70X_2 - 8.76X_3 - 435.7,$$

which yields predictions of 86.7, 91.7, 62.0, 68.0, and 81.7 for the five subjects, and a sum of squared errors of prediction of 130.7.

Table 2.1

Multiple Regression Analyses of Data Set 1

| Predictors included | $b_0$ | $b_1$ | $b_2$ | $b_3$ | $\hat{Y}_1$ | $\hat{Y}_2$ | $\hat{Y}_3$ | $\hat{Y}_4$ | $\hat{Y}_5$ | $\sum e^2$ | $R$ | $F^a$ |
|---|---|---|---|---|---|---|---|---|---|---|---|---|
| 1 | −102.0 | 1.80 | 0 | 0 | 81.6 | 85.2 | 79.8 | 65.4 | 78.0 | 535.2 | 546 | 1.271 |
| 1, 2 | −67.3 | 2.03 | −3.21 | 0 | 85.3 | 82.9 | 70.4 | 63.8 | 87.6 | 333.1 | 750 | 1.287 |
| 1, 2, 3 | −435.7 | 5.18 | 1.70 | −8.76 | 86.7 | 91.7 | 62.0 | 68.0 | 81.7 | 130.7 | 910 | 1.611 |

[a] The basis for this column will be explained in Section 2.3.

Table 2.1 summarizes the results of our regression analysis of Data Set 1. Note that as we add predictors, we obtain increasingly better fit to the $Y$s, as measured by our criterion, the sum of the squared errors. Note also that the weight assigned $X_1$ in the regression equation was positive and moderately large for all three equations, but the weight assigned $X_2$ shifted drastically from $-3.20$ (larger in absolute value than $b_1$) when only $X_1$ and $X_2$ were being used, to 1.70 (about $\frac{1}{3}$ as large as $b_1$) when all three predictor variables were employed. Finally, note that $X_3$ was the most important contributor to the regression equation employing all three predictors, even though its correlation with the outcome variable is only $-.065$, that is, used by itself it would account for only about four-tenths of one percent of the variance in $Y$ scores. These shifts in sign and magnitude of a given regression coefficient as the context in which that variable is imbedded (that is, the other predictors employed) is changed are quite common in applications of multiple regression analysis. They should not be surprising, since the contribution variable $j$ makes to predicting $Y$ will of course depend on how much of the information it provides about $Y$ (as reflected in $r_{jy}^2$) is already provided by the other predictors (as reflected partially in its correlations with the other $X$s). Social psychologists among readers of this Primer will recognize a very close parallel between this general phenomenon of multiple regression analysis and Wishner's (1960) analysis of changes in the "centrality" of a given personality trait (that is, the magnitude of the effect on $S$'s impression of the person being described of changing his reported position on this one trait) as the nature of the other traits mentioned in the description is varied. The Venn diagram used in answering Real Data Quiz I, Problem II.B.1, may also be of help. Now let us consider our second criterion for deriving the $b$s.

## 2.2.2  Maximum Correlation Criterion

It would be nice to have an overall measure of how good a job the regression equation derived from a multiple regression analysis does of fitting the observed $Y$s. A fairly obvious measure is the Pearson product–moment correlation between predicted and observed values $r_{YY}$. This has been computed for each of the three sets of predictions generated for Data Set 1, and is listed in the column headed $R$ in Table 2.2. (The reader will, of course, wish to verify these computations.) This also provides an overall measure of the degree of relationship between the set of predictor variables and the outcome variable, since it is the correlation between $Y$ and a particular linear combination of the $X$s. We naturally wish to be as optimistic as possible in our statement about the strength of this relationship, so the question arises as to whether the values of $b_0$, $b_1$, $b_2$, etc. can be improved upon in terms of the value of $r_{YY}$ to which they lead. Intuitively, the answer would seem to be " No," since we know that there is an extremely close relationship between the magnitude of $r$ and the adequacy of prediction in the univariate case. However, to put this hunch on firmer grounds, let us ask what would have happened had we used the "optimism criterion" (choose $b$s which make $r_{YW}$ as large as possible, where $W = b_0 + b_1 X_1 + \cdots$) to derive the $b$s.

The maximum-$R$ criterion is of no help to us in the univariate case, since we know that any linear transformation of the $X$ scores will leave the correlation between $X$ and $Y$ unaffected. There may be a deep logical (or philosophical) lesson underlying this fact, but it is more likely simply an indication that prediction optimization was used as a criterion in developing the formula for the Pearson $r$. It should also be clear that the maximum-$R$ criterion will be of no value in picking a value of $b_0$, since the correlation between $W = b_1 X_1 + b_2 X_2 + \cdots + b_m X_m$ and $Y$ is unchanged by addition or subtraction of any constant whatever. However, one of the properties of univariate regression equations is that they always predict that a subject who scores at the mean of one variable will also score at the mean of the other variable. Thus, if we can use the maximum-$R$ criterion to pick values of $b_1$, $b_2$, etc., $b_0$ can be computed from the requirement that $\hat{Y} = \bar{Y}$ when $W = \bar{W}$, that is,

$$\bar{Y} = \bar{W} = b_1 \bar{X}_1 + b_2 \bar{X}_2 + \cdots + b_m \bar{X}_m + b_0.$$

※ Let us start with the simplest case, namely the two-predictor case. We know that

$$r_{WY} = \sum (W - \bar{W})(Y - \bar{Y}) / \sqrt{\sum (W - \bar{W})^2 \sum (Y - \bar{Y})^2},$$

and that

$$\bar{W} = \sum (b_1 X_1 + b_2 X_2)/N = b_1 \bar{X}_1 + b_2 \bar{X}_2,$$

whence

$$w = W - \bar{W} = b_1(X_1 - \bar{X}_1) + b_2(X_2 - \bar{X}_2) = b_1 x_1 + b_2 x_2.$$

Thus,

$$r_{WY} = \sum (b_1 x_1 + b_2 x_2) y / \sqrt{\sum (b_1 x_1 + b_2 x_2)^2 \sum y^2}$$
$$= (b_1 \sum x_1 y + b_2 \sum x_2 y) / \sqrt{[b_1^2 \sum x_1^2 + 2 b_1 b_2 \sum x_1 x_2 + b_2^2 \sum x_2^2] \sum y^2}.$$

Since the value of $r$ is unaffected by linear transformations of $Y$, we could apply a transformation to $Y$ [namely, converting to $z_y/(N-1)$] which would guarantee that $\sum y^2 = 1$, thereby simplifying our arithmetic somewhat. Also, we know that requiring that $R_{Y \cdot X_1, X_2} = r_{WY}$ be as large as possible is not enough to uniquely specify $b_1$ and $b_2$, since the correlation of $b_1 X_1 + b_2 X_2$ with $Y$ is identical to the correlation of $c(b_1 X_1 + b_2 X_2) = (cb_1)X_1 + (cb_2)X_2$ with $Y$. We therefore need to put some additional restriction on the permissible values of $b_1$ and $b_2$. We might as well choose values which simplify our mathematical work. Square roots are bothersome when taking derivatives, so we will simply require that

$$\sum w^2 = b_1^2 \sum x_1^2 + 2b_1 b_2 \sum x_1 x_2 + b_2^2 \sum x_2^2$$

be equal to unity. Thus, we are left with the problem of maximizing $b_1 \sum x_1 y + b_2 \sum x_2 y$, subject to the restriction that $\sum w^2 = 1$. The method of Lagrangian multipliers is ideal for this purpose. If you are not presently conversant with Lagrangian multipliers, either turn to Digression 1 for a review, or take the following derivation on faith. (In either case, you should by all means test the end result by trying out other weights which satisfy the side condition.)

※ Our side restriction is that $g(b_1, b_2) = \sum (b_1 x_1 + b_2 x_2)^2 - 1 = 0$. The function whose derivatives we must find is, then

$$L = (b_1 \sum x_1 y + b_2 \sum x_2 y) - \lambda[\sum (b_1 x_1 + b_2 x_2)^2 - 1],$$

whence

$$dL/db_1 = \sum x_1 y - \lambda[2b_1(\sum x_1^2) + 2b_2 \sum x_1 x_2]$$

and

$$dL/db_2 = \sum x_2 y - \lambda[2b_2(\sum x_2^2) + 2b_1 \sum x_1 x_2].$$

Setting these two derivatives equal to zero produces the two simultaneous equations,

$$2\lambda b_1 \sum x_1^2 + 2\lambda b_2 \sum x_1 x_2 = \sum x_1 y;$$

and

$$2\lambda b_1 \sum x_1 x_2 + 2\lambda b_2 \sum x_2^2 = \sum x_2 y.$$

Comparison of these two equations with the equations which resulted when we derived $b_1$ and $b_2$ from the best-prediction criterion shows that the two pairs of equations are identical if $\lambda = \frac{1}{2}$, and indeed substituting in that value for $\lambda$ and the previously computed expressions for $b_1$ and $b_2$ can be readily shown to satisfy the above two equations.

We have thus shown, for the one- and two-predictor cases, that the same weights which minimize the sum of the squared errors of prediction also yield a linear combination of the original variables which correlates more highly with $Y$ than any other linear combination of the predictor variables. We could proceed to the case where three predictors are involved, but it will prove considerably easier if instead we pause to show how matrix algebra can be used to simplify our task.

### 2.2.3   The Utility of Matrix Algebra

Looking back over the formulas we have developed so far, we can see some patterns which are suggestive of the form the multiple regression problem and solution might take in the general case of $m$ predictors. For instance, it is fairly clear (and we gave some reasons earlier for believing) that $b_0$ will always be computed from the value of the other coefficients as $b_0 = \overline{Y} - b_1 \overline{X}_1 - b_2 \overline{X}_2 - \cdots - b_m \overline{X}_m$. Similarly, the pattern of coefficients involved in the simultaneous equations which must be solved for the values of $b_1, b_2, \ldots, b_m$ seems to be very closely related to the sample covariances, $s_{ij}$ and $s_{iy}$, as follows:

$$s_1^2 b_1 + s_{12} b_2 + s_{13} b_3 + \cdots + s_{1m} b_m = s_{1y}$$

$$s_{12} b_1 + s_2^2 b_2 + s_{23} b_3 + \cdots + s_{2m} b_m = s_{2y}$$

$$s_{13} b_1 + s_{23} b_2 + s_3^2 b_3 + \cdots + s_{3m} b_m = s_{3y}$$

$$\vdots \qquad \vdots \qquad \vdots \qquad \vdots \qquad \vdots \quad \vdots$$

$$s_{1m} b_1 + s_{2m} b_2 + s_{3m} b_3 + \cdots + s_m^2 b_m = s_{my}.$$

What is *not* clear, however, is what will be the general form of the expressions for $b_1$, $b_2$, etc. which result from solving these equations. It seems strange that the apparent regularity of the equations would not be paralleled by predictability of their solutions. This is precisely the sort of situation which leads a mathematician to suspect that a new level of operating is needed, an algebra designed to handle arrays of numbers or symbols as a whole, rather than having to deal with each symbol or number separately. Such a " higher-order " algebra has been developed, and is known as *matrix algebra*.

If you are not already familiar with matrix algebra, you will wish to turn to Digression 2 for a simplified review of this important tool. First, however, we shall try to show why the operations on sets of numbers described there are so relevant to multivariate statistics. Basically, there are three advantages to matrix algebra:

(1)   it *summarizes* expressions and equations very compactly;
(2)   it facilitates our *memorizing* these expressions; and
(3)   it greatly *simplifies* the procedures for deriving solutions to multivariate problems.

The first function is illustrated by the following summary of the solution for the general, $m$-predictor case:

$$b_0 = \overline{Y} - \overline{\mathbf{X}}\mathbf{b}, \quad \mathbf{b} = \mathbf{S}_x^{-1}\mathbf{s}_{xy}; \qquad R^2 = (\mathbf{b}'\mathbf{s}_{xy})^2/(\mathbf{b}'\mathbf{S}_x \mathbf{b} s_y^2) = \mathbf{b}'\mathbf{s}_{xy}/s_y^2.$$

This is a bit more compact than the expressions derived earlier for, say, the three-predictor case, and it is many orders of magnitude less complex than would be, say, the expressions for the 30-variable case written in " single-symbol " form.

The second, mnemonic function of matrix algebra is in part a direct consequence of the first, since it is always easier to remember a compact expression than one requiring dozens (or hundreds) of symbols. This advantage is greatly enhanced, however, by the fact that the matrix expressions usually turn out to be such direct and obvious generalizations of the formulas for the univariate (in the present context, single-predictor) case. For example, in multiple regression, the set of $m$ $b$s corresponding to the $m$ predictors are computed from the matrix equation $\mathbf{b} = \mathbf{S}_x^{-1}\mathbf{s}_{xy}$, where $\mathbf{S}_x$ is the covariance *matrix* of the $m$ predictors (the obvious generalization of $s_x^2$); $\mathbf{s}_{xy}$ is a column vector listing the covariance of each predictor variable with the outcome variable (the obvious generalization of $s_{xy}$); and the $-1$ exponent represents the operation of *inverting* a matrix, which is analogous to the process of inverting (taking the reciprocal of) a single number.

To see how matrix algebra accomplishes the third function (that of simplifying derivations), let us review the reasoning which produced the matrix-form solution for the multiple regression problem.

First, we want to minimize

$$\sum (Y - \hat{Y})^2 = \sum (y - \hat{y})^2 = \sum y^2 - 2 \sum y\hat{y} + \sum \hat{y}^2 = \mathbf{y}'\mathbf{y} - 2\mathbf{b}'\mathbf{s}_{xy} + \mathbf{b}'\mathbf{S}_x\mathbf{b},$$

whence

$$dE/d(\mathbf{b}) = 0 - 2\mathbf{s}_{xy} + 2\mathbf{S}_x\mathbf{b} = 0 \qquad \text{iff} \quad \mathbf{S}_x\mathbf{b} = \mathbf{s}_{xy}.$$

Since the $b$s are what we are after, it would be convenient if we could isolate the vector $\mathbf{b}$ on the left-hand side by "dividing" both sides by the matrix $\mathbf{S}_x$. The closest matrix analog of division is multiplication by the inverse of a matrix (just as division by a scalar is equivalent to multiplication by its inverse). It can be shown that there exists a matrix $\mathbf{S}_x^{-1}$, the inverse of $\mathbf{S}_x$, which has the property that $\mathbf{S}_x^{-1}\mathbf{S}_x = \mathbf{S}_x\mathbf{S}_x^{-1} = \mathbf{I}$, the identity matrix (which has ones on the main diagonal and zeros elsewhere), whence

$$\mathbf{S}_x^{-1}\mathbf{S}_x\mathbf{b} = \mathbf{S}_x^{-1}\mathbf{s}_{xy}; \qquad \text{whence} \quad \mathbf{b} = \mathbf{S}_x^{-1}\mathbf{s}_{xy}.$$

[Actually, as pointed out in Digression 2, $\mathbf{S}_x^{-1}$ exists only if $\mathbf{S}_x$ has what is called full rank, that is, if and only if $|\mathbf{S}_x| \neq 0$. However, whenever $\mathbf{S}_x$ is singular, that is, has a zero determinant, one or more of the predictors can be expressed as an exact linear combination of the other predictors, so that this redundant predictor(s) can be eliminated from the regression procedures by deleting the corresponding row and column of $\mathbf{S}_x$ and the corresponding member of $\mathbf{s}_{xy}$, thus eliminating the singularity without losing any predictive ability.]

The above derivation is a great deal simpler than attempting to solve explicitly for the relationship between each $b_j$ and particular elements of $\mathbf{s}_{xy}$ and $\mathbf{S}_x$. It is also much more general than the single-symbol approach, since it holds for all sizes of the set of predictor variables. Of course, if you wish an explicit algebraic expression for the relationship between each $b_j$ and particular elements of $\mathbf{S}_x$ and $\mathbf{s}_{xy}$, you will find the derivational labor involved in developing explicit expressions for $\mathbf{S}_x^{-1}$ of the same order of magnitude (though always somewhat less, because of the straightforwardness of the matrix inversion process) as solving the single-symbol equations. However, such expressions are seldom needed for any but the simplest cases, say $m \leq 3$, where the full formulas may be useful in testing computer programs. It is, of course, to readily available matrix inversion computer programs that we will turn when numerical solutions for the $b$s are desired in a problem of practical size.

Before we digress to a discussion of the use of "canned" computer programs, however, let us use the matrix algebra introduced in Digression 2 to make two more derivations. First, the extremely complex maximum-$R$ criterion for deriving the $b$s:

$$R = \mathbf{b's}_{xy}/\sqrt{\mathbf{b'S}_x\mathbf{bs}_y^2}.$$

Noting that when $m = 1$,

$$\mathbf{b'S}_x\mathbf{b} = b_1^2 s_x^2 = \frac{(\sum xy)^2/\sum x^2}{N-1} = \frac{(\sum xy/\sum x^2)\sum xy}{N-1} = b_1\frac{\sum xy}{N-1} = \mathbf{b's}_{xy},$$

and wishing to maintain compatibility with the $b$s derived from the prediction accuracy criterion, we take as our side condition $\mathbf{b'S}_x\mathbf{b} - \mathbf{b's}_{xy} = 0$. Thus, we wish to maximize $R = \sqrt{\mathbf{b's}_{xy}}$, subject to the specified constraint. (Remember that for purposes of the derivation we can ignore $s_y$, since it is unaffected by our choice of $b$s, and we can at any rate apply a standard-score transformation to the $Y$s to obtain $s_y = 1$.) However, we might as well maximize $R^2$ instead, and thereby simplify our math. Using the method of Lagrangian multipliers (cf. Digression 1) gives

$$L = \mathbf{b's}_{xy} - \lambda(\mathbf{b'S}_x\,\mathbf{b} - \mathbf{b's}_{xy}) = \mathbf{b's}_{xy}(1 + \lambda) - \lambda(\mathbf{b'S}_x\,\mathbf{b});$$

whence

$$dL/d\mathbf{b} = \mathbf{s}_{xy}(1 + \lambda) - 2\lambda\mathbf{S}_x\,\mathbf{b} = 0 \qquad \text{iff} \quad \mathbf{S}_x\,\mathbf{b} = (1 + \lambda)\mathbf{s}_{xy}/2\lambda.$$

This last equation seems quite reminiscent of our side condition. In fact, premultiplying both sides of the equation by $\mathbf{b}'$ gives us

$$\mathbf{b'S}_x\,\mathbf{b} = [(1 + \lambda)/2\lambda]\mathbf{b's}_{xy},$$

which is consistent with the side condition if and only if $\lambda = 1$. Plugging this value of lambda back into the equation we had before the postmultiplication by $\mathbf{b}'$ gives

$$\mathbf{S}_x\,\mathbf{b} = \mathbf{s}_{xy}, \qquad \text{whence} \quad \mathbf{b} = \mathbf{S}_x^{-1}\mathbf{s}_{xy},$$

the same values derived from the accuracy criterion.


### 2.2.4   Independence of Irrelevant Parameters

Second, we wish to confirm the statement that the procedures for estimating the $b$s satisfy the condition of "independence of irrelevant parameters." We need a more precise way of expressing this condition. Let us assume that we have a set of expressions for computing the $b$s from the observed values on $Y$, or $y$. These will be of the form

$$b_j = k_{1j}y_1 + k_{2j}y_2 + \cdots + k_{Nj}y_N,$$

or, in matrix notation, $\mathbf{b} = \mathbf{Ky}$, where each row of $\mathbf{K}$ contains the coefficients for computing the corresponding $b$ from the $y$s. Now, we also have a theoretical expression for each $y$, namely

$$y_i = \beta_1 x_{i1} + \beta_2 x_{i2} + \cdots + \beta_m x_{im} + \varepsilon_i,$$

that is, $\mathbf{y} = \mathbf{x\beta} + \mathbf{\varepsilon}$ in matrix form. We wish to estimate $\mathbf{b} = \mathbf{Ky} = \mathbf{Kx\beta} + \mathbf{K\varepsilon}$. Now since $\varepsilon$ is a vector of (presumed) random errors, we can assume that the expected value of $\mathbf{K\varepsilon} = \mathbf{0}$, the null vector, and concentrate on $\mathbf{Kx\beta}$. A few seconds of reflection should make it clear that one way of ensuring that the expression for each $b_j$ involves only its corresponding $\beta_j$ is to ask that $\mathbf{Kx}$ be equal to the identity matrix, that is, $\mathbf{Kx} = \mathbf{I}$. However, the expression for $\mathbf{b}$ is

$$\mathbf{b} = \mathbf{S}_x^{-1}\mathbf{s}_{xy} = (\mathbf{x'x})^{-1}(\mathbf{x'y}) = [(\mathbf{x'x})^{-1}\mathbf{x'}]\mathbf{y} = \mathbf{Ky};$$

whence

$$\mathbf{K} = (\mathbf{x'x})^{-1}\mathbf{x'} \quad \text{and} \quad \mathbf{Kx} = (\mathbf{x'x})^{-1}(\mathbf{x'x}) = \mathbf{I},$$

as asserted. For example, in the two-predictor case applied to Data Set 1 we have

$$\mathbf{K} = (\mathbf{x'x})^{-1}\mathbf{x'} = \frac{1}{1375}\begin{bmatrix} 20 & -5 \\ -5 & 70 \end{bmatrix}\begin{bmatrix} 2 & 4 & 1 & -7 & 0 \\ -1 & -1 & 3 & 0 & -3 \end{bmatrix}$$

$$= \frac{1}{1375}\begin{bmatrix} 45 & 75 & 5 & -140 & 15 \\ -80 & 50 & 205 & 35 & -210 \end{bmatrix}.$$

Applying this $\mathbf{K}$ matrix to the theoretical expressions for the $y$s gives

$$E(b_1) = [45(2\beta_1 - \beta_2) + 75(4\beta_1 + \beta_2)$$
$$+ 5(\beta_1 + 3\beta_2) - 140(-7\beta_1) + 15(-3\beta_2)]/1375 = \beta_1;$$

and

$$E(b_2) = [-80(2\beta_1 - \beta_2) + 50(4\beta_1 + \beta_2)$$
$$+ 205(\beta_1 + 3\beta_2) + 35(-7\beta_1) - 210(-3\beta_2)]/1375 = \beta_2;$$

where, for present purposes, $E(b)$ stands for the expected value of $b$. Note, however, that the estimates $b_1 = -y_4/7$ and $b_2 = -y_5/3$ also satisfy the criterion; thus requiring independence of irrelevant parameters is insufficient to *derive* the $b$s uniquely. One potentially useful offshoot of this demonstration arises from the recognition that $\mathbf{K}$ is completely independent of the $y_i$s. Thus if the same set of predictor variables were to be related to each of several outcome measures, it might prove efficient to compute $\mathbf{K}$ and then simply postmultiply it by each of the different $\mathbf{y}$ vectors to compute the several different regression equations.

## 2.3   Relating the Sample Equation to the Population Equation

Let us consider for a moment the four-predictor case applied to Data Set 1. Let us assume that our five $S$s had scores of 1, 2, 3, 4, and 5 on $X_4$. Thus $X_4$ has correlations of $-.567$, $-.354$, $-.424$, and $-.653$ with $X_1, X_2, X_3$, and $Y$, respectively. It thus has the highest correlation with the outcome variable of any of the predictor variables, so that we might anticipate that it would contribute greatly to the regression equation and lower $\sum e^2$ considerably from its three-predictor value of 130.7. However, even though $X_4$ would account for about 43% of the variation in $Y$ even if it were used by itself, it also shares a great deal of variance with the other three predictors. Thus, the information it provides about $Y$ is somewhat redundant, and might contribute relatively little to our ability to predict $Y$ over what the first three predictors already provide. To find out which factor is more important, we carry out the regression analysis, finding that

$$(\mathbf{x'x})^{-1} = \begin{bmatrix} .12364 & .15273 & -.23636 & .12000 \\ .15273 & .27636 & -.34545 & .16000 \\ -.23636 & -.34545 & .54545 & -.20000 \\ .12000 & .16000 & -.20000 & .24000 \end{bmatrix},$$

whence

$$\mathbf{b'} = [2.3761 \quad -2.03586 \quad -4.09066 \quad -5.60]$$

and the vector of predicted $y$s is

$$\mathbf{y'} = [17.99 \quad 8.98 \quad -16.00 \quad -9.94 \quad -1.00],$$

whence $\sum e^2 = .0021$ and $R = 1.00$. In other words, inclusion of this very arbitrary predictor variable has produced a near-perfect fit between predicted and observed scores on $Y$. It is important to ask whether this result was a lucky accident due to the particular values chosen for $X_4$, or whether instead the fact that $\sum e^2$ is not *exactly* zero is due solely to inaccuracy in numerical computations. The latter turns out to be the case, as the reader may verify by trying out another set of values for $X_4$. No matter what numbers are selected, you will find that, except for round-off errors, the regression analysis procedures for predicting five subjects' $Y$ scores on the basis of their scores on four predictor variables, regardless of what numbers are used for $X_4$ (or, for that matter, $X_1$, $X_2$, and $X_3$) will lead to a multiple $R$ of 1.0. This should lead us to suspect that our $R$ of 1 is telling us more about the efficiency of the optimization process than about any underlying relationship between these variables and $Y$ in the population from which the data were "sampled." It has been shown (for example, Wishart, 1931) that the expected value of a sample coefficient of multiple correlation based on $N$ subjects and $m$ predictors *none of which have any true relationship to the outcome variable* in terms of their population correlations with $Y$, is equal to $m/(N-1)$. This also clearly establishes that the sample value of $R^2$ is a biased estimate of the corresponding population coefficient. Olkin and Pratt (1958) have developed an unbiased estimate of $R_{pop}^2$, the squared population multiple correlation coefficient, the first two terms of which are given in Eq. (2.3):

$$\hat{R}_{pop}^2 = 1 - \left(\frac{N-3}{N-m-1}\right)\left[(1-R^2) + \left(\frac{2}{N-m+1}\right)(1-R^2)^2\right]. \qquad (2.3)$$

The problem, of course, is that our optimization procedure has no way of knowing what portion of the variation among the variables is truly representative of the population, and what portion represents "wild" behavior of the particular sample. It therefore utilizes *all* variation in picking values for $b_1$, $b_2$, etc. This is very reminiscent of two similar situations in elementary statistics:

(1)   the need to use $N - 1$, rather than $N$, in the denominator of $s_x^2$ in order to compensate for the fact that in taking deviations about $\bar{X}$ rather than about the (unfortunately unknown) true population value of the mean, we guarantee that $\sum x^2$ will be smaller than $N\sigma_x^2$, and

(2)   the fact that the Pearson $r$ between two variables when only two observations on their relationship are available must be either $+1$ or $-1$ (or undefined if the two scores for either variate are identical).

The geometric interpretation of this latter fact is the statement that a straight line (the line representing the regression equation) can always be passed through any two points (the two points on the scatter diagram which summarize the two Ss' scores on $X$ and $Y$). A similar interpretation holds for the two-predictor case in multiple regression. We now have a three-dimensional scattergram, and perfect fit consists of passing a plane through all points on this three-dimensional scattergram—which, of course, is always possible when there are only three points. In the general case, we will have an $(m + 1)$-dimensional scatter plot for which we seek the best fitting $m$-dimensional figure. Perfect fit will always be possible when the number of points on the scatter diagram (the number of subjects) is less than or equal to $m + 1$.

The case where $N \leq m + 1$ represents the most extreme capitalization on chance. There will in all cases, however, be a need to assess how much information the regression equation computed for our sample of data provides about the *population* regression equation. If we have selected an appropriate model for the data, so that the residuals show no systematic relationship to, for example, the magnitude of the predicted value of $Y$ or the order in which the observations were drawn, and if $Y$ is normally distributed while the $X_{i,j}$s are fixed constants for each subject, then it can be shown (as friendly mathematical statisticians have done) that our estimates of the true population variance of residual scores and of the true population variance of predicted scores are independent of each other and each follow a chi-square distribution, while the $b_j$s are normally distributed. We can thus, for instance, take $F = R^2(N - m - 1)/[(1 - R^2)m]$ as a test statistic to be compared with the critical value of the $F$-distribution with $m$ and $N - m - 1$ degrees of freedom in deciding whether the sample provides strong enough evidence to reject the null hypothesis of no true relationship between the predictor variables and the outcome variable.

If the $X_j$s are *not* fixed constants, but are themselves random variables which together with $Y$ are distributed in accordance with the multivariate normal distribution, the distributions of our sample estimates of the entries of the covariance matrix follow a Wishart distribution (a multivariate generalization of the chi-square distribution) and the $b_j$s follow a complex distribution which is *not* exactly multivariate normal. However, the statistical tests which apply in this case (known as the "variance component" model, in contrast to the "fixed" model discussed in the preceding paragraph) are identical to those derived from the fixed model, as is demonstrated by the fact that Winer (1971) and Morrison (1967) arrive at the same formulas working from the assumptions of the fixed and variance-component model, respectively. What is not as clear are the consequences for inferential uses of multiple regression of having some of the $X_j$s fixed constants and others random variables, which is the case known as the "mixed" model. Inferential procedures for the mixed model in analysis of variance—which, as we shall see in Chapter 5, can be viewed as a special case of multiple regression in which the predictor variables are all dichotomous, group-membership variables—are well known, and generally involve using different "error terms" for tests involving only the fixed $b$s than for those involving regression coefficients associated with random variables. The more general case (the mixed model with other than dichotomous variables) does not appear to have been extensively studied—perhaps because multiple regression has typically been applied in correlational research involving measured, rather than manipulated, variables, in which context fixed variables (implying zero error of measurement in assigning subjects scores on that predictor variable) are rarely encountered.

Table 2.2 summarizes the various significance tests which are available to us if our initial assumption of random residuals is tenable. In any study one should plot the residuals and examine them for gross departures from randomness. (This will, however, be a truly meaningful procedure only if a large number of subjects, say at least 30 more than the number of predictor variables employed, is used.) Draper and Smith (1966), Daniel and Wood (1971), Anscombe (1973), and Mosteller and Tukey (1968) have all argued convincingly for the usefulness of graphical methods in detecting departures of the data from the assumptions of the linear regression model, and

*Table 2.2*

*Summary of Significance Tests for Multiple Regression[a]*

| Source | df | Sum of squares (SS) | Mean square (MS) |
|---|---|---|---|
| Total | $N - 1$ | $\sum y^2 = \mathbf{y'y}$ | $s_y^2 = \sum y^2/(N - 1)$ |
| Reduction in $\sum y^2$ due to first $k$ predictors | $k$ | $(\mathbf{y'x})\mathbf{b}^*$, where the nonzero elements of $\mathbf{b}^*$ are computed on basis of first $k$ predictors | $\mathbf{y'xb}^*/k$ |
| Reduction in $y^2$ due to all $m$ predictors | $m$ | $(\mathbf{y'x})\mathbf{b} = \sum [b_j(x_j y)]$ | $\mathbf{y'xb}/m = R^2 \sum y^2/m$ |
| Improvement in fit due to addition of last $m - k$ predictors | $m - k$ | $(\mathbf{y'x})\mathbf{b} - (\mathbf{y'x})\mathbf{b}^*$ | $(\mathbf{y'x})(\mathbf{b} - \mathbf{b}^*)/(m - k)$ |
| Hypothesis that a single $\beta_j$ has a true population value $\alpha$ | 1 | $(b_j - \alpha)^2/d_{jj}$, where $d_{jj}$ is the $j$th main diagonal entry of $(\mathbf{x'x})^{-1}$ | Same as SS |
| Contrast among $\beta$s, $\mathbf{a'\beta} = \sum a_j \beta_j = 0$, where $\sum a_j = 0$ | 1 | $(\mathbf{a'b})^2/V$, where $V = \mathbf{a'(x'x)}^{-1}\mathbf{a}$, which equals $\mathbf{a'a} = \sum a_j^2$ when the predictors are uncorrelated | Same as SS |
| Deviation from prediction = error = residual | $N - m - 1$ | $\sum e^2 = \sum (y - \hat{y})^2$ $= \sum y^2 - (\mathbf{y'x})\mathbf{b}$ $= \sum y^2(1 - R^2)$ | $\sum e^2/(N - m - 1)$ $= \sum y^2(1 - R^2)/(N - m - 1)$ |

[a] All tests consist of dividing MS for that source of variance by the MS for deviation from prediction and comparing the resulting value to the critical value for your chosen significance level of an $F$-distribution having the specified degrees of freedom. The overall test of significance of regression can be written as $F = [(N - m - 1)/m][R^2/(1 - R^2)]$.

suggest a number of techniques which serve to enhance the resolving power of these graphical approaches.

✻ In particular, Draper and Smith (1966) demonstrate that the usual sum of squares for residual from regression is actually composed of two components: error variance and systematic departures from linearity. Where there are several "repeat" observations (sets of subjects having exactly the same scores on all predictor variables), the sum of squared deviations about a mean $Y$ score within each of these sets of repeat observations can be used to compute a mean square for error which is independent of systematic departures from the model being employed, which in turn can be used to test the statistical significance of the departures from the model, using

$$F = \text{MS}_{\text{dep}}/\text{MS}_{\text{err}} \, ,$$

where $\text{MS}_{\text{err}}$ is the mean square for error just described and

$$\text{MS}_{\text{dep}} = (\text{SS}_{\text{res}} - \text{SS}_{\text{err}})/(\text{df}_{\text{res}} - \text{df}_{\text{err}}).$$

The major drawback to this test is that it is often difficult to find *any* repeat observations in a study where the predictors are organismic variables not subject to experimental control.

✻ Daniel and Wood (1971) suggest a "factorial" approach to developing the most adequate model before considering a multiple regression analysis complete. Once possible sources of improvement in the fit of the model have been identified, primarily by graphical means, a series of $2^s$ MRAs is conducted, where $s$ is the number of such possible sources of improvement and the $2^s$ MRA "runs" through the computer consist of all possible combinations of the presence or absence of each improvement in a particular run. These sources of improvement might include deleting one or more "outliers" (observations which do not seem to "belong" on the same scatter diagram as the others), including the square or the cube of one of the predictor variables or a cross-product of two of the predictors as additional predictor variables, performing a square root or logarithmic transformation on the predicted variable $Y$, etc. This would seem to be an excellent procedure for an exploratory study. However, three cautions need to be considered before (or while) adopting this factorial approach:

(a) The usual significance levels for tests conducted on the overall $R^2$, on individual regression coefficients, etc. do not apply to the final regression equation which results from this factorial exploration, since those probability levels were derived under the assumption that this is a single, preplanned analysis. This is, of course, no problem if the researcher is not interested in hypothesis testing, but only in model development.

(b) The number of runs can become prohibitively large if the researcher is not rather selective in generating possible sources of improvement.

(c) This approach to model development should *not* be confused with the development of mathematical models of behavior from a set of axioms about the processes which generate the observed behavior (cf. Atkinson, Bower, and Crothers, 1965, and/or Estes, 1957, for discussion of this latter approach). Daniel and Wood's factorial approach is essentially a systematization of the "curve-fitting" approach to the development of theories of learning which proved to be such a dead end. There are, for any single relationship such as that between a specific set of predictors and a single outcome variable, an almost limitless number of quite different looking functions relating the two which nevertheless provide very nearly equally good fits to the data. Moreover, a function developed on purely empirical grounds without an underlying axiomatic justification cannot be generalized to situations involving slightly different sets of predictors or different experimental conditions.

Where $Y$ and/or the $X_j$ are not normally distributed, the multivariate equivalent of the central limit theorem so crucial in univariate statistics assures us that the statistical tests outlined in Table 2.2 remain valid for large $N$ (cf., for example, Ito, 1969). Unfortunately, there have been few empirical sampling studies of how large an $N$ is needed for this robustness of multiple regression analysis under violations of normality assumptions to evidence itself. We can expect the overall test of the hypothesis that the population value of $R$ is zero will display considerably more robustness than tests on individual regression coefficients. The author recommends interpreting these latter tests with great caution if the $X$s and the $Y$s have grossly nonnormal distributions and $N - m$ is less than about 50.

Other authors have been even more pessimistic. Cooley and Lohnes (1971) cite Marks's (1966) recommendation, on the basis of a series of computer-generated cross-validation studies (in which a prediction equation is derived from one sample and then tested for predictive ability on a second random sample), that the sample regression coefficients be ignored altogether in developing a prediction equation, replacing them with the simple correlation between each predictor $i$ and $Y$, $r_{iy}$, whenever $N$ is less than 200. Cooley and Lohnes consider this recommendation overly pessimistic, though they do not provide an alternative "critical value" of $N$. Actually, the sample sizes used by Marks were 20, 80, and 200, so that his data would really only support a recommendation that the regression weights not be taken seriously unless $N > 80$—not too different from the author's earlier $N - m > 50$ stipulation. More importantly, Marks's findings—which include higher predictor-criterion correlations using unity (1.0) as the weight for each predictor than from the regression coefficients computed on a previous sample in 65% of the cases—seem to be at odds with the well-known theorem (cf., for example, Draper and Smith, 1966, p. 59) that for data which fit the assumptions of normality and linearity the sample regression coefficient $b_i$ provides an unbiased estimate of the corresponding population parameter $\beta_i$, and has a sampling distribution with lower variance than any other unbiased estimator which is linearly related to the $Y$s.

These findings have nevertheless been supported by further Monte Carlo and empirical studies summarized by Dawes and Corrigan (1974). In particular, Schmidt (1971) found that for population correlation matrices which involved "suppressor" variables (variables whose correlations with the criterion are negligible, but which receive sizeable weights in the population regression equation), unit weights outperformed least-squares regression weights unless the ratio of $N$ to $m$ was 25 or more, while for his matrices which did include such "suppressor variables," a ratio of 15 to 1 was necessary. This suggests that the crucial factor determining whether regression weights will outperform unity weights (or simple predictor–criterion correlations) is how different the population regression weights are from the *a priori* weights of unity or $r_{iy}$. For instance, Dawes and Corrigan (1974) point out that Marks's (1966) simulations were subject to the condition that "the partial correlation between any two predictors, partialling out the criterion variable, was zero." It can readily be shown that this implies that the true population regression weight for predictor variable $X_i$ will be equal to $r_{iy}/[\sum_j r_{jy}^2 + (1 - r_{iy}^2)]$, i.e., that the population regression weights are very nearly directly proportional to the simple predictor–criterion correlations. Marks thus inadvertently "loaded" his Monte Carlo runs in favor of $r_{iy}$ as the weighting factor—and probably unity weights as well, depending on how variable his population values of $r_{iy}$ were. (Marks's paper is unavailable at this writing except in abstract form.)

Perhaps the best remedy for a tendency to read more from a set of regression coefficients than their reliability warrants would be to construct a confidence interval about each one, that is, $b_j \pm t_{.05}\sqrt{d_{jj}\mathrm{MS}_{\mathrm{res}}}$, where $t_{.05}$ is the 97.5th percentile of the *t*-distribution having $N - m - 1$ degrees of freedom. (This set of *m* individual confidence intervals is of course only an approximation to the joint confidence interval or confidence ellipsoid for all *m* regression coefficients considered simultaneously. Consult the work of Morrison, 1967, for computational procedures for constructing such joint confidence intervals.)

※ It may help to reduce the level of faith necessary to accept the tests in Table 2.2 if we spend some time discussing the "reasonableness" of one of these tests, namely the test of the hypothesis that $\beta_j = \alpha$. Now, $b_j = j$th element of $(\mathbf{x}'\mathbf{x})^{-1}\mathbf{x}'\mathbf{y}$ = a linear function of *y*, where the weighting coefficients are given by the *j*th row of $\mathbf{K} = (\mathbf{x}'\mathbf{x})^{-1}\mathbf{x}'$. However, each *y* is an independently sampled observation from the population of *y* scores, so the variance of $b_j$ should be given by $(\sum_{i=1}^N k_{j,i}^2)\sigma_y^2$, which we can of course estimate by

$$(\sum k_{j,i}^2)\mathrm{MS}_{\mathrm{res}} = (\mathbf{k}_j' \mathbf{k}_j)\mathrm{MS}_{\mathrm{res}},$$

where $\mathbf{k}_j'$ is the *j*th row of $\mathbf{K} = \mathbf{e}'\mathbf{K}$, where $\mathbf{e}$ is a vector whose only nonzero entry occurs in the *j*th position. Thus the variance of $b_j$ is

$$(\mathbf{e}'\mathbf{KK}'\mathbf{e})\mathrm{MS}_{\mathrm{res}} = \mathbf{e}'(\mathbf{x}'\mathbf{x})^{-1}\mathbf{x}'\mathbf{x}(\mathbf{x}'\mathbf{x})^{-1}\mathbf{e} = \mathbf{e}'(\mathbf{x}'\mathbf{x})^{-1}\mathbf{e}$$

$$= d_{jj},$$

where the last equation on the first line is the *j*th main diagonal entry of $(\mathbf{x}'\mathbf{x})^{-1}$. Thus $(b_j - \alpha)^2/(d_{jj}\mathrm{MS}_{\mathrm{res}})$ is directly analogous to the familiar expression $[(x - \mu_x)/s_x]^2 = t^2 = F$. A highly similar argument can be made for the reasonableness of the test for a contrast among the $\beta$s, since a linear combination (for example, a contrast) of the *b*s is in turn a linear combination of the *y*s, so that the variance of $\mathbf{a}'\mathbf{b}$ is estimated by

$$\mathrm{var}(\mathbf{a}'\mathbf{b}) = \mathbf{a}'(\mathbf{x}'\mathbf{x})^{-1}\mathbf{x}'\mathbf{x}(\mathbf{x}'\mathbf{x})^{-1}\mathbf{a} = \mathbf{a}'(\mathbf{x}'\mathbf{x})^{-1}\mathbf{a}.$$

Finally, note that the entries for degrees of freedom (df) in Table 2.2 assume that the *m* predictors are not linearly related, that is, cannot be related perfectly by a linear function. More generally, we should replace *m* by the rank of the matrix $(\mathbf{x}'\mathbf{x})$.

Of course, where $\mathbf{x}'\mathbf{x}$ is of less than full rank we cannot obtain its inverse, so that we must resort to one of the techniques discussed in Section 3.2, namely dropping one of the redundant predictors or working instead with the subjects' scores on the principal components (cf. Chapter 6) of $\mathbf{x}'\mathbf{x}$.

## 2.4   Computer Programs for Multiple Regression

At the beginning of Section 2.3 the inverse of a $4 \times 4$ matrix was displayed. This is well within the lower range of regression analyses in terms of the number of predictors apt to be employed in a problem of practical importance, but already it is a considerable strain on the patience and accuracy of a researcher armed only with a desk calculator. Using the cofactor approach requires, as an *initial* step, computing the determinants of 10 different $3 \times 3$ matrices. Using the row-operation approach requires 12 additions of a multiple of one row to another row, and provides multitudinous possibilities for numerical errors, including round-off losses and just plain arithmetic goofs (reversing a sign, for instance). Fortunately, computers are now so generally available, together with "canned" programs designed to handle most commonly desired statistical analyses, that there is no longer a need for the many involved discussions of methods to reduce the labor of hand computations and provide intermediate checks on accuracy which fill early textbooks on multivariate statistics. There are, however, some precautions and words of advice with respect to the use of computer programs—as well as descriptions of some of the most useful programs available nationally—in Digression 4, to which you should now turn.

## 2.5   Some General Properties of Covariance Matrices

One of the steps involved in conducting an $m$-predictor multiple regression analysis is the inversion of the $m \times m$ matrix of the covariances (or cross-products, or intercorrelations) of the predictor variables. As is proved in Section D2.11 in Digression 2, we may eliminate the usual succeeding step of postmultiplication by $\mathbf{x}'\mathbf{y}$ (or $\mathbf{s}_{xy}$, if we inverted the covariance matrix) if we are willing to undertake the inversion of the slightly larger $(m + 1) \times (m + 1)$ augmented covariance matrix which includes the covariance of each predictor with the outcome variable as its $(m + 1)$st row and column. Specifically, it is proved in Digression 2 that

$$d_{ii} = [s_i^2(1 - R_{i \cdot c}^2)]^{-1};$$

whence

$$R_{i \cdot c}^2 = 1 - 1/(s_i^2 d_{ii}) \qquad \text{and} \qquad b_{i,j \cdot c} = -d_{ij}/d_{ii},$$

where $d_{ij}$ is the entry in the $i$th row and $j$th column of $\mathbf{S}^{-1}$, the inverse of the covariance matrix of a set of variables $= (N - 1)c_{ij}$, where $c_{ij}$ is the $(i,j)$th entry of $(\mathbf{x}'\mathbf{x})^{-1}$; $R_{i \cdot c}^2$ is the squared coefficient of multiple correlation between variable $i$ and all variables other than $i$ that are represented in $\mathbf{S}$; and $b_{i,j \cdot c}$ is the regression coefficient corresponding to variable $j$'s contribution to the regression equation which best predicts scores on variable $i$ from scores on $\mathbf{c}$, the set of variables other than $i$ involved in the regression equation. The expression for $d_{ii}$ in terms of $R_{i \cdot c}^2$ provides yet another insight into why $c_{ii}$ appears in the denominator of the test of significance for the single sample regression coefficient $b_i$ in Table 2.2. Note that

$$c_{ii} = d_{ii}/(N - 1) = 1/[\sum x_i^2 (1 - R_{i \cdot c}^2)],$$

the reciprocal of the portion of $\sum x_i^2$ which is unique to variable $i$, not being shared with (predictable from) the other predictor variables. The more "unique variance" $X_i$ contributes to the system, the smaller will be $c_{ii}$ and thus the larger will be the $F$-ratio for this test of significance of $X_i$'s contribution.

This relationship provides an opportunity for "milking" some information from canned regression analysis computer programs that is not ordinarily provided explicitly by such programs. Any such program will provide a significance test of the difference between $b_i$ and a hypothesized value of zero. BMDO2R employs the $F$-test outlined in Table 2.2, while other programs (for example, U6603H, a program written at the University of New Mexico and based on readily available subroutines from IBM's Scientific Subroutine Package) report a $t$-test which is simply the square root of the $F$-ratio. Thus, for instance, the entry for predictor variable $i$ under the $F$-test column of the output from BMD02R would be equal to

$$F_i = b_i^2/(c_{ii} \, \mathrm{MS}_{\mathrm{err}});$$

whence

$$F_i = (N - 1)b_i^2/(\mathrm{MS}_{\mathrm{err}} \, d_{ii}) = (N - 1)b_i^2 s_i^2 (1 - R_{i \cdot c}^2)/\mathrm{MS}_{\mathrm{err}};$$

whence

$$R_{i \cdot c}^2 = 1 - F_i \, \mathrm{MS}_{\mathrm{err}}/[b_i^2 s_i^2 (N - 1)].$$

Thus, although the multiple correlation between each predictor and the other predictor variables is not usually reported by "canned" programs, it can readily be obtained from values which *are* reported.

The ready availability of all this additional information, while very convenient for, for example, item analysis where the predictors are scores on individual items of some attitude scale or battery of tests, also raises the spectre of the multiple comparison problem in yet another guise. It might be tempting to obtain the inverse of the covariance matrix of some large set of variables, use the main diagonal entries of this inverse to compute $R^2$ for each variable predicted from the other variables, and test each $R^2$ for significance. It would therefore be useful to have available, before we start this "fishing expedition," a test of the overall null hypothesis that none of the off-diagonal entries in the *population* covariance matrix differ from zero. This is equivalent to testing the hypothesis that the true population correlation between each pair of variables is zero. Bartlett (1954) has provided such a test, namely

$$\chi^2 = -[N - 1 - (2m + 5)/6] \ln |\mathbf{R}| \qquad \text{with} \quad m(m - 1)/2 \text{ df},$$

where $m$ is the number of variables involved in the correlation matrix $\mathbf{R}$, and $\ln |\mathbf{R}|$ represents the natural logarithm of the determinant of $\mathbf{R}$. That $|\mathbf{R}|$ should be involved in this test is intuitively reasonable, since the smaller $|\mathbf{R}|$ is, the closer the system is to a perfect linear relationship among the variables (the direct opposite of no relationship, scores on any one being predictable from a linear combination of the other) and the more closely $-\ln|\mathbf{R}|$ approaches positive infinity.

Lawley (1940) has shown that if $N$ is large or the sample coefficients small, a good approximation to this chi-square test is provided by replacing $\ln|\mathbf{R}|$ with $\sum\sum_{i<j}r_{ij}^2$, the sum of the above-diagonal entries of $\mathbf{R}$.

## 2.6   Measuring the Importance of the Contribution of a Single Variable

One of the issues which often arises in applications of multiple regression analysis is that of the relative importance of the contribution of a particular variable to the predictive ability represented by $R^2$. This turns out to be closely related to the issue of interpreting the regression coefficients in an application of multiple regression analysis. Both of these issues are discussed by Darlington (1968), on whose paper a number of the comments in this section are based.

The crucial question to ask in assessing the contribution of a given variable is "Why?" Is this measure of importance to be used in deciding whether to retain the variable in the regression equation? Or is it to be used instead in assessing which among a number of variables are the most important determinants of scores on the outcome measure?

In the first case, the obvious measure to use is the drop in $R^2$ which results when that variable is eliminated from the regression equation. This measure is referred to by Darlington as the "usefulness" of the variable. Looking back at Table 2.2, we can see that this can be written as

$$U_i = R^2 - R^{*2} = (\mathbf{x}'\mathbf{y})(\mathbf{b} - \mathbf{b}^*)/\sum y^2,$$

where $R^{*2}$ is the squared multiple $R$ obtained when variable $i$ is omitted and $\mathbf{b}^*$ is the vector of regression coefficients in which the $i$th position is set to 0 and the $m-1$ other $b_j$s are computed on the basis of the first $m-1$ predictors only. This would appear to require two separate multiple regression analyses. However, it is possible to show (proof available from the author) that $U_i$ also equals $b_{z_i}^2(1 - R_{i \cdot c}^2)$, where $b_{z_i}$ is the normalized regression coefficient for predictor $i$. That is, it is the regression coefficient used in the equation for predicting $z$-scores on $Y$ from $z$-scores on the predictors $= b_i(s_i/s_y)$, and $R_{i \cdot c}$ is the multiple correlation between variable $i$ and the remaining $m-1$ predictor variables. In Section 2.5 we demonstrated that $R_{i \cdot c}^2$ could be computed from the $i$th main diagonal entry of $\mathbf{C} = \mathbf{S}^{-1}$, or indirectly from the $t$- or $F$-ratio obtained in testing the hypothesis that $b_i = 0$. Thus the output of any of the common computer programs for multiple regression analysis provides all the information needed to compute $U_i$ for each predictor variable.

The second goal is much more ambiguous, and often inappropriate. Many researchers who ask this question are really interested in assessing the value of the variable in a very different context than that in which the prediction equation was derived. For instance, Steele and Tedeschi (1967) attempted to develop an index based on the payoffs used in a two-person, two-choice experimental game which would be predictive of the overall percentage of cooperative choices $C$ made in that game. Their approach was to generate 210 different indices, select from among these the 15 which correlated most highly with $C$, compute the corresponding regression equation, and then compare $b_z^2$ for the various predictors. If, as their discussion implies, they were interested in generating a single best predictor of $C$, this procedure was clearly inappropriate. Such a goal would seem to dictate selection of that variable the absolute value of whose simple correlation with $C$ was highest. Putting the various variables into a regression analysis was irrelevant to the goal of selecting a single best predictor, since a regression coefficient provides information about that variable's contribution only within the highly specific context of the other $m - 1$ predictors. As we have seen in Data Set 1, a variable having a very low correlation with $Y$ considered by itself may receive quite high weight in a regression analysis involving two other predictors, and *vice versa*. Also a variable which is very important in the three-predictor case may receive near-zero weight when a fourth predictor is added.

Given that we have computed the regression equation for the set of predictors we actually plan to employ, there are still several possible measures of the relative importance of the various predictors in that equation. The most straightforward is $b_{z_i}^2$, since it indicates how variable $z$-scores on $b_i z_i$ will be, and thus how much influence scores on $X_i$ will have in determining the ranking of subjects on $\hat{Y}$. However, the squares of the $z$-score regression coefficients do not provide an additive "partition" of $R^2$, since

$$R^2 = \sum b_{z_i}^2 + \sum\sum_{i<j} b_{z_i} b_{z_j} r_{ij} .$$

In other words, $b_{z_i}^2$ ignores the contributions variable $X_i$ makes through its correlations with the other predictors. If we arbitrarily assign one-half of a pair of predictors' "joint contribution" $2b_i b_j r_{ij}$ to each of the two predictors, we obtain

$$b_{z_i}^2 + \left(\sum_{j \neq i} b_{z_j} r_{ij}\right) b_{z_i} = b_{z_i}\left(\sum_{\text{all } j} b_{z_j} r_{ij}\right)$$

—which, as it turns out, also equals $b_{z_j} r_{iy}$—as a measure of importance which *does* provide an additive partition of $R^2$. However, it lacks any operational interpretation like that given above for $U_i$ (the drop in $R^2$ which would result if $X_i$ were dropped from the regression equation), which interpretation makes $U_i$ a serious candidate for a measure of importance.

Cooley and Lohnes (1971) make perhaps the most radical suggestion in their recommendation that $r_{iy}$ be used as the measure of the importance of $X_i$ to the endeavor of predicting scores on $Y$. They base this recommendation primarily on the fact that this simple correlation with $Y$ is proportional to that variable's "loading on," that is, its correlation with, $\hat{Y}$. $\hat{Y}$ is the linear composite of the $X$s which the regression analysis proposes as indicative of the true relationship between $Y$ and the $X$s and is thus similar in that respect to a factor in factor analysis. Traditionally the emphasis in interpreting factors has been on examining those original variables which correlate most highly with (have the highest loadings on) a given factor (cf. Chapters 6 and 7). It can be argued, however, that Cooley and Lohnes' argument should be reversed, and that the often rather striking discrepancy between $b_{z_i}^2$ and $r_{iy}^2$ should serve as evidence that the emphasis in interpreting factors would more appropriately be put on factor score coefficients (corresponding to $b_{z_i}$) than on the indirect criteria of factor loadings (corresponding to $r_{iy}$).

Finally, perhaps the most appropriate resolution to the issue of importance of contributions is to recognize that the relationship summarized by $R^2$ is a relationship between $Y$ and the *set* of predictor variables as a whole. Rather than attempting to decompose this relationship, and assigning "credit" to the original variables, it might be more rewarding theoretically to try to interpret this new composite variable $\hat{Y}$ substantively.

## DEMONSTRATION PROBLEM

Each of the remaining chapters of the present text will provide two kinds of data for you to use in practicing the techniques presented in that chapter: real data about which some researcher really cared, and a highly simplified set of data to be used as the basis of a demonstration problem which asks you to do a large number of analyses by hand which serve to illustrate that the heuristic explanations of what these techniques "really" do actually hold up when you "plug into" the various formulas. The demonstration problem is very important to a full understanding of, in this case, multiple regression.

| Subject | $X_1$ | $X_2$ | $X_3$ | $Y$ |
|---------|-------|-------|-------|-----|
| A | 6 | 3 | 2 | 4.2 |
| B | 5 | 6 | 0 | 4.0 |
| C | 4 | 2 | 9 | 5.0 |
| D | 7 | 1 | 5 | 4.1 |
| E | 3 | 8 | 9 | 6.4 |

**A.**  Do each of the following by hand (and/or with the help of a desk calculator).

**1.**  Construct the cross-product matrix for the predictor variables, and the vector of cross-products of the predictor variables with $Y$.

**2.**  Use the information from Problem 1 to find the regression coefficients for predicting:

(a)  $Y$ from $X_1$;     (b)  $Y$ from $X_1$ and $X_2$;     (c)  $Y$ from $X_1, X_2$, and $X_3$.

**3.**  Calculate $R^2$ and $R$ for Problems 2a–2c.

**4.**  Construct summary tables for Problems 2a–2c and use the tables to decide whether $\rho$ (the true population multiple correlation coefficient) can "reasonably" (say, with $\alpha = .05$) be assumed to be zero.

**5.**  For Problem 2c, calculate $\hat{Y}$ for each subject. In addition, calculate $\sum \hat{Y}/N, \sum (\hat{Y} - \bar{Y})^2$, and the Pearson product–moment correlation coefficient for the correlation between $\hat{Y}$ and $Y$. To what already computed statistic does each of these correspond?

**6.** For Problem 2c, calculate $e_i = Y_i - \hat{Y}_i$ for each subject. Also calculate $\sum e$, $\sum (e_i - \bar{e})^2$, and the correlation between $Y_i$ and $e_i$. To what previously calculated statistic does each of these correspond?

**7.** Calculate the inverse of the $3 \times 3$ covariance matrix of the predictor variables.

**8.** Use the results of Problem 7 to calculate the multiple correlation between

(a) $X_3$ vs $X_1$ and $X_2$;  (b) $X_2$ vs $X_1$ and $X_3$;  (c) $X_1$ vs $X_2$ and $X_3$.

**9.** Write the regression equations for Problems 8a–8c.

**10.** For Problems 2a–2c test the hypothesis that the population value of $\beta_i$ (the "weight" or "loading" of variable 1 in the regression equation) is zero.

**11.** Test the significance of the increase in $R^2$ (and thus in predictive ability) which arises from adding variable 2.

**12.** Test the significance of the increase in $R^2$ which results from adding variable 3 to the first two prediction variables.

**13.** Test the hypothesis that $\beta_1$ and $\beta_3$ are identical in the three-predictor equation.

**14.** Plot $Y$ against $\hat{Y}$ and examine for departures from linearity.

**B.** Do each of the following problems by hand and/or desk calculator.

**1.** Calculate the intercorrelations among $X_1$, $X_2$, $X_3$, and $Y$.

**2.** Test the hypothesis that the true population value of each of the six intercorrelations is zero.

**3.** On the basis of the results of Problem A2 (that is, the raw-score coefficients), predict what the corresponding regression coefficients will be in each case for predicting $z$-scores on $Y$ from $z$-scores on the $X$-variables.

**4.** Now carry out the multiple regression analyses of Problem A2, beginning in each case with the intercorrelation matrix, rather than with the covariance matrix.

**5.** Repeat Problems A4–A14, starting in each case with the intercorrelations, rather than the covariances.

**C.** Use the computer (for example, program BMDO2R or SPSS's REGRESSION subprogram) for each of the following problems.

**1.** Conduct as many of the analyses of parts A and B as possible. Do the analyses of Problems A8 and A9 in two ways:
   (a) Using the computer for each multiple regression analysis, specifying a new dependent variable and a new pair of predictor variables for each "run," and
   (b) Using only the information provided by the computer's analysis of $Y$ predicted from $X_1$, $X_2$, and $X_3$.

**2.** Perform the additional analyses needed to test the significance of the increase in $R^2$ provided by adding a quadratic component to the regression analysis in Problem A2b, that is, by adding $X_1^2$ as a predictor variable. Conduct the significance test.

**3.** Why were you not asked to compare the $R^2$ obtained when $X_1^2$, $X_2^2$, and $X_3^2$ are added as predictor variables?

## Answers

**A.**

**1.**

|   | $x_1$ | $x_2$ | $x_3$ | $y$ |   | $x_1$ | $x_2$ | $x_3$ | $y$ |
|---|---|---|---|---|---|---|---|---|---|
| A | 1 | −1 | −3 | −.54 | | 10 | −13 | −15 | −5.4 |
| B | 0 | 2 | −5 | −.74 | | | 34 | 1 | 7.1 |
| C | −1 | −2 | 4 | .26 | | | | 66 | 13.0 |
| D | 2 | −3 | 0 | −.64 | | | | | 4.072 |
| E | −2 | 4 | 4 | 1.66 | | | | | |
| Mean | 5 | 4 | 5 | 4.74 | | | | | |

Cross-product matrix     Cross-product vector

**2.** (a) $b_1 = -5.4/10 = -.54$;    $b_0 = \overline{Y} - b_1\overline{X} = 4.74 - (-.54)(5) = 7.44$.

(b) $\begin{bmatrix} b_1 \\ b_2 \end{bmatrix} = \begin{bmatrix} 10 & -13 \\ -13 & 34 \end{bmatrix}^{-1} \begin{bmatrix} -5.4 \\ 7.1 \end{bmatrix} = \frac{1}{171}\begin{bmatrix} 34 & 13 \\ 13 & 10 \end{bmatrix}\begin{bmatrix} -5.4 \\ 7.1 \end{bmatrix} = \frac{1}{171}\begin{bmatrix} -91.3 \\ 0.8 \end{bmatrix} = \begin{bmatrix} -.534 \\ .005 \end{bmatrix}$;

$b_0 = 4.74 - 5(-.534) - 4(.005) = 7.390$.

(c) $(\mathbf{x'x})^{-1} = \frac{1}{4016}\begin{bmatrix} 2243 & 843 & 497 \\ 843 & 435 & 185 \\ 497 & 185 & 171 \end{bmatrix}$;   $\begin{bmatrix} b_1 \\ b_2 \\ b_3 \end{bmatrix} = \frac{1}{4016}\begin{bmatrix} 334.1 \\ 941.3 \\ 852.7 \end{bmatrix} = \begin{bmatrix} .083 \\ .234 \\ .212 \end{bmatrix}$.

$b_0 = 4.74 - 5(.083) - 4(.234) - 5(.212) = 2.329$

**3.** (a) $R^2 = r_{1y}^2 = .846^2 = -.54(-5.4)/s_y^2 = 2.916/4.072 = .716$.

(b) $R^2 = [-.5.4(-.534) + 7.1(.005)]/4.072 = 2.916/4.072 = .716$;    $R = .846$.

(The computer program's greater accuracy reveals that $R$ in the first case is .84623, while it is .84629 in the two-predictor case.)

(c) $R^2 = [(-5.4)(.083) + 7.1(.234) + .212(13.0)]/4.072 = .975$;

$R = .988$.

**4.** Summary tables. (None of the $F$s are statistically significant.)

| Source | df | SS | MS | F | df | SS | MS | F | df | SS | MS | F |
|---|---|---|---|---|---|---|---|---|---|---|---|---|
| Regression | 1 | 2.916 | 2.916 | 7.57 | 2 | 2.916 | 1.458 | 2.52 | 3 | 3.976 | 1.325 | 13.68 |
| Residual | 3 | | .386 | | 2 | 1.156 | .577 | | 1 | .097 | .097 | |
| Total | 4 | 4.072 | | | 4 | 4.072 | | | 4 | 4.073 | | |

**5.** Predicted scores

| Subject | $.083x_1$ | $.234x_2$ | $.212x_3$ | $y'$ | $Y'$ | $\varepsilon$ |
|---|---|---|---|---|---|---|
| A | .083 | −.234 | −.636 | −.787 | 3.953 | .247 |
| B | 0 | .468 | −1.060 | −.592 | 4.148 | −.148 |
| C | −.083 | −.468 | .848 | .297 | 5.037 | −.037 |
| D | .166 | −.702 | 0 | −.536 | 4.204 | −.104 |
| E | −.166 | .936 | .848 | 1.618 | 6.358 | .042 |

$$\hat{Y} = 23.7/5 = 4.74 = \bar{Y}; \qquad \sum(\hat{Y} - \bar{Y})^2 = 3.965 = SS_{regr};$$

$$r_{\hat{Y}Y} = [(-.787)(-.54) + (-.592)(-.74) + \cdots]/\sqrt{(3.965)(4.072)}$$

$$= .986 = R.$$

**6.** See Problem A5 for error scores.

$$\sum e_i = .489 - .489 = 0.000;$$

$$\sum(e_i - \bar{e}_i)^2 = \sum e_i^2 = .0969 = SS_{res}; \qquad r_{Ye} = .160 = \sqrt{1 - R^2} \neq 0.$$

Thus our estimates of the residuals are *not* completely independent of the *Ss'* scores on *Y*.

**7.** $S_x^{-1} = 4(x'x)^{-1}$. Note that $(x'x)^{-1}$ was given in the answers to Problem 2.

**8.**

| $i$ | $s_i^2 c_{ii}$ | $1/(s_i^2 c_{ii})$ | $R_{i \cdot c}^2$ | $R_{i \cdot c}$ |
|---|---|---|---|---|
| (c) 1 | $2.5(2243/1004) = 5.585$ | .179 | .821 | .908 |
| (b) 2 | $8.5(435/1004) = 3.68$ | .272 | .728 | .853 |
| (a) 3 | $16.5(171/1004) = 2.81$ | .356 | .644 | .802 |

**9.** (a) $x_3 = -(497/171)x_1 - (185/171)x_2 = -2.91x_1 - 1.082x_2;$
(b) $x_2 = -(843/435)x_1 - (185/435)x_3 = -1.94x_1 - .425x_3;$
(c) $x_1 = -(843/2243)x_2 - (497/2243)x_3 = -.375x_2 - .222x_3.$

**10.** (a) Same as overall $F$, ns.
(b) $F = b_1^2/(d_{ii} MS_{res}) = (.534)^2/[(34/171)(.57781)] = 2.48 = 1.575^2$, ns.
(c) $F = (.083)^2/[(2243/4016)(.097)] = 27.7/220.9 = .1259 = .355^2$, ns.

**11.** Increase is very close to zero. Not worth testing.

**12.** $F = (3.976 - 2.916)/.097 = 10.91$ with 1 and 1 df, ns.

**13.** $H_0 : \beta_1 - \beta_3 = 0$. The sum of squares for contrast,
$SS_{contr} = (.212 - .083)^2/.3536 = .0471.$
$F = SS_{contr}/MS_{res} = .0471/.097 < 1$, ns.

**14.** No nonlinearity apparent.

**B.**

**1.** $R = DSD$, where $D$ is a diagonal matrix whose entries are the reciprocals of the standard deviations of the variables, whence

$$R = \begin{bmatrix} 1 & -.706 & -.584 & -.846 \\ & 1 & .021 & .604 \\ & & 1 & .793 \\ & & & 1 \end{bmatrix}.$$

**2.** $\chi^2 = -(4 - 13/6) \ln|R| = -(11/6) \ln|R|$. However, $|AB| = |A| \cdot |B|$, so

$$|R| = |D|^2 \cdot |S_x| = 388.96/(10 \cdot 34 \cdot 66 \cdot 4.072) = .0443;$$

whence

$$\chi^2 = (11/6)(5.459) = 10.01 \quad \text{with 6 df, ns.}$$

The approximate test gives

$$\chi^2 = (11/6) \sum r^2 = 4.67 \quad \text{with 6 df, ns.}$$

**3.** $\hat{Y} = b_0 + b_1 X_1 + b_2 X_2 + \cdots + b_m X_m;$     $\hat{Y} = \bar{Y} = b_1 x_1 + b_2 x_2 + \cdots + b_m x_m;$

$\hat{z}_y = (Y' - \bar{Y})/s_y = (b_1/s_y)x_1 + (b_2/s_y)x_2 + \cdots + (b_m/s_y)x_m.$

However,   $z_i = x_i/s_i;$   whence   $x_i = s_i z_i;$   whence

$\hat{z}_y = b_1(s_1/s_y)z_1 + b_2(s_2/s_y)z_2 + \cdots + b_m(s_m/s_y)z_m = b_1^* z_1 + b_2^* z_2 + \cdots + b_m^* z_m$

where $b_i^* = b_i(s_i/s_y).$

| Predictors | $b_1^*$ | $b_2^*$ | $b_3^*$ |
|---|---|---|---|
| $z_1$ | −.846 | — | — |
| $z_1, z_2$ | −.836 | .014 | — |
| $z_1, z_2, z_3$ | .130 | .676 | .853 |

**4.**   Same results as in Problem A2.

**5.**   Problem A4:   $SS_{regr} = (N - 1)R^2,$   $SS_{res} = (N - 1)(1 - R^2),$   $SS_{tot} = N - 1 = 4$ in each case; F unchanged.

   Problem A5:   Mean $\hat{z} = 0;$   $\sum \hat{z}^2 = (N - 1)R^2;$   $r_{\hat{z}z} = R.$

   Problem A6:   Mean error $= 0;$   sum of squared errors $= (N - 1)(1 - R^2).$

   Problem A7:   $\mathbf{R}^{-1} = (\mathbf{DS}_x\mathbf{D})^{-1} = \mathbf{D}^{-1}\mathbf{S}_x^{-1}\mathbf{D}^{-1} = \begin{bmatrix} 5.590 & 3.87 & 3.07 \\ & 3.68 & 2.17 \\ & & 2.81 \end{bmatrix}.$

   Problem A8:   Same multiple Rs, but obtained much more simply as $1 - 1/c_{ii}.$

$$\hat{z}_3 = -1.091z_1 - .772z_2.$$

(Note that the coefficients in each case are simply $s_i/s_y$ times previous coefficients.)

   Problem A10:   Same Fs.

   Problems A11–A12:   Same Fs, simpler arithmetic.

   Problem A13:   $F = [(.130 - .853)^2/2.26]/.0239 = 9.68$     with   3 and 1 df;     ns.
   *Not* the same as for covariance-based bs.

   Problem A14:   Same shape, though different units,   origin.

## C.

**1.** Comparable results. Only "new" analysis is using program output from the three-predictor case to do Problem A8, as follows:

$$R_{i \cdot c}^2 = 1 - [s_i^2 c_{ii}(N - 1)]; \qquad t_i = b_i/(MS_{res} c_{ii});$$

whence

$R_{i \cdot c}^2 = 1 - t_i^2 MS_{res}/[b_i^2 s_i^2 (N - 1)]$
   $= 1 - 7.4(.0969)/4 = 1 - .177 = .823$     for   $i = 1;$
   $= 1 - [2.288/(.2344)(2.915)]^2(.09685/4) = 1 - .272 = .728$     for   $i = 2;$
   $= 1 - .355 = .645$     for   $i = 3.$

(Substitute F for $t^2$ if the program you used reports F-tests.) There is no simple way, however, to obtain the regression coefficients (answer to Problem A9) from the given information.

**2.** Adding only $X_1^2$ gives $R = .985$; adding only $X_2^2$, gives $R = .960$. Neither $F$ significant.

**3.** This would have produced negative degrees of freedom and $R = 1.00$, since you would be using five predictors on only five $S$s.

## SOME REAL DATA AND A QUIZ THEREON

As part of a demonstration of Thurstone and Likert scaling techniques (cf. Selltiz, Jahoda, Deutsch, and Cook, 1959), students in the author's social psychology lab course served both as judges and as pilot subjects in the development of a scale to measure the position of students on a liberal–conservative continuum. The primary issue in comparing the lab students' ratings of how conservative various items are with the student's own conservatism scores is whether (as Thurstone contended when he first developed Thurstone scaling techniques) the scale value of an item is unaffected by the positions of the judges used. The data are given in Data Set 2.

If you are reading through this Primer with others—for example, as part of a class—and if your local computing facility has a shared file system such as IBM's RAX (for "remote access") system, you may wish to pool your efforts in punching and verifying these data in a file accessible to all of you.

## REAL DATA QUIZ I

**I.** Conduct each of the following analyses of the Conservatism Data.

**A.** Using a computer program which employs a step-down multiple regression procedure, test the null hypothesis that the judges' responses to (judgments of) statements 3, 5, 6, 12, 18, and 27 were unrelated to their own scores on the Conservatism scale. Which individual variables contribute significantly to the regression equation? Test the hypothesis that the simple sum of the judges' judgments of statements 3, 5, 6, 12, 18, and 27 is significantly related to their Conservatism scale score. Using a $t$-value of 1.96 for each successive step, conduct a "step-down" stepwise multiple regression analysis of these data.

**B.** Using BMDO2R or a similar program and treating all six predictors as "free variables," conduct a "step-up" regression in such a way that you are certain that all six predictors are included. Now force a step-up to include variables in the order 3, 5, 6, 12, 18, 27 and then in the order 27, 18, 12, 6, 5, 3. Any interesting differences? Now do a step-up in which all predictors are free variables and the $F$s for inclusion and deletion are both 1.0, while the tolerance level is 50%. Any comments?

**II.** Answer the following "thought questions."

**A.** (Answerable on the basis of intuition about the nature of multiple regression.)

**1.** Can we, simply by examining the intercorrelation matrix, determine a minimum value of multiple $R$?

**2.** In conducting an item analysis of a questionnaire, each item is typically correlated with "total scale score," which equals the sum of the scores on the separate items. Could we instead

*Data Set 2*

*Ratings of Conservatism of Statement Number, Multiplied by 10[a]*

| Subject | 3 | 5 | 6 | 12 | 18 | 27 | Conservatism score[b] |
|---|---|---|---|---|---|---|---|
| A | 15 | 15 | 65 | 70 | 70 | 100 | 61 |
| B | 10 | 10 | 100 | 50 | 50 | 65 | 72 |
| C | 20 | 30 | 100 | 70 | 30 | 70 | 61 |
| D | 23 | 19 | 83 | 46 | 53 | 84 | 56 |
| E | 10 | 15 | 80 | 15 | 25 | 80 | 67 |
| F | 10 | 40 | 70 | 100 | 40 | 100 | 87 |
| G | 25 | 15 | 85 | 60 | 45 | 70 | 48 |
| H | 15 | 12 | 10 | 30 | 50 | 100 | 65 |
| I | 20 | 20 | 70 | 50 | 50 | 80 | 65 |
| J | 10 | 20 | 80 | 60 | 40 | 90 | 68 |
| K | 37 | 13 | 76 | 66 | 45 | 50 | 70 |
| L | 30 | 50 | 10 | 10 | 50 | 40 | 63 |
| M | 57 | 32 | 78 | 76 | 42 | 87 | 69 |
| N | 40 | 30 | 80 | 50 | 40 | 70 | 54 |
| O | 10 | 20 | 90 | 70 | 30 | 82 | 96 |
| P | 30 | 42 | 82 | 34 | 82 | 44 | 67 |
| Q | 30 | 15 | 75 | 50 | 75 | 94 | 76 |
| R | 15 | 30 | 80 | 20 | 50 | 50 | 71 |
| S | 20 | 30 | 95 | 40 | 35 | 100 | 86 |
| T | 15 | 35 | 100 | 65 | 95 | 85 | 73 |
| U | 05 | 27 | 85 | 72 | 40 | 87 | 53 |
| V | 80 | 95 | 100 | 80 | 50 | 27 | 59 |
| W | 01 | 09 | 84 | 36 | 49 | 69 | 57 |
| X | 20 | 11 | 100 | 49 | 30 | 60 | 70 |
| Y | 20 | 10 | 70 | 40 | 70 | 80 | 54 |
| Z | 90 | 15 | 90 | 69 | 35 | 80 | 52 |
| 1 | 20 | 40 | 95 | 80 | 40 | 35 | 53 |

[a] Ratings were made on a 10-point scale, 1.0–10.0. (Not all $S$s stayed within this range.) 1.0 was labeled "Extremely liberal," and 10.0 was labeled "Extremely conservative."

[b] The conservatism score was derived from $S$'s agreement or disagreement with the same statements he had previously rated for degree of conservatism. The most conservative possible score was 138; the most liberal possible score was 23. The statements which correspond to the numbered columns are:

3. "The University of New Mexico should hire a full-time counselor to counsel legal avoidance of the draft."

5. "Birth control information and equipment should be available at the university health center to any interested coed."

6. "Police should have unlimited authority to apply force when necessary."

12. "Union shops are opposed to the American principle of individual freedom."

18. "Most people who are on welfare are not able to support themselves."

27. "The campus humor magazine, The Juggler, was obscene and not suited for a sophisticated college population."

do a multiple regression of the items on total scale score and use the overall magnitude of multiple $R$ as an overall indication of internal consistency, with the $b$s indicating which items should be retained?

3.  What other suggestion for analysis of internal consistency do you have?

4.  What differences between *internal consistency* and *validity* does all this suggest?

**B.**  (Requiring some analysis, or at least doodling with the data.)

1.  Can we use the intercorrelation matrix to put an upper bound on $R$?

2.  Can we use a variable's correlation (Pearson $r$) with the outcome variable as a guide to how high a weight it will receive in the regression equation employing all $m$ predictors to predict $y$?

3.  How adequately can we check for curvilinearity by examining separately each of the scattergrams of $y$ plotted against one of the predictor variables? (In other words, how useful is the Index–Plot option of BMDO2R?)

4.  What is the most adequate visual check for curvilinearity? How can this be verified statistically?

5.  Does curvilinearity of the relationship between one or more predictors and $Y$ imply curvilinearity of the relationship between $\hat{Y}$ and $Y$?

6.  Does curvilinearity of the relationship between $\hat{Y}$ and $Y$ imply that the relationship between $Y$ and at least one of the $X$s is curvilinear?

7.  Can you develop some simple formulas for the $\beta$-weights and for the overall value of multiple $R$ in the case where the predictor variables are uncorrelated with each other? Would you share these formulas with the author, please?

*Answers*

**I.**

**A.**  Own score $= 47.62 - .1701I_3 + .2394I_5 + .0672I_6 - .0373I_{12} - .0655I_{18} + .2313I_{27}$.

$\qquad$ $F = .916$ $\quad$ with $\quad$ 6 and 20 df, $\quad$ ns.

Clearly $I_5, I_{27}, I_3$ contribute most. The quickest way to check the simple sum (or any other linear combination of the original variables) for significance of relationship to $Y$ is to compute the correlation between this combined variable and compare this figure to the multiple $R$ which would have been required to achieve significance. Since the multiple $R$ represents the ultimate in capitalizing on chance (it is the correlation you obtain for the very best linear combination of the original variable), and since the overall $F$-test for significance takes this capitalization on chance into account, therefore, whether *a priori* or *post hoc*, any linear combination of the original variables you can think up which exceeds this critical value is statistically significant. This is essentially the "union–intersection principle" which we will use in Manova. In the present case, an $F$ of 2.71 is needed for significance at the .05 level, whence

$$R^2(N - m - 1)/[(1 - R^2)(m)] = 2.71$$

gives us the critical value of $R$ at the .05 level for these degrees of freedom, namely $R = .670$. We could compute the correlation of the sum with $Y$ and compare this to .670, but since multiple $R$ for our sample did not exceed .670, we know that this particular linear combination will not either.

**B.**   The first three requested analyses produce the same end result as part A. The differences arise in the speed with which $R$ "builds up." The following table depicts the buildup for each of the requested analyses:

*Order of Addition of Variables*

| Step | "Free" Order | R | 1, 2, ..., 6 Order | R | 27, 18, ..., 3 Order | R | Free with $Fs = 1$; tolerance = .50 Order | R |
|------|------|------|------|------|------|------|------|------|
| 1 | 6 | .293 | 1 | .275 | 6 | .293 | 6 | .293 |
| 2 | 1 | .352 | 2 | .306 | 5 | .299 | 1 | .3517 |
| 3 | 2 | .439 | 3 | .315 | 4 | .300 | 2 | .439 |
| 4 | 3 | .452 | 4 | .330 | 3 | .311 | No variable has $F$ to |  |
| 5 | 5 | .461 | 5 | .337 | 2 | .378 | enter greater than 1, |  |
| 6 | 4 | .464 | 6 | .464 | 1 | .464 | so analysis stops here. |  |

**II.**

**A.**

**1.**   Clearly $R$ must be at least as large as the absolute value of the correlation between $Y$ and that single $X$ variable having the highest (in absolute magnitude) Pearson $r$ with $Y$, since the optimization routine could always use regression coefficients which were zero except for that particular $X$ variable.

**2.**   No; or, at least, not to any useful purpose. Since total scale score is just the sum of the individual item scores, the regression equation will of course simply have coefficients of 1.0 for each variable, resulting in a multiple $R$ of 1.0, regardless of what items were used.

**3.**   Take the multiple $R$ between each item and the rest of the items (excluding itself) as indicative of how closely related to the other items it is. This is easily computed from the main diagonal entries of the intercorrelation matrix, and gives us the most optimistic possible picture of the degree of internal consistency.

**4.**   The suggestion in answer 3 employs only the intercorrelation matrix of the scores on the items of the scale and uses no information at all about how the items correlate with some external criterion. Thus internal consistency and validity are logically (and often empirically) independent.

**B.**

**1.**   Consideration of a set-theoretic interpretation of correlation in terms of "common variance" suggests that $R^2$ must be less than or equal to $\sum r_{iy}^2$. We can draw the various variables as overlapping circles, with the overlapping area representing the amount of variance variables $i$

and $j$ "share," namely $s_{ij} = s_i^2 r_{ij}^2 = s_j^2 (r_j^2)$ (see Fig. 2.1). It seems fairly clear that the portion of its variance which $Y$ shares with the entire set of variables $X_1, X_2, \ldots, X_m$ (which equals $R^2$) must be less than or equal to the sum of the portions shared separately with $X_1, X_2, \ldots, X_m$, since any overlap between $X_i$ and $X_j$ in the variance they share with $Y$ would be "counted" twice in the simple sum $\sum r_{iy}^2$. The overlap interpretation suggested above strongly suggests that we should be able to prove that

$$ s_{y \cdot x_1, x_2, \ldots, x_m}^2 = s_{1y} + s_{2y \cdot 1, 2} + \cdots + s_{my \cdot 1, 2, \ldots, m-1}, $$

where $s_{iy \cdot c}$ is the covariance of $x_i$ with $y$, with the effects of the variables in the set $\mathbf{c}$ "partialled out," where of course the covariance so corrected will be less than the covariance between $x_i$ and $y$ before the predicted scores (predicted from linear regression equations involving the variables in $\mathbf{c}$) have been subtracted.

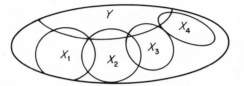

*Figure 2.1.* Venn diagram of correlations among $Y$ and four predictors.

**2.** We certainly cannot take $r_{iy}$ as a minimum value. For instance, in the two-predictor equation for the $z$-score form of the Demonstration Problem, $r_{2y} = .604$, but $b_2^*$ only $= .014$. Similarly, it cannot be taken as a maximum value, since for example in the three-predictor case $r_{2y} = .604$ and $r_{3y} = .793$, but $b_2^*$ (the regression coefficient in the $z$-score equation) $= .676$ and $b_3^* = .853$. However, if all the predictor variables are uncorrelated, the regression coefficients in the $z$-score regression equation are $b_i^* = r_{iy}$. (See Quiz Problem B7.)

**3.** Not very adequately. It is, for instance, possible for the relationships of $X_1$ and $X_2$ to be curvilinear, while the relationship between $Y'$ and $Y$ is perfectly linear. For instance, we can have $Y = X_1 + X_1^2 = X_1 + X_2$ if $X_2 = X_1^2$. In this case $Y = X_2^{1/2} + X_2$, so that the separate relationship of $X_1$ to $Y$ and $X_2$ to $Y$ are *both* curvilinear, while the relationship of $Y$ to $X_1$ and $X_2$ considered together is perfectly linear.

**4.** A plot of $Y'$ vs $Y$. If this is linear, then $Y = a + cY'$ is an adequate description of the data, whence

$$ Y = a + c(b_0 + b_1 X_1 + b_2 X_2 + \cdots + b_m X_m) = b_0' + b_1' X_1 + b_2' X_2 + \cdots + b_m' X_m $$

is an adequate description of the data (no $X^2$ terms are needed), where $b_0' = a + cb_0$ and $b_i' = cb_i$.

**5.** No. See Quiz Problem B3.

**6.** Yes. If $Y = a + cY' + dY'^2$, then

$$ Y = a + cY' + d(b_0^2 + b_1^2 X_1^2 + \cdots + b_m^2 X_m^2 + 2 \sum \sum b_i b_j X_k X_j). $$

Thus if no $X_i^2$ terms or $X_i X_j$ terms were required in the regression equation for predicting $Y$, $d$ in the univariate regression of $Y$ on $Y'$ would vanish.

**7.**   Yes. In that case, $\mathbf{S}_x$ is a diagonal matrix whose entries are the variances of the predictor variables, whence

$$\mathbf{S}_x^{-1} = \begin{bmatrix} 1/s_1^2 & & & \bigcirc \\ & 1/s_2^2 & & \\ & & \ddots & \\ \bigcirc & & & 1/s_m^2 \end{bmatrix} \quad \text{and} \quad \mathbf{b} = \mathbf{S}_x^{-1}\mathbf{s}_{xy} = \begin{bmatrix} s_{1y}/s_1^2 \\ s_{2y}/s_2^2 \\ \vdots \\ s_{my}/s_m^2 \end{bmatrix}$$

and

$$R^2 = (\mathbf{b}'\mathbf{s}_{xy})^2/s_y^2 = [\sum (s_{iy}/s_i)^2]/s_y^2 = \sum [s_{iy}/(s_i s_y)]^2 = \sum r_{iy}^2 .$$

# 3 | HOTELLING'S $T^2$
## Tests on One or Two Mean Vectors

This chapter describes a rather direct generalization of Student's $t$-test to situations involving more than one outcome variable. The general approach, as in all of the multivariate techniques we shall study in this Primer except factor analysis, is to compute a combined score for each subject which is simply a linear combination of his scores on the various outcome measures. The weighting factors employed in obtaining this combined score are those which make the univariate $t$-test conducted on the combined score as large as possible. Just as Student's $t$-test can be used either to test a single sample mean against some hypothetical value or to test the difference between two independent sample means and a hypothesized difference (usually zero) between the means of the populations from which they were drawn, so is there a single-sample and a two-sample version of Hotelling's $T^2$ (the statistical procedure we shall discuss in this chapter), corresponding to testing a mean outcome *vector* or the difference between two vectors of mean scores on dependent measures against corresponding population mean vectors. As we shall see, if the dependent measures are sampled from a multivariate normal distribution, the sampling distribution of Hotelling's $T^2$ (which is simply the square of the maximum possible univariate $t$ computed on any linear combination of the various outcome measures) has the same shape as the $F$-distribution. Now for the details.

### 3.1 Single-Sample $t$ and $T^2$

Consider Data Set 3, which reports the number of mutually cooperative (CC), unilaterally cooperative (CD), and unilaterally competitive (DC) outcomes in a 40-trial Prisoner's Dilemma game for each of 48 subject pairs. (The DD column will be explained later in this chapter.)

These data were gathered in the first of a series of replications and extensions of Deutsch's (1960) article designed to assess the effects of directive versus nondirective motivating instructions on behavior in experimental games (highly structured, two-person, two-choice interaction situations). See the work of Harris, Flint, and Everitt (1970) for details.

Ignoring for the moment differences between conditions, we can treat the data as representing 48 subject pairs and compute an overall mean for each outcome measure,

$$(\overline{CC}, \overline{CD}, \overline{DC}) = (11.23, 6.92, 8.33).$$

*Data Set 3*

*Results of Deutsch Replication 1*

| Motivating orientation | Experimenter | Subject pair | Number of responses | | | |
|---|---|---|---|---|---|---|
| | | | CC | CD | DC | DD |
| Competitive | DR | A | 0 | 1 | 7 | 32 |
| | | B | 4 | 9 | 8 | 19 |
| | HC | C | 11 | 12 | 9 | 8 |
| | | D | 5 | 9 | 7 | 19 |
| | RC | E | 15 | 11 | 5 | 9 |
| | | F | 8 | 9 | 8 | 15 |
| | RK | G | 4 | 9 | 10 | 17 |
| | | H | 5 | 8 | 11 | 16 |
| | RS | I | 2 | 15 | 2 | 21 |
| | | J | 4 | 6 | 6 | 24 |
| | SI | K | 12 | 14 | 4 | 10 |
| | | L | 1 | 5 | 7 | 27 |
| | | $\sum$ | 71 | 108 | 84 | 217 |
| Individualistic | DR | M | 1 | 4 | 11 | 24 |
| | | N | 6 | 5 | 6 | 23 |
| | HC | O | 9 | 8 | 10 | 13 |
| | | P | 1 | 6 | 8 | 25 |
| | RC | Q | 0 | 8 | 6 | 26 |
| | | R | 2 | 6 | 7 | 25 |
| | RK | S | 17 | 6 | 11 | 6 |
| | | T | 2 | 6 | 6 | 26 |
| | RS | U | 9 | 2 | 7 | 22 |
| | | V | 9 | 11 | 7 | 13 |
| | SI | W | 10 | 14 | 9 | 7 |
| | | X | 10 | 4 | 10 | 16 |
| | | $\sum$ | 76 | 80 | 98 | 226 |
| No motivating orientation | DR | Y | 10 | 9 | 11 | 10 |
| | | Z | 3 | 5 | 2 | 30 |
| | HC | AB | 11 | 10 | 9 | 10 |
| | | BC | 5 | 8 | 12 | 15 |
| | RC | CD | 6 | 9 | 12 | 13 |
| | | DE | 9 | 5 | 11 | 15 |
| | RK | EF | 11 | 12 | 7 | 10 |
| | | FG | 3 | 5 | 13 | 19 |
| | RS | GH | 11 | 8 | 8 | 13 |
| | | HI | 7 | 5 | 20 | 8 |
| | SI | IJ | 6 | 10 | 8 | 16 |
| | | JK | 17 | 7 | 14 | 2 |
| | | $\sum$ | 99 | 93 | 127 | 161 |

Data Set 3 (continued)

| Motivating orientation | Experimenter | Subject pair | Number of responses | | | |
|---|---|---|---|---|---|---|
| | | | CC | CD | DC | DD |
| Cooperative | DR | KL | 30 | 0 | 10 | 0 |
| | | LM | 14 | 10 | 11 | 5 |
| | HC | MN | 18 | 1 | 17 | 4 |
| | | NO | 34 | 0 | 4 | 2 |
| | RC | OP | 16 | 1 | 20 | 3 |
| | | PQ | 40 | 0 | 0 | 0 |
| | RK | QR | 31 | 3 | 4 | 2 |
| | | RS | 39 | 1 | 0 | 0 |
| | RS | ST | 19 | 12 | 6 | 3 |
| | | TU | 2 | 11 | 9 | 18 |
| | SI | UV | 37 | 3 | 0 | 0 |
| | | VW | 13 | 9 | 10 | 8 |
| | | $\Sigma$ | 293 | 51 | 91 | 45 |

We would like to be certain that Ss are not merely choosing C or D at random. If they were choosing purely randomly, we would expect 10 CCs in a 40-trial game. We can therefore test one implication of our null hypothesis of random responding by conducting a $t$-test on the difference between our observed mean number of CCs (11.23) and our hypothesized value (10.0). Thus we have

$$H_0 : \mu_{\text{CC}} = 10.0$$

$$t = (\bar{X} - \mu_0)/s_{\bar{X}} = (\bar{X} - \mu_0)/\sqrt{s^2/N} = (11.23 - 10.0)/\sqrt{(5196.5/47)(1/48)}$$

$$= 1.23/1.515 = .812 \quad \text{with} \quad N - 1 = 47 \text{ df.}$$

This is clearly too small a $t$-value to permit rejection of our null hypothesis, since $t$-values of .8 or greater in absolute magnitude occur over half the time when $H_0$ is true.

Of course our hypothesis of random responding has implications for the other three outcomes, too. We would expect, under that hypothesis, that CD and DC would each equal 10.0 also. We could conduct a separate $t$-test on each of the four outcome measures and compare each to a table of critical values (upper percentiles) of Student's $t$-distribution with 47 degrees of freedom. This, however, leads us straight into the multiple comparison problem. If we adopt a significance level (Type I error probability) of .05 for each $t$-test, the probability of falsely identifying at least one of the three means as significantly different from 10.0 will be somewhere between .05 (the probability if all four measures are perfectly correlated with each other) and .143 $(= 1 - (1 - \alpha)^k = 1 - .95^3$, the experimentwise error rate if the four measures are mutually independent). We thus need to find a way of controlling our $\alpha$-level for the set of three comparisons as a whole directly. The approach which Hotelling discovered and which was later subsumed as a special case of Roy's *union–intersection* principle involves

(i)   representing any comparison of one or more of the sample means with one or more of the corresponding hypothesized means as a univariate $t$-ratio computed on a single combined variable which is a linear combination of the original outcome measures, and

(ii)   taking as the test statistic for such comparisons the maximum possible value that such a "combined $t$-ratio" could attain for *any* selection of coefficients for the linear combining rule, and selecting a critical value of that statistic which guarantees occurrence of maximum $t$-ratios exceeding that level only $\alpha \times 100\%$ of the time when $H_0$ is correct.

The development of this multivariate significance test proceeds as follows:

(a)   We define a new variable

$$\mathbf{W} = a_1 \mathbf{X}_1 + a_2 \mathbf{X}_2 + \cdots + a_p \mathbf{X}_p = \mathbf{Xa},$$

where $\mathbf{X}_j$ is an $N$-element column vector giving each of the $N$ subjects' score on dependent measure (outcome variable) $j$; $\mathbf{X} = [\mathbf{X}_1 \mathbf{X}_2 \cdots \mathbf{X}_p]$ is an $N \times p$ data matrix whose $i$th row gives subject $i$'s scores on each of the outcome variables; $\mathbf{a}$ is a $p$-element column vector giving the weights by which the dependent measures are to be multiplied before being added together; and $\mathbf{W}$ is therefore an $N$-element column vector whose $i$th entry is subject $i$'s score on the new outcome variable $W$.

(b)   Our null hypothesis is that $\mu_1 = \mu_{10}; \mu_2 = \mu_{20}; \ldots; \mu_p = \mu_{p0}$ are all true. If one or more of these equalities is false, the null hypothesis is false. This hypothesis can be expressed in matrix form as

$$\boldsymbol{\mu} = \begin{bmatrix} \mu_1 \\ \mu_2 \\ \vdots \\ \mu_p \end{bmatrix} = \begin{bmatrix} \mu_{10} \\ \mu_{20} \\ \vdots \\ \mu_{p0} \end{bmatrix} = \boldsymbol{\mu}_0 , \tag{3.1}$$

and it implies that $\mu_W = \mathbf{a}'\boldsymbol{\mu}_0$ .

(c)   The variance of a linear combination of variables can readily be expressed as a linear combination of the variances and covariances of the original variables, whence

$$s_W^2 = a_1^2 s_1^2 + a_2^2 s_2^2 + \cdots + a_p^2 s_p^2 + 2a_1 a_2 s_{12} + 2a_1 a_3 s_{13} + \cdots + 2a_{p-1} a_p s_{p-1,p}$$

$$= \mathbf{a}'\mathbf{Sa},$$

where $\mathbf{S}$ is the covariance matrix of the outcome variables. Thus the univariate $t$-ratio computed on the combined variable $W$ is given by

$$t(\mathbf{a}) = (\mathbf{a}'\bar{\mathbf{X}} - \mathbf{a}'\boldsymbol{\mu}_0)/\sqrt{\mathbf{a}'\mathbf{Sa}/N} = \mathbf{a}'(\bar{\mathbf{X}} - \boldsymbol{\mu}_0)/\sqrt{\mathbf{a}'\mathbf{Sa}/N}, \tag{3.2}$$

where the notation $t(\mathbf{a})$ emphasizes the fact that the value of $t$ depends on (is a function of) the particular coefficients selected for $\mathbf{a}$. Now, we are clearly interested in the absolute magnitude of $t$, irrespective of sign. Since absolute values are a tremendous inconvenience mathematically, we shall seek instead to maximize the square of $t(\mathbf{a})$, which is of course equal to

$$t^2(\mathbf{a}) = [\mathbf{a}'(\bar{\mathbf{X}} - \boldsymbol{\mu}_0)]^2/(\mathbf{a}'\mathbf{Sa}/N) = N\mathbf{a}'(\bar{\mathbf{X}} - \boldsymbol{\mu}_0)(\bar{\mathbf{X}} - \boldsymbol{\mu}_0)'\mathbf{a}/(\mathbf{a}'\mathbf{Sa}). \tag{3.3}$$

It should be fairly clear that multiplication of the coefficients in $\mathbf{a}$ by a constant will have no effect on $t^2(\mathbf{a})$. In other words, the elements of $\mathbf{a}$ are defined only up to a change of units, and we must add a side condition on the possible choices of **a** in order that the vector which maximizes $t^2(\mathbf{a})$ may be uniquely defined. The most convenient such choice is the condition that $\mathbf{a'Sa}$, the variance of $W$, be equal to 1. This reduces our problem to that of maximizing the numerator of $t^2(\mathbf{a})$ subject to the constraint that the denominator be equal to unity. The method of Lagrangian multipliers comes into play here. The reader may wish to review the use of Lagrangian multipliers (Digression 1) and the procedures for differentiation of matrix expressions (Digression 2), so that he may follow the derivation outlined below. Alternately, he may accept this derivation on faith and proceed directly to Eqs. (3.4).

Our side condition is expressed as $\mathbf{a'Sa} - 1 = 0$, whence we need to maximize

$$h(\mathbf{a}) = N\mathbf{a'}(\bar{\mathbf{X}} - \boldsymbol{\mu}_0)(\bar{\mathbf{X}} - \boldsymbol{\mu}_0)'\mathbf{a} - \lambda(\mathbf{a'Sa} - 1).$$

Differentiating $h(\mathbf{a})$ with respect to **a**—which is equivalent to differentiating it with respect to each $a_j$ separately and then putting the $p$ different derivatives into a single column vector—we obtain

$$dh(\mathbf{a})/d\mathbf{a} = 2N(\bar{\mathbf{X}} - \boldsymbol{\mu}_0)(\bar{\mathbf{X}} - \mu_0)'\mathbf{a} - 2\lambda\mathbf{Sa},$$

which equals the zero vector if and only if

$$[N(\bar{\mathbf{X}} - \boldsymbol{\mu}_0)(\bar{\mathbf{X}} - \boldsymbol{\mu}_0)' - \lambda\mathbf{S}]\mathbf{a} = 0, \tag{3.4a}$$

whence

$$[N\mathbf{S}^{-1}(\bar{\mathbf{X}} - \boldsymbol{\mu}_0)(\bar{\mathbf{X}} - \boldsymbol{\mu}_0)' - \lambda\mathbf{I}]\mathbf{a} = 0. \tag{3.4b}$$

Thus our problem is reduced to that of finding the characteristic roots and vectors (see Digression 2) of the matrix $N\mathbf{S}^{-1}(\bar{\mathbf{X}} - \boldsymbol{\mu}_0)(\bar{\mathbf{X}} - \boldsymbol{\mu}_0)'$. Note that this matrix will generally *not* be symmetric.

Digression 4 includes a description of a "canned" computer program, HEINV, which can be of help in computing $T^2$. HEINV takes as input any two symmetric matrices, **H** and **E**, and computes the characteristic roots and vectors of the product matrix $\mathbf{E}^{-1}\mathbf{H}$. However, use of this program represents a considerable degree of "overkill," since, as we shall show, the matrix whose characteristic roots we seek is of rank 1, so that it has only a single nonzero root, which is equal to the trace of the matrix and which can be shown to equal

$$T^2 = N(\bar{\mathbf{X}} - \boldsymbol{\mu}_0)'\mathbf{S}^{-1}(\bar{\mathbf{X}} - \boldsymbol{\mu}_0). \tag{3.5}$$

※ Equation (3.5) can be proved by a "brute force" approach in which single-symbol expressions for the matrices are actually multiplied together and then rearranged. However, it is much easier to make use of a few well-known properties of rank and of characteristic roots. First, we know that the rank of the product of two matrices is $\leq$ the product of the ranks of the separate matrices. However, $(\bar{\mathbf{X}} - \boldsymbol{\mu}_0)$ is a column vector and therefore obviously of rank 1 (or zero, if the sample vector identically equals the hypothesized population vector), and therefore $(\bar{\mathbf{X}} - \boldsymbol{\mu}_0)(\bar{\mathbf{X}} - \boldsymbol{\mu}_0)'$ must have rank 1, and so must $\mathbf{S}^{-1}(\bar{\mathbf{X}} - \boldsymbol{\mu}_0)(\bar{\mathbf{X}} - \boldsymbol{\mu}_0)'$. Next, we know that the number of nonzero characteristic roots of a matrix is equal to the rank of that matrix, so that there must be a single nonzero root in the present case. However, putting this information into the Laplace expansion for generating the characteristic equation gives

$$(-\lambda)^p + (-\lambda)^{p-1}S_1 = 0,$$

whence $\lambda = S_1$, the trace of $NS^{-1}(\bar{\mathbf{X}} - \boldsymbol{\mu}_0)(\bar{\mathbf{X}} - \boldsymbol{\mu}_0)'$, that is, the sum of its main diagonal entries, but the $i$th main diagonal entry (mde) of $\mathbf{S}^{-1}(\bar{\mathbf{X}} - \boldsymbol{\mu}_0)(\bar{\mathbf{X}} - \boldsymbol{\mu}_0)'$ is the product of the $i$th row of $\mathbf{S}^{-1}$ and the $i$th column of

$$(\bar{\mathbf{X}} - \boldsymbol{\mu}_0)(\bar{\mathbf{X}} - \boldsymbol{\mu}_0)' = \boldsymbol{\Delta\Delta'}, \qquad \text{where} \quad \boldsymbol{\Delta'} = (\Delta_1, \Delta_2, \ldots, \Delta_p) = (\bar{\mathbf{X}} - \boldsymbol{\mu}_0)'.$$

However, the $i$th column of $\boldsymbol{\Delta\Delta'}$ is just $\Delta_i\boldsymbol{\Delta}$. Now, the position of a scalar in a chain of matrix multiplications is immaterial, so we can write

$$i\text{th mde of } NS^{-1}\boldsymbol{\Delta\Delta'} = N(i\text{th row of } \mathbf{S}^{-1})\Delta_i\boldsymbol{\Delta} = N\Delta_i(i\text{th row of } \mathbf{S}^{-1})\boldsymbol{\Delta};$$

whence

$$\lambda = \text{trace of } NS^{-1}\boldsymbol{\Delta\Delta'} = N\sum_{i=1}^{p}[\Delta_i(i\text{th row of } \mathbf{S}^{-1})\boldsymbol{\Delta}] = N\boldsymbol{\Delta'}\mathbf{v},$$

where $\mathbf{v}$ is a column vector whose $i$th entry is the vector product of the $i$th row of $\mathbf{S}^{-1}$ postmultiplied by $\boldsymbol{\Delta}$, that is, $\mathbf{v} = \mathbf{S}^{-1}\boldsymbol{\Delta}$, whence

$$\lambda = N\boldsymbol{\Delta'}\mathbf{S}^{-1}\boldsymbol{\Delta}.$$

Moreover, premultiplication of both sides of Eq. (3.4a) by $\mathbf{a}'$ shows that

$$\lambda = N\mathbf{a}'\boldsymbol{\Delta\Delta'}\mathbf{a}/\mathbf{a}'\mathbf{Sa} = \max_{\mathbf{a}} t^2(\mathbf{a}) = T^2,$$

so that $\lambda = T^2 = N\boldsymbol{\Delta'}\mathbf{S}^{-1}\boldsymbol{\Delta}$ as asserted.

Equation (3.5) should be relatively easy to remember, since it is such a direct generalization of the formula for the univariate single-sample $t$. Specifically, the univariate $t$-formula can be written as

$$t = (\bar{X} - \mu_0)/s_{\bar{X}} = (\bar{X} - \mu_0)/(s/\sqrt{N});$$

whence

$$t^2 = (\bar{X} - \mu_0)^2/(s^2/N) = N(\bar{X} - \mu_0)(\bar{X} - \mu_0)/s^2$$
$$= N(\bar{X} - \mu_0)(s^2)^{-1}(\bar{X} - \mu_0);$$

so that Eq. (3.5) is obtained from the formula for $t^2$ by replacing $s^2$ with $\mathbf{S}$ and $(\bar{X} - \mu_0)$ with $(\bar{\mathbf{X}} - \boldsymbol{\mu}_0)$, keeping in mind the matching up of dimensions of various matrices needed to yield a single number as the end result of applying (3.5).

Yet another procedure for computing $T^2$ is available. Using the results in Digression 2 with respect to determinants of partitioned matrices, it can be shown that

$$T^2 = (N-1)\left(\frac{|\mathbf{A} + N(\bar{\mathbf{X}} - \boldsymbol{\mu}_0)(\bar{\mathbf{X}} - \boldsymbol{\mu}_0)'|}{|\mathbf{A}|} - 1\right)$$
$$= \frac{|\mathbf{S} + N(\bar{\mathbf{X}} - \boldsymbol{\mu}_0)(\bar{\mathbf{X}} - \boldsymbol{\mu}_0)'|}{|\mathbf{S}|} - 1, \qquad (3.6)$$

where $\mathbf{A} = \mathbf{x}'\mathbf{x} = (N-1)\mathbf{S}$ is the cross-product matrix of deviation scores on the $p$ outcome measures. Either of these two determinantal formulas avoids the need to invert $\mathbf{S}$.

(d)  The statistical significance of the observed value of $T^2$ must now be determined. $T^2$ is the square of a univariate $t$-ratio computed on a linear combination of the outcome measures. We know that if scores on $X$ are sampled from a normal distribution, the square of a $t$-ratio whose variance estimate is based on $N - 1$ degrees of freedom has the same frequency distribution as an $F$-distribution with 1 and $N - 1$ degrees of freedom. However, this distribution is based on the assumption that a single variable, defined in advance of collecting the data on purely *a priori* grounds, is being tested. Instead, in the present situation we have generated a new variable, largely *post hoc* in nature, which is a linear combination of the various outcome measures whose coefficients were deliberately chosen so as to "inflate" $t^2$ to the greatest extent possible. Hotelling was able to prove that if scores on $X_1, X_2, \ldots, X_p$ are sampled from a single multivariate normal distribution, the effects of all this capitalization on chance can be exactly compensated for by the simple expedient of multiplying the observed value of $T^2$ by a constant $(N - p)/[p(N - 1)]$, which is always less than 1.0 except when $p$ and the constant both equal 1, and then comparing the resulting statistic with the $F$-distribution having $p$ and $N - p$ degrees of freedom. That is,

$$F = \frac{N - p}{p(N - 1)} T^2 \qquad (3.7)$$

is distributed as $F$ with $p$ and $N - p$ degrees of freedom. This expression assumes that we have available $N$ independently sampled outcome vectors (one for each subject), whence each variance or covariance estimator is based on $N - 1$ degrees of freedom. If we use instead variance–covariance estimators based on fewer degrees of freedom (for example, if we compute each deviation score separately about the mean of the *subgroup* to which that subject belongs, rather than about the grand mean across all subjects, whence we have only $N - k$ degrees of freedom, where $k$ is the number of subgroups), the constant multiplier becomes $(\text{df} - p + 1)/(p \cdot \text{df})$, where df is the number of degrees of freedom going into each estimate of $s_i^2$ or $s_{ij}$, and will usually be the number by which each sum of squared deviations or sum of cross-products of deviation scores is divided in computing **S**. In this case, the degrees of freedom for our $F$-test become $p$ and $\text{df} - p + 1$.

(e)  If $T^2$ exceeds the critical value for the chosen level of significance, we must now specify which variable or combination of variables is "responsible" for the rejection of $H_0$. The most complete, though seldom the simplest, answer is that that linear combination of the original variables which was actually used to maximize $t^2$—known as the "discriminant function"—is the variable (dimension?) on which the observed data differ most sharply from our null hypothesis. We shall usually, however, wish to test other, more natural or more theoretically interesting, linear combinations of the variables (for example, each original variable by itself) for significant departure from the value implied by our null hypothesis vector. Any of these comparisons which were planned in advance employ the univariate $t$-test and the corresponding critical values in a table of the $t$-distribution. *Post hoc* tests are also conducted by computing the appropriate univariate $t$-ratio on the appropriate combination of the outcome measures, but the resulting value of such a *post hoc* $t$ is compared with that value of $T$ ($\sqrt{T^2}$) which would have just barely achieved statistical significance. This criterion ensures us that no matter how carefully we search for a combining rule which produces a large *t*-value, we shall have at most an $\alpha \times 100\%$ chance of making a Type I error (rejecting a true null hypothesis) in *any* comparison.

Conversely, if we are interested primarily in testing the overall significance of the differences between our observed sample means and the hypothesized population means, we may be able to avoid computing a covariance matrix (let alone inverting it) at all. Clearly $T^2$ must be greater than or equal to the square of the largest of the univariate *t*-ratios. If, therefore, any one of these univariate *t*-ratios exceeds $\sqrt{T_{\text{crit}}^2}$, so will $T^2$. If $\max(t_i^2) < T_{\text{crit}}^2$, we must of course compute $T^2$, since it still may exceed $T_{\text{crit}}^2$. Use of this "quick and dirty" test, however, even if successful, foregoes information on the discriminant function, that is, on which particular linear combination of the original variables deviates most drastically from the value implied by our overall null hypothesis.

***Example 3.1***    Applying all this to our original problem, that of testing the hypothesis that the mean response vector for *S*s in the Deutsch replication study arose through random sampling from a population in which the true mean response vector is (10, 10, 10), we obtain

$$\bar{\mathbf{X}} = \tfrac{1}{48}(539, 332, 400) = (11.229, 6.917, 8.333),$$

$$\mathbf{\Delta}' = (1.229, -3.083, -1.667),$$

$$\mathbf{X'X} = \begin{array}{c} \\ \text{CC} \\ \text{CD} \\ \text{DC} \end{array} \begin{array}{ccc} \text{CC} & \text{CD} & \text{DC} \\ \left[\begin{array}{ccc} 11{,}249 & 2875 & 3835 \\ 2875 & 3036 & 2732 \\ 3835 & 2732 & 4216 \end{array}\right], \end{array}$$

$$N\bar{\mathbf{X}}\bar{\mathbf{X}}' = \begin{bmatrix} 6052.5 & 3728.1 & 4491.7 \\ 3728.1 & 2296.3 & 2766.7 \\ 4491.7 & 2766.7 & 3333.3 \end{bmatrix},$$

$$\mathbf{A} = \mathbf{X'X} - N\bar{\mathbf{X}}\bar{\mathbf{X}}' = \begin{bmatrix} 5196.5 & -853.1 & -656.7 \\ -853.1 & 739.7 & -34.7 \\ -656.7 & -34.7 & 882.7 \end{bmatrix},$$

$$N\mathbf{\Delta}\mathbf{\Delta}' = 48 \begin{bmatrix} 1.229 \\ -3.083 \\ -1.667 \end{bmatrix} \overset{(1.229 \quad -3.083 \quad -1.667)}{\begin{bmatrix} 72.5 & -181.9 & -98.3 \\ -181.9 & 456.2 & 246.7 \\ -98.3 & 246.7 & 133.3 \end{bmatrix}},$$

$$\mathbf{A} + N\mathbf{\Delta}\mathbf{\Delta}' = \begin{bmatrix} 5269.0 & -1035.0 & -755.0 \\ -1035.0 & 1196.0 & 212.0 \\ -755.0 & 212.0 & 1016.0 \end{bmatrix},$$

$$|\mathbf{A} + N\mathbf{\Delta}\mathbf{\Delta}'| = 5269 \begin{vmatrix} 1196 & 212 \\ 212 & 1016 \end{vmatrix} + \begin{vmatrix} -1035 & 212 \\ -755 & 1016 \end{vmatrix} 1035 - 755 \begin{vmatrix} -1035 & 1196 \\ -755 & 212 \end{vmatrix}$$

$$= 5269(1, 170, 192) + 1035(-891, 500) - 755(683, 560)$$

$$= 4.727 \times 10^9,$$

$$|\mathbf{A}| = 5196.5(651, 729.10) + 853.1(-775, 818.86) - 656.7(515, 363.56)$$

$$= 2.3864 \times 10^9,$$

$$T^2 = 47[(4.727/2.386) - 1] = (.9808)47 = 46.10,$$

$$F = (48 - 3)/[47(3)]T^2 = 15(.9808) = 14.712$$

with 3 and 45 df, $p < .001$.

Since we used the determinantal formula (3.6) to compute $T^2$, we must return to the matrix equations (3.4) to compute the vector of coefficients which produced the maximum $t^2$, whence we have

$$[N(\bar{\mathbf{X}} - \mathbf{\mu}_0)(\bar{\mathbf{X}} - \mathbf{\mu}_0)' - 46.10\mathbf{S}]\mathbf{a} = [N\Delta\Delta' - (46.10/47)\mathbf{A}]$$

$$= \begin{bmatrix} -5024.5 & 654.9 & 545.8 \\ 654.9 & -269.3 & 280.7 \\ 545.8 & 280.7 & -732.5 \end{bmatrix}\begin{bmatrix} a_1 \\ a_2 \\ a_3 \end{bmatrix} = \mathbf{0};$$

whence our discriminant function, in nonnormalized form, is given by $\mathbf{a}' = (.358, 1.913, 1)$. In other words—as the reader will wish to verify himself—computation of the combined variable $W = .358CC + 1.913CD + DC$ for each subject, followed by a $t$-test of the difference between $\bar{W}$ and $10 \sum a_i = 10(3.271)$, will yield a $t$-ratio of $\sqrt{46.10} = 6.79$; and no other combined variable can be found which will discriminate as strongly as (produce a $t$-ratio whose absolute value is greater than) this particular choice. This discriminant function is not, however, a particularly simple one, so we shall wish to conduct some tests on simpler linear combinations of the variables. For this purpose, we first need to compute $t_{\text{crit}}$. With 3 and 45 degrees of freedom, we would need an $F$ of 2.82 to achieve statistical significance at the .05 level. This corresponds to a $T^2$ of $47(3)/(48 - 3)$ times $F_{\text{crit}} = 8.836$, whence $t_{\text{crit}} = 2.97$. Testing each separate mean outcome against 10.0, we obtain

$$t_{\text{CC}} = .812,$$

as from our earlier computation;

$$t_{\text{CD}} = -3.08/\sqrt{739.7/[47(48)]} = -5.59,$$

which is statistically significant by our experimentwise criterion; and

$$t_{\text{DC}} = -1.67/.626 = -2.66,$$

which would have been significant had this been an *a priori* comparison, but is not significant by our *post hoc* criterion. Thus the only single variable of these three which can be reliably discriminated from our hypothesized mean value of 10.0 is CD. Note that the correspondence between these individual $t$s and the contribution of the different variables to the discriminant function is far from perfect—a result which should surprise no one who has been through the discussion of multiple regression in Chapter 2.

## 3.2   Linearly Related Outcome Variables

The reader has probably guessed from the brief discussion of the study which generated Data Set 3 that the DD column of that data set represents a fourth measure of the outcome of the game, namely the number of mutually competitive choices made during the 40 trials of the game. This conjecture is correct. It is also true that our null hypothesis of random, 50–50 responding implies that $\mu_{DD}$, too, = 10.0. Why, then, was DD not included in the above analysis? Fundamentally, because it adds no new information beyond that already provided by CC, CD, and DC. Since each pair played 40 trials, CC + CD + DC + DD = 40, whence knowledge of a pair's "scores" on the first three measures permits us to predict with 100% certainty the value of DD for the pair. As might be anticipated (and as you should verify) the true dimensionality (rank) of the complete 4 × 4 covariance matrix is only 3, whence $S^{-1}$ is undefined, and our computational procedures break down.

This problem will crop up whenever there is a perfect linear relationship among the outcome measures—that is, whenever one or more of the dependent variables can be written as a linear combination of the others. In such a case there is a linear dependence among the rows and columns of $A$ and $S$, so that $|A| = |S| = 0$. Going back to an earlier step in the derivation of $T^2$, we find that we need a solution to the matrix equation $[N\Delta\Delta' - \lambda S]a = 0$. We cannot in the case of linear dependence use the usual trick of multiplying through by $S^{-1}$, since it does not exist. However, it can be shown, with a bit of algebra, that if $\lambda$ and $a$ are solutions to this equation, then $\lambda$ (same value) and $a^* = [a_i + c_i a_p]$ are solutions to the matrix equation

$$[N\Delta^*\Delta^{*\prime} - \lambda S^*]a^* = 0,$$

where the asterisk (*) indicates that that vector or matrix has had its last ($p$th) row and/or column deleted, and where

$$X_p = c_1 X_1 + c_2 X_2 + \cdots + c_{p-1} X_{p-1} + a_0.$$

We can now multiply through by $S^{*-1}$ (assuming that $S$ is of rank $p - 1$; if it is of rank $p - r$, we must delete $r$ of the dependent variables) and proceed to obtain the same formulas as before.

Thus whenever the rank of $S$ is less than $p$, we must delete dependent variables until the reduced system yields a reduced covariance matrix with a nonzero determinant, and then apply the usual $T^2$ formulas to this reduced system of outcome measures. It does not matter which of the measures is deleted, so long as it is one of the dependent variables involved in the perfect linear relationship. For instance, retaining DD but deleting DC in the present example leads to a discriminant function of

$$-.642CC + .913CD - DD = -.642CC + .913CD - (40 - CC - CD - DC)$$

$$= .358CC + 1.913CD + DC,$$

which is obviously equivalent to our original discriminant function and which in fact leads to precisely the same value of $T^2$.

Tatsuoka (1972) has suggested that linear dependence in **A** and **S** be handled by first performing a principal component analysis (cf. Chapter 6 or Section 1.2.11) of the data, then conducting $T^2$ (or Manova, or multiple regression) on the Ss' scores on the principal components. The number of these principal components (PCs) will equal the rank of the covariance matrix, so that no problem arises in computing the inverse of the covariance matrix of scores on PCs. The resulting discriminant function expressed in terms of scores on PCs can then be converted back to a discriminant function expressed in terms of scores on the original variables by substituting for each PC its definition in terms of the original variables.

The only argument Tatsuoka advances for preferring this rather involved procedure over simple deletion of one or more original variables is that the resulting discriminant function will usually assign nonzero weights to each of the $p$ original variables. This desideratum can be achieved much more simply, however, by using the variable deletion method to obtain an initial discriminant function and then using the known linear relationship between the deleted variable and the original variables to generate any one of the infinitely many equivalent expressions which assign nonzero weights to all variables. For instance, in the present case,

$$.358\text{CC} + 1.913\text{CD} + \text{DC} = (.358 + b)\text{CC} + (1.913 + b)\text{CD} + (1 + b)\text{DC}$$

$$+ b\text{DD} - 40b$$

for any choice of $b$.

A word of caution should be inserted here about the dangers of relying on $|\mathbf{S}|$, rather than knowledge of the constraints operating on the data, to reveal linear dependencies among the outcome variables. A very small error in computing the entries of **S** or **A** can lead to a computed determinant which appears greatly different from zero. For instance, if the entry for $\sum X_{\text{CD}} X_{\text{DD}}$ in $\mathbf{X'X}$ is changed from 4637 to 4647, $\sum x_{\text{CD}} x_{\text{DD}}$ changes from 148.1 to 158.1, and the resulting determinant of **A** is $2.345 \times 10^{10}$, which is unlikely to be confused with zero. The problem is that the $+10$ error in this one entry enters into all four-term products involving that entry which are used in computing $|\mathbf{A}|$. Near-dependencies like the example can be made more detectable by dividing the determinant of any cross-product or covariance matrix by the product of the main diagonal elements (the maximum possible value of its determinant) and casting a suspicious eye on the possibility that $|\mathbf{A}|$ or $|\mathbf{S}|$ is truly zero if this ratio is less than $10^{-2}$. (The use of .01 as the criterion is arbitrary.) In the present case, the product of the main diagonal elements of **A** is approximately $12.4 \times 10^{12}$, and the suggested ratio is approximately $.19 \times 10^{-2}$.

## 3.3   Two-Sample $t$ and $T^2$

Returning to Data Set 3, we might wish to test the effects of our various instructional sets on the performance of our subjects. For instance, we could ask whether the no motivating orientation (NMO) and cooperative (COOP) conditions differ significantly in the frequency of mutually cooperative outcomes. This calls for the old, familiar $t$-test for the difference between two independent sample means. Thus we have

$$H_0 : \mu_1 = \mu_2; \quad \text{that is,} \quad \mu_{x_1-x_2} = \mu_1 - \mu_2 = 0.$$

$$t = [(\bar{X}_1 - \bar{X}_2) - 0]/\sqrt{s_c^2(1/N_1 + 1/N_2)} \quad \text{with} \quad N_1 + N_2 - 2 \text{ df}; \quad (3.8)$$

where $s_c^2 = (\sum x_1^2 + \sum x_2^2)/(N_1 + N_2 - 2)$ is our best available estimate of the presumed common value of the variance of the two populations from which samples 1 and 2 were drawn. Thus in the present situation

$$t_{\text{NMO-COOP}} = (8.25 - 24.42)/\sqrt{[(180.25 + 1662.95)/22](2/12)}$$

$$= -16.17/\sqrt{1843.20/132} = -4.33 \quad \text{with} \quad 22 \text{ df}, \quad p < .001.$$

However our "eyeball analysis" of the data led us to concentrate on the difference in CCs as the most dramatic one, so that we would still like to have an overall test of the difference between the two sample mean vectors. The development of the appropriate multivariate test follows that of the single-sample $T^2$ point for point. The matrix analog of $\bar{X}_1 - \bar{X}_2$ is $\bar{\mathbf{X}}_1 - \bar{\mathbf{X}}_2$, while the matrix analog of $\sum x_i^2$ is $\mathbf{A}_i$, so that $s_c^2$ becomes

$$\mathbf{S}_c = (\mathbf{A}_1 + \mathbf{A}_2)/(N_1 + N_2 - 2).$$

Thus the two-sample $T^2$ can be computed from any of the following formulas:

$$T^2 = [N_1 N_2/(N_1 + N_2)](\bar{\mathbf{X}}_1 - \bar{\mathbf{X}}_2)'\mathbf{S}_c^{-1}(\bar{\mathbf{X}}_1 - \bar{\mathbf{X}}_2) \quad (3.9)$$

$$= \text{single nonzero root of} \quad [N_1 N_2/(N_1 + N_2)](\bar{\mathbf{X}}_1 - \bar{\mathbf{X}}_2)(\bar{\mathbf{X}}_1 - \bar{\mathbf{X}}_2)'\mathbf{S}_c^{-1} \quad (3.10)$$

$$= \frac{|\mathbf{S}_c + N_1 N_2(\bar{\mathbf{X}}_1 - \bar{\mathbf{X}}_2)(\bar{\mathbf{X}}_1 - \bar{\mathbf{X}}_2)'/(N_1 + N_2)|}{|\mathbf{S}_c|} - 1. \quad (3.11)$$

The application of formula (3.9) to the present comparison proceeds as follows:

|  |  | NMO |  |  |  |  | COOP |  |  |
|---|---|---|---|---|---|---|---|---|---|
|  | CC | CD | DC | DD |  | CC | CD | DC | DD |
| CC | 997 | 803 | 1076 | 1084 | CC | 8817 | 784 | 1602 | 517 |
| CD | 803 | 783 | 949 | 1135 | CD | 784 | 467 | 420 | 369 |
| DC | 1076 | 949 | 1557 | 1498 | DC | 1602 | 420 | 1159 | 459 |
| DD | 1084 | 1135 | 1498 | 2673 | DD | 517 | 369 | 459 | 455 |

$\mathbf{X'X} =$ (left matrix above), (right matrix above),

$$\bar{\mathbf{X}} = (99, 93, 127, 161)/12 \qquad (293, 51, 91, 45)/12,$$

$$\mathbf{A} = \begin{bmatrix} 180.25 & 35.75 & 28.25 & -244.24 \\ 35.75 & 62.25 & -35.25 & -112.75 \\ 28.25 & -35.25 & 212.92 & -205.90 \\ -244.24 & -112.75 & -205.90 & 512.93 \end{bmatrix} \begin{bmatrix} 1662.95 & -461.23 & -618.9 & -581.75 \\ -461.23 & 250.25 & 33.25 & 177.75 \\ -618.9 & 33.25 & 468.92 & 117.75 \\ -581.75 & 177.75 & 117.75 & 286.25 \end{bmatrix},$$

$$\mathbf{S}_{\text{pool}} = (\mathbf{A}_1 + \mathbf{A}_2)/22 = \begin{bmatrix} 83.782 & -19.340 & -26.848 & -37.545 \\ -19.340 & 14.205 & -.091 & 2.955 \\ -26.843 & -.091 & 30.992 & -4.007 \\ -37.545 & 2.955 & -4.007 & 36.326 \end{bmatrix}.$$

After deleting DD (the fourth row and column of **S**) we have

$$\mathbf{S}^{-1} = \begin{bmatrix} .029432 & .040235 & .025614 \\ .040235 & .12540 & .035223 \\ .025614 & .035223 & .054559 \end{bmatrix}; \quad \mathbf{\Delta} = \bar{\mathbf{X}}_1 - \bar{\mathbf{X}}_2 = \begin{bmatrix} -16.167 \\ 3.500 \\ 3.000 \end{bmatrix};$$

and

$$\mathbf{\Delta}'\mathbf{S}^{-1}\mathbf{\Delta} = [.029432(261.372) + .12540(12.25) + .054559(9) + .19697(93.451)]$$

$$+ 2[-.040235(56.5845) - .025614(48.5) + .035223(10.5)]$$

$$= 9.720 - 2(3.149) = 3.422;$$

whence

$$T^2 = 6(3.422) = 20.532;$$

and

$$F = (N_1 + N_2 - p - 1)T^2/[p(N_1 + N_2 - 2)]$$

$$= (20/66)T^2 = 6.222 \quad \text{with} \quad p \quad \text{and} \quad N_1 + N_2 - p - 1 \text{ df,}$$

that is, with 3 and 20 degrees of freedom, $p < .01$. Thus we can reject the hypothesis that all three population mean differences are zero.

In yet another direct analogy to the one-sample $T^2$, we can conduct all the specific comparisons of the two groups in terms of linear combinations of the outcome measures and be assured that the probability of any one or more of these comparisons leading to a false rejection of the hypothesis of no true difference is at most $\alpha$, simply by selecting as our critical value of $|t|$ for all such comparisons $\sqrt{T^2_{\text{crit}}}$, where $T^2_{\text{crit}}$ is that value of $T^2$ which would have just barely led to rejection of the overall null hypothesis. Thus in the present case we have

$$(20/66)T^2_{\text{crit}} = 3.10$$

for significance at the .05 level, whence

$$T^2_{\text{crit}} = 3.3(3.10) = 10.23.$$

Thus our criterion for the statistical significance of any *post hoc* comparisons is that $|t|$ for that comparison must equal or exceed $\sqrt{10.23} = 3.20$. For instance, testing each separate variable gives us

$$t_{\text{CC}} = -16.167\sqrt{6.0/83.782} = -4.33, \quad p < .05;$$

$$t_{\text{CD}} = 2.27, \text{ ns}; \quad t_{\text{DC}} = 1.32, \text{ ns}; \quad \text{and} \quad t_{\text{DD}} = 3.93, \quad p < .05.$$

Of course, had we been interested solely in significance tests, we could have computed $\sqrt{T^2_{\text{crit}}}$ and compared it to each of the univariate $t$s, thereby discovering that $t_{\text{CC}}$ exceeds this value and thus, *a fortiori*, so must $T^2$. We could therefore omit computation of **S**, $\mathbf{S}^{-1}$, etc. However, in so doing we would lose information about the discriminant function, that is, that particular linear combination of our three measures which maximally discriminates between the two groups. For the present data, the discriminant function (ignoring DD) is .842CC + .345CD + .415DC.

## 3.4 Profile Analysis

The overall $T^2$ test for two samples "lumps together" two sources of difference between the two groups' response vectors ("profiles"): a difference in the *level* of the two curves, and differences in the *shapes* of the two curves. Figures 3.1a and 3.1b illustrate, respectively, a pair of groups which differ only in mean response level and a pair of groups which differ only in shape. Methods which analyze these two sources of difference separately and in addition provide a simple test of the flatness of the

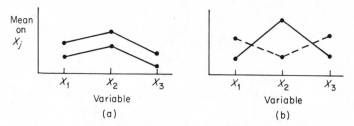

Figure 3.1. Response vectors for groups differing only in (a) level or (b) slope.

combined or pooled profile for the two groups are known as *profile analysis*. A profile analysis of the response vectors for the two groups involves tests of three separate null hypotheses:

### (a) Parallelism Hypothesis

$H_{01}$:  The profiles for the two groups are parallel (have the same shape), that is,

$$\boldsymbol{\mu}_{\text{slope, 1}} = \begin{bmatrix} \mu_{1,1} & -\mu_{2,1} \\ \mu_{2,1} & -\mu_{3,1} \\ \vdots \\ \mu_{p-1,1} & -\mu_{p,1} \end{bmatrix} = \begin{bmatrix} \mu_{1,2} & -\mu_{2,2} \\ \mu_{2,2} & -\mu_{3,2} \\ \vdots \\ \mu_{p-1,2} & -\mu_{p,2} \end{bmatrix} = \boldsymbol{\mu}_{\text{slope, 2}};$$

that is,

$$\boldsymbol{\mu}_{\text{slope, 1}} - \boldsymbol{\mu}_{\text{slope, 2}} = \mathbf{0}.$$

The equivalence of the two forms of the null hypothesis can be seen from the fact that if the two profiles are truly parallel, then the slope of each line segment making up that profile will be the same for both groups.

### (b) Levels Hypothesis

$H_{02}$:  The profiles for the two groups are at the same mean level, that is,

$$\mu_{W,1} = \mu_{\sum x_{i,1}} = \mu_{1,1} + \mu_{2,1} + \cdots + \mu_{p,1} = \mu_{1,2} + \cdots + \mu_{p,2} = \mu_{W,2};$$

that is,

$$\mu_{W,1} - \mu_{W,2} = 0.$$

This simply tests the hypothesis that the mean of the means (actually, the mean sum) of the separate variables is identical for the two groups. If the parallelism hypothesis

Comedel

is true, then this is equivalent to a test of the hypothesis that the difference between the group means on any variable (this distance being the same for all variables) is truly zero.

### (c)   Flatness Hypothesis

$H_{03}$:   The "pooled" profile for the two groups combined is perfectly flat, that is, the combined means are all equal to the same value; that is,

$$\mu_{\text{slope}} = \begin{bmatrix} \mu_1 & - \mu_2 \\ \mu_2 & - \mu_3 \\ \vdots \\ \mu_{p-1} - \mu_p \end{bmatrix} = \mathbf{0}.$$

collum effect

(The absence of subscripts on the $\mu_j$s indicates that these are *combined* means, pooled across both groups.) The final version of $H_{03}$ simply takes advantage of the fact that a flat profile implies that all line-segment slopes are truly zero.

These three tests are analogous to the three standard tests of a two-way univariate analysis of variance in which treatments (groups) correspond to rows, and response measures (dependent variables) correspond to columns, whence $H_{02}$ corresponds to a test of the row main effect; $H_{03}$, to a test of the column main effect; and $H_{01}$, to a test of the interaction between rows and columns. Thus in profile analysis, as in two-way Anova, the "interaction" (parallelism) test takes precedence, with a significant departure from parallelism implying that

(a)   the two groups must be compared separately on each outcome measure, since the magnitude and/or direction of the difference between the two groups varies, depending on which variable is considered, and

(b)   nonsignificant departures from the equal levels $H_{02}$ hypothesis and/or flatness $H_{03}$ hypothesis are essentially noninterpretable, since the significant interaction between groups and measures implies that both are significant sources of variance, the magnitude of their influence depending, however, on the level of the other source of variation in responses.

These three tests are *statistically* independent, but the last two are usually irrelevant substantively when parallelism is rejected.

In order that profile analysis give meaningful results, it is necessary that the various outcomes be measured in meaningful units. Formally, a statement about some empirical finding is meaningful if its truth or falsehood is unaffected by supposedly irrelevant transformations of the data. In the present context, the shape of the pooled profile is affected by a change of origin (addition or subtraction of a constant), while parallelism can be destroyed or guaranteed by changing the units of the different measures. Thus the investigator who uses profile analysis must be able to justify his particular choice of origin and units for each measure. (The overall $T^2$-test for two samples is unaffected by any linear transformation of any of the dependent variables.)

Computationally, the parallelism and flatness hypotheses are handled by converting the original set of $p$ outcome measures to $p - 1$ *slope* measures, where slope measure $i = X_i - X_{i+1}$ is the difference between two "adjacent" dependent variables. Parallelism is then handled by conducting a two-sample $T^2$-test on the difference between the slope measures for the two groups, while flatness is tested by conducting a single-sample $T^2$-test of the hypothesis that all of the grand means (combined means, combining the data for both groups) of the various slope measures are equal to zero. The equal levels hypothesis is tested by a univariate $t$-test on the sum of the $p$ original measures.

These two transformations (to slope measures and to sums of scores) could be carried out on each single subject, but it is generally much easier to take advantage of the simple relationships between the coefficients used in constructing a linear combination of scores and the mean and variance of that combined score, namely

$$\bar{X}_W = w_1 \bar{X}_1 + w_2 \bar{X}_2 + \cdots + w_p \bar{X}_p = \mathbf{w}'\bar{\mathbf{X}} \qquad \text{and} \qquad s_W^2 = \mathbf{w}'\mathbf{S}\mathbf{w}.$$

Thus our three significance tests become

(i)   $H_{01}$:   $\boldsymbol{\mu}_{\text{slope},\,1} = \boldsymbol{\mu}_{\text{slope},\,2}$   is tested by

$$T^2 = N_1 N_2 (\bar{\mathbf{X}}_{\text{slope},\,1} - \bar{\mathbf{X}}_{\text{slope},\,2})' \mathbf{S}_{\text{slope}}^{-1} (\bar{\mathbf{X}}_{\text{slope},\,1} - \bar{\mathbf{X}}_{\text{slope},\,2})/(N_1 + N_2),$$

whence

$$F = (N_1 + N_2 - p)/[(p-1)(N_1 + N_2 - 2)]T^2$$

$$\text{with} \quad p - 1 \quad \text{and} \quad N_1 + N_2 - p \text{ df};$$

where

$$\bar{\mathbf{X}}_{\text{slope},\,i} = \begin{bmatrix} \bar{X}_{1,i} & - \bar{X}_{2,i} \\ \bar{X}_{2,i} & - \bar{X}_{3,i} \\ \vdots \\ \bar{X}_{p-1,i} - \bar{X}_{p,i} \end{bmatrix} \qquad \text{for} \quad i = 1, 2;$$

and

$$\mathbf{S}_{\text{slope}} = \mathbf{S}_c \text{ computed on the slope measures} = \mathbf{CSC}',$$

where

$$\mathbf{C} = \begin{bmatrix} 1 & -1 & 0 & \cdot & \cdot & \cdot & 0 & 0 \\ 0 & 1 & -1 & \cdot & \cdot & \cdot & 0 & 0 \\ \cdot & & & \cdot & \cdot & \cdot & & \cdot \\ 0 & 0 & 0 & \cdot & \cdot & \cdot & 1 & -1 \end{bmatrix}.$$

Alternately,

$$\mathbf{S}_{\text{slope}} = \begin{bmatrix} s_1^2 + s_2^2 - 2s_{12} & s_{12} + s_{23} - s_2^2 - s_{13} & \cdots & s_{1,p-1} + s_{2p} - s_{2,p-1} - s_{1p} \\ & s_2^2 + s_3^2 - 2s_{23} & \cdots & s_{2,p-1} + s_{3p} - s_{3,p-1} - s_{2p} \\ & & \ddots & \\ & & & s_{p-1}^2 + s_p^2 - 2s_{p-1,p} \end{bmatrix}$$

that is,

the $(r, t)$th entry of $\mathbf{S}_{\text{slope}}$ = the covariance of $(X_r - X_{r+1})$     with     $(X_t - X_{t+1})$

$$= s_{rt} + s_{r+1,t+1} - s_{r,t+1} - s_{t,r+1}.$$

(ii) $H_{02} : \mu_{W,1} = \mu_{W,2}$ is tested by

$$t = (\bar{X}_{W,1} - \bar{X}_{W,2})/[s_W^2(1/N_1 + 1/N_2)] \qquad \text{with} \quad N_1 + N_2 - 2 \text{ df.}$$

where $\bar{X}_{W,i} = \bar{X}_{1,i} + \bar{X}_{2i} + \cdots + \bar{X}_{p,i}$ for $i = 1, 2$; $s_W^2$ is the sum of all entries in $\mathbf{S}_c$.

(iii) $H_{03} : \boldsymbol{\mu}_{\text{slope}} = \mathbf{0}$ is tested by

$$T^2 = (N_1 + N_2)\bar{\bar{\mathbf{X}}}_{\text{slope}} \mathbf{S}_{\text{slope}}^{-1} \bar{\bar{\mathbf{X}}}_{\text{slope}}$$

whence

$$F = (N_1 + N_2 - p)T^2/[(p - 1)(N_1 + N_2 - 2)]$$

$$\text{with} \quad p - 1 \quad \text{and} \quad N_1 + N_2 - p \text{ df;}$$

where $\bar{\bar{\mathbf{X}}}_{\text{slope}}$ is a column vector of "adjacent" differences in sample grand means,

$$\bar{\bar{\mathbf{X}}}_{\text{slope}} = \begin{bmatrix} \bar{\bar{X}}_1 & - \bar{\bar{X}}_2 \\ \bar{\bar{X}}_2 & - \bar{\bar{X}}_3 \\ \vdots \\ \bar{\bar{X}}_{p-1} & - \bar{\bar{X}}_p \end{bmatrix}, \qquad \text{where} \quad \bar{\bar{X}}_j = (N_1 \bar{X}_{j,1} + N_2 \bar{X}_{j,2})/(N_1 + N_2).$$

As an illustration, let us conduct a profile analysis of the difference between the NMO and COOP groups of Data Set 2. If we include all four dependent measures in our analysis, we will thereby destroy the independence of the tests of the levels and parallelism hypotheses, since the four means for each group must add to 40 (the number of trials) and any true difference on one measure must be "compensated for" by a difference in the opposite direction on one or more other dependent variables, thus ensuring nonparallelism. We will therefore carry out the analysis on CC, CD, and DC, omitting DD. (The overall $T^2$ was conducted on this "reduced" matrix.)

## (a) Parallelism Test

$$\mathbf{S}_{\text{slope}} = \begin{bmatrix} 136.677 & -6.788 \\ -6.788 & 45.379 \end{bmatrix}; \qquad \mathbf{S}_{\text{slope}}^{-1} = \frac{1}{6141} \begin{bmatrix} 45.379 & 6.788 \\ 6.788 & 136.667 \end{bmatrix};$$

$$\bar{\mathbf{X}}_{\text{slope}, 1} - \bar{\mathbf{X}}_{\text{slope}, 2} = \frac{1}{12} \begin{bmatrix} 236 \\ 6 \end{bmatrix};$$

whence

$$T_{\text{par}}^2 = [\tfrac{1}{2}(12)/144][236(45.379) + 36(136.667) + 2(6)(236)(6.788)]/6141$$

$$= 17.15;$$

whence

$$F = (21/44)T^2 = 8.19 \qquad \text{with} \quad 2 \text{ and } 21 \text{ df}, \qquad p < .01.$$

We can thus reject the hypothesis of parallel response vectors for the two groups. We might wish to stop right here, since the interaction between treatment effect and measure thereof—especially a "criss-cross" interaction of the sort we have here, where the difference between the two groups is of different sign for different dependent variables—makes the interpretation of the levels and flatness hypotheses difficult. In this case, we would proceed to make specific comparisons between the two groups with respect to single dependent variables or groups of outcome variables. We would, of course, use the critical value of the overall two-sample $T^2$ as our criterion for assessing statistical significance of these specific comparisons in order to avoid the inflated Type I error rate simple repeated application of univariate $t$-tests would entail.

Alternately, we might wish to carry out the other two tests (which are, after all, well defined and statistically independent of the parallelism test), cautioning our readers that the interpretation of the flatness and levels tests is complicated by the presence of interaction between groups and measures. Since the purpose of this example is primarily didactic, we shall adopt this latter option and go through the computations for the other two profile analysis hypotheses.

### (b)   Levels Test

$$t = (116/12)/\sqrt{82.4/6} = 2.60 \qquad \text{with} \quad 22 \text{ df,} \qquad p < .05;$$

that is, the extremely high mean for the COOP group on CC is sufficient to make its overall mean across all three variables significantly higher than the NMO group's overall mean, despite the fact that the COOP group's means fall below the NMO means for the other two outcome measures.

### (c)   Flatness Hypothesis

$$\bar{\bar{\mathbf{X}}} = \frac{1}{24}\begin{bmatrix} 392 \\ 144 \\ 218 \end{bmatrix} = \begin{bmatrix} 16.333 \\ 6.000 \\ 13.250 \end{bmatrix}; \qquad \bar{\bar{\mathbf{X}}}_{\text{slope}} = \begin{bmatrix} 10.333 \\ -7.250 \end{bmatrix};$$

whence

$$T^2_{\text{flat}} = (24/6141)[10.333^2(45.38) + 7.25^2(136.7) - (14.5)(10.33)(6.788)]$$

$$= 24(10{,}996)/6141 = 43.0;$$

$$F = (21/44)T^2 = 20.5 \qquad \text{with} \quad 2 \text{ and } 21 \text{ df,} \qquad p < .01;$$

that is, the "pooled" profile is definitely not flat, though it is a pretty good bet that the NMO profile does not differ significantly from a horizontal line.

## 3.5   Discriminant Analysis

The clearer the distinction a particular measure makes between two groups, the more useful that measure is apt to be in deciding to which group an as yet unclassified subject belongs. The discriminant function—that linear combination of the original variables which yields the highest possible $t$-ratio for the difference between the two groups—therefore seems a logical candidate for such applications. However, a slightly more general version of discriminant analysis is available through multivariate analysis of variance, so we will postpone further discussion of discriminant analysis to Chapter 4.

## 3.6   Assumptions Underlying $T^2$

### 3.6.1   The Assumption of Equal Covariance Matrices

The procedure of averaging together the covariance matrices for groups 1 and 2 before carrying out a $T^2$-analysis of differences between the two groups involves the implicit assumption that the differences between $S_1$ and $S_2$ simply represent random fluctuations about a common population covariance matrix $\Sigma$. Logically, our null hypothesis includes both the hypothesis that $\mu_1 = \mu_2$ and that $\Sigma_1 = \Sigma_2$. (The second hypothesis is usually thought of as an assumption upon whose correctness the validity of the first hypothesis—the one of real interest to the investigator—depends, rather than as part of $H_0$.) Rejection of $H_0$ thus *could* be due to the fact that $\Sigma_1 \neq \Sigma_2$, rather than (or in addition to) a nonnull difference between $\mu_1$ and $\mu_2$. However, studies by statisticians have shown that $T^2$, like the univariate $t$, is much more sensitive to differences in means than to differences in variances (and covariances, in the case of $T^2$). In particular, Ito and Schull (1964) have shown that the true significance level of the $T^2$-test is unaffected by discrepancies between $\Sigma_1$ and $\Sigma_2$, so long as $N_1 = N_2$ is fairly large. For small and unequal sample sizes, it is handy to have a separate test of the equality of the covariance matrices of the populations from which groups 1 and 2 were drawn. A multivariate analog of Bartlett's test for homogeneity of variance (a standard tool in the kit of the user of univariate analysis of variance) provides such a test. Namely,

$$H_0 : \Sigma_1 = \Sigma_2 = \cdots = \Sigma_k$$

is tested by one of two tests. Each requires calculation of

$$M = n \ln |S_c| - \sum_j (n_j \ln |S_j|),$$

where $n_i$ is the degrees of freedom for the variance–covariance estimates in $S_j$ and $n = \sum n_j$. If $p$ (the number of dependent variables) and $k$ (the number of independent groups) are both $\leq 5$ and each sample size is $\geq 20$, the chi-square approximation can be used, namely

$$\chi^2 = M/C$$

is distributed as $\chi^2$ with $\frac{1}{2}(k-1)(p)(p+1)$ degrees of freedom, where

$$1/C = 1 - (2p^2 + 3p - 1)/[6(p+1)(k-1)](\sum n_j^{-1} - n^{-1})$$
$$( = 1 - (2p^2 + 3p - 1)(k+1)/[6(p+1)kn] \qquad \text{when} \quad n_1 = n_2 = \cdots = n_k).$$

In any other situation, the $F$-approximation must be used, namely

$$F = M/b$$

is distributed as $F$ with $f_1$ and $f_2$ degrees of freedom, where

$$b = f/[1 - A_1 - (f_1/f_2)]; \qquad f_1 = \tfrac{1}{2}(k-1)p(p+1); \qquad f_2 = (f_1 + 2)/(A_2 - A_1^2);$$
$$A_1 = 1 - 1/C;$$

and

$$A_2 = (p - 1)(p + 2)[\sum (1/n_j^2) - 1/n^2]/[6(k - 1)]$$

$$[= (p - 1)(p + 2)(k^2 + k + 1)/(6k^2n^2) \text{ for equal sample sizes}].$$

The $F$-approximation is excellent so long as none of the covariance matrices is based on fewer than 9 degrees of freedom. For extremely small sample sizes an extremely complex series solution for the exact distribution of the test statistic $M$ must be employed (cf. Box, 1949).

### 3.6.2   Known Covariance Matrix

In some situations we have *a priori* grounds for specifying, in advance of examining the data, the entries in the population variance–covariance matrix. The first test conducted in this chapter, that of the hypothesis that the grand mean response vector, computed across all four groups, = (10, 10, 10, 10) is an example of such a situation, since the hypothesis of random selection of $C$ or $D$ on each trial with probability $= \frac{1}{2}$ implies specific values of the $\sigma_{ij}$ as well as the $\mu_j$, namely $\sigma_j = 40(3/16) = 7.5$ and $\sigma_{ij} = 40(-1/16) = -2.5$. (The fact that the observed variances are uniformly larger than these hypothesized values suggests that individual differences in choice probability are inflating the response variabilities. One should be careful in applying formulas for the mean and variance of a multinomial distribution to situations where the assumption of those formulas that all Ss have the same generating probabilities is unlikely to be valid.)

The formulas for $T^2$ are very easily corrected to known-covariance formulas simply by substituting $\Sigma$ for $S$ or $S_c$. The resulting value of $T^2$ is then referred (without postmultiplication by any constant other than 1.0) to the chi-square table with $p$ degrees of freedom. For instance, in the single-sample case

$$T^2 = N(\bar{X} - \mu_0)' \Sigma^{-1}(\bar{X} - \mu_0)$$

is distributed as chi-square with $p$ degrees of freedom; and in the two-sample case

$$T^2 = N_1 N_2 (\bar{X}_1 - \bar{X}_2)' \Sigma^{-1}(\bar{X}_1 - \bar{X}_2)/(N_1 + N_2)$$

is distributed as chi-square with $p$ degrees of freedom. Applying the single-sample formula to our hypothesis of completely random responding from all 48 subject pairs gives

$$\chi^2 = 48 \times [1.23 - 3.09 - 1.67] \begin{bmatrix} .2 & .1 & .1 \\ .1 & .2 & .1 \\ .1 & .1 & .2 \end{bmatrix} \begin{bmatrix} 1.23 \\ -3.09 \\ -1.67 \end{bmatrix}$$

$$= [.2(13.84) + .1(-6.16)]48 = 48(2.14) = 102.2 \quad \text{with} \quad 3 \text{ df}.$$

Actually, had we simply applied the same formula as is used for the estimated covariance–matrix case, keeping in mind that we have an arbitrarily large number of degrees of freedom for the covariance matrix, we would have obtained identical results, since the $F$-distribution with $\infty$, $n_2$ degrees of freedom is identical to the distribution of $\chi^2/n_2$ with $n_2$ degrees of freedom.

### 3.6.3   The Assumption of Multivariate Normality

The derivation of the distribution of $T^2$ makes use of the assumption that the vectors of outcome variables are sampled from a multivariate normal distribution. We might suspect from the multitude of empirical sampling experiments which have demonstrated the remarkable robustness of Student's $t$-test (when 2-tailed tests are employed) against violations of the normality assumption, that violations of multivariate normality would have similarly small effects on the validity of $T^2$-tests unless especially "wild" distributions are employed. However, little is known about the robustness of $T^2$, except that for sufficiently large sample sizes, computed $T^2$-values do conform to the $F$-distribution, no matter what the shape of the parent population. Empirical sampling studies to determine how large "sufficiently large" is for the $T^2$-test are unfortunately lacking.

## 3.7   Single-Symbol Expressions for Simple Cases

We give the single-symbol expressions for the two-variable, three-variable, and uncorrelated-variable cases for both the single-sample and two-sample $T^2$-tests. The reader will probably wish to test his understanding of $T^2$ by deriving these formulas himself. The reader may also have noticed that $T^2$-analogs for only two of the three $t$-test procedures usually described in introductory statistics texts have been discussed explicitly in this chapter. The generalizations of the single-sample $t$ and the $t$-test for independent means have been presented, but there has been no separate section on a generalization of the $t$-test for correlated means. The reason for this is that this third application of Student's $t$-test is equivalent to a flatness test—which you will recall is a test of the hypothesis that the "profile" of population means of the $p$ dependent variables is flat, whence every difference between two "adjacent" means is zero—conducted on a two-variable outcome vector. In this case $\mathbf{S}^*$ is the single entry

$$s_1^2 + s_2^2 - 2s_{12} = s_{X_1 - X_2}^2 = s_d^2,$$

so that

$$T_{\text{flat}}^2 = N\mathbf{dS}^{*-1}\mathbf{\bar{d}}' = Nd^2/s_d^2$$

is the square of the usual test statistic for the $t$-test for correlated means.

**(a)  $p = 2$**

$$T^2 = \frac{K(d_1^2 s_2^2 - 2d_1 d_2 s_{12} + d_2^2 s_1^2)}{s_1^2 s_2^2 - s_{12}^2} = \frac{K(d_1^2 + d_2^2 - 2r_{12} d_1 d_2)}{s^2(1 - r_{12}^2)}$$

when $s_1^2 = s_2^2 = s^2$;

$$F = (\text{df} - 1)T^2/(2\ \text{df});$$

$$a_1 = \frac{s_{12} T^2 - K d_1 d_2}{K d_1^2 - T^2 s_1^2} a_2 = \frac{K d_2^2 - T^2 s_2^2}{s_{12} T^2 - K d_1 d_2} a_2.$$

**(b)** $p = 3$

$$T^2 = \frac{K\left[\begin{array}{c} d_1^2(s_2^2 s_3^2 - s_{23}^2) + d_2^2(s_1^2 s_3^2 - s_{13}^2) + d_3^2(s_1^2 s_2^2 - s_{12}^2) \\ + 2Kd_1 d_2(s_{13}s_{23} - s_{12}s_3^2) + d_1 d_3(s_{12}s_{23} - s_{13}s_2^2) + d_2 d_3(s_{12}s_{13} - s_{23}s_1^2) \end{array}\right]}{s_1^2 s_2^2 s_3^2 + 2s_{12}s_{13}s_{23} - s_1^2 s_{23}^2 - s_2^2 s_{13}^2 - s_3^2 s_{12}^2}$$

$$= \frac{K\left[\begin{array}{c} d_1^2(1 - r_{23}^2) + d_2^2(1 - r_{13}^2) + d_3^2(1 - r_{12}^2) + 2d_1 d_2(r_{13}r_{23} - r_{12}) \\ + 2d_1 d_3(r_{12}r_{23} - r_{13}) + 2d_2 d_3(r_{12}r_{13} - r_{23}) \end{array}\right]}{s^2[1 + 2r_{12}r_{13}r_{23} - r_{23}^2 - r_{13}^2 - r_{12}^2]}$$

when variances are equal;

$$F = (\text{df} - 2)T^2/(3 \, \text{df});$$

$$a_1 = \frac{L_{12}L_{23} - L_{22}L_{13}}{L_{11}L_{22} - L_{12}^2} a_3; \qquad a_2 = \frac{L_{12}L_{33} - L_{13}L_{23}}{L_{13}L_{22} - L_{12}L_{23}} a_3;$$

where $L_{ij} = Kd_i d_j - T^2 s_{ij}$.

**(c) Uncorrelated Variables**

$$T^2 = K[(d_1/s_1)^2 + (d_2/s_2)^2 + \cdots + (d_p/s_p)^2] = t_1^2 + t_2^2 + \cdots + t_p^2,$$

where $t_j$ is the univariate $t$ computed on $X_j$ alone;

$$F = (\text{df} - p + 1)T^2/(p \, \text{df});$$

$$a_j = (d_j/d_p)(s_p^2/s_j^2)a_p.$$

In the above formulas, $K = N$ for the single-sample $T^2$, and $N_1 N_2/(N_1 + N_2)$ for the two-sample $T^2$; $d_j = \bar{X}_j - \mu_{j0}$ for the single-sample $T^2$, and $d_j = \bar{X}_{j1} - \bar{X}_{j2}$ (the difference between the sample means for variable $j$) for the two-sample $T^2$; and df is the degrees of freedom for each covariance or variance estimate, and will usually be $N - 1$ for the single-sample $T^2$, and $N_1 + N_2 - 2$ for the two-sample $T^2$.

## DEMONSTRATION PROBLEM

**A. Hypothetical data.** Consider the following set of data.

| Group | Subject | | $X_1$ | $X_2$ | $X_3$ |
|---|---|---|---|---|---|
| 1 | 1 | | 3 | 4 | 11 |
| | 2 | | 9 | 8 | 10 |
| | 3 | | 3 | 9 | 6 |
| | | $\bar{X}_1$ | 5.0 | 7.0 | 9.0 |
| 2 | 4 | | 10 | 9 | 11 |
| | 5 | | 11 | 9 | 10 |
| | 6 | | 6 | 9 | 12 |
| | | $\bar{X}_2$ | 9.0 | 9.0 | 11.0 |
| | | $\bar{\bar{X}}$ | 7.0 | 8.0 | 10.0 |

1. Compute a pooled covariance matrix based on deviations about *group* means (rather than about the grand means). This yields an estimate of the common covariance matrix which is not affected by differences in population means.

2. Using the pooled covariance matrix computed in Problem 1, test the null hypothesis that the population grand mean profile is (10, 10, 10).

3. Compute the discriminant function corresponding to the test in Problem 2 and compute a score for each S on this discriminant function. Now conduct a univariate *t*-test on the difference between the mean of this discriminant function and the population mean implied for this combined score by the null hypothesis adopted in Problem 2. Compare with your results in Problem 2.

4. Now test the null hypothesis that the population mean profiles for groups 1 and 2 are identical, that is, that populations 1 and 2 do not differ in their means on any of the three dependent measures.

5. Compute a score for each S on the discriminant function corresponding to the test in Problem 4. Now conduct a univariate *t*-test on the difference in mean discriminant scores for the two groups. Compare with your results in Problem 4.

6. Using *post hoc* significance criteria, test the hypotheses that the two groups are sampled from populations having identical means on
(a)  $X_1$,     (b)  $X_2$,     (c)  $X_1 + X_2 + X_3$,     (d)  $4.0X_1 + 2.0X_2 + 2X_3$.
The last test uses the observed mean differences as weighting factors.

7. Conduct a profile analysis of these data. Compare the results of the flatness test with the results of the test in Problem 2, which also assumes a flat profile. In addition, test the significance of the difference in grand means on $X_1$ and $X_2$.

**B.  Some real data.**   All of the following refer to Data Set 3.

1. Use Hotelling's $T^2$ to test the significance of the difference between the mean outcome vectors for the NMO and the COMP groups.

2. Compute the discriminant function for the difference between NMO and COMP. Compute each subject-pair's score on the discriminant function D.

3. Conduct a univariate *t*-test on the difference between the NMO and COMP groups in mean D (discriminant function) score. Compare the results of Problem 1.

4. Test the statistical significance of each of the following sources of difference between the NMO and COMP groups:
(a)  CC;     (b)  CD − DC;     (c)  CC + CD;     (d)  CC + DD;
(e)  a simplified (rounded-off) version of the discriminant function.
Indicate in each case both the conclusion you would reach if this were a *post hoc* comparison and the conclusion employing an *a priori* criterion.

5. Conduct a profile analysis of the difference between NMO and COMP outcome vectors.

*Answers*

**A.**

1. $S_c = \dfrac{1}{4} \begin{bmatrix} 38 & 6 & 1 \\ 6 & 14 & -11 \\ 1 & -11 & 16 \end{bmatrix} = \begin{bmatrix} 9.5 & 1.5 & .25 \\ 1.5 & 3.5 & -2.75 \\ .25 & -2.75 & 4.0 \end{bmatrix}.$

**2.**  $\bar{\mathbf{X}}' = (7, 8, 10);$     $\Delta' = (-3, -2, 0);$

$$\mathbf{S}_c + N\Delta\Delta' = \mathbf{S}_c + 6\begin{bmatrix} 9 & 6 & 0 \\ 6 & 4 & 0 \\ 0 & 0 & 0 \end{bmatrix} = \begin{bmatrix} 63.5 & 37.5 & .25 \\ 37.5 & 27.5 & -2.75 \\ .25 & -2.75 & 4.0 \end{bmatrix};$$

$|\mathbf{S}_c| = 49.875;$     $|\mathbf{S}_c + N\Delta\Delta'| = 826.50.$

Thus, using the determinantal formula,

$$T^2 = (826.5/49.875) - 1 = 15.571.$$

Then

$$F = (4 - 3 + 1)/[3(4)]T^2 = T^2/6 = 2.595 \qquad \text{with} \quad 3 \text{ and } 2 \text{ df,} \qquad \text{ns.}$$

It is always a good idea to check your work by applying an alternate computational procedure. Using Eq. (3.5), we obtain

$$\mathbf{S}_c^{-1} = \begin{bmatrix} .12907 & -.13409 & -.10025 \\ & .76065 & .53133 \\ & & .62515 \end{bmatrix},$$

whence

$$N\Delta'\mathbf{S}_c^{-1}\Delta = 6[9(.12907) + 4(.76065) + 12(-.13409)] = 6(2.59515) = 15.5709.$$

Almost identical results are obtained by employing HEINV with $\mathbf{H} = N\Delta\Delta'$ and $\mathbf{E} = \mathbf{S}_c$. (HEINV gave a $T^2$ value of 15.5710.)

**3.**  It is probably most convenient to work with $[N\Delta\Delta' - \lambda\mathbf{S}]\mathbf{a} = \mathbf{0}$, rather than carrying out the premultiplication by $\mathbf{S}^{-1}$. At any rate, in nonnormalized form the discriminant function is $.15625X_1 + 1.46874X_2 + X_3$. Any other set of coefficients which preserves the *ratios* of the above figures (that is, their relative magnitude) is correct. The $D$ scores are 17.344, 23.156, 19.687, 25.781, 24.937, 26.156 for subjects $1, \ldots, 6$, respectively. The resulting $t$-test gives

$$t = (22.844 - 26.250)/\sqrt{.74493} = -3.946 = \sqrt{T^2}.$$

Note that as an *a priori* test (which it certainly is not), this $t$-ratio would have been statistically significant. Recall, too, that we are using a within-groups estimate of $\sigma_c^2$.

**4.**  $\Delta' = (4, 2, 2),$  whence

$$\mathbf{S}_c + \tfrac{3}{2}\Delta\Delta' = \begin{bmatrix} 33.5 & 13.5 & 12.25 \\ 13.5 & 9.5 & 3.25 \\ 12.25 & 3.25 & 10.0 \end{bmatrix},$$

and the determinantal formula gives

$$T^2 = (655.5/49.875) - 1 = 12.143,$$

whence

$$F = T^2/6 = 2.024 \qquad \text{with} \quad 3 \text{ and } 2 \text{ df,} \qquad \text{ns.}$$

HEINV and Eq. (3.5) agree.

**5.**  The resulting $t$ should be equal to the square root of $T^2$. The discriminant function is $.0170X_1 + .7321X_2 + .6810X_3$.

**6.** By far the quickest way to compute these $t$-ratios is to make use of the two-sample version of Eq. (3.2), whence

$$t^2(\mathbf{c}) = \tfrac{3}{2}(\mathbf{c}'\Delta)^2/(\mathbf{c}'\mathbf{S}_c\mathbf{c}),$$

where $\mathbf{Xc}$ is the set of combined scores. Thus

$$t^2(1, 0, 0) = \tfrac{3}{2}(16/9.5) = 2.53, \qquad \text{ns};$$
$$t^2(1, 1, 1) = \tfrac{3}{2}(64/15) = 6.4, \qquad \text{ns};$$
$$t^2(0, 1, 0) = \tfrac{3}{2}(4/3.5) = 1.71, \qquad \text{ns};$$
$$t^2(4, 2, 2) = \tfrac{3}{2}(576/188) = 4.596, \qquad \text{ns}.$$

We knew, of course, that none of these comparisons would be significant by the *post hoc* criterion, but it is worth noting how *much* better the discriminant function does (on this sample of data) than any of the others. Note, too, how unimportant $X_1$ is in the discriminant function.

**7. (a) Parallelism**

$$\mathbf{S}^* = \begin{bmatrix} 10 & -5 \\ -5 & 13 \end{bmatrix}; \qquad \Delta'_{\text{slope}} = (2, 0); \qquad \mathbf{S}^{*-1} = \frac{1}{105}\begin{bmatrix} 13 & 5 \\ 5 & 10 \end{bmatrix};$$

$$T^2 = \tfrac{3}{2}(52/105) = 78/105; \qquad F = 3T^2/8 < 1.$$

**(b) Levels** This is essentially the test in Problem 6c, with

$$F = 6.4 \qquad \text{with} \quad 1 \text{ and } 4 \text{ df}, \qquad \text{ns}.$$

**(c) Flatness**

$$\frac{1}{105}\begin{bmatrix} 13 & 5 \\ 5 & 10 \end{bmatrix}\begin{bmatrix} 1 \\ 2 \end{bmatrix}$$

$$T^2 = 6[1 \quad 2] \qquad\qquad = 6(73/105) = 4.17;$$

$$F = (3/8)T^2 = 1.56 \qquad \text{with} \quad 2 \text{ and } 3 \text{ df}, \qquad \text{ns}.$$

This is *not* the same as the test against $(10, 10, 10)$, since the overall level is unspecified. It is closer to (but not identical to) a test of the hypothesis that $\mu' = (8.67, 8.67, 8.67)$.

Testing the significance of the difference between the grand means on variables 1 and 2 yields an $F$ of $6(1)^2/10 = .6$, nonsignificant.

**B.**

**1.** $\mathbf{A}_c = \begin{bmatrix} 417.17 & 149.75 & 14.25 \\ 149.75 & 226.25 & -79.25 \\ 14.25 & -79.25 & 282.92 \end{bmatrix};$  $\mathbf{A}_c + 6\Delta\Delta' = \begin{bmatrix} 449.84 & 132.25 & 64.42 \\ 132.25 & 235.63 & -106.13 \\ 64.42 & -106.13 & 359.96 \end{bmatrix};$

$$\Delta' = [-2.333 \quad 1.250 \quad -3.583]; \qquad N_1 N_2/(N_1 + N_2) = 6;$$

$$|\mathbf{A}_c| = .17355 \times 10^8; \qquad |\mathbf{A}_c + 6\Delta\Delta'| = .24005 \times 10^8;$$

$$T^2 = 22[(.24005/.17355) - 1] = 8.4287;$$

$$F = (22 - 3 + 1)/[3(22)]T^2 = (20/66)T^2 = 2.554 \qquad \text{with} \quad 3 \text{ and } 20 \text{ df}, \qquad \text{ns}.$$

**2.** $D = .75565CC - .68658CD + DC.$

**3.** $t = \sqrt{T^2} = 2.904.$

**4.** For *post hoc* comparisons, $T^2_{\text{crit}} = 3.3(3.10) = 10.23$, whence $t_{\text{crit}} = 3.20$. Of course, since the overall test failed, we know that none of these specific comparisons will be significant on a *post hoc* basis. For *a priori* comparisons, a squared *t*-ratio of $2.074^2 = 4.30$ is required for significance at the .05 level.

(a) $t^2(1, 0, 0, 0) = 132[(2.333)^2/417.17] = 1.722.$

(b) $t^2(0, 1, -1, 0) = 132[(4.833)^2/667.670] = 4.618,$     $p < .05$   if *a priori*.

(c) $t^2(1, 1, 0, 0) = .164.$     (d)   $t^2(1, 0, 0, 1) = 2.049.$

(e) $t^2(1, -1, 1, 0) = 8.374.$

Note the (perfectly legitimate) inclusion of DD in test (d), despite its apparently not having been involved in the overall test. This is a result of the linear dependency between DD and the other three outcome measures.

**5.** Profile analysis is an example of a case where it *does* make a difference how we decide to handle a linear dependency among the outcome measures. For instance, in the present case, the fact that the four measures must sum to 40 guarantees that a levels test on the full four-variable outcome vector would yield a *t*-ratio of exactly zero. It further forces a correlation between the parallelism test and the overall $T^2$ if all variables are included in the former test, since a difference between the population means of any variable or combination of variables forces the population mean profiles to be nonparallel. To keep the results from being absolutely trivial, the profile analysis was conducted on the first three outcome measures only.

**(a) Parallelism**

$$S_{\text{slope}} = \frac{1}{22}\begin{bmatrix} 343.92 & -170.0 \\ -170.0 & 667.67 \end{bmatrix}; \qquad \Delta'_{\text{slope}} = \frac{1}{12}[-43 \quad 58];$$

$$T^2 = 6\Delta'S^{-1}\Delta = 7.049; \qquad F = (21/44)T^2 = 3.364 \qquad \text{with} \quad 2 \text{ and } 21 \text{ df}, \qquad \text{ns.}$$

**(b) Levels**

$$t^2(1, 1, 1, 0) = 132[(-4.667)^2/1095.84] = 2.624, \qquad \text{ns.}$$

**(c) Flatness**

grand mean slope vector $= \Delta'_{\bar{\bar{x}}} = \dfrac{1}{24}[31 \quad 10];$

$$T^2 = 24[31 \quad 10]S^{-1}_{\text{slope}} \frac{1}{576}\begin{bmatrix} 31 \\ 10 \end{bmatrix} = 3.569;$$

$$F = (21/44)T^2 = 1.703 \qquad \text{with} \quad 2 \text{ and } 21 \text{ df}, \qquad \text{ns.}$$

# 4 MULTIVARIATE ANALYSIS OF VARIANCE

## Differences among Several Groups on Several Measures

The use of Hotelling's $T^2$ is limited to comparisons of only two groups at a time. Many studies, such as the one which provided the data of Data Set 3 (pp. 68–69), involve more than just two independent groups of subjects. Just as doing a series of six univariate $t$-tests to test all the differences on a single variable among four experimental groups would be inappropriate because of the inflation of Type I error this produces, so would a series of six $T^2$ analyses of the differences in mean response vectors among these same groups be just as inappropriate for the same reason. The answer to the problem of multiple experimental groups in the univariate case is (univariate) analysis of variance (Anova); the solution to the multiple comparison problem when there are two or more outcome measures as well as two or more groups is multivariate analysis of variance (Manova). Before we discuss Manova, it might be well to review the basic rationale and computational formulas of Anova.

## 4.1 One-Way (Univariate) Analysis of Variance

### 4.1.1 The Overall Test

Let us say that we are interested in assessing differences among the four motivating instruction conditions of Data Set 3 (see pp. 68–69) in the number of mutually competitive outcomes (DDs) achieved by a pair in a 40-trial game. The least interesting of the possible true "states of nature" which might underlie these data is the possibility that there are no real differences due to motivating instructions, the differences among the sample means having arisen solely through random sampling from populations having identical means. Thus, before we proceed with any more detailed examination of the means on DD, we must test the null hypothesis that

$$\mu_{\text{NMO}} = \mu_{\text{IND}} = \mu_{\text{COMP}} = \mu_{\text{COOP}} .$$

More generally, the null hypothesis can be stated as

$$H_0 : \mu_1 = \mu_2 = \cdots = \mu_k , \tag{4.1}$$

93

where $k$ is the number of groups. If this null hypothesis were true, then *on the average* we would obtain $\bar{X}_1 = \bar{X}_2 = \cdots = \bar{X}_k$, though any particular set of samples would produce pairwise differences between sample means differing somewhat from zero as a consequence of random fluctuation. Our first task is to select a single statistic which will summarize how far our sample means depart from the equality implied by $H_0$. An attractive candidate, and the one we shall adopt, is

$$s_{\bar{X}}^2 = \sum (\bar{X}_j - \bar{\bar{X}})^2/(k - 1),$$

the sample variance of the $k$ means. In addition to the obvious reasons for selecting $s_{\bar{X}}^2$ (its sensitivity to any difference between sample means, the algebraic tractability of variances, etc.), it also is equal to one-half the mean squared difference between the $k(k - 1)/2$ pairs of sample means. For instance, the variance of the four $\overline{\text{DD}}$s for Data Set 3 is

$$\tfrac{1}{3}[(18.81 - 13.52)^2 + \cdots + (3.75 - 13.52)^2] = 48.14,$$

while the mean squared pairwise difference is

$$\tfrac{1}{6}[(18.81 - 18.09)^2 + (18.09 - 13.41)^2 + \cdots + (18.81 - 3.75)^2] = 96.31.$$

The second thing we need to ask is whether the magnitude of the discrepancies among the sample means (as measured by the variance of the $\bar{X}$s) is consistent with our assumption of identical population means. A basis for comparison is provided by the well-known relationship between the variance of a population and the variance of samples drawn from that population, namely that the variance of the means is

$$\sigma_{\bar{X}}^2 = \sigma^2/n, \tag{4.2}$$

where $n$ is the size of the sample on which each mean is based. Thus if we can obtain an estimate of the variance which we assume to be common to the $k$ populations from which our samples were drawn, we could thereby also obtain an estimate of how large we would expect the variance of our sample means to be if there were in fact no true differences in population means. [Any true difference between any one or more pairs of population means will of course inflate the variance of the sample means beyond the figure computed from Eq. (4.2).] We can of course obtain such an estimate by looking at the variability within each group of subjects, and pooling these $k$ estimates of $\sigma^2$ into a single best estimate, just as we pool our two available estimates in the univariate $t$-test for the difference between two independent means. It would then seem informative to compare the ratio of our direct estimate of the variance of sample means drawn from our four populations with our indirect estimate of this same variance arrived at through an examination of the within-group variances in conjunction with the assumption of no true differences among the means of the four populations. In other words, we compute

$$F = \frac{s_{\bar{X}}^2}{s_{\text{w}}^2/n} = \frac{\text{direct estimate of } \sigma_{\bar{X}}^2}{\text{estimate of } \sigma_{\bar{X}}^2 \text{ assuming } H_0}, \tag{4.3}$$

where

$$s_w^2 = [\sum (X_1 - \bar{X}_1)^2 + \sum (X_2 - \bar{X}_2)^2 + \cdots + \sum (X_k - \bar{X}_k)^2]/(N - k);$$
$$s_{\bar{X}}^2 = [(\bar{X}_1 - \bar{\bar{X}})^2 + (\bar{X}_2 - \bar{\bar{X}})^2 + \cdots + (\bar{X}_k - \bar{\bar{X}})^2]/(k - 1);$$

and $n$ is the (common) sample size, that is, the number of subjects in each of our $k$ independent groups, so that the total number of observations is $N = nk$.

Statisticians have computed for us how often $F$-ratios of various sizes would occur if, in fact, the null hypothesis were correct and the additional assumptions of homogeneity of variance and normality of parent populations are met. We can thus compare the value of $F$ (the ratio of the variability we observe among our sample means to the variability we would expect to observe, were $H_0$ correct) we obtain to tabled percentiles of the $F$-distribution. We need to specify two degree-of-freedom parameters for this comparison: the degrees of freedom going into our direct estimate of $\sigma_{\bar{X}}^2$ ($k - 1$, the number of numbers used in computing $s_{\bar{X}}^2$ minus one degree of freedom "lost" in taking deviations about $\bar{\bar{X}}$ instead of $\mu$) and the number of degrees of freedom going into our indirect estimate ($N - k = n_1 - 1 + \cdots + n_k - 1$, the sum of the degrees of freedom going into the various group variances). If our computed value of $F$ is greater than the "critical value" tabled for an $F$ with $k - 1$ and $N - k$ degrees of freedom at the chosen significance level, $H_0$ is rejected; otherwise, it is not rejected.

Computationally, Eq. (4.3) is inconvenient to work with, since each $\bar{X}_j$, as well as $\bar{\bar{X}}$, can be expected to involve decimals, and since repetitive subtractions are not easily worked into cumulations of cross-products on a desk calculator. For computational purposes it has become traditional to reverse the intuitively obvious comparison procedure, using the variability of the sample means to estimate $\sigma^2 = n\sigma_{\bar{X}}^2$ and comparing this indirect estimate of $\sigma^2$ with its direct estimate $s_w^2$. The computational version of (4.3) is thus

$$F = \frac{\sum n_j(\bar{X}_j - \bar{\bar{X}})^2/(k - 1)}{\sum \sum (X_j - \bar{X}_j)^2/(N - k)} = \frac{[\sum (T_j^2/n_j) - (\sum \sum X)^2/N]/(k - 1)}{[\sum \sum X^2 - \sum (T_j^2/n_j)]/(N - k)}$$

$$= \frac{SS_b/(k - 1)}{SS_w/(N - k)} = \frac{MS_b}{MS_w} \tag{4.4}$$

with $k - 1$ and $N - k$ degrees of freedom, where $T_j = \sum_u X_{ju}$ is the sum of the observations in group $j$. The results of our one-way Anova are conventionally reported in a summary table (Table 4.1). The computational formulas of (4.4) are of course simply raw-score formulas for the two kinds of variances computed in (4.3).

Table 4.1

Summary Table of Anova on Dependent Variable

| Source | df | SS | MS | F |
|---|---|---|---|---|
| Between groups | $k - 1$ | $\sum (T_j^2/n_j) - (\sum \sum X)^2/N$ | $SS_b/(k - 1)$ | $MS_b/MS_w$   (4.5) |
| Within groups (error) | $N - k$ | $\sum \sum X^2 - \sum (T_j^2/n_j)$ | $SS_w/(N - k)$ | |
| Total | $N - 1$ | $\sum \sum X^2 - (\sum \sum X)^2/N$ | | |

### 4.1.2   Specific Comparisons

Rejection of our overall $H_0$ simply tells us that something other than chance fluctuation is generating the differences among our $k$ group means. It is the purpose of a *specific comparison* procedure to allow us to specify in more detail the source of our significant overall $F$. There are many alternative procedures for conducting such specific comparisons. Rather than repeating the comprehensive analyses of the various approaches by Games (1971) and Kirk (1968), the present text focuses on two approaches: Scheffé's contrast method and Bonferroni critical values.

Scheffé's contrast method has the following properties:

(i)   It tests any and all hypotheses of the form $c_1 \mu_1 + c_2 \mu_2 + \cdots + c_k \mu_k = 0$, that is, all *linear contrasts* among the population means. All alternative methods (except the Bonferroni approach to be described next) confine themselves to pairwise comparisons, that is, comparisons involving only two groups at a time. It has been the author's experience that truly illuminating descriptions of the relationships among three or more groups usually involve more than pairwise comparison of group means. For example, the differences among the four instructional sets of Data Set 3 in terms of mean number of DD outcomes can most naturally be summarized in terms of the difference between the COOP group and the other three (corresponding to the $H_0$ that $\mu_{\text{COOP}} - \mu_{\text{NMO}}/3 - \mu_{\text{COMP}}/3 - \mu_{\text{IND}}/3 = 0$), the difference between the NMO group and the other two ($\mu_{\text{NMO}} - \frac{1}{2}\mu_{\text{IND}} - \frac{1}{2}\mu_{\text{COMP}}$), and the difference between the Individualistic and Competitive groups ($\mu_{\text{IND}} - \mu_{\text{COMP}}$).

(ii)   The procedure supplies significance tests which are easily adjusted— through multiplication by a constant—to handle either *a priori* or *post hoc* comparisons. That these two types of comparisons require different significance criteria should be obvious. The probability that the COOP group would score higher than the other three groups by chance is, for instance, considerably lower than the probability that *some* one of the four groups would have the highest sample mean. Similarly the probability that this prespecified (*a priori*) difference would achieve a specific magnitude is considerably lower than the probability that *some* after-the-fact (*post hoc*) contrast would reach that same magnitude. Each *a priori* comparison is tested by comparing an obtained $F$-value, defined by

$$F_{\text{contr}} = \frac{(\sum c_j \bar{X}_j)^2}{\sum (c_j^2/n_j)(\text{MS}_{\text{w}})} \tag{4.6}$$

$$\left\{ = \frac{n(\sum c_j \bar{X}_j)^2}{(\sum c_j^2)(\text{MS}_{\text{w}})} = \frac{(\sum c_j T_j)^2}{n(\sum c_j^2)(\text{MS}_{\text{w}})} = \frac{\text{SS}_{\text{contr}}}{\text{MS}_{\text{w}}} = \frac{\text{MS}_{\text{contr}}}{\text{MS}_{\text{w}}} \right.$$

$$\left. \text{when } n_1 = n_2 = \cdots = n_k = n \right\},$$

with the critical value at the chosen level of significance for an $F$ with 1 and $N - k$ degrees of freedom. Each *post hoc* $F_{\text{contr}}$ is compared to $k - 1$ times the critical value for an $F$ with $k - 1$ and $N - k$ degrees of freedom. This *post hoc* procedure has the property that the probability of *any* false rejection of a null hypothesis is *at most* $\alpha$

(the chosen significance level), no matter how many of the infinite number of possible contrasts are computed. Further, the overall $F$ is statistically significant if and only if at least one linear contrast can be found which reaches statistical significance by the *post hoc* criterion, though it is not at all uncommon to find that no single pairwise difference between any two group means is significant even though the overall $F$ is.

(iii)   When all sample sizes are equal, then any set of $k - 1$ *independent* contrasts completely partitions the between-groups sum of squares, in the sense that the sum of the $k - 1$ $SS_{contr}$s is equal to $SS_b$. Scheffé's method thus has enormous descriptive value in specifying the percentage of the total variation among the groups which is attributable to each of $k - 1$ independent sources of differences among the $k$ groups. Two contrasts are independent if and only if the sum of the cross-products of their respective coefficients equals zero, that is, $c_1'c_2 = 0$. It is convenient for descriptive purposes to choose independent contrasts, but this should never get in the way of testing those contrasts among the means which are of greatest theoretical import to the researcher.

To illustrate the Scheffé procedures we have outlined for one-way Anova, let us apply them to the DD measure only in Data Set 3. We obtain

$$\sum\sum X = 649; \qquad \sum\sum X^2 = 12{,}505;$$
$$\tfrac{1}{12}\sum T_j^2 = 10{,}509.25; \qquad C = 649^2/48 = 8775.02;$$

$SS_{contr\,1} = [217 + 226 + 161 - 3(45)]^2/(12 \cdot 12)$

$\qquad = 1527.51 \qquad$ for the COOP group versus the other three groups;

$SS_{contr\,2} = [217 + 226 - 2(161)]^2/(12 \cdot 6)$

$\qquad = 203.35 \qquad$ for NMO versus COMP and IND;

and

$SS_{contr\,3} = (217 - 226)^2/(12 \cdot 2)$

$\qquad = 3.38 \qquad$ for COMP versus IND;

whence we have a summary table (Table 4.2). Note that the "NMO versus COMP,

Table 4.2

*Summary Table for Effects of Instructions on Frequency of DD Outcomes*

| Source | df | SS | MS | F |
|---|---|---|---|---|
| Between groups | 3 | 1734.23 | 578.08 | 12.745[a] |
| COOP versus other 3 | 1 | 1527.51 | | 33.7[a] |
| NMO versus COMP, IND | 1 | 203.35 | | 4.49 |
| COMP versus IND | 1 | 3.38 | | < 1 |
| Within groups | 44 | 1995.75 | 45.36 | |
| Total | 47 | 3729.98 | | |

[a] $p < .01$ by *post hoc* criterion.

IND" contrast would be considered statistically significant *if* it were specified before the data were collected. If, however, the thought of making this particular comparison arose only after seeing the data, it cannot be considered statistically significant. Note also that $F$-ratios less than 1 can only be the result of chance fluctuation or of failure to take account of a constraint on the data, such as matching of groups. $F$-ratios lower than 1 thus need not be computed in detail. (If there are several very low $F$s, however, the researcher should check his computations very carefully and examine the possibility that his sampling or measurement procedures may impose restrictions on the variability of the data.)

The issue of how best to conduct multiple comparisons is still a highly controversial one, as a glance at any collection of readings on statistical issues (for example, Kirk, 1968) will verify. In particular, it is clear that the procedure outlined thus far, which entails complete freedom to conduct any *a priori* tests using the same significance level as if that test were the only one conducted, while requiring that *post hoc* tests employ a critical value which sets the probability that *any* one of the (infinitely many) *possible* comparisons leads to a false rejection of a null hypothesis at $\alpha$, has some disadvantages. For example, if the researcher simply predicts in advance that every treatment will lead to a distinctly different effect, so that no two of the corresponding population means are identical, then he can claim that *any* specific comparison he chooses to make is logically *a priori*. The Bonferroni procedure provides a solution to this problem.

The Bonferroni approach is based on the *Bonferroni inequality*, which states that

$$\alpha_{\text{set}} \leq \sum_{i=1}^{b} \alpha_i, \tag{4.7}$$

where $\alpha_i$ is the Type I error rate (pertaining to the probability of falsely rejecting the null hypothesis, given that it is true) for the test of the $i$th of a set of $b$ null hypotheses, and $\alpha_{\text{set}}$ is the probability that one or more or the $b$ null hypotheses in the set will be rejected, given that all are true. If the $b$ tests involved were statistically independent of each other (as for instance is true of orthogonal contrasts), then $\alpha_{\text{set}}$ would precisely equal $1 - \prod_{i=1}^{b} (1 - \alpha_i)$, which is very close to, but less than $\sum_{i=1}^{b} \alpha_i$, for small $\alpha_i$. Any correlations among the tests in the set lower $\alpha_{\text{set}}$, so that (4.7) holds for correlated tests as well as for orthogonal comparisons. The Bonferroni approach adjusts the critical values of the individual tests so as to yield individual Type I error rates $\alpha_i$ which sum to the desired *setwise error rate* $\alpha_{\text{set}}$. If the set contains all null hypotheses of any possible interest to the researcher, then $\alpha_{\text{set}} = \alpha_{\text{ew}}$, the *experiment-wise error rate*. (Cf. Games, 1971, and/or Kirk, 1968, for a discussion of the issue of selection of error rate.)

If the tests are highly intercorrelated, $\alpha_{\text{set}}$ may be considerably lower than $\sum \alpha_i$—it does in fact equal $\alpha_i$ when the tests are perfectly correlated. The Bonferroni approach thus provides conservative tests of statistical significance, but it nevertheless often yields less stringent critical values (and thus more powerful tests) than use of the Scheffé *post hoc* criterion, especially when $b$, the total number of comparisons, is small. It also has the advantage of allowing the researcher to set lower $\alpha_i$s for (and thus provide more powerful tests of) the hypotheses of greatest theoretical or practical importance.

The Bonferroni approach has the disadvantage of often requiring critical values for significance levels not included in readily available tables of the percentiles of sampling distributions. The desired critical values will usually be those for the extreme tails of Student's $t$-distribution. (An $F$-ratio computed on a Scheffé contrast can of course be converted to a $t$-ratio with degrees of freedom equal to the denominator degree of freedom of the $F$-test by the simple expedient of taking its square root.) Myers (1972, Table A-12) provides a table of the critical values of $t$ for significance levels $\alpha/b$, where $\alpha = .01, .05$, or $.10$ and $b$ ranges from 2 to 5000, and where the degrees of freedom range from 5 to 120. (Comparison with other available tables suggests that Myers's entries should be considered accurate only to two decimal places, rather than the four provided.) Appendix B of the present Primer includes a computer subroutine which computes critical values of $t$ for any choice of $\alpha$ and degrees of freedom. Finally, the reader should be warned that an approximation to the critical values of $t$ provided by Kirk (1968) can lead to rather poor approximations for low values of degrees of freedom or $\alpha$. For instance, Kirk's formula approximates $t_{.0005}(10)$ as 6.59, versus the true value of 7.28; and yields a value of 5.92 for $t_{.01}(5)$, versus the true value of 5.60.

The major disadvantage of the Bonferroni approach is its restriction to prespecified sets of comparisons, which thereby reduces its utility for *post hoc* exploration of the obtained data. Of course this is not a disadvantage to the researcher who is interested only in testing a limited number of specific hypotheses generated on theoretical grounds, or who finds that one of the following two sets of comparisons includes all those he would consider meaningful:

(a)   all pairwise comparisons of the $k$ means, for a total of $\binom{k}{2} = k(k-1)/2$ comparisons [The notation $\binom{n}{m}$ stands for the number of possible combinations of $m$ things selected from a set of $n$ things, order ignored.];

(b)   all comparisons of the unweighted means of two subsets of the means, leading to a total of

$$\frac{1}{2}\sum_{i=1}^{k-1}\sum_{j=1}^{k-i}\binom{k}{i}\binom{k-i}{j} = (3^k - 2^{k+1} + 1)/2,$$

where $k$ is the number of groups. Note that this set includes tests of hypotheses such as $\mu_4 = (\mu_1 + \mu_2 + \mu_3)/3$, but does not include the standard tests for linear, quadratic, cubic, etc. trends.

It is especially important to note that $b$, the size of the set of comparisons whose individual $\alpha_i$ sum to $\alpha_{set}$, is equal to the number of *possible* comparisons, not just the number the researcher actually carries out. Thus, for instance, a decision to test the largest mean against the smallest involves a $b$ of $k(k-1)/2$, not 1, since it implicitly requires examination of *all* pairs of means in order to select the largest pairwise difference.

Note too that choice of the Bonferroni approach to controlling the experiment-wise error rate for multiple comparisons is inconsistent with use of the overall $F$, $F_{ov} = MS_a/MS_w$, as a test of the overall hypothesis of no differences among the population means, since this $F$-test considers all sources of differences among the means, including some kinds of comparisons which the researcher has declared to be of no conceivable interest to him. Thus the overall null hypothesis for the user of the Bonferroni approach should be rephrased as "Each of the $b$ null hypotheses (in the set of comparisons of interest) is true," which hypothesis is rejected at the $\alpha_{ew}$-level of significance if one or more of the $b$-tests leads to rejection of its corresponding null hypothesis at the $\alpha_i$- (usually $= \alpha_{ew}/b$) level of significance.

Finally, note that specialized multiple comparison procedures have been developed for certain sets of prespecified contrasts, such as comparisons of each of the $k - 1$ experimental groups' means with the single control group's mean (Dunnett's test) or examination of all possible pairwise differences (Tukey's test). These specialized procedures provide more powerful tests than the Bonferroni procedure when they are applicable, since their critical values take into account the nature of the intercorrelations among the various comparisons. (The reader is again referred to the work of Kirk, 1968; or Games, 1971, for details of these specialized procedures.) On the other hand, the Bonferroni approach is more flexible, generalizing readily to situations involving multiple dependent variables.

For instance, if the researcher is willing to confine himself to examination of single dependent variables, eschewing all consideration of the emergent variables represented by linear combinations of the original variables, the test of the overall null hypothesis in a situation where Manova would normally be applied can be accomplished instead by performing $p$ separate univariate analyses, taking $F_{\alpha/p}(df_{eff}, df_w)$ as the critical value for each such test, where $\alpha$ is the desired significance level for the overall test. Similarly he could specify a set of $b \cdot d$ contrasts encompassing all $b$ of the relevant comparisons among the means on any of the $d$ linear combinations of the dependent measures he deems worth consideration, taking $\alpha_{ew}/(b \cdot d)$ as the significance level for each such comparison. Such a procedure, however, seems recklessly wasteful of the information which a true multivariate analysis could provide.

The Bonferroni approach can be illustrated with the same data used to illustrate Scheffé procedures, namely the three contrasts among the four means on DD from Data Set 3. From the summary table provided earlier (Table 4.2), we have $F$s of 33.7, 4.49, and $< 1$ for tests of

$$\mu_{COOP} = (\mu_{NMO} + \mu_{IND} + \mu_{COMP})/3, \qquad \mu_{NMO} = (\mu_{IND} + \mu_{COMP})/2,$$

and

$$\mu_{IND} = \mu_{COMP},$$

respectively. If we had specified these in advance as the only comparisons of any conceivable interest to us, and if we had considered them of equal importance, then we would use a percomparison error rate $\alpha_i$ of $.05/3 = .0167$ for each test, yielding a critical value of $F_{contr}$ of 6.09 (using the TINV subroutine of Appendix B.4 and squaring the result). (This assumes that we have selected an experimentwise error rate of .05.) If, on the other hand, we had actually chosen these comparisons after seeing the data as the most interesting or interpretable of all comparisons from the set $b$ described on page 99, then the size of this set is $b = (3^4 - 2^5 + 1)/2 = 25$, and the appropriate critical value for each comparison is $3.292^2 = 10.84$. Scheffé's *post hoc* criterion is $3(2.82) = 8.46$, which is smaller than the Bonferroni critical value, despite being based on the infinitely large set of *all* possible contrasts.

(Be sure that you understand why this is possible.) Finally, if we anticipate performing the same kind of *post hoc* exploration of the means separately for each of the four
outcome measures, we would thereby be contemplating 100 possible comparisons
and would therefore require an $F$ of $3.912^2 = 15.30$ for significance. (As we shall see
later, the Manova equivalent of Scheffé's *post hoc* criterion, taking into account all
possible contrasts on all possible linear combinations of the dependent measures,
yields a critical value of 15.44.)

The basic principle underlying the Bonferroni approach is sufficiently simple that
the reader should be able to adapt it quite readily to any multiple comparisons
problem he encounters in which he can specify the size of the set of all "relevant"
comparisons. Because of the conservative nature of the Bonferroni procedure, he
should, however, consider the possibility that test procedures based on the infinitely
large set of *all* possible comparisons may provide less stringent critical values than
the Bonferroni approach.

## 4.2   One-Way Multivariate Analysis of Variance

After being exposed to multiple regression and to Hotelling's $T^2$, it should come
as no surprise to the reader that Manova is handled by reducing each subject's scores
on $p$ variables to a single number—a simple linear combination of his scores on those
original $p$ variables. Heuristically, one-way Manova consists of a search for that
linear combination of the variables which maximally discriminates among the $k$
groups in the sense of producing the largest possible univariate $F$-ratio, followed by
comparison of this largest possible univariate $F$ to a critical value appropriate to
such a statistic, taking into account the extreme capitalization on chance involved in
finding it.

Formally, it should be fairly clear from the reader's familiarity with the relationship between the variance of a linear combination of variables and the variances of
the component variables that

$$F(\mathbf{a}) = \mathrm{MS}_{b,V}/\mathrm{MS}_{w,V} = \frac{\sum n_j(\bar{V}_j - \bar{\bar{V}})^2/(k-1)}{\sum\sum(V_j - \bar{V}_j)^2/(N-k)} = \frac{\mathbf{a'Ha}}{\mathbf{a'Ea}}\left(\frac{N-k}{k-1}\right), \quad (4.8)$$

where $\mathrm{MS}_{b,V}$ is the mean square between groups for variable (dependent measure) $V$;
$\mathrm{MS}_{w,V}$ the within-group mean square for dependent variable $V$;

$$V = a_1 X_1 + a_2 X_2 + \cdots + a_p X_p = \mathbf{Xa};$$

$\mathbf{H}$ is the between-group covariance matrix,

$$= \begin{bmatrix} \mathrm{SS}_{b,1} & \mathrm{SS}_{b;1,2} & \mathrm{SS}_{b;1,3} & \cdots & \mathrm{SS}_{b;1,p} \\ \mathrm{SS}_{b;1,2} & \mathrm{SS}_{b;2} & \mathrm{SS}_{b;2,3} & \cdots & \mathrm{SS}_{b;2,p} \\ \mathrm{SS}_{b;1,3} & \mathrm{SS}_{b;2,3} & \mathrm{SS}_{b,3} & \cdots & \mathrm{SS}_{b;3,p} \\ \vdots & \vdots & \vdots & \vdots & \vdots \\ \mathrm{SS}_{b;1,p} & \mathrm{SS}_{b;2,p} & \mathrm{SS}_{b;3,p} & \cdots & \mathrm{SS}_{b,p} \end{bmatrix},$$

where

$$\mathrm{SS}_{b;r,s} = \sum n_j(\bar{X}_{j,r} - \bar{\bar{X}}_r)(\bar{X}_{j,s} - \bar{\bar{X}}_s) = \sum(T_{j,r}T_{j,s}/n_j) - T_r T_s/N$$

where $T_{j,r}$ is the sum of observations on variable $r$ in group $j$ and $T_r$ the sum of all observations in all groups on variable $r$; $\mathbf{E}$ is the within-group covariance matrix, the matrix whose $(r, s)$th entry is

$$SS_{w;r,s} = \sum\sum (X_{j,r} - \bar{X}_r)(X_{j,s} - \bar{X}_s) = \sum\sum X_r X_s - \sum (T_{j,r} T_{j,s}/n_j).$$

The task is to choose an $\mathbf{a}$ such that $F(\mathbf{a})$ is maximized. However maximizing $F(\mathbf{a})$ is equivalent to maximizing $\mathbf{a'Ha/a'Ea}$, which is just $F(\mathbf{a})$ with the bothersome constant "stripped off." Further, we know that the univariate $F$-ratio is unaffected by linear transformations of the original data, so that we are free to put a side condition on $\mathbf{a}$ so that it may be uniquely defined. The most convenient restriction is $\mathbf{a'Ea} = 1$, whence, applying the method of Lagrangian multipliers (Digression 1) and the matrix differentiation procedures discussed in Digression 2, we obtain

$$L = \mathbf{a'Ha} - \lambda\mathbf{a'Ea};$$

whence

$$dL/d\mathbf{a} = 2\mathbf{Ha} - 2\lambda\mathbf{Ea}$$
$$= \mathbf{0} \quad \text{iff} \quad [\mathbf{H} - \lambda\mathbf{E}]\mathbf{a} = \mathbf{0} \tag{4.9}$$
$$\text{iff} \quad [\mathbf{E}^{-1}\mathbf{H} - \lambda\mathbf{I}]\mathbf{a} = \mathbf{0}. \tag{4.10}$$

However, (4.9) is precisely the kind of matrix equation characteristic roots and vectors (cf. Digression 2) were designed to solve. We need only solve the determinantal equation, $|\mathbf{E}^{-1}\mathbf{H} - \lambda\mathbf{I}| = 0$, for $\lambda$, use this value to solve (4.10) for $\mathbf{a}$, and then plug into (4.8) to obtain the desired maximum possible $F$. However, $\mathbf{E}^{-1}\mathbf{H}$ will in general have a rank greater than or equal to 2 (usually it will equal either $k - 1$ or $p$, whichever is smaller), so that $\mathbf{E}^{-1}\mathbf{H}$ will have more than one characteristic root and associated vector. Which shall we take as our desired solution, and what meaning do the other roots and vectors have?

Returning to (4.9) and premultiplying both sides of that equation by $\mathbf{a}'$, we find that each $\lambda$ satisfying (4.9) and (4.10) also satisfies the relationship

$$\mathbf{a'Ha} - \lambda\mathbf{a'Ea} = 0,$$

whence

$$\lambda = \mathbf{a'Ha/a'Ea} = (k - 1)F(\mathbf{a})/(N - k); \tag{4.11}$$

that is, each characteristic root is equal (except for a constant multiplier) to the univariate $F$-ratio resulting from using the coefficients of its associated characteristic vector to obtain a single combined score for each $S$. Since this is what we are trying to maximize, the *largest* characteristic root is clearly the solution to our maximization problem, with the characteristic vector associated with this largest root giving the linear combining rule used to obtain this maximum possible $F(\mathbf{a})$. The $i$th largest characteristic root is the maximum possible univariate $F$-ratio obtainable from any linear combination of the $p$ original variables which is uncorrelated with the first $i - 1$ "discriminant functions," the characteristic vector associated with the $i$th largest root giving the coefficients used to obtain this maximum possible $F$.

The distribution of the largest characteristic root (which will, of course, be the basis of our decision as to whether to reject the overall null hypothesis of identical group means on each of the $p$ variables, that is, that $\mu_1 = \mu_2 = \cdots = \mu_p$, where $\mu_j$ is a population mean response *vector* for condition $j$) is rather complicated. Tables are, however, available through the efforts of Heck (1960) and Pillai (1965, 1967) and are provided in Appendix A. Three degree-of-freedom parameters must be specified, namely

$$s = \min(\mathrm{df}_{\mathrm{eff}}, p); \qquad m = (|\mathrm{df}_{\mathrm{eff}} - p| - 1)/2; \qquad n = (\mathrm{df}_{\mathrm{err}} - p - 1)/2;$$

$$(4.12)$$

where $\mathrm{df}_{\mathrm{eff}}$ is the degrees of freedom for the effect being tested ($= k - 1$ for one-way Manova); and $\mathrm{df}_{\mathrm{err}}$ is the degrees of freedom for the appropriate error term ($= N - k$ for one-way Manova). When $s = 1$, $\lambda_{\mathrm{max}}$ has an $F$-distribution (as might have been anticipated from the fact that in such cases Manova reduces either to one-way univariate Anova or to Hotelling's $T^2$); specifically,

$$s = 1 \Rightarrow F = \frac{2n + 2}{2m + 2} \lambda_{\mathrm{max}} \qquad \text{with} \quad 2m + 2 \text{ and } 2n + 2 \text{ df.} \qquad (4.13)$$

When $s = 2$ or larger, we must compute $\theta = \lambda_{\mathrm{max}}/(\lambda_{\mathrm{max}} + 1)$ and compare this to the critical value read from the appropriate table in Appendix A.

Specific comparisons following a significant overall greatest characteristic root test are handled by the same reasoning (Roy's union intersection principle) that provided *post hoc* procedures in Hotelling's $T^2$, combined with the contrast methods we recommended for one-way univariate Anova. Since $\lambda_{\mathrm{max}}$ represents the most favorable picture of the differences among the groups on all $p$ variables, any univariate $F$ on a specific variable or linear combination of variables which is larger than that value of $F_{\mathrm{max}} = (N - k)\lambda_{\mathrm{max}}/(k - 1)$, which would just barely have led to rejection of the overall null hypothesis, must be considered statistically significant at at least the same level employed in testing the overall $H_0$. The overall null hypothesis is rejected if and only if there exists at least one linear combination of the original variables which would have yielded an $F$-ratio significant by this *post hoc* criterion. There is thus no need to run any specific *post hoc* comparison if the overall null hypothesis (based on the greatest characteristic root test) did not achieve statistical significance. *A priori* tests should of course always be run.

We can perform a Scheffé-type contrast among the means using any single variable or linear combination of variables as the dependent measure. However, the critical value to be used in testing the significance of such a contrast will depend on whether (a) the coefficients for this particular contrast among the groups and (b) the coefficients for this particular linear combination of the outcome measures were selected on *a priori* grounds or only after examining the data, as indicated in Table 4.3.

The first column of Table 4.3 needs no further explanation, since it amounts to applying Scheffé's contrast method (discussed in Section 4.1) to the linear combination of outcome measures just as if it were the single outcome measure selected for a univariate Anova. The lower right-hand corner represents a relatively straightforward extension of Scheffé's contrast method to the multivariate case. It, like the univariate application, is so designed that the overall test is significant if and only if

Table 4.3

Critical Values for Contrasts Performed on Linear Combinations of Variables[a]

| | | Was this linear combination of the outcome measures selected on *a priori* grounds? | |
| | | Yes, *a priori* | No, *post hoc* |
|---|---|---|---|
| Was this particular contrast among the groups selected on *a priori* grounds? | Yes, *a priori* | $F_\alpha(1, \mathrm{df_{err}})$ | $\dfrac{p \cdot \mathrm{df_{err}}}{\mathrm{df_{err}} - p + 1} F_\alpha(p, \mathrm{df_{err}} - p + 1)$ |
| | No, *post hoc* | $\mathrm{df_{eff}} \cdot F_\alpha(\mathrm{df_{eff}}, \mathrm{df_{err}})$ | $\mathrm{df_{eff}} \cdot F_{\mathrm{crit}}$ |

[a] $F_{\mathrm{crit}} = \mathrm{df_{err}} \lambda_{\mathrm{crit}}/\mathrm{df_{eff}} = \mathrm{df_{err}} \theta_{\mathrm{crit}}/[(1 - \theta_{\mathrm{crit}})\mathrm{df_{eff}}]$, where $\theta_{\mathrm{crit}}$ is the critical value at the $\alpha$-level of significance of the greatest characteristic root distribution having the degree-of-freedom parameters specified in Eq. (4.12).

*some* contrast among the means on *some* linear combination of the outcome measures would yield an $F$-ratio exceeding the specified *post hoc* critical value. Roy and Bose (1953, summarized in Morrison, 1967) established similar, but more general critical values in the context of simultaneous confidence intervals.

The $(1 - \alpha)$-level confidence interval developed by Roy and Bose (1953) can, when applied to a contrast on a linear combination of variables, be expressed as

$$\mathbf{c'\bar{X}a} - \sqrt{\sum c_j^2(\mathbf{a'Ea})\lambda_{\mathrm{crit}}/n} < \mathbf{c'\mu a} < \mathbf{c'\bar{X}a} + \sqrt{\sum c_j^2(\mathbf{a'Ea})\lambda_{\mathrm{crit}}/n},$$

where $n$ is the number of subjects per group; $\mathbf{c}$ the vector of $k$ contrast coefficients; $\mathbf{a}$ the vector of $p$ weights defining the linear combination of the dependent measures; $\mathbf{E}$ the error matrix; $\mathbf{\bar{X}} = [\bar{X}_{j,r}]$ a $k \times p$ matrix of group means on dependent measures; $\mathbf{\mu}$ a $k \times p$ matrix of the corresponding population means; $\mathbf{c'\mu a}$ therefore the contrast among the population means on the specified combined variable for which we are constructing the confidence interval; and $\lambda_{\mathrm{crit}} = \theta_{\mathrm{crit}}/(1 - \theta_{\mathrm{crit}})$ is the $\alpha$-level critical value of the greatest characteristic root (gcr) of $\mathbf{E}^{-1}\mathbf{H}$; and $\theta_{\mathrm{crit}}$ the critical value of the gcr of $(\mathbf{H} + \mathbf{E})^{-1}\mathbf{H}$, that is the critical value tabled in Table A5. By way of illustration, the 95% confidence interval about

$$\mu_{\mathrm{COOP,W}} - (\mu_{\mathrm{NMO,W}} + \mu_{\mathrm{IND,W}} + \mu_{\mathrm{COMP,W}})/3,$$

where $\mathrm{W} = \mathrm{CC} + \mathrm{DC}$ (cf. Demonstration Problem A4 at the end of this chapter), is computed as

$$\mathbf{c'\bar{X}a} = [384 - (226 + 174 + 155)/3]/12 = 16.583.$$

$$s = 3; \qquad m = [\,|3 - 3| - 1]/2 = -.5; \qquad n = (44 - 3 - 1)/2 = 20;$$

whence

$$\lambda_{\mathrm{crit}} = .260/.740 = .35135.$$

$$\sum c_j^2(\mathbf{a'Ea})\lambda_{\mathrm{crit}}/n = (12/9)(2064.85)(.35135)/12 = 80.609;$$

whence

$$16.583 - \sqrt{80.609} < \mu_{\mathrm{COOP,\,W}} - (\mu_{\mathrm{NMO,\,W}} + \mu_{\mathrm{IND,\,W}} + \mu_{\mathrm{COMP,\,W}})/3$$
$$< 16.583 + \sqrt{80.609}$$

$$7.604 < \mathbf{c'\mu a} < 25.561.$$

It is important to understand the degree of protection we obtain with this 95% confidence interval. If we were to run a large number of experiments and follow up each with *post hoc* confidence intervals of the type that we just constructed, constructing as many as we found interesting (including, if we had the time and the fortitude, all possible such contrasts), then in at most 5% of these experiments would one or more of our statements about the range within which the population values of the various contrasts lie, be incorrect. These are thus extremely conservative confidence intervals, in that we can construct as many of them as we please, on whatever grounds we choose, and still be 95% confident (in the sense just described) that *all* of these confidence intervals are correct statements.

In the case where an *a priori* contrast is to be performed on a combined variable which was constructed only after examining the data (for example, the discriminant function), the critical value is essentially that for a $T^2$-analysis performed by treating the *a priori* contrast as representing the difference between two weighted means: one represented by the positive coefficients in the contrast and the other represented by the negative coefficients. Note that the discriminant function, which maximizes the overall *F*-ratio, will generally *not* be the combined variable which maximizes the *F*-ratio for this particular contrast. A separate maximization procedure could be carried out for each contrast, but then the resulting $SS_{contr}$ s would no longer add up to the total sum of squares for the subjects' scores on the discriminant function. This is highly similar to the problem arising in higher-order Manova (Section 4.8) in deciding whether to conduct a separate maximization and greatest characteristic root test for each effect, or to select instead the discriminant function from a one-way Manova which ignores the factorial structure of the groups, followed by a higher-order univariate Anova treating subjects' scores on this single discriminant function as the dependent measure.

To give a specific example, consider the problem of following up a Manova on Data Set 3 with various specific contrasts on a variety of linear combinations of the outcome measures. (Remember that here $p = 3$ because of the linear dependence among the four outcome measures.) For these four groups,

$$df_{eff} = 3; \qquad df_{err} = 44;$$

$$s = 3; \qquad m = (|3 - 3| - 1)/2 = -\tfrac{1}{2}; \quad \text{and} \quad n = (44 - 3 - 1)/2 = 20.$$

If we had specified in advance that we would compare $(\overline{CD} - \overline{DC})$, the mean value of the combined variable CD − DC, for COOP versus the other three groups, the resulting *F*-ratio for this contrast on this combined variable would be compared to $F_{.05}(1, 44) = 4.06$. If we had specified CD − DC in advance as a theoretically relevant variable, but decided to compare the COOP and NMO groups only after examining our results, the appropriate critical value would be $3 \cdot F_{.05}(3, 44) = (2.82) = 8.46$. If instead we specified the COOP versus others contrast in advance, but used the (clearly *post hoc*) discriminant function as our dependent measure, we would compare the *F*-ratio for the contrast to $(3 \cdot 44/42)F_{.05}(3, 42) = 8.96$. Finally, if we decided after examining the data to compare the COOP and NMO groups in terms of their means on the discriminant function, the critical value would be $3(44/3) \times (.260/.740) = 15.44$—.260 being the .05-level critical value of the gcr statistic having the specified degree-of-freedom parameters.

If the researcher is uninterested in linear combinations of his original variables and will consider only comparisons involving a single dependent measure, he may find that the Bonferroni approach described in Section 4.1.2 provides more powerful tests of his preplanned comparisons than does the gcr-based approach described in this section.

## 4.3    Multiple Profile Analysis

The situations in which we might wish to employ multiple profile analysis are the same as those in which "simple" profile analysis, which was discussed in Chapter 3, is appropriate, except that multiple profile analysis can be applied to distinctions among three or more groups. All variables must be expressed in meaningful, nonarbitrary units and the researcher must be interested in distinguishing between differences in shape and differences in overall level of response, averaged across outcome measures. The three null hypotheses to be tested are the same, and the test procedures are also the same, once allowances for the presence of more than two independent groups have been made. Specifically, the *parallelism* hypothesis (that the shapes of the $k$ profiles are mutually identical, that is, that the difference between each pair of adjacent dependent measures is the same for all $k$ groups) is tested by conducting a one-way Manova on slope measures. This can be done either by actually computing each subject's slope scores $d_i = x_i - x_{i+1}$, and proceeding with standard one-way Manova procedures from there, or by using the formulas developed in Chapter 3 for the relationship between $\mathbf{S}$ computed on all $p$ original variables and $\mathbf{S}_{\text{slope}}$ computed on slope scores. The *levels* hypothesis is tested by conducting a one-way *univariate* Anova on $W = \sum X_i$, the sum of each subject's scores on the $p$ dependent measures. This test is of course most conveniently carried out by employing formula (4.8) with $\mathbf{a}' = (1, 1, \ldots, 1)$. As pointed out in discussing two-group profile analysis in Chapter 3, the levels hypothesis is difficult to interpret when the parallelism hypothesis has been rejected. The *flatness* hypothesis (that the "pooled" profile is flat, that is that the $p$ grand means on the various dependent measures are identical) is tested by conducting a single-sample $T^2$ on the vector of grand mean slope measures (that is, the vector of differences between adjacent grand means).

Specifically, the hypotheses and their tests are as follows.

### (a)    Parallelism Hypothesis

$$H_{01}: \begin{bmatrix} \mu_{1,1} - \mu_{1,2} \\ \mu_{1,2} - \mu_{1,3} \\ \vdots \\ \mu_{1,p-1} - \mu_{1,p} \end{bmatrix} = \begin{bmatrix} \mu_{2,1} - \mu_{2,2} \\ \mu_{2,2} - \mu_{2,3} \\ \vdots \\ \mu_{2,p-1} - \mu_{2,p} \end{bmatrix} = \cdots = \begin{bmatrix} \mu_{k,1} - \mu_{k,2} \\ \mu_{k,2} - \mu_{k,3} \\ \vdots \\ \mu_{k,p-1} - \mu_{k,p} \end{bmatrix};$$

that is,

$$\boldsymbol{\mu}_{\text{slope},1} = \boldsymbol{\mu}_{\text{slope},2} = \cdots = \boldsymbol{\mu}_{\text{slope},k},$$

where $\mu_{j,v}$ is the mean on dependent measure $v$ of population from which group $j$ was sampled, and $\boldsymbol{\mu}_{\text{slope},j}$ is a column vector of the $p - 1$ differences between means on "adjacent" dependent measures for population $j$. This hypothesis is tested by computing

$$\lambda_{par} = \text{greatest characteristic root of } (\mathbf{E}^*)^{-1}\mathbf{H}^*,$$

where $\mathbf{E}^*$ and $\mathbf{H}^*$ are error and hypothesis matrices, respectively, computed on the $p-1$ difference scores $(X_1 - X_2), (X_2 - X_3), \ldots, (X_{p-1} - X_p)$. Repeating the rather extensive computational procedures required to compute $\mathbf{H}$ and $\mathbf{E}$ matrices can be avoided by noting that

$$e_{i,j}^* = e_{i,j} + e_{i+1,j+1} - e_{i,j+1} - e_{i+1,j}$$

and

$$h_{i,j}^* = h_{i,j} + h_{i+1,j+1} - h_{i,j+1} - h_{i+1,j}.$$

In other words, the $(i,j)$th entry of the "reduced" or "starred" or "slope" matrix is obtained by adding together the main diagonal entries and then subtracting the off-diagonal entries of the $2 \times 2$ submatrix whose upper left-hand corner is the $(i,j)$th entry of the original, $p \times p$ hypothesis or error matrix.

The degrees-of-freedom parameters for this test are

$$s = \min(k-1, p-1); \qquad m = (|k-p|-1)/2; \qquad n = (N-k-p)/2.$$

### (b)  Levels Hypothesis

$$H_{02}: \mu_{W,1} = \mu_{W,2} = \cdots = \mu_{W,k},$$

where $\mu_{W,j}$ is the mean for population $j$ of the sum of all dependent measures, and thus is equal to $\mu_{j,1} + \mu_{j,2} + \cdots + \mu_{j,p}$. This hypothesis is tested by comparing

$$F = (SS_{b,W}/SS_{w,W})[(N-k)/(k-1)],$$

where $SS_{Wb,} = \sum\sum h_{ij}$ is the sum of all the entries in the original, $p \times p$ hypothesis matrix, and $SS_{w,W} = \sum\sum e_{ij}$ is the sum of all the entries in $\mathbf{E}$, the original $p \times p$ error matrix to the $F$-distribution having $k-1$ and $N-k$ degrees of freedom.

### (c)  Flatness Hypothesis

$$H_{03}: \bar{\mu}_1 = \bar{\mu}_2 = \bar{\mu}_3 = \cdots = \bar{\mu}_p,$$

where $\bar{\mu}_v = (\mu_{1,v} + \mu_{2,v} + \cdots + \mu_{k,v})/k$ is the arithmetic mean of the means on dependent variable $v$ of the $k$ populations from which the $k$ groups were sampled. This hypothesis is tested by computing

$$T_{\text{flat}}^2 = N(N-k)\bar{\bar{\mathbf{X}}}'_{\text{slope}}(\mathbf{E}^*)^{-1}\bar{\bar{\mathbf{X}}}_{\text{slope}},$$

where

$$\bar{\bar{\mathbf{X}}}_{\text{slope}} = \begin{bmatrix} \bar{\bar{X}}_1 - \bar{\bar{X}}_2 \\ \bar{\bar{X}}_2 - \bar{\bar{X}}_3 \\ \vdots \\ \bar{\bar{X}}_{p-1} - \bar{\bar{X}}_p \end{bmatrix}$$

is a $(p - 1)$-element column vector of differences between "adjacent" grand means (averaged across groups); that is,

$$\bar{\bar{X}}_v = (n_1 \bar{X}_{1,v} + n_2 \bar{X}_{2,v} + \cdots + n_k \bar{X}_{k,v})/N.$$

Then

$$F = \frac{(N - k - p + 2)}{[(N - k)(p - 1)]} T^2_{\text{flat}}$$

is distributed as $F$ with $p - 1$ and $N - k - p + 2$ degrees of freedom.

## 4.4   Multiple Discriminant Analysis

One of the by-products of a one-way Manova is the set of coefficients which produce the largest possible univariate $F$-ratio when they are taken as the coefficients of a linear function of the original variables. This linear combination of the original $p$ variables is the discriminant function (more accurately, the *first* or primary *discriminant function*) for these $k$ groups, and is a logical candidate for use in classifying some new subject whose classification is unknown into one of the $k$ groups. His score on the primary discriminant function is computed and compared with the mean scores on this discriminant function for each of the $k$ groups. If the true population proportions of the various kinds of persons who constitute these $k$ groups are identical, and if the "cost" of making an error of classification is identical for all of the various kinds of error that might be made, then $S$ is assigned to (given the same classification as) that group whose mean discriminant score is closest to $S$'s discriminant score. Where population proportions differ and/or not all errors are equally costly, the "cutoff point" for deciding between group $i$ and group $j$ is moved closer to the mean discriminant score for the more common group or away from the group, classification into which is more costly (for example, the "mentally ill" group or the "habitual criminal" group). (See Morrison, 1967, Chapter 5, or Tatsuoka, 1971a, for further details.)

There is an added complication in multiple discriminant analysis, in that there are $s = \min(p, k - 1)$ discriminant functions, not just one. The $i$th discriminant function is that linear combination of the original variables which maximally discriminates among the groups (in the sense that it produces the highest possible univariate $F$-ratio), subject to the condition that it be uncorrelated with the first $i - 1$ discriminant functions. The $i$th discriminant function is of course given by the characteristic vector associated with the $i$th largest characteristic root of $\mathbf{E}^{-1}\mathbf{H}$, and need be considered an indication of a true source of differences among the groups only if it is statistically significant. The first degree-of-freedom parameter for testing the significance of the $i$th characteristic root is $s = \min(k - i, p - i + 1)$; the other two degree-of-freedom parameters remain the same. [Harris (1974a) points out that the resulting tests of the significance of the second, third, ..., $s$th roots are conservative.]

## 4.5   Greatest Characteristic Roots versus Wilks's Lambda in Manova

The general technique of reducing Manova to a univariate problem by computing a univariate $F$ on the "optimal" linear combination of the $p$ original variables led us naturally to significance tests based on the magnitude of the greatest characteristic root of $\mathbf{E}^{-1}\mathbf{H}$. The original development of multivariate analysis of variance (Wilks, 1932) took the alternative approach of computing the ratio of the "likelihood" of the data under the assumption of population mean vectors identical to the corresponding sample mean vectors, to the likelihood of the observed data under the assumption of no differences among population mean vectors. This leads to *Wilks's lambda* statistic, also referred to as the $U$-statistic (thereby avoiding Greek letters),

$$\Lambda = U = \frac{|\mathbf{E}|}{|\mathbf{H} + \mathbf{E}|} = \frac{1}{|\mathbf{E}^{-1}\mathbf{H} + \mathbf{I}|} = \left( \prod_{i=1}^{s} \lambda_i^* \right)^{-1},$$

where $\lambda_i^*$ is the $i$th characteristic root of $\mathbf{E}^{-1}\mathbf{H} + \mathbf{I}$. However, $\lambda_i^*$ is a solution to the matrix equation

$$[(\mathbf{E}^{-1}\mathbf{H} + \mathbf{I}) - \lambda_i^*\mathbf{I}]\mathbf{a} = \mathbf{0},$$

that is of

$$[\mathbf{E}^{-1}\mathbf{H} - (\lambda_i^* - 1)\mathbf{I}]\mathbf{a} = \mathbf{0},$$

where $\lambda_i = \lambda_i^* - 1$ is the $i$th characteristic root of $\mathbf{E}^{-1}\mathbf{H}$. Thus greatest characteristic root tests and Wilks's lambda criterion are closely related, with

$$U = \left[ \prod_{i=1}^{s} (\lambda_i + 1) \right]^{-1} = \prod_{i=1}^{s} (1 - \theta_i) \tag{4.14}$$

where $\lambda_i$ is defined as above.

Most texts and most computer programs employ Wilks's lambda criteria for significance tests in Manova. The reasons for this preference are at least fourfold:

(a)   historical precedence.

(b)   while the exact distribution of $U$ is extremely complex, fairly good approximations to this distribution in terms of readily available chi-square and $F$-table entries exist.

(c)   under certain circumstances, especially for data where the successive characteristic roots of $\mathbf{E}^{-1}\mathbf{H}$ are nearly equal, statistical tests based on $U$ are more powerful than greatest characteristic root (gcr) tests.

(d)   determinants are easier to compute than are characteristic roots.

The present Primer has concentrated on the latter approach to significance testing because of its greater heuristic and didactic value and because:

(a)   It leads directly and naturally to multiple comparison procedures on both an *a priori* and an *ad hoc* basis.

(b)   It makes the connection between discriminant analysis and Manova much clearer, whereas texts employing $U$-statistics have to introduce gcr procedures as a separate body of techniques for discriminant analysis.

(c)   The power of gcr tests is greater than that for likelihood-ratio (lr) tests when the first characteristic root is much larger than the other roots, and the two approaches have equivalent power for large sample sizes.

(d)   Closely related to point (b), the *source* of any statistically significant differences among the groups is much more readily described (via the statistically significant discriminant function or functions) when gcr techniques are employed.

About all that can be done to follow up a statistically significant $\Lambda$ is to perform univariate analyses of variance on each separate dependent measure, with all the potential for compounded Type I error rates that entails and with no hope of detecting crucial *combinations* of variables.[1] At any rate, the empirical fact is that $U$-statistics are much more common in canned computer programs and in current research literature than are gcr statistics, so it is important to know how to compute significance levels for such statistics.

The degrees of freedom for $U$ ($= \Lambda$) are $u$, $g$, and $n$ where $u$ is the rank of $\mathbf{E}$, the error matrix and usually is equal to the number of dependent (outcome) variables; $g$ is the rank of $\mathbf{H}$, the hypothesis matrix, and is equal to $k - 1$ (number of groups $- 1$) for the overall test in one-way Manova. More generally, this corresponds to the degree of freedom for the effect being tested in any Manova; $n = N - r$, where $r$ is the rank of the design matrix, usually equal to $k$. In other words, $n$ corresponds to $df_w$, the degrees of freedom on which each entry in $\mathbf{E}$ or $\mathbf{S}$ (the pooled covariance matrix) is based. Tables of the distribution of the $U$-statistic are not readily available, though Anderson (1958) provides formulas to use in estimating the probability of an obtained $U$-statistic when various special cases apply. By far the best known of these is Bartlett's (1947) chi-square approximation based on the natural logarithm of $U$. Specifically,

$$\chi^2 = -m(\ln U), \tag{4.15}$$

where $m = N - (u - g) - (g + u + 1)/2$ and $\ln U$ is the natural logarithm of $U$, is distributed as chi-square with $gu$ degrees of freedom when $N$ is large. As Anderson (1958) points out, $\chi^2_{ug}$ is only the first term in an asymptotic expansion which includes the percentiles of $\chi^2_{ug+4}$ and of $\chi^2_{ug+8}$ and which provides the exact probability of a

---

[1] Gabriel (1969, 1970) has suggested (as a special case of a broad class of *simultaneous test procedures* (stp) he discusses in the 1969 paper) that individual contrasts among the groups on specified linear combinations of the original variables be tested against the same critical value as employed for the overall likelihood ratio, that is, that we reject the hypothesis that $\mathbf{c'\mu a} = \mathbf{0}$ if and only if the "stripped $F$-ratio," $R(\mathbf{c}, \mathbf{a}) = SS_{contr}/SS_{err}$ computed for that specific comparison is sufficiently large that $1/(1 + R) < U_{crit}$, where $U_{crit}$ is the critical value of the overall $U$-statistic. However, Gabriel also points out that the power of the resulting tests of specific contrasts (which Gabriel refers to as tests of the "atomic hypotheses" of which the overall null hypothesis is the intersection) is uniformly lower than the power of the equivalent gcr-based simultaneous test procedure outlined in Section 4.2. Moreover, the likelihood-ratio-based simultaneous test procedure is not *coherent* (to use Gabriel's term), in that the overall null hypothesis may be rejected on the basis of a low value of $U$ even though *no* statistically significant specific contrast among the groups on any linear combination of the variables exists. The lr-based simultaneous test procedure thus need not be given serious consideration as a multiple comparison procedure.

value of chi-square as large as or larger than the observed value, given that the null hypothesis is true. Rao (1959) has developed an approximation to the distribution of $U$, based on the $F$-distribution, which Jones (1966) states is more accurate than the chi-square approximation of Eq. (4.15), though not as accurate as the full chi-square approximation. Rao's $F$-approximation involves comparing

$$F = \frac{1 - U^{1/s}}{U^{1/s}}\left(\frac{h}{ug}\right) \qquad (4.16)$$

with an $F$-distribution having $ug$ and $h$ degrees of freedom, where

$$s = 1 \qquad\qquad \text{if} \quad u^2 + g^2 = 5,$$

$$= \sqrt{\frac{u^2 g^2 - 4}{u^2 + g^2 - 5}} \qquad \text{otherwise},$$

and

$$h = \left(n - \frac{u - g + 1}{2}\right)s - \frac{ug - 2}{2}.$$

Rao's approximate $F$-statistic is reported by the BMDX69 computer program for Manova (cf. Digression 4), and provides an exact test when $u$ or $g = 1$.

More recently Schatzoff (1966a) has developed an algorithm for computing exact significance levels for $U$ and has used it to generate a table of correction factors by which $\chi^2_{ug}$ is to be multiplied in order to provide an exact critical value for $U$. Schatzoff's algorithm is too complicated for hand calculation, and his tables are restricted to $ug \leq 70$ and $u$ or $g$ even. Wall (1968) has extended these tables.

To return briefly to the issue of whether to employ $U$ or gcr tests: The most damning evidence against the gcr approach is probably Schatzoff's (1966b) study, in which he showed for a wide range of population **E** and **H** matrices (from which a computer selected numerous samples) that the gcr test had considerably less power than $U$. The greater power comes from the fact that, as Eq. (4.14) makes clear, $U$ takes into account *all* of the $s$ characteristic roots of $\mathbf{E}^{-1}\mathbf{H}$, that is all sources of differences among the groups. The gcr tests, however, consider only those differences which can be summarized in a single "dimension" or combined variable. As also follows from Eq. (4.14), the chi-square approximation to the distribution of $U$, $\chi^2 = -m \ln U$, can be decomposed into additive components, since

$$\chi^2 = -m \ln U = \sum_{i=1}^{s}[m \ln(1 + \lambda_i)] = \sum[-m \ln(1 - \theta_i)], \qquad (4.17)$$

where $\theta_i$ is the gcr statistic associated with the $i$th largest characteristic root. This strongly suggests that the chi-square approximation *could* be used to test the significance of the individual discriminant functions—including that associated with the gcr.

This makes it tempting to speculate that the individual components of $U$ would be chi-quare distributed and thus provide significance tests for the individual discriminant functions. In fact, Tatsuoka (1972) and Kendall and Stuart (1961) each assert that, under the overall null hypothesis, $-M \ln(1 - \theta_i) = M \ln(1 + \lambda_i)$ is distributed as a chi-square variate having $p + q - 2i + 1$ degrees of freedom. However, as pointed out to the present author by W. N. Venables (personal communication, 1974), Lancaster (1963) proved this assertion to be false with respect to $\lambda_1$, thus forcing Kendall and Stuart (1968) to retract their earlier assertion. Harris (1974a) reports that Monte Carlo runs based on the case $s = 10$, $m = 0$, $n = 100$ yielded true Type I error rates for the test of $\lambda_1$ based on the first component of $-M \ln U$ of 61 percent for a nominal $\alpha$ of .05 and 24 percent for a nominal $\alpha$ of .01! The problem is that the individual *population* roots do indeed generate distributions of $M \ln(1 + \lambda)$ having the specified characteristics, but there is no way of knowing which of the $p$ *population* roots has generated the $i$th largest *sample* value of $\lambda_i$.

This attempt (abortive though it is) to use the individual components of $U$ to test the individual sample values of $\lambda_i$ does, however, make it clear that those cases where $U$ leads to rejection of the overall null hypothesis but the gcr test does not are precisely the cases where no single discriminant function discriminates significantly among the various groups. In such a case we are in the unenviable state of knowing *something* beyond random fluctuation is "going on" in our data, but being unable to specify what that "something" is, since no statement about differences among the groups on any single variable or linear combination of variables can be supported by a significance test. This state of limbo can be avoided (ostrich-like?) by concentrating on the gcr approach.

There is no objection to use of Wilks's lambda statistics for testing the overall null hypothesis, so long as the discriminant functions are also computed and tested individually for significance. What *is* objectionable on logical grounds, however, is the procedure most authors of multivariate texts (for example, Cooley and Lohnes, 1971; Tatsuoka, 1971c, 1972; Overall and Klett, 1972) recommend for testing the significance of the individual discriminant functions. (A similar procedure for testing the separate pairs of canonical variates is suggested by each of these authors.) This procedure tests the *residuals* after removing the effects of the first, second, ..., $(s - 1)$st discriminant functions. This is accomplished by including only the higher values of $i$ in the summation of Eq. (4.17). As soon as the residual after removing $r \geq 1$ characteristic roots fails to achieve significance, it is claimed that each of the first $r$ discriminant functions is statistically significant. In particular, the first discriminant function is never tested directly, the argument being that if the overall test is significant, then at least the first discriminant function must also be significant. This is analogous to claiming that if we obtain a significant overall $F$ in a one-way Anova, then at least one of the pairwise differences (in particular, the largest mean minus the smallest) must be statistically significant—a claim which is not true.

In addition to the fundamental logical flaw of treating statistical significance of $U$ as equivalent to statistical significance of $\lambda_1$, this residual-testing procedure is based on invalid assertions about the sampling distributions of the residual components of $U$. Since, for instance, $M \ln(1 + \lambda_i)$ does *not* have the chi-square distribution asserted for it, $-M \ln U - M \ln(1 + \lambda_1)$ will not, under the null hypothesis, have the distribution which almost all texts advocating use of Wilks's lambda (for example, Cooley and Lohnes, 1971; Tatsuoka, 1971a; Overall and Klett,

1972; Press, 1972; and Kshirsagar, 1972) claim for it. Thus partitioned-$U$ tests—whether of individual roots or of the residuals remaining after the effects of preceding roots have been "removed"—must be abandoned. See Harris (1974a,b) for a fuller discussion of these points. This proscription applies equally, of course, to partitioned-$U$ tests developed for canonical correlation (cf. Section 1.2.9 and Chapter 5), of which Manova is a special case.

Summarizing the last few paragraphs crudely: The gcr approach "speaks up" (rejects the overall null hypothesis) only if it can then go on to provide information about the source(s) of the differences among the groups, while Wilks's lambda speaks up even when there's nothing useful to say. These points are elaborated on by Harris (1973b).

<h3 style="text-align:center">※4.6   Practical Computation of Characteristic Roots</h3>

Most Manova problems of practical size will require use of a computer program such as those described in Digression 4. Formulas for $s = 2$, $s = 3$, and uncorrelated cases are provided in Section 4.7. However, the reader could conceivably have occasion for conducting a "hand" analysis of a problem for which $s \geq 4$. If the straightforward approach to finding characteristic roots and then their vectors is adopted, the researcher will find himself faced with the horrendous task of finding the roots of a fourth or higher degree polynomial. An alternate procedure (which is actually the one employed by most computer subroutines) is available in which the characteristic vectors are computed first, and the corresponding characteristic roots are inferred from them.

This approach makes use of the fact that characteristic vectors are defined only up to a scale unit, so that any $p$-element column vector can be represented, given appropriate choices of scale for each of the characteristic vectors, as the sum of the $p$ characteristic vectors of $A$. Symbolically,

$$\mathbf{x}_0 = \mathbf{a}_1 + \mathbf{a}_2 + \cdots + \mathbf{a}_p,$$

where $\mathbf{a}_i$ is the $i$th characteristic vector of $\mathbf{A}$. Premultiplication by $\mathbf{A}$ gives

$$\mathbf{A}\mathbf{x}_0 = \mathbf{A}\mathbf{a}_1 + \mathbf{A}\mathbf{a}_2 + \cdots + \mathbf{A}\mathbf{a}_p.$$

However, by definition of a characteristic vector,

$$\mathbf{A}\mathbf{a}_i = \lambda_i \mathbf{a}_i,$$

so that we have

$$\mathbf{x}_1 = \mathbf{A}\mathbf{x}_0 = \lambda_1 \mathbf{a}_1 + \lambda_2 \mathbf{a}_2 + \cdots + \lambda_p \mathbf{a}_p;$$

and, by repeated application of $\mathbf{x}_{i+1} = \mathbf{A}\mathbf{x}_i$,

$$\mathbf{x}_n = \mathbf{A}^n \mathbf{x}_0 = \lambda_1^n \mathbf{a}_1 + \lambda_2^n \mathbf{a}_2 + \cdots + \lambda_p^n \mathbf{a}_p;$$

and

$$\lambda_1^{-n} x_n = \mathbf{a}_1 + \left(\frac{\lambda_2}{\lambda_1}\right)^n \mathbf{a}_2 + \cdots + \left(\frac{\lambda_p}{\lambda_1}\right)^n \mathbf{a}_p.$$

However, $\lambda_1$ is the largest root, so for large values of $n$, *any* trial vector premultiplied by the $n$th power of $\mathbf{A}$ will be very nearly equal to a constant times the first characteristic vector of $\mathbf{A}$. In practice, $\mathbf{A}$, $\mathbf{A}^2$, $\mathbf{A}^4$, $\mathbf{A}^8$, ... are computed by successive matrix squarings to obtain a high power of $\mathbf{A}$ before performing the premultiplication of $\mathbf{x}_0$ (which can be any nontrivial, $p$-element vector), $\mathbf{A}^n$ is then squared and the resulting $\mathbf{x}_{2n}$ is compared to $\mathbf{x}_n$ (after dividing each by the sum of its elements to improve comparability), the process being stopped when $\mathbf{x}_n$ and $\mathbf{x}_{2n}$ agree to within the desired limits of accuracy; at which point either of the normalized $\mathbf{x}_n$ and $\mathbf{x}_{2n}$ vectors can be taken as equal to $\mathbf{a}_1$; or the ratios of paired elements within $\mathbf{x}_n$ can be compared to the corresponding ratios in $\mathbf{x}_{2n}$ to see if agreement is reached. The ratio between $\mathbf{x}_{2n}$ and $\mathbf{x}_n$ *before* dividing each by the sum of its elements gives $\lambda^n$; or we can compute $\lambda$ from $\mathbf{A}$ and $\mathbf{a}$ by using the relationship $\mathbf{A}\mathbf{a} = \lambda\mathbf{a}$.

If characteristic vectors and roots beyond the first are desired, a *residual* matrix

$$\mathbf{A} - \lambda_1 \mathbf{a}_1 \mathbf{a}_1' - \lambda_2 \, \mathbf{a}_2 \, \mathbf{a}_2' - \cdots - \lambda_{i-1} \mathbf{a}_{i-1} \mathbf{a}_{i-1}'$$

is computed based on the first $i - 1$ (normalized) characteristic roots and vectors. The iterative process is then applied to this residual matrix to obtain $\mathbf{a}_i$ and thus $\lambda_i$.

A computational example may be helpful. To keep it simple, but also to illustrate that there is nothing about the iterative method which requires a symmetric matrix, let us take the matrix product which arises in testing the main effect of visual contact for the (CD, DC) outcome vector in the game study used to illustrate higher-order Manova in Section 4.8, namely

$$\mathbf{E}_{p(vc)}^{-1}\mathbf{H}_v = \begin{bmatrix} .947 & -.146 \\ -.146 & .625 \end{bmatrix}^{-1} \begin{bmatrix} .211 & .174 \\ .174 & .143 \end{bmatrix} \times 10^{-2}$$

$$= \begin{bmatrix} .27566 & .22720 \\ .34279 & .28187 \end{bmatrix} \times 10^{-2} = \mathbf{A}$$

in the discussion below. From the special formula provided in Table 4.1 for Manova when $p = 2$, we "predict" in advance of beginning our iterative procedure that

$$\lambda =$$

$$\frac{\left\{ \begin{array}{l} (.211)(.625) + (.143)(.947) - 2(.174)(-.146) \\ \pm \sqrt{[(.211)(.625) - (.143)(.947)]^2 + 4[(.174)(.947) - (.211)(-.146)] \times [(.174)(.625) - (.143)(-.146)]} \end{array} \right\} \times 10^{-2}}{2[(.947)(.625) - (.146)^2]}$$

$$= \frac{.318104 + \sqrt{.101425}}{2(.570559)} = (.558, -.0001) \times 10^{-2}.$$

(From the fact that there were only two levels of the visual contact factor, we know that $s = 1$ and there should be only one nonzero root, so the second value is attributable to roundoff errors.)

Trying the iterative approach, we make use of the fact that

$$\begin{bmatrix} a & b \\ c & d \end{bmatrix}^2 = \begin{bmatrix} a^2 + bc & b(a + d) \\ c(a + d) & d^2 + bc \end{bmatrix}$$

to help us compute

$$\mathbf{A}^2 = \begin{bmatrix} .15382 & .12667 \\ .19112 & .15733 \end{bmatrix}; \qquad \mathbf{A}^4 = \begin{bmatrix} .04787 & .03941 \\ .05947 & .04909 \end{bmatrix}; \ldots;$$

$$\mathbf{A}^{16} = \begin{bmatrix} .43515 & .35875 \\ .54137 & .44632 \end{bmatrix}; \qquad \mathbf{A}^{32} = \begin{bmatrix} .38357 & .31623 \\ .47720 & .39342 \end{bmatrix}.$$

Note that, until we have found **a**, it is only the relative magnitudes of the entries in these matrices that need concern us, so that constant multipliers of the form " $\times 10^{-36}$ " have been omitted from the listed matrices.

Now, a rather simple trial vector would be $\mathbf{x}_0' = (1, 1)$, since postmultiplication by this vector simply gives us the row sums of the matrix it multiplies. Thus

$$\mathbf{x}_{16} = \mathbf{A}^{16}\mathbf{x}_0 = \begin{bmatrix} .79390 \\ .98769 \end{bmatrix} \quad \text{and} \quad \mathbf{x}_{32} = \mathbf{A}^{32}\mathbf{x}_0 = \begin{bmatrix} .69980 \\ .87062 \end{bmatrix}.$$

However, the two ratios of $a_1$ to $a_2$ are each .80379, so we can stop our iteration here, concluding that $\mathbf{a}_1'$ (the first discriminant function) $= (.80379, 1)$ in nonnormalized form. Now, returning to the basic definition of characteristic roots and vectors, we use the fact that $\mathbf{A}\mathbf{a} = \lambda\mathbf{a}$ to compute $\lambda_1$, as

$$\mathbf{A}\mathbf{a}_1 = \begin{bmatrix} .44877 \\ .55740 \end{bmatrix} \times 10^{-2} = \lambda_1 \begin{bmatrix} .80379 \\ 1.0 \end{bmatrix},$$

whence our two estimates of $\lambda_1$ are .5574 and .5583 (each $\times 10^{-2}$), which average to .558, in good agreement with the value computed from Table 4.3. Since the original matrices were rounded off to three significant digits, the slight failure of proportionality between $\mathbf{A}\mathbf{a}_1$ and $\mathbf{a}_1$ should not bother us.

## 4.7   Simple Cases of Manova

We now give some formulas for simple cases of Manova. Note that when both **H** and **E** are diagonal matrices—which corresponds to having mutually uncorrelated dependent measures—the greatest characteristic root is equal to the largest univariate $F$.

In what follows,

$$e_{ij} = \text{the } (i, j)\text{th entry of } \mathbf{E}; \qquad h_{ij} = \text{the } (i, j)\text{th entry of } \mathbf{H};$$

and each $\lambda$ must be "converted" either to

$$F = \frac{n + 1}{m + 1}\lambda \qquad \text{with} \quad 2m + 2 \quad \text{and} \quad 2n + 2 \text{ df} \quad (\text{when} \quad s = 1)$$

or to

$$\theta = \lambda/(\lambda + 1)$$

with degrees-of-freedom parameters $s$, $m$, and $n$ in order to test for statistical significance.

**(a)   $p = 2$**

$$\lambda = \frac{\left| h_{11}e_{22} + h_{22}e_{11} - 2h_{12}e_{12} \atop \pm \sqrt{(h_{11}e_{22} - h_{22}e_{11})^2 + 4(h_{12}e_{11} - h_{11}e_{12})(h_{12}e_{22} - h_{22}e_{12})} \right|}{2(e_{11}e_{22} - e_{12}^2)}$$

$$= \frac{R_1 + R_2 \pm \sqrt{(R_1 - R_2)^2 + 4h_{12}^2/(e_{11}e_{22})}}{2} \qquad \text{when} \quad e_{12} = 0,$$

where $R_i$ is the "stripped" univariate $F$-ratio computed on dependent measure $i$, that is,

$$R_i = \frac{k - 1}{N - k} F_i.$$

**(b)   $p = 3$**

$\lambda$ is obtained from the cubic equation

$$\lambda^3 - (1/d)\left(\sum_{i=1}^{3} \sum_{j=1}^{3} c_{ij}h_{ij}\right)\lambda^2 + (1/d^2)\left(\sum_{i=1}^{3} \sum_{j=1}^{3} g_{ij}\right)\lambda + D = 0$$

where

$$c_{ij} = e_{ik}e_{jk} - e_{ij}e_{kk}; \qquad c_{ii} = e_{jj}e_{kk} - e_{jk}^2;$$

$$g_{ij} = (h_{kk}h_{ij} - h_{ik}h_{jk})(c_{kk}c_{ij} - c_{ik}c_{jk});$$

$$d = e_{11}e_{22}e_{33} + 2e_{12}e_{13}e_{23} - \sum_{i=1}^{3} e_{ii}e_{jk}^2;$$

and

$$\begin{aligned}
D = (1/d^3)\{ &f_1[l_1(h_{22}h_{33} - h_{23}^2) + l_2(h_{22}h_{13} - h_{13}h_{23}) + l_3(h_{12}h_{33} - h_{13}h_{23})] \\
+ &f_2[l_1(h_{13}h_{23} - h_{12}h_{33}) + l_2(h_{11}h_{23} - h_{12}h_{13}) + l_3(h_{13}^2 - h_{11}h_{33})] \\
+ &f_3[l_1(h_{12}h_{23} - h_{22}h_{13}) + l_2(h_{12}^2 - h_{11}h_{22}) + l_3(h_{11}h_{23} - h_{12}h_{13})]\}
\end{aligned}$$

where

$$f_i = \sum_{j=1}^{3} h_{ij}c_{ij};$$

$$l_1 = c_{22}c_{33} - c_{23}; \qquad l_2 = c_{22}c_{13} - c_{13}c_{23}; \qquad l_3 = c_{33}c_{12} - c_{13}c_{23}.$$

(See Digression 3 for procedures for solving cubic equations.)

When $e_{12} = e_{13} = e_{23} = 0$, the above reduces to

$$\lambda^3 - \left(\sum R_i\right)\lambda^2 + \left[\sum_{i=j}\sum\left(R_iR_j - \frac{h_{ij}^2}{e_{ii}e_{jj}}\right)\right]\lambda$$

$$- R_1R_2R_3 - \frac{2h_{12}h_{13}h_{23} - h_{11}h_{23}^2 - h_{22}h_{13}^2 - h_{33}h_{12}^2}{e_{11}e_{22}e_{33}}.$$

If in addition $k = 3$, so that $s$ (the rank of $\mathbf{E}^{-1}\mathbf{H}$) = 2,

$$\lambda = \frac{\left(\sum R_i\right)^2 \pm \sqrt{\left(\sum R_i\right)^2 - 4S_2}}{2}$$

where

$$S_2 = (R_1 + R_2 + R_3)^2 - 4(R_1R_2 + R_1R_3 + R_2R_3) + 4\sum_i\sum_j\frac{h_{ij}^2}{e_{ii}e_{jj}}.$$

Also, if $k = 2$, then the single nonzero characteristic root is given by

$$\lambda = \frac{k-1}{N-k}T^2 = \sum R_i.$$

### (c)   Outcome Scores and Corresponding Group Means Uncorrelated

If

$$h_{ij} = 0, \qquad e_{ij} = 0 \qquad \text{for any} \quad i \neq j,$$

then

$$\lambda_i = R_i \qquad \text{and} \qquad a_i' = [0, 0, \ldots, 1, \ldots, 0],$$

that is, the characteristic roots are simply the "stripped" univariate $F$s, and the characteristic vector associated with $\lambda_i$ has zeros everywhere except in the $i$th position. Thus the greatest characteristic root is the largest $F$ computed on any *single* outcome measure with the corresponding "discriminant function" simply being that most discriminating single variable. Note, however, that it is not sufficient for $\mathbf{E}$ alone to be diagonal—as the rather complex expressions for the $e_{ij} = 0$ cases in (a) and (b) above clearly demonstrate. The point is that both $\mathbf{E}$ *and* $\mathbf{H}$ are covariance matrices whose off-diagonal entries reflect the intercorrelations among the dependent measures, and each could in fact be converted to an intercorrelation matrix by dividing its $(i, j)$th entry by the square roots of its $i$th and $j$th main diagonal entries: $\mathbf{E}$ reflects the tendency for an individual subject who scores high on one dependent measure to also score high on other dependent measures, while $\mathbf{H}$ reflects the tendency for those groups which have high *mean scores* on one measure to also have high means on other measures. Thus $\mathbf{E}/(N-k)$ is a *within-groups* covariance matrix, while $\mathbf{H}/(k-1)$ is a *between-groups* covariance matrix. Grice (1966) has pointed out that these two sources of correlation are logically—and often empirically—quite distinct. Thus, for instance, the chronic level of anxiety of the subjects who are assigned to a given treatment may be not at all useful in predicting individual differences in response to the treatment, while applying different treatments to each of several groups may produce *mean* levels of anxiety which are highly predictive of the ordering of the groups in terms of *mean* responseness to the various treatments. All of which is simply to say that the researcher will seldom encounter situations in which both $\mathbf{H}$ and $\mathbf{E}$ are uncorrelated—even when the variables have deliberately been submitted to a set of transformations, such as principal components analysis, specifically designed to yield uncorrelated variables. The chances of being able to employ the special formulas for uncorrelated variables (that is for $\mathbf{H}$ and $\mathbf{E}$ both diagonal) are even slimmer when we consider the usually very small number of degrees of freedom $(k - 1)$ on which each of the "covariance" estimates in $\mathbf{H}$ is based. A sample correlation coefficient based on only two pairs of observations must have a value of either $+1$ or $-1$ (or be undefined if either $s_1^2$ or $s_2^2 = 0$), so that if only two groups are employed, each off-diagonal entry in $\mathbf{H}$ (which entry is the sum across groups of the cross-products of deviations of the means on two measures from the grand means for those two measures) will be either $+\sqrt{h_{ii}h_{jj}}$ or $-\sqrt{h_{ii}h_{jj}}$. When $k > 2$, the expected value of the correlation between the $k$ sample means of variable $i$ and the $k$ sample means of variable $j$ is still $\sqrt{1/(k-1)}$ even if the true correlation between population means is zero. Thus, unless $k$ is quite large, it is extremely unlikely that $\mathbf{H}$ will even come close to being a diagonal matrix.

## 4.8   Higher-Order Manova

So far we have discussed only one-way multivariate analysis of variance—the obvious, straightforward generalization of one-way univariate Anova. More complex univariate Anova designs—$n$-way factorial, mixed, repeated measures—also have their counterparts in multivariate Anova. To describe the multivariate extensions in detail would require that we first describe the univariate models—an undertaking for which an entire textbook is required and for which purpose several excellent ones (for example, Winer, 1971) exist. What will be done instead is to

(a)   describe the procedure which would be followed in extending any univariate Anova design to the multivariate (multiple dependent measure) case, and

(b)   discuss a general class of designs—those involving repeated measures or "within-subjects" designs—which are, as pointed out in Chapter 1, commonly treated by univariate Anova techniques though they are usually more properly handled by Manova.

Univariate Anova designs vary considerably in their complexity, in the number of different "error terms" employed in constructing $F$-ratios, etc. However, the end result of the effort involved in selecting an appropriate univariate Anova model for a particular study will be a *summary table* listing each source of variance, its associated degrees of freedom, and which other entry in the table is to serve as the error term (denominator) for the $F$-ratio testing the null hypothesis that this particular component of variance is truly zero in the population. The procedures for extending this design to take into account differences among the groups on all $p$ dependent variables are

(a)   Construct a cross-product matrix (more accurately, a matrix of sums of cross-products of deviation scores) for each effect you wish to test and each error term you will use, that is for each row of the univariate Anova summary table. Each $r$th main diagonal entry of the $\mathbf{H}_i$ matrix corresponding to a particular effect $i$ will be the sum of squares for that effect computed on the basis of subjects' scores on dependent measure $r$. Each $(r, s)$th off-diagonal entry in $\mathbf{H}_i$ or $\mathbf{E}_i$ is computed by substituting a product of $S$'s score on outcome measures $r$ and $s$ for every occurrence in the single-variable computational formula of a square of $S$'s score on a single variable. Thus, for instance,

$$\sum \frac{T_j^2}{n_j} \quad \text{becomes} \quad \sum \frac{T_{j,r} T_{j,s}}{n_j}$$

and

$$(\bar{X}_j - \bar{\bar{X}}_j)^2 \quad \text{becomes} \quad (\bar{X}_{j,r} - \bar{\bar{X}}_r)(\bar{X}_{j,s} - \bar{\bar{X}}_s).$$

Computer program BMDX69 provides an efficient means of computing these matrices.

(b)   For every

$$F\text{-ratio} = \frac{\mathrm{MS}_{\mathrm{eff}}}{\mathrm{MS}_{\mathrm{err}}} = \frac{\mathrm{SS}_{\mathrm{eff}}/\mathrm{df}_{\mathrm{eff}}}{\mathrm{SS}_{\mathrm{err}}/\mathrm{df}_{\mathrm{err}}}$$

computed in the univariate analysis, compute the corresponding matrix product $\mathbf{H}_{\mathrm{eff}} \mathbf{E}_{\mathrm{err}}^{-1}$ or the determinantal ratio $|\mathbf{E}_{\mathrm{err}}|/|\mathbf{H}_{\mathrm{eff}} + \mathbf{E}_{\mathrm{err}}|$. The matrix product can be the basis of a gcr test *or* of an lr test while the determinantal form yields only Wilks's lambda criterion for the hypothesis of no true population differences corresponding to that effect in the summary table for *any* of the $p$ outcome measures.

(c)   For each source (row) of the summary table which yielded rejection of the overall, multivariate null hypothesis, perform specific *post hoc* comparisons among the various experimental groups on the various dependent measures and linear combinations thereof. (Of course, specific *a priori* hypotheses about the pattern of means on a specific variable or combination of variables can always be tested whether the corresponding multivariate test yielded statistical significance or not.) This *post hoc* analysis will almost always include the univariate $F$-ratio for that effect computed separately on each outcome measure—which univariate $F$s are of course computed directly from the main diagonal entries of $\mathbf{H}_{\mathrm{eff}}$ and $\mathbf{E}_{\mathrm{err}}$ via the equation

$$F_i = \frac{\mathrm{df}_{\mathrm{err}}\, h_{ii}}{\mathrm{df}_{\mathrm{eff}}\, e_{ii}}.$$

In addition, univariate $F$s for the effects of that source on any linear combinations of the outcome measures can be computed via the equation

$$F(a) = \frac{\mathrm{df}_{\mathrm{err}}\, \mathbf{a}'\mathbf{H}\mathbf{a}}{\mathrm{df}_{\mathrm{eff}}\, \mathbf{a}'\mathbf{E}\mathbf{a}}.$$

In particular, if the characteristic vectors of $\mathbf{E}_{\mathrm{err}}^{-1}\mathbf{H}_{\mathrm{eff}}$ have been computed in (b), each of these whose corresponding characteristic root exceeds the criterion for multivariate statistical significance should be "interpreted" in terms of a simpler pattern of weights, and that interpretation tested by a univariate $F$-ratio on the specified linear combination via the $F(\mathbf{a})$ formula. For instance, $.71X_1 + .49X_2 + .92X_3 + .63X_4$

might be interpreted as representing "essentially" the sum of the four measures, whence $F(\mathbf{a})$ for $\mathbf{a} = [1, 1, 1, 1]$ would be computed, while $.30X_1 - .82X_2 + .25X_3 + .31X_4$ could be interpreted as the difference between variable two and the average of the other three, implying an $\mathbf{a}$ of $[1, -3, 1, 1]$. If available (as it will be if the greatest characteristic root approach to significance testing has been adopted), the *post hoc* criterion for significance, rather than the critical values of the (univariate) $F$-distribution, should be used in all of these "followup" $F$-tests. Even if, because of the tedium involved in computing inverses or because of the unavailability of computer programs employing gcr statistics, $U$-statistics have been used for the multivariate significance tests, the gcr criterion for *post hoc* $F$-ratios can be computed from the gcr tables in Appendix A via the formula

$$F_{crit} = \frac{df_{err}}{df_{eff}} \frac{\theta_{crit}}{1 - \theta_{crit}},$$

where $\theta_{crit}$ is that value required for significance at the chosen significance level for the applicable degrees of freedom, namely

$$s = \min(df_{eff}, p), \qquad m = \frac{|df_{eff} - p| - 1}{2}, \qquad n = \frac{df_{err} - p - 1}{2}.$$

(d)  As the final stage of the analysis, applicable only if $df_{eff} > 1$, each univariate $F$ on a single variable or a combination of variables which is found in step (c) to be statistically significant by the *post hoc* criterion is further explained through the use of Scheffé-type contrasts on the differences among the means on that particular original variable or combined variable of the various experimental groups. Again, *post hoc* significance criteria are preferred, each $F_{contr}$ being compared to $df_{eff} F_{crit}$, where $F_{crit}$ is the *post hoc* significance criterion computed for *post hoc* overall $F$s on the effect in step (c).

It has been the author's experience that steps (a)–(d) provide a highly efficient means of "screening out" the multitude of chance fluctuations which would otherwise crop up if the unfortunately common alternative procedure of simply conducting $p$ different univariate Anova were followed. Moreover, the recommended steps also provide a logical procedure for arriving at a relatively compact summary of the sources of differences among the various groups on the various outcome measures—including descriptions in terms of important linear combinations of these variables which might never have been detected via univariate approaches. There are, however, at least two alternate approaches to conducting a higher-order Manova which should be mentioned. First, if the dependent variables are commensurable, that is measured in common, meaningful units, the researcher may wish to begin the analysis of each row of the summary table with a profile analysis, rather than a single, overall significance test. Second, the reader will have noted that the discriminant function used in testing, say, the main effect of factor $C$ is *not* in general the same discriminant function used in testing the $A \times B$ interaction, the gcr maximization procedure being free to pick that linear combination of variables which maximizes the univariate $F$-ratio for the particular effect being tested. The researcher may therefore prefer to ignore the factorial structure of his groups initially and simply conduct a one-way Manova on the differences among all $k$ experimental groups to test the overall null hypothesis, and then perform a higher-order *univariate* Anova on the resulting discriminant function or a simpler interpretation thereof—using, of course, gcr *post hoc* significance levels, rather than the usual $F$-table critical values. Such an approach is roughly equivalent to conducting a test of the total between-group sum of squares in a higher-order Anova before beginning the finer-grained tests of main effects and interactions of the various factors. The primary advantage in the multivariate case is that the combined outcome measure used to assess the various main effects and interactions is the same for all such tests. Its primary disadvantage is a loss of power in each of these tests, since the combined measure which maximally discriminates among all $k$ experimental groups may be very different from the combined variable which maximally discriminates among the levels of factor $A$.

It may be helpful to give an example of higher-order Manova. In the fall of 1972, students in the Social Psychology Lab course at the University of New Mexico conducted a study on the effect on behavior in a Prisoner's Dilemma game, if a pair of subjects could (or could not) see each other's faces and they exchanged (did not exchange) explicit messages about their intentions prior to each trial. The design was thus a 2 × 2 factorial with both factors (visual presence or not and communication or not) fixed. (The reader is not necessarily expected to understand the terminology of higher-order Anova at this point. Come back to this example after you have had a course in Anova designs other than one-way, or after you have read a text in the area.) The summary table for a univariate analysis of, say, the proportion of the 30 trials which yielded mutually cooperative outcomes (the number of such trials, divided by 30) looks as shown in Table 4.4. As can be gathered from this table, there were a total of 76 pairs of subjects, 19 in each of the four experimental conditions.

*Table 4.4*

| Source | df | SS | MS | F |
|---|---|---|---|---|
| Visual contact (V) | 1 | .0122 | .0122 | .3369 |
| Communication (C) | 1 | .0270 | .0270 | .7461 |
| V × C | 1 | .0159 | .0159 | .4404 |
| P(VC) | 72 | 2.6075 | .0362 | |
| Total | 75 | 2.6626 | | |

Computer program BMDX69 was used to generate the cross-product matrices and to provide likelihood-ratio tests of the significance of each effect, yielding results which included the cross-product matrix (based on four dependent measures) for each of the effects in Table 4.4 (except the Total row):

$$
\mathbf{H_V} = \begin{bmatrix}
.01220 & & & \\
.00507 & .00211 & & \\
.00418 & .00174 & .00143 & \\
-.02153 & -.00894 & -.00738 & .03798
\end{bmatrix};
$$

$$
\mathbf{H}_C = \begin{bmatrix}
.02702 \\
.00762 & .00215 \\
-.01252 & -.00353 & .00580 \\
-.02200 & -.00620 & .01020 & .01792
\end{bmatrix};
$$

$$
\mathbf{H}_{VC} = \begin{bmatrix}
.01595 \\
-.02031 & .02586 \\
.01350 & -.01719 & .01143 \\
-.00926 & .01179 & -.00784 & .00537
\end{bmatrix};
$$

and

$$
\mathbf{E}_{P(VC)} = \begin{bmatrix}
2.60751 \\
-.54722 & .94746 \\
-.51449 & -.14619 & .62542 \\
-1.54555 & -.25407 & .03526 & 1.76417
\end{bmatrix}.
$$

Each of these matrices is of course symmetrical, so that only the below-diagonal and main-diagonal entries need be presented. The program also output the means for each group on each outcome measure, as summarized in Table 4.5.

*Table 4.5*

*Mean Proportion of Total Responses Accounted for by Each Outcome*

| Group | CC | CD | DC | DD |
|---|---|---|---|---|
| Isolated, | | | | |
|   no communication | .14563 | .21047 | .19479 | .44916 |
|   communication | .21232 | .18421 | .20184 | .40163 |
| Face-to-face, | | | | |
|   no communication | .19995 | .18410 | .22800 | .38763 |
|   communication | .20868 | .23163 | .18600 | .37374 |

Finally, the program included a Manova summary table (Table 4.6). Dependent variables 2, 3, and 4 are CD, DC, and DD, respectively. Only 3 of the 4 are included because their sum must add to 1.00000, so that one of them must be omitted from any multivariate analysis to avoid zero determinants. The results would have been identical had we included CC, DC, and CD, instead.

*Table 4.6*

*Multivariate Analysis of Variance*

Dependent variables:   2   3   4

| Source | Log (generalized variance) | U-Statistic | Degrees of freedom | Approximate F-statistic | Degrees of freedom |
|---|---|---|---|---|---|
| V | −0.00690 | 0.975452 | 3  1  72 | 0.5872 | 3   70.00 |
| C | −0.01300 | 0.981422 | 3  1  72 | 0.4417 | 3   70.00 |
| VC | 0.01360 | 0.955661 | 3  1  72 | 1.0826 | 3   70.00 |
| P(VC) | −0.03175 | | | | |

It may be instructive to show how the cross-product matrices were constructed. We cannot compute $\mathbf{E}$ without access to the original scores, but the various $\mathbf{H}$ matrices require only knowledge of the group mean vectors. Thus, for example, the formula for the sum of squares for the communication main effect for a single variable is $\sum_{j=1}^{2} n_j(\bar{X}_j - \bar{X})^2$, where $n_j$ is the number of subjects at the $j$th level of the communication factor ($=$ twice the number of subjects in a single group), $\bar{X}_j$ is the mean score on the outcome variable of the subjects at that level of the C factor, and $\bar{\bar{X}}$ is the grand mean across all subjects of the dependent measure. Thus, for CC, this sum of squares is

$$38(.17297 - .19164)^2 + 38(.21050 - .19164)^2 = .027019,$$

which appears, in rounded-off form, in the first main diagonal position of $\mathbf{H}_C$. If we now wish to compute, say, the (CC, DC) entry—first column, third row—of $\mathbf{H}_C$, we simply replace each squared deviation term in the previous formula with the product of a deviation score on CC and a deviation score on DC, giving

$$h_{CC,DC} = 38(.17279 - .19164)(.21139 - .20266)$$

$$+ 38(.21050 - .19164)(.19392 - .20266)$$

$$= -.012507,$$

which is fairly close to the listed value of $-.01252$. Of course, we would not actually use these deviation-score formulas if we were computing the cross-product matrices from "scratch," but would instead use more convenient (and more accurate) computational versions. When in doubt, however, it is always best to go back to the deviation-score formulas, which minimize the chances of a *major* goof. Note, too, that it *is* possible to get a negative entry in an *off*-diagonal cell of one of the $\mathbf{H}$ or $\mathbf{E}$ matrices. Your strongly conditioned aversions to negative sums of squares can, however, be safely transferred to the main diagonal entries of these matrices. (Be sure you understand why.) The reader would find it instructive to reconstruct each of the above $\mathbf{H}$ matrices from the mean vectors given above.

Given that we have succeeded in constructing the cross-product matrices (or, more probably, have let the computer do so), the likelihood-ratio test for each effect in the table is computed from

$$U = |\mathbf{E}|/|\mathbf{H} + \mathbf{E}|.$$

Thus, for the main effect of Visual contact the program computes

$$U = .968988/.99312 = .9757,$$

with

$$u = \text{rank of } \mathbf{E} = 3; \qquad g = \text{rank of } \mathbf{H} = 1; \qquad \text{and} \qquad n = df_{err} = 72.$$

(Note that labels for the $u$, $g$, and $n$ columns are not provided in the BMDX69 printout.) For $g = 1$, the "approximate" $F$-test is exact, as well as agreeing perfectly with the gcr test (cf. Section 4.5). For this test,

$$s = \sqrt{(9 - 4)/(10 - 5)} = 1; \qquad h = (72 - 3/2) \cdot 1 - (3 - 2)/2 = 70;$$

whence

$$F = (70/3)[(1 - U)/U] = .58112 \quad \text{with} \quad 3 \text{ and } 70 \text{ df.}$$

Note that, even though our value of $U$ agreed to within .0002 with the program's answer, the resulting discrepancy in the approximate $F$ was .0061. We could hardly have hoped for greater accuracy in $U$, given the rounded-off matrices with which we began. We could, however, have obtained more significant digits (given the computer's insistence on printing out only five decimal positions) simply by multiplying each of our original observations by 10 (thus multiplying each entry of $\mathbf{E}$ or $\mathbf{H}$ by 100). This may sound tedious, but it actually only requires changing the input format statement to (31X, 4F5.0). We must not overdo this "trick," however, since the program's print format statement allows for only six positions (including sign) left of the decimal point, with numbers exceeding this limit being replaced by a field of (highly uninformative) asterisks.

Note, too, that Table 4.6 includes $U$, $F$, and the various degree of freedom parameters. It also includes the natural logarithm of $|\mathbf{E}|$, as well as $\ln|\mathbf{H}_{\text{eff}} + \mathbf{E}|$ for each of the other rows of Table 4.6. These logarithms appear in the "Log(generalized variance)" column, though the relevance of the concept of generalized variance to $|\mathbf{H} + \mathbf{E}|$ is unclear, since $(\mathbf{H} + \mathbf{E})$ is *not* equal to the overall covariance matrix except in one-way Manova.

At this point, we really should stop, since the $F$-statistics indicate that the null hypothesis is still a strong contender for each effect. However, we will go through the gcr approach for demonstration's sake. The major reason for doing so in the present case would be to obtain information about what linear combination of the outcome measures maximally discriminates among the levels of each factor. (This example should also help allay the fears of those who look on multivariate statistics as another of those complicated techniques by which researchers can make *anything* come out significant.) To carry out the gcr analysis, we can submit the cross-product matrices produced by BMDX69 to a computer program such as HEINV to obtain the greatest (in this case only) characteristic root and associated vector of $\mathbf{E}^{-1}\mathbf{H}_{\text{eff}}$. For example, $\lambda$ for the V main effect is .02502 with $s = 1$; $m = \frac{1}{2}$; and $n = 34$; whence

$$F = (70/3)\lambda = .5837 \quad \text{with} \quad 3 \text{ and } 70 \text{ df.}$$

The other $F$s yielded by the gcr approach (all with 3 and 70 degrees of freedom) are .4440 for the C main effect and 1.0795 for the V × C interaction. The univariate $F$-test for the V main effect is maximized by choosing

$$.426CC + .549CD + .719DC$$

as our combined variable, while

$$.608CC + .708CD - .359DC$$

maximizes the C main effect, and

$$.287CC - .420CD + .861DC$$

maximizes the V × C interaction. If we are troubled by this inconstancy of just what we are talking about in analyzing each effect, we can instead perform a four-group

one-way Manova and then perform a univariate analysis on the resulting discriminant function. Thus in the present case we submit

$$\mathbf{H}_{between} = \mathbf{H}_V + \mathbf{H}_C + \mathbf{H}_{VC}$$

and the same $\mathbf{E}_{P(VC)}$ as before to a program like HEINV, discovering thereby that for
the one-way Manova $\lambda = .0519$ (there are now a second and a third root of .0357 and
.0124, which we shall ignore), whence $\theta = .0493$ with $s = 3$, $m = -\frac{1}{2}$, and $n = 36$,
nonsignificant. (The critical value for $\theta$ at the .05 level is .150.) The coefficients of the
discriminant function are given by

$$\mathbf{d}' = (.413, .252, .875);$$

that is,

$$D = .413CC + .252CD + .875DC.$$

The univariate Anova on this discriminant function is most simply carried out by
realizing that

$$F_{eff} = 72(\mathbf{d}'\mathbf{H}_{eff}\,\mathbf{d})/(\mathbf{d}'\mathbf{E}\mathbf{d}).$$

The results of this Anova are summarized in Table 4.7. The appropriate critical value
for each of these comparisons is $72(3/70)F_{.05}(3, 70) = 7.344$, rather than the univariate critical value of 2.78. (Be sure you understand why.) Note, too, that the discriminant function derived from the one-way Manova does *not* do a better job (in terms of
the resulting $F$-ratio) than does the discriminant function derived from the higher-
order Manova for that effect. (Be sure you understand why it *cannot*.) Rao's approximate $F$-statistic does *not* equal the univariate $F$ computed on the discriminant
function for that effect. For comparison, the $F_{max}$ for each effect is $72 \cdot \lambda_{eff} = 1.801$,
1.370, and 3.331 for the three rows of Table 4.7.

*Table 4.7*

*Summary Table for Anova on Discriminant Function from
One-Way Manova*

| Source | df | SS | MS | F |
|--------|-----|--------|------------|-------|
| V | 1 | .00815 | Same as SS | 1.354 |
| C | 1 | .000828 | Same as SS | .138 |
| V × C | 1 | .01106 | Same as SS | 1.837 |
| P(VC) | 72 | .43354 | .00602 | |

## 4.9   Within-Subject Univariate Anova versus Manova

An increasingly common experimental design is the *within-subjects* or *repeated
measures* design, in which one or more factors of the design are manipulated in such a
way that each subject receives all levels of that factor or factors. The primary advantage of this approach is that it provides control for individual differences among the
Ss in their responsiveness to the various experimental treatments, and thus provides
more powerful tests of the effects of the within-subjects factor(s) when such intersub-

ject variability is high. (It is also necessary that $S$'s responses across trials be sufficiently highly correlated to offset the loss of degrees of freedom in the error term.) On the other hand, it also introduces intercorrelations among the means on which the tests of the main effect of and interactions with the within-subjects factor are based. The usual univariate $F$-ratios used to test these effects provide therefore only approximate significance tests, the resulting $p$-values being accurate only if highly restrictive assumptions about the nature of the intercorrelations among $Ss$' responses to the various levels of the within-$S$ factor(s) are met. Specifically, it must be assumed that the variance of the difference in response to treatment levels $i$ and $j$ is the same for all possible pairs of treatment levels. (The significance tests are therefore exact when there are only two levels of the within-$S$ factor, as is the case in the standard $t$-test for correlated means.) These significance tests could more appropriately be handled by treating $S$'s responses to the $w$ levels of the within-$S$ factor $W$ as a $w$-element outcome vector and then applying Manova techniques. The main effect of $W$ would then be tested by a flatness test on the grand mean outcome vector (cf. Section 4.3 on Multiple Profile Analysis), and interactions with between-$S$ factors would be reflected in the parallelism test within the profile analysis of each separate term in the Manova summary table.

Huynh and Feldt (1970) prove that a necessary and sufficient condition for the assumption of homogeneity of treatment-difference variances (the h.o.t.d.v. assumption) to be met is that all contrasts among the levels of $W$—that is, all linear combinations $\sum c_j \overline{T}_j$ such that $\sum c_j = 0$, where $\overline{T}_j$ is the mean of all responses at the $j$th treatment level of $W$—have identical population variances, which makes it clear that testing each such contrast against a single variance estimate ($\text{MS}_{S \times W}$) is legitimate. These authors further show that the h.o.t.d.v. assumption is met if and only if $\mathbf{C\Sigma C'} = k\mathbf{I}$, i.e., is proportional to an identity matrix, whence the characteristic roots of $\mathbf{C\Sigma C'}$ are all identical, where $\mathbf{C}$ is a $(p-1) \times p$ matrix whose rows are any $(p-1)$ orthogonal vectors of contrast coefficients which have been normalized to unit length. $\mathbf{C}$ is thus easily constructed by selecting any $(p-1)$ orthogonal contrasts among the treatment means and dividing each by $\sqrt{\sum c_j^2}$, the square root of the sum of the squares of the coefficients for that contrast (that is, for that row of $\mathbf{C}$). For instance, we may always take as the $i$th row of $\mathbf{C}$ a row vector having zeros in the first $i-1$ positions, $p-i$ in the $i$th position, and $-1$ in the remaining $p-i$ positions. The h.o.t.d.v. assumption is then tested by computing

$$M + |\mathbf{CSC'}|/[\text{tr } \mathbf{CSC'}/p - 1]^{p-1},$$

where tr $\mathbf{A}$ is the trace of $\mathbf{A}$ (the sum of its main diagonal elements). In addition, we compute $f = [p(p-1)/2] - 1$ and $d = 1 - (2p^2 - 3p + 3)/[6(p-1)df]$, and then compare $-df\, d \ln M$ to the chi-square distribution having $f$ degrees of freedom. (The term "df" here is the number of degrees of freedom on which $\mathbf{S}$, the sample covariance matrix for the $p-1$ treatment levels and thus our best estimate of $\mathbf{\Sigma}$, is based.)

Several current design texts, such as Winer (1971) and Kirk (1968) provide a test due to Box (1950) of the assumption of *compound symmetry*, that is, that all main diagonal elements of $\mathbf{\Sigma}$ are equal to a common value and all of its off-diagonal elements are equal to a second, lower common value. Compound symmetry is, however (despite Kirk's claim to the contrary), only a *sufficient* and not a *necessary* condition for the h.o.t.d.v. assumption to be met.

Greenhouse and Geisser (1959) have shown that an upper bound to the probability of the observed data given the null hypothesis of no main effect of, or interaction with, the within-subject factor (and thus a conservative test of any effect involving that factor) is provided by dividing both numerator and denominator degrees of freedom for any test involving the within-subjects factor by $p - 1$, where $p$ is the number of response measures per subject, before looking up the critical value for that $F$-ratio in the tables of the $F$-distribution. Computational details of within-subjects univariate Anova, together with expressions for both standard tests (which apply when h.o.t.d.v. holds) and conservative tests, are provided by Winer (1971).

The above considerations suggest two important questions:

(a)  Why does anyone ever conduct within-subjects univariate Anova, when so many—and so infrequently tested—assumptions must be made which are *not* required by Manova?

(b)  Why does anyone ever use Manova to analyze within-subjects designs, when the computationally simpler univariate Anova procedure is available?

In answer to the first question, it should be pointed out that there are many situations—especially those in which the within-subjects factor is a trials (repeated measures) factor, with the same dependent measure being used to assess the outcome of each trial—in which the assumptions of within-subjects Anova, including compound symmetry, seem entirely reasonable. However, the major compromise implicit in these two questions is between computational simplicity and statistical power. The formulas for within-subjects univariate Anova are indeed orders of magnitude simpler than the computational procedures for Manova. Anova, unlike Manova, can be quite reasonably and (with appropriate checks) quite accurately conducted on a desk calculator, since the computational formulas involve nothing more complex than sums of squares of individual scores or of cell totals. In particular, they do not require matrix inversion. Also if no attempt is made to verify the compound symmetry assumption, there is the additional advantage of avoiding computation of **E** and **E***. On the other hand, performing the within-$S$ Anova without evidence supportive of the compound symmetry assumption requires use of the conservative tests, which may entail considerable loss of power as compared to the Manova approach. (Actually, when compound symmetry of the population from which **E*** is sampled *does* hold, within-$S$ Anova provides more powerful tests of the within-$S$ effects than Manova, since the latter procedure must correct for its more extreme capitalization on chance in the general case.) The present text favors incorporating the within-$S$ factor into the Manova outcome vector *except* in a repeated-measures design employing several measures of the outcome of each experimental treatment. In this situation, inclusion of the within-subjects factor(s) would multiply the length of the outcome vector for each subject by $w$ (the number of levels of the within-subjects factor) and would often bring the total number of variables in the outcome vector close to the total number of subjects, thereby producing a very small value of $n$ (one of the degree-of-freedom parameters).

## DEMONSTRATION PROBLEM

### A.   Computations

All of the following problems refer to Data Set 3 (pp. 68–69), which involved four groups of 12 subject-pairs each, with each group receiving a different motivational set (COOPerative, COMPetitive, INDividualistic, or No Motivating Orientation) prior to beginning a 40-trial Prisoner's Dilemma game. Six different experimenters were employed, with each E running two complete replications of the experiment. (This experimenter factor is ignored in the first five problems.) The outcome measures of interest here are the number of mutually cooperative (CC), unilaterally cooperative (CD), unilaterally competitive (DC), and mutually competitive (DD) outcomes. Recall that there is a perfect linear relationship among these four measures (their sum must equal 40), so that we can only include three of the four variables in any multivariate test, and $p = 3$ in computing degree-of-freedom parameters for such a test. You may find it instructive to work one or more of the problems first omitting DD and then omitting one of the other measures.

**1.** Repeat Problems B1 and B2 of the Demonstration Problem in Chapter 3 using the formulas for Manova.

**2.** Using Manova, test the null hypothesis of no differences among the population mean outcome vectors for these four experimental conditions.

**3.** Compute the discriminant function $D_i$ for each subject-pair $i$.

**4.** Conduct a one-way Anova on the differences among the four conditions in discriminant function scores. Compare with Problem 2. Use Scheffé's contrast method to "decompose" the differences among the groups in mean discriminant function scores into the following three sources of variability:

(a)   COOP versus the other three groups;
(b)   NMO versus IND and COMP;    (c)   IND versus COMP.
What is the appropriate *a priori* criterion for each of these contrasts? What is the appropriate *post hoc* criterion?

**5.** Test the statistical significance of the differences among the four groups in terms of each of the five linear combinations of outcome variables mentioned in Problem B4 of the Demonstration Problem in Chapter 3. Further decompose the differences among the groups on each of these combined variables into the three sources of variability mentioned in Problem 4. What are the appropriate *a priori* and *post hoc* criteria for each of these tests?

**6.** Using BMDX69 and HEINV, conduct a Manova of these data, including experimenter as a variable, with motivating instruction now treated as a within-experimenter variable. (*Hint:* the design and index cards will be as follows:
INDEX  6     2 4
DESIGN  ERM     E, R(E), M, $EM.)
Conduct this analysis once omitting DD and once omitting DC. Compare.

### B.   Thought questions

**1.** It is not uncommon for an overall Manova to indicate significant differences among the various groups, and yet to find that none of the obvious comparisons among the groups reach statistical significance by the *post hoc* criterion. How are we to interpret such a situation?

**2.** What would the formulas for Manova look like if we knew that our dependent measures were uncorrelated? What does this suggest about the possibility of preceding Manova with a principal component or factor analysis of our dependent measures?

**3.** In general, Manova produces $k - 1$ discriminant functions. If two or more of these are statistically significant sources of between-group differences, could we do a better job of discriminating among the groups by employing a linear combination of the various statistically significant discriminant functions? How are we to use the second, third, etc. discriminant functions?

## Answers

**A.**

**1.** These should lead to the same ultimate $F$-ratio as did the $T^2$-analyses. In addition, if $F_{max} = (df_{err}/1)\lambda_{max}$ is computed from the Manova results, it will equal $T^2$. Also, the discriminant function should be the same.

**2.** $\lambda_{max} = 1.3775$ with $s = 3$, $m = -\frac{1}{2}$, $n = 20$; whence $\theta = 1.3775/2.3775 = .262$ at the .05 level and .325 at the .01 level, so that we may soundly reject $H_0$. The other roots are .116 and .054, nonsignificant.

**3.** $D_i = .832CC_i - .159CD_i + .532DC_i$, which seems close to the number of cooperative responses made by member B of the pair, $CC + DC$.

**4.** The overall $F$-ratio $= 20.20$ with 3 and 44 degrees of freedom, which equals $F_{max} = (44/3)\lambda_{max}$. Applying Scheffé's contrast method to the discriminant function scores, we obtain sums of squares of 1934.3 for COOP versus others, 91.2 for NMO versus the other two, and 10.8 for IND versus COMP. These add to 2036.3, the total SS for the between-groups factor. Note that the COOP versus others contrast accounts for 95.0% of the total between-groups sum of squares. Considering the selection of the discriminant function $D$ as our dependent measure as a *post hoc* selection, we must use the second column of Table 4.3 to compute significance levels for the contrasts, whence the critical value for each contrast will be 8.96 if it is considered *a priori* (which each of these contrasts actually was) or 15.44 if considered *post hoc*. The obtained $F$s for contrast were 57.58, 2.71, and .320, so that the COOP versus others contrast is significant by either criterion, while the other two do not reach significance by either criterion.

**5.** The results of all these analyses can be summarized as follows:

SS and $F$ for:

| Source | CD − DC | CC + CD | CC + DD | CC + DC | CC |
|---|---|---|---|---|---|
| Between-groups | 208.33(2.06) | 1826.39(11.15) | 263.00(2.99) | 2700.24(19.18) | 2819.7(17.4) |
| COOP versus others | 58.78(1.74) | 1771.00(32.42) | 186.78(6.37) | 2475.06(52.74) | 2782.6(51.5) |
| NMO versus other 2 | 76.06(2.26) | 33.35( .61) | 68.06(2.32) | 210.13( 4.48) | 36.1( 0.7) |
| IND versus COMP | 73.50(2.18) | 22.04( .40) | 8.17( .28) | 15.04( .32) | 1.0( .02) |
| Within-groups | 1482.59 | 2403.58 | 1289.27 | 2064.85 | 2376.7 |

The critical value for the overall (between-groups) $F$ for each combination of variables is 2.82 if that combination was selected on *a priori* grounds but 5.15 if it is considered *post hoc*. Presuming that the selection of these dependent measures is considered *a priori* (in contrast to the definitely *post hoc* discriminant function analyzed in Problem 4), the critical value for each specific comparison is 4.06 if the contrast is considered *a priori* and 8.46 if it is a *post hoc* comparison of the means.

This is essentially an "expanded" subjects × treatments design, but with experimenters taking the role of subjects (statistically) and with additional information obtained through having each experimenter run two replications of the experimental conditions, so that information is available to test the null hypothesis of no differences among experimenters in the absolute level of the responses they elicit (*E* main effect) or in the pattern of means across the four conditions computed from their data (the *E* × *M* interaction). The summary table for, say, the CC outcome measure *should* therefore look as follows:

| Source | df | SS | MS | F |
|---|---|---|---|---|
| E | 5 | 250.60 | 50.12 | < 1 |
| R(E) | 6 | 228.62 | 38.10 | |
| M | 3 | 2819.73 | 939.91 | 14.10 |
| E × M | 15 | 697.64 | 46.51 | < 1 |
| R × M(E) | 18 | 1199.81 | 66.66 | |

The *F*-ratio for the main effect of motivating instructions is statistically significant at the .05 level by the Greenhouse and Geisser conservative test (df = 1 and 5).

Note that *M* is tested against *E* × *M*, *E* against *R(E)*, and *E* × *M* against *R* × *M(E)*. In fact, however, BMDX69 uses *R* × *M(E)* as the error term for *all* of these tests. This is no problem for the univariate analyses, since the user need only use the listed MSs to obtain the correct *F*-ratios. However, the multivariate analysis is not so easily corrected, since (as pointed out in Section D4.4.4), use of the wrong error term leads to computation of the wrong determinants for the numerator and denominator of the *U*-statistic, and there is no simple way to get from the listed information to the correct *U*-statistic. Instead, the cross-product matrices printed by BMDX69 are input to HEINV to obtain characteristic roots and vectors for the appropriate tests. (*U*-statistics can of course be computed from the characteristic roots.) Thus, for instance, the test of the *E* × *M* interaction is accomplished by using the "CROSS PRODUCT MATRIX OF COMPONENT EM" as the **H** matrix and the "CROSS PRODUCT MATRIX OF COMPONENT RM(E)" as the **E** matrix in a run of the HEINV program (deleting one row and column of each matrix, for example the DD row and column, in order to handle the linear dependence among the outcome measures), resulting in a $\lambda_{max}$ of 1.376 (the others are .528 and .325) with degree-of-freedom parameters $s = 3$, $m = (|15 - 3| - 1)/2 = 5.5$, and $n = (18 - 3 - 1)/2 = 7$. The critical value for $\theta$ at the .05 level is .780, versus our obtained value of $1.376/2.376 = .579$, nonsignificant. The results of all three of the multivariate tests, together with the univariate tests, can be presented in the following summary table:

| Source | df | CD SS | F | DC SS | F | DD SS | F | (CC, CD, DC, DD) $\lambda_{max}$ | $\theta$ | s | m | n |
|---|---|---|---|---|---|---|---|---|---|---|---|---|
| E | 5 | 68.92 | < 1 | 17.42 | < 1 | 274.60 | 1.38 | 1.751 | .636 | 3 | .5 | 1.0 |
| R(E) | 6 | 92.75 | | 174.25 | | 238.12 | | | | | | |
| M | 3 | 146.50 | 3.82 | 89.17 | 1.42 | 1734.23 | 10.4[a] | 4.371 | .814 | 3 | −.5 | 5.5 |
| E × M | 15 | 186.25 | < 1 | 227.08 | < 1 | 523.64 | < 1 | 1.376 | .579 | 3 | 5.5 | 7.0 |
| RM(E) | 18 | 245.24 | | 374.74 | | 959.37 | | | | | | |

[a] $p < .05$ if considered *a priori* with respect to selection of dependent measure.

Note that the CC column has been omitted from the above table only to conserve space. Note, too, that the results of this analysis with respect to the effect of motivating instructions are quite comparable to the one-way Manova ignoring experimenter differences. The critical value for $s = 3$, $m = -.5$, and $n = 5.5$ is a $\theta$ of .725 at the .01 level.

The results of the multivariate tests (gcr or maximum likelihood) would of course come out exactly the same whether CC, CD, DC, or DD was the measure deleted from the cross-product matrices.

**B.**

**1.** Whatever is going on in our data is simply nonobvious. So long as we have used the gcr test, we know that there will be at least *one* contrast among the group means for *some* linear combination of the outcome measures will exceed $F_{\text{crit}}$. In particular, the $F$ for contrast which results when we take the discriminant function as our outcome measure and use contrast coefficients proportional to the difference between each group mean (on the discriminant function) and the grand mean across all groups, will exactly equal the stripped $F$-ratio corresponding to $\lambda_{\text{max}}$, that is, will equal $\text{df}_{\text{err}} \cdot \lambda_{\text{max}}$. If our multivariate test barely reached significance, it may well be that only something very close to this (usually rather arbitrary looking) contrast will reach significance by the *post hoc* criterion.

If we use likelihood-ratio ($U$) statistics, however, there is no guarantee that any (even the first) of the discriminant functions is statistically significant, so that we might in fact come up with a significant $U$-statistic but have nothing else to say, that is, be unable to specify any contrast on any linear combination of outcome measures which is statistically significant. Such cases are, of course, those in which the gcr test would *not* yield significance, and are the basis of the $U$-statistic's generally greater power.

**2.** See the comments in Section 4.7.

**3.** We could *not* do any better by using a linear combination of the discriminant functions, since the resulting combined variable will still be a linear combination of the original measures, and if this linear combination had been better than the first discriminant function, it would have been used as the first discriminant function. Put another way, the optimal linear combination of the various discriminant functions will turn out to be one in which all but the first discriminant function receive zero weights. This does not, however, rule out the possibility that a *nonlinear* combining rule would yield more accurate sorting of subjects into groups than the first discriminant function. For instance, we might put $S$ into that group to which he is "closest" in the discriminant function space. In other words, we would compute, for each group $j$, the measure

$$d_{uj} = \sqrt{(d_{1u} - \bar{d}_{1j})^2 + (d_{2u} - \bar{d}_{2j})^2 + \cdots + (d_{su} - \bar{d}_{sj})^2},$$

where $d_{iu}$ is subject $u$'s score on discriminant function $i$ and $\bar{d}_{ij}$ is the mean of group $j$ on discriminant function $i$. We would then put the subject into that group for which $d_{uj}$ was smallest. It would be interesting to explore the conditions under which this procedure does indeed provide better discrimination among the groups than does the first discriminant function.

# 5 | CANONICAL CORRELATION
## Relationships between Two Sets of Variables

In Chapter 2 we discussed multiple regression as a means of measuring the degree of relationship between a set of predictor variables and a *single* predicted (outcome) variable. In this chapter we shall discuss a technique known as *canonical correlation* for assessing the degree of relationship between a *set* of $p$ predictor variables and a *set* of $q$ outcome variables. (Actually, canonical correlation is a perfectly symmetric technique in which the distinction between the predictor set and the outcome set is not mirrored by any difference in statistical treatment of the two sets, and the distinction thus need not be made.) Our approach to multiple regression was to obtain a "combined predictor," a linear combination of scores on the original $p$ predictor variables $X_1, X_2, \ldots, X_p$, and correlate this combined variable with our single $Y$-score. In canonical correlation, we obtain *two* linear combinations: one combination of the $p$ predictor variables and one combination of the $q$ outcome measures $Y_1, Y_2, \ldots, Y_q$. Naturally we take as our coefficients for these linear combinations those vectors $\mathbf{a}$ and $\mathbf{b}$ (of length $p$ and $q$, respectively) which make the Pearson product–moment correlation between the two combined variables, $u = \mathbf{a}'\mathbf{X}$ and $v = \mathbf{b}'\mathbf{Y}$, as large as possible. The value of the maximum possible Pearson $r$ is known as the *canonical correlation* $R_c$ between the two sets of variables, and $u$ and $v$ are known as *canonical variates*, while $\mathbf{a}$ and $\mathbf{b}$ constitute the two sets of *canonical coefficients*.

## 5.1 Derivation of Formulas for Computing Canonical $R$s

※ The mathematical procedure for obtaining $\mathbf{a}$, $\mathbf{b}$, and $R_c$ is facilitated somewhat if we recognize that the correlation between $u$ and $v$ is given by

$$r(\mathbf{a}, \mathbf{b}) = (\mathbf{a}'\mathbf{S}_{xy}\,\mathbf{b})/\sqrt{(\mathbf{a}'\mathbf{S}_x\,\mathbf{a})(\mathbf{b}'\mathbf{S}_y\,\mathbf{b})}. \tag{5.1}$$

To eliminate the mathematically inconvenient square root we seek a method of maximizing $r^2$, knowing that this will maximize the absolute value of $r$ as well. (Note that the sign of $r$ is arbitrary since this sign can be reversed by multiplying either $u$ or $v$ by $-1$.) To simplify our "maximanda" further, while assuring the uniqueness of $\mathbf{a}$ and $\mathbf{b}$ in the face of the well-known invariance of the Pearson $r$ under linear transformations of either of the variables being correlated we establish the side conditions that $\mathbf{a}'\mathbf{S}_x\mathbf{a}$ (the variance of our linear combination of predictor variables) $= \mathbf{b}'\mathbf{S}_y\mathbf{b}$ (the variance of the linear combination of outcome variables) $= 1$. Applying the methods of Lagrangian multipliers and matrix differentiation (cf. Digressions 1 and 2), we find that

$$L = (\mathbf{a}'\mathbf{S}_{xy}\,\mathbf{b})^2 - \lambda(\mathbf{a}'\mathbf{S}_x\,\mathbf{a} - 1) - \theta(\mathbf{b}'\mathbf{S}_y\,\mathbf{b} - 1);$$

whence

$$dL/d\mathbf{a} = 2(\mathbf{a}'\mathbf{S}_{xy}\,\mathbf{b})\mathbf{S}_{xy}\,\mathbf{b} - 2\lambda\mathbf{S}_x\,\mathbf{a},$$

and

$$dL/d\mathbf{b} = 2(\mathbf{a}'\mathbf{S}_{xy}\,\mathbf{b})\mathbf{S}'_{xy}\,\mathbf{a} - 2\theta\mathbf{S}_y\,\mathbf{b}. \tag{5.2}$$

Setting the derivatives (5.2) equal to zero leads to a homogenous system of two simultaneous matrix equations, namely

$$-\lambda\mathbf{S}_x\,\mathbf{a} + (\mathbf{a}'\mathbf{S}_{xy}\,\mathbf{b})\mathbf{S}_{xy}\,\mathbf{b} = 0,$$

$$(\mathbf{a}'\mathbf{S}_{xy}\,\mathbf{b})\mathbf{S}'_{xy}\mathbf{a} - \theta\mathbf{S}_y\,\mathbf{b} = 0. \tag{5.3}$$

If we premultiply the first equation by $\mathbf{a}'$ and the second equation by $\mathbf{b}'$ we obtain (keeping in mind that $\mathbf{a}'\mathbf{S}_{xy}\mathbf{b}$ is a scalar)

$$-\lambda\mathbf{a}'\mathbf{S}_x\,\mathbf{a} + (\mathbf{a}'\mathbf{S}_{xy}\,\mathbf{b})^2 = 0,$$

and

$$(\mathbf{a}'\mathbf{S}_{xy}\,\mathbf{b})^2 - \theta\mathbf{b}'\mathbf{S}_y\,\mathbf{b} = 0;$$

whence, keeping in mind our side conditions,

$$\lambda = \theta = r^2(\mathbf{a}, \mathbf{b}).$$

In other words, each of our Lagrangian multipliers is equal to the desired maximum value of the squared correlation between $u$ and $v$. For purposes of simplifying (5.3) we note that $(\mathbf{a}'\mathbf{S}_{xy}\,\mathbf{b}) = \sqrt{\lambda}$ since by our side condition $\mathbf{b}'\mathbf{S}_y\,\mathbf{b} = \mathbf{a}'\mathbf{S}_x\,\mathbf{a} = 1$, whence Eqs. (5.3) become

$$-\sqrt{\lambda}\mathbf{S}_x\,\mathbf{a} + \mathbf{S}_{xy}\,\mathbf{b} = 0,$$

$$\mathbf{S}'_{xy}\,\mathbf{a} - \sqrt{\lambda}\mathbf{S}_y\,\mathbf{b} = 0. \tag{5.4}$$

However, a nontrivial solution of Eqs. (5.4) exists only if the determinant of the coefficient matrix vanishes. Using the results of Digression 2 on partitioned matrices, we can express this condition in one of two alternative ways:

$$\left| -\lambda\mathbf{S}_x + \mathbf{S}_{xy}\,\mathbf{S}_y^{-1}\mathbf{S}'_{xy} \right| = 0; \quad \text{or} \quad \left| -\lambda\mathbf{S}_y + \mathbf{S}'_{xy}\,\mathbf{S}_x^{-1}\mathbf{S}_{xy} \right| = 0. \tag{5.5}$$

The roots of (5.5) are the characteristic roots of the matrices $\mathbf{S}_x^{-1}\mathbf{S}_{xy}\mathbf{S}_y^{-1}\mathbf{S}'_{xy}$ and $\mathbf{S}_y^{-1}\mathbf{S}'_{xy}\mathbf{S}_x^{-1}\mathbf{S}_{xy}$. The characteristic vector associated with the gcr of the first matrix gives the canonical coefficients for the left-hand set of variables, while the characteristic vector associated with the gcr of the second matrix gives the canonical coefficients for the right-hand set of variables, and the gcr of either matrix gives the squared canonical correlation between the two sets of variables.

The largest characteristic root (that is, the square of the canonical correlation between the two sets) is tested for significance by comparing it directly (no division by $1 + \lambda$ is required) to the Heck or Pillai-type charts of Appendix A with degree-of-freedom parameters

$$s = \min(p, q); \qquad m = (|p - q| - 1)/2; \qquad \text{and} \qquad n = (N - p - q - 2)/2. \quad (5.6)$$

There are, in general, $s$ nonzero roots of $S_x^{-1}S_{xy}S_y^{-1}S_{xy}'$. Just as with Manova, the "extra" roots have an interpretation quite similar to our interpretation of the largest root. The pair of characteristic vectors associated with the $i$th largest characteristic root represents linear combinations of the $X$s and the $Y$s, respectively, which correlate as highly with each other as possible, subject to the conditions that they be uncorrelated with the first $i - 1$ canonical variates. These successive roots thus describe $s$ uncorrelated ways in which the two sets of measures relate to each other, $s$ "factors" they have in common. The investigator will of course wish to concentrate his interpretative efforts on those pairs of canonical variates which he is sure represent more than just random fluctuation in the data. The procedure for testing the $i$th canonical $R$ and associated pair of canonical variates for statistical significance is identical to that for testing the first, largest canonical $R$, *except* that the first degree-of-freedom parameter $s_i$ now equals $\min(p - i + 1, q - i + 1)$. The other two parameters, $m$ and $n$, remain the same. However, as pointed out in Section 4.4, the tests of all but the first of these canonical $R$s are conservative.

As might be suspected from our knowledge of the Pearson $r$, the value of the canonical $R$ between two sets of variables is unaffected by linear transformations of the variables within either set. In particular, precisely the same canonical $R$ (or canonical $R$s) is (are) obtained when the variables are first transformed to standard-score form, which of course results in the substitution of correlation matrices for the covariance matrices in the formulas we have developed so far. The numerical values of **R**-derived canonical coefficients will be different from the **S**-derived coefficients, but the difference simply reflects the same combined variable described either in terms of raw scores (**S**-derived coefficients) or in terms of standard scores (**R**-derived coefficients).

### 5.1.1  Heuristic Justification of Canonical Formulas

The formulas given above for canonical $R$ and the canonical coefficients look—and in fact are, as we shall see when we discuss explicit formulas for simple cases—quite complicated. However, the product matrices whose characteristic roots and vectors must be obtained are highly patterned. The matrix used to obtain the canonical variates for the left-hand set of variables (the $X$-variables) begins with the inverse of the matrix of covariances of the $X$s, $S_x^{-1}$, and the matrix used to obtain the canonical variates for the right-hand set of variables begins with $S_y^{-1}$. Both products involve both inverses ($S_x^{-1}$ and $S_y^{-1}$) and both between-set covariance matrices ($S_{xy}$ and $S_{xy}' = S_{yx}$) that make up the complete $(p + q) \times (p + q)$ covariance matrix, with inverses and between-set covariance matrices alternating in such a way as to maintain conformability of the dimensions, that is, in such a way that the indices also alternate (assuming $\mathbf{S}'_{xy}$ to be rewritten as $\mathbf{S}_{yx}$).

More importantly, the products "make sense" in terms of the multiple regression formulas with which we are already familiar. For instance, $\mathbf{s}'_{xy_j}\mathbf{S}_x\mathbf{s}_{xy_j}$, where $\mathbf{s}_{xy_j}$ is the $j$th column of $\mathbf{S}_{xy}$, is clearly equal to $s_j^2(R_{y_j \cdot x}^2)$, that is, to that portion of the variation among scores on $Y_j$ which is predictable from the optimum linear combination of the $X$s, that is, to the variance of the predicted scores on $Y_j$ due to the (linear) effects of the $X$s. Thus $\mathbf{S}'_{xy}\mathbf{S}_x^{-1}\mathbf{S}_{xy}$ represents the $q \times q$ covariance matrix of the $q$ residual variates. (The stepwise multiple regression algorithms, discussed in Section D4.4.3, on programs for multiple regression use computation of just such matrices.) Thus $\mathbf{S}_y^{-1}\mathbf{S}'_{xy}\mathbf{S}_x^{-1}\mathbf{S}_{xy}$ is a matrix analog of $(s_y^2 R_{y \cdot x}^2)/s_y^2 = R_{y \cdot x}^2$, and is indeed equal to that value when there is only one $Y$ variable. To get more explicit about the relationship of the matrix product defining the canonical variates for the $Y$s, consider the task of predicting a linear combination of the $Y$s, $\mathbf{b}'\mathbf{Y}$, on the basis of knowledge of the $X$-scores. From Chapter 2, we know that

$$s_W^2 R_{W \cdot X}^2 = \mathbf{s}'_{xw}\, \mathbf{S}_x^{-1}\mathbf{s}_{xw},$$

where $W = \mathbf{b}'\mathbf{Y}$. From by now familiar formulas for the variances and covariances of linear combinations of variables, we know that

$$s_W^2 = \mathbf{b}'\mathbf{S}_y\, \mathbf{b} \qquad \text{and} \qquad \mathbf{s}_{xw} = \mathbf{S}_{xy}\, \mathbf{b},$$

whence

$$(\mathbf{b}'\mathbf{S}_y\, \mathbf{b})R_{\mathbf{b}'\mathbf{Y} \cdot X}^2 = \mathbf{b}'\mathbf{S}'_{xy}\, \mathbf{S}_x^{-1}\mathbf{S}_{xy}\, \mathbf{b},$$

whence

$$\mathbf{b}'[\mathbf{S}_y R_{\mathbf{b}'\mathbf{Y} \cdot X}^2 - \mathbf{S}'_{xy}\, \mathbf{S}_x^{-1}\mathbf{S}_{xy}]\mathbf{b} = 0,$$

which would certainly be satisfied if

$$[\mathbf{S}_y R_{\mathbf{b}'\mathbf{Y} \cdot X}^2 - \mathbf{S}'_{xy}\, \mathbf{S}_x^{-1}\mathbf{S}_{xy}]\mathbf{b} = \mathbf{0},$$

that is, if

$$[\mathbf{S}_y^{-1}\mathbf{S}'_{xy}\, \mathbf{S}_x^{-1}\mathbf{S}_{xy} - R_{\mathbf{b}'\mathbf{Y} \cdot X}^2\, \mathbf{I}]\mathbf{b} = \mathbf{0},$$

which is precisely the relationship which holds when we have found a solution to the matrix equation which defines $R_c^2$ and the corresponding canonical variate. A perfectly symmetric argument establishes that the matrix equation defining the canonical variates for the set of $X$ variables implies an equality between canonical $R$ and the multiple $R$ resulting from predicting a particular linear combination of the $X$ variables (one of the left-hand canonical variates) from knowledge of scores on the $Y$ variables. Thus the process of finding characteristic roots and vectors of $\mathbf{S}_y^{-1}\mathbf{S}'_{xy}\mathbf{S}_x^{-1}\mathbf{S}_{xy}$ and $\mathbf{S}_x^{-1}\mathbf{S}_{xy}\mathbf{S}_y^{-1}\mathbf{S}'_{xy}$ amounts to a partitioning of the variance of each set predicted by knowledge of scores on variables in the other set (that is, the variance shared by the two sets of variables) into portions attributable to $s$ different pairs of linear combinations of the variables in the two sets.

### 5.1.2  Simple Cases of Canonical Correlations

We give explicit scalar (single-symbol as opposed to matrix form) formulas for "simple" cases in which canonical analysis might be appropriate. Note that because of the extreme complexity of the single-symbol expressions which result from multiplying together four matrices, two of which are inverses of the original matrices, available to the user, the $p$ or $q = 3$ case has not been included. The reader will find it much easier to employ the matrix operations for computing $R_c^2$ and the canonical variates, rather than trying to "plug into" single-symbol expressions for any case beyond $p = q = 2$.

**(a)  $q = 1$**

$$R_c^2 = \mathbf{s}'_{xy}\mathbf{S}_x^{-1}\mathbf{s}_{xy} = R_{y \cdot x}^2$$

is the squared multiple correlation between the single $Y$-score and the $p$, $X$-scores;

$$\mathbf{a} = \mathbf{S}'_x\mathbf{S}_{xy}^{-1}$$

gives the regression coefficients for predicting $Y$ from the $X$s; $\mathbf{b}$ is undefined, since multiplication of the single $Y$-score by any constant has no effect on its correlation with any other variable.

A parallel expression holds, of course, for $p = 1$.

**(b)  $p = q = 2$**

$$R_c^2 = \tfrac{1}{2}[m_s \pm \sqrt{m_s^2 - 4(r_{13}r_{24} - r_{14}r_{23})^2},$$

where

$$m_s = \frac{\left[\begin{array}{c} r_{13}^2 + r_{14}^2 + r_{23}^2 + r_{24}^2 - 2r_{13}r_{14}r_{34} - 2r_{23}r_{24}r_{34} \\ - 2r_{13}r_{23}r_{12} - 2r_{14}r_{24}r_{12} + 2r_{12}r_{34}(r_{13}r_{24} + r_{14}r_{23}) \end{array}\right]}{(1 - r_{12}^2)(1 - r_{34}^2)}$$

$$= [R_{1 \cdot 3,4}^2 + R_{2 \cdot 3,4}^2 - 2r_{12}C_1]/(1 - r_{12}^2)$$

$$= [R_{3 \cdot 1,2}^2 + R_{4 \cdot 1,2}^2 - 2r_{34}C_2]/(1 - r_{34}^2);$$

$$C_1 = [r_{13}r_{23} + r_{14}r_{24} - r_{34}(r_{13}r_{24} + r_{23}r_{14})]/(1 - r_{34}^2);$$

$$C_2 = [r_{13}r_{14} + r_{23}r_{24} - r_{34}(r_{13}r_{24} + r_{23}r_{14})]/(1 - r_{12}^2);$$

and

$$R_{i \cdot j,k}^2 = [r_{ij}^2 + r_{ik}^2 - 2r_{ij}r_{ik}r_{jk}]/(1 - r_{jk}^2)$$

is the squared multiple correlation of variable $i$ predicted from variables $j$ and $k$.
The canonical variates are defined by

$$a_1/a_2 = (r_{12}R_{2 \cdot 3,4}^2 - C_1)/[R_{1 \cdot 3,4}^2 - r_{12}C_1 - (1 - r_{12}^2)R_c^2]$$

$$= [(1 - r_{12}^2)R_c^2 + r_{12}C_1 - R_{2 \cdot 3,4}^2]/(C_1 - r_{12}R_{1 \cdot 3,4}^2);$$

and

$$b_1/b_2 = (r_{34}R_{4\cdot 1,2}^2 - C_2)/[R_{3\cdot 1,2}^2 - r_{34}C_2 - (1 - r_{34}^2)R_c^2]$$
$$= [(1 - r_{34}^2)R_c^2 + r_{34}C_2 - R_{4\cdot 1,2}^2]/(C_2 - r_{34}R_{3\cdot 1,2}^2).$$

When $r_{12} = r_{34} = 0$ (that is, when the variables within each set are uncorrelated), these expressions become

$$R_c^2 = \tfrac{1}{2}[r_{13}^2 + r_{14}^2 + r_{23}^2 + r_{24}^2$$
$$\pm \sqrt{(r_{13}^2 + r_{14}^2 + r_{23}^2 + r_{24}^2)^2 - 4(r_{13}r_{24} - r_{14}r_{23})^2}];$$
$$a_1/a_2 = (r_{13}r_{23} + r_{14}r_{24})/(R_c^2 - r_{13}^2 - r_{14}^2)$$
$$= (R_c^2 - r_{23}^2 - r_{24}^2)/(r_{13}r_{23} + r_{14}r_{24});$$

and

$$b_1/b_2 = (r_{13}r_{14} + r_{23}r_{24})/(R_c^2 - r_{13}^2 - r_{23}^2)$$
$$= (R_c^2 - r_{14}^2 - r_{24}^2)/(r_{13}r_{14} + r_{23}r_{24}).$$

**(c)  Variables within Each Set Uncorrelated, That Is, $R_x = R_y = I$**

$R_c^2$ is the characteristic root of $\mathbf{R}_{xy}\mathbf{R}_{xy}'$ or of $\mathbf{R}_{xy}'\mathbf{R}_{xy}$, where of course

$$\mathbf{R}_{xy}\mathbf{R}_{xy}' = \begin{bmatrix} \sum_{j=1}^{q} r_{x_1y_j}^2 & \sum_{j=1}^{q} r_{x_1y_j}r_{x_2y_j} & \cdots & \sum_{j=1}^{q} r_{x_1y_j}r_{x_py_j} \\ \sum_{j=1}^{q} r_{x_1y_j}r_{x_2y_j} & \sum_{j=1}^{q} r_{x_2y_j}^2 & \cdots & \sum_{j=1}^{q} r_{x_2y_j}r_{x_py_j} \\ \vdots & \vdots & \vdots & \vdots \\ \sum_{j=1}^{q} r_{x_1y_j}r_{x_py_j} & \cdots & \cdots & \sum_{j=1}^{q} r_{x_py_j}^2 \end{bmatrix}$$

and $\mathbf{R}_{xy}'\mathbf{R}_{xy}$ is a $q \times q$ matrix whose $(i,j)$th entry is the sum of the cross-products of columns $i$ and $j$ of $\mathbf{R}_{xy}$. The characteristic vectors of $\mathbf{R}_{xy}\mathbf{R}_{xy}'$ yield the canonical variates for the left-hand set of variables, while the characteristic vectors of $\mathbf{R}_{xy}'\mathbf{R}_{xy}$ yield the right-hand canonical variates.

### 5.1.3  Example of a Canonical Analysis

Data collected in the fall of 1970 by Professor Douglas P. Ferraro (Ferraro and Billings, in press, a, b) on drug use among undergraduates at the University of New Mexico will serve to illustrate the use of canonical analysis in uncovering interesting relationships between two sets of variables. The items on this anonymous question-naire included a number of questions on general drug use (from cigarettes to "hard" narcotics such as heroin), as well as a set of questions explicitly directed toward marijuana usage and a number of questions tapping general background variables (age, sex, etc.). These data thus afford the opportunity to investigate (among other things) the extent to which various constellations of background variables are predictive of drug use. Focusing on marijuana for the moment, Table 5.1 reports the correlations between eight background variables (sex, age, class in school, parents' annual income, degree of religious activity, political activity, degree of agreement with parents on Vietnam, and degree of agreement with parents on premarital sex and abortion) and six questions pertaining to marijuana usage (how often the respondent had smoked marijuana, the age when he had first smoked it, how frequently he currently smoked it, whether he would use it in the future if he had the chance, whether he feels pot use should be restricted to those 18 years of age or older, and whether he feels marijuana use should be legalized). The between-set

*Table 5.1*

*Correlations of Background Variables with Marijuana Questions*[a]

| Background variables | Marijuana questions | | | | | |
|---|---|---|---|---|---|---|
| | Nmarij | FstMarij | NowUseMj | WldUseMj | RestrMj | LegalMj |
| Sex | −.186 | .174 | −.171 | .144 | .034 | −.079 |
| Age | −.080 | .202 | −.083 | .057 | −.107 | −.052 |
| Class | −.005 | .144 | −.026 | −.008 | −.052 | .043 |
| Pannuinc | .101 | −.102 | .109 | −.084 | .009 | .052 |
| ReligAct | .389 | −.319 | .325 | −.349 | .100 | .329 |
| YrPolAct | −.150 | .117 | −.136 | .125 | −.033 | −.132 |
| AgrPVN | −.174 | .134 | −.162 | .206 | −.067 | −.195 |
| AgrPSex | −.261 | .246 | −.200 | .272 | −.072 | −.308 |

[a] The questions and the available responses were as follows:

Sex:   0 = Male, 1 = Female.

Age:   0 = ≤ 17, 1 = 18, 2 = 19, etc.

Class:   0 = Freshman, 1 = Sophomore, . . . , 4 = Graduate student.

Pannuinc (Parents' annual income):   0 = < \$5000, 1 = \$5–10,000, . . . , 4 = > \$25,000.

ReligAct ("How active are you in your religion?"):   0 = Very active, 1 = Active, 2 = Barely active, 3 = Inactive.

YrPolAct ("How active are you politically?"):   0 = Very active, . . . , 3 = Inactive.

AgrPVN ("Do you agree with your parents on Vietnam?"):   0 = Strongly disagree, 1 = Generally disagree, 2 = Do not know, 3 = Generally agree, 4 = Strongly agree.

AgrPSex ("Do you agree with your parents on premarital sex and abortion?"): 0 = Strongly disagree, . . . , 4 = Strongly agree.

NMarij ("How many times have you used marijuana?"):   0 = Never, 1 = 1 time, 2 = 2–4 Times, 3 = 5–14 Times, 4 = 15–30 Times, 5 = More than 30 times.

FstMarij ("When did you first use marijuana?"):   1 = Elementary school, grades 1–6; 2 = Junior high school, grades 7–9; 3 = High school, grades 10–12; 4 = College, fresh year; . . . ; 7 = College, senior year; 8 = Have never used.

NowUseMj ("How often do you currently use marijuana?"):   0 = Do not use marijuana, 1 = Less than once per month, 2 = 1–4 Times per month, 3 = 1–3 Times per week; 4 = 4–6 Times per week; 5 = 7 or More times per week.

WldUseMj ("If you had the opportunity in the future, would you use marijuana?"):   0 = Yes, 1 = No.

RestrMj ("If marijuana were legalized, should it be restricted to those 18 years of age or older?"):   0 = Yes, 1 = No.

LegalMj ("Do you feel that society should legalize marijuana at this time?"):   0 = No, 1 = Undecided, 2 = Yes.

correlations reported in Table 5.1 are of course insufficient as the basis for a canonical analysis, since we must consider the within-set correlations as well. Table 5.1 is provided so as to give the reader some feeling for the interpretive chore involved in the usual approach of examining these 48 correlations one at a time.

Table 5.2 reports the six coefficients of canonical correlation, together with the corresponding pairs of canonical variates. The overall null hypothesis is hardly in doubt. Since these data came from the responses of 2675 students's responses to the anonymous questionnaire, we have $m = (|8 - 6| - 1)/2 = \frac{1}{2}$ and $n = (2675 - 16)/2 = 1329.5$, with $s$ being 6 for $R_c^2$, 5 for $R_2^2$, etc. Taking $n = 1000$ as the basis for a conservative critical value, we see from the tables in Appendix A that each of the first three coefficients of canonical correlation are statistically significant at the .01 level. The first pair of canonical variates seem to be identifying a tendency for young, politically active but religiously inactive respondents who disagree with their parents on the topics of premarital sex and abortion to have tried marijuana at an early age, to have used marijuana frequently in the past, to be in favor of legalizing its use, and to be willing to try marijuana again in the future should the opportunity present itself. Of course, this first canonical correlation is only .551, with the specified linear combination of the background variables accounting for only about 30% of the total variation in the specified combination of the responses to the marijuana questions. The second canonical $R$ of .344 seems to suggest primarily that older, upperclass respondents have used marijuana extensively in the past but began this use late in

*Table 5.2*

*Canonical Analysis of Background Variables versus Marijuana Questions*

| | $i$: | 1 | 2 | 3 | $4^a$ | $5^a$ | $6^a$ |
|---|---|---|---|---|---|---|---|
| | $R_i^2$: | .3035 | .1182 | .0254 | .0062 | .0042 | .0008 |
| | $R_i$: | .551 | .344 | .159 | .079 | .065 | .028 |
| Sex | | −.188 | .185 | .719 | | | |
| Age | | −.309[b] | .473 | −.322 | | | |
| Class | | .045 | .616 | .155 | | | |
| Pannuinc | | .085 | −.078 | −.385 | | | |
| ReligAct | | .623 | .180 | −.117 | | | |
| YrPolAct | | −.258 | .031 | .011 | | | |
| AgrPVN | | −.194 | −.123 | −.023 | | | |
| AgrPSex | | −.395 | .002 | −.504 | | | |
| Nmarij | | .295 | 1.815 | −.581 | | | |
| FstMarij | | −.185 | 1.920 | −.228 | | | |
| NowUseMj | | .061 | −.508 | −.662 | | | |
| WldUseMj | | −.244 | −.209 | −.157 | | | |
| RestrMj | | .067 | −.147 | .491 | | | |
| LegalMj | | .414 | .270 | .723 | | | |

[a] Coefficients have been omitted since $R_i^2$ is nonsignificant.

[b] Italicized coefficients are those which were emphasized in developing substantive interpretations of the canonical variates.

their academic career. The third $R_i$ of .159 indicates that young females from low-income families who disagree with their parents on the issues of premarital sex and abortion tend not to have used marijuana much in the past nor to be using it at present, but nevertheless to be more in favor of legalizing marijuana for everyone, 18 or not, than most other respondents.

It must be kept in mind that there is some loss in shared variance when we go from the canonical variates as actually computed by the canonical analysis to the "simplified" canonical variates implied by the above verbal descriptions. For instance, the correlation between

$$(-\text{Age} + \text{ReligAct} - \text{YrPolAct} - \text{AgrPSex})$$

and

$$(\text{Nmarij} - \text{FstMarij} + \text{LegalMj} - \text{WldUseMj})$$

is only .539 vs .551 for the first pair of canonical variates. Similarly, the second and third pairs of "simplified" canonical variates yield correlations of .285 (vs .344) and .148 (vs .159).

At any rate, the three sources of relationship between the two sets of variables uncovered by the canonical analysis seem to be tapping important trends which would have been difficult to uncover through examination of the individual pairwise correlations. The first canonical correlation, with its suggestion that "activist" students who have "tuned out" religion are especially likely to use marijuana, is particularly interesting in light of theories which predict that drug users should be primarily dropouts who have been discouraged by the difficulty of political reform and have turned their interests inward.

## 5.2   Relationships to Other Statistical Techniques

All of the techniques we have talked about in preceding chapters can be seen as special cases of canonical correlation. When one of the two sets of variables consists entirely of group membership dummy variables, one-way Manova results. When $s = 1$, conducting a canonical analysis on the two "sets" of data is equivalent to conducting a multiple regression of the set having two or more variables as predictors of the single variable in the other set. Within a canonical analysis in which both groups have two or more variables, two multiple regression analyses are "imbedded." If we compute $u$ (the canonical variate for the left-hand set of variables) for each $S$ and then conduct a multiple regression analysis of the relation between $u$ and the right-hand set of variables, the resulting set of regression coefficients will be identical to the coefficients which define $v$. Similarly, a multiple regression analysis of $v$ predicted from the $p$ variables of the left-hand set produces $u$ as the best combination of those $p$ variables for predicting $v$. This suggests an iterative (and unfortunately very inefficient) procedure for finding $u$ and $v$ by regression analyses involving successive "guesses" as to the coefficients defining $u$ and $v$.

For those readers who are not easily satisfied by the hand-waving approach to statistics, a proof of the relationship between canonical analysis and Manova is given in Section 5.2.1. This section may of course be skipped without serious damage to the reader's understanding of multivariate statistics. The relationship between canonical analysis and multiple regression when $s = 1$ is detailed in Sections 5.1.2a–c and can easily be derived from the formulas given in Chapter 2. The relationship between canonical analysis and multiple regression in the more general case was presented mathematically in Section 5.1.1 on the heuristic justification of canonical correlation formulas, and is demonstrated in the Demonstration Problem at the end of this chapter.

## ※ 5.2.1  Proof of Relationship between Canonical Analysis and Manova

As you will recall, there is a close relation between multiple regression and univariate one-way analysis of variance. Since one-way multivariate analysis of variance (Manova) is an extension of univariate Anova to the case where more than one dependent variable is employed, and since canonical analysis (Canona) is an extension of multiple regression, along the same lines, we might expect that a simple relation would hold between Canona and one-way Manova. This turns out to be the case.

We begin by defining "dummy" variables, namely

$$x_{ij} = \begin{cases} 1 & \text{if individual } i \text{ is a member of group } j \\ 0 & \text{otherwise.} \end{cases}$$

(We will use $x$s to represent independent variables, which we will arbitrarily consider "set 1" variables.) Then

$$\mathbf{S}_{11} = 1/(N-1) \begin{bmatrix} N_1 & & & \\ & N_2 & \bigcirc & \\ & \bigcirc & \ddots & \\ & & & N_k \end{bmatrix}, \quad \text{and} \quad \mathbf{S}_{11}^{-1} = (N-1) \begin{bmatrix} 1/N_1 & & & \\ & 1/N_2 & \bigcirc & \\ & \bigcirc & \ddots & \\ & & & 1/N_k \end{bmatrix}$$

(where $k$ is the number of groups; the index $j$ will be used to refer to a specific group; $p$ is the number of dependent measures; the index $h$ will be used to refer to a specific variable).

So far the development is identical to the approach we would take in establishing the relationship between multiple regression and one-way Anova. Now, however, things get more complicated, since our $y$-"vector" is now a matrix with a separate column for values on each of the response measures. In other words

$$\mathbf{y} = \begin{bmatrix} y_{11} & y_{12} & y_{13} & \cdots & y_{1p} \\ y_{21} & y_{22} & y_{23} & \cdots & y_{2p} \\ \vdots & \vdots & \vdots & \vdots & \vdots \\ y_{N1} & y_{N2} & y_{N3} & \cdots & y_{Np} \end{bmatrix}, \quad \text{where} \quad y_{ih} = Y_{ih} - \bar{\bar{Y}}_h,$$

that is, $y_{ih}$ is individual $i$'s score on variable $h$, expressed as a deviation from the grand mean of all scores on that variable, ignoring the division into groups. Thus

$$
S_{22} = y'y/(N-1) = \begin{bmatrix} \sum_{i=1}^{N} y_i^2 & \sum y_{i1} y_{i2} & \cdots & \sum y_{i1} y_{ip} \\ & \sum y_{i2}^2 & \cdots & \sum y_{i2} y_{ip} \\ & \vdots & \vdots & \cdots & \sum y_{ip}^2 \end{bmatrix} /(N-1),
$$

that is, $S_{22}$ is a $p \times p$ matrix whose main diagonal elements represent the total variability across all observations from all groups on each of the $p$ response variables, and whose off-diagonal elements represent the total covariance between each pair of response measures. However, as we all know by now, each overall sum of squares or sum of cross-products is made up of two components: variation or covariation *within* the experimental groups and variation or covariation *between* the group means. More formally,

$$
\sum_{i=1}^{N} (Y_{ij} - \bar{Y}_h)^2 = \sum_{j=1}^{k} \sum_{i=1}^{N_j} (Y_{ijh} - \bar{Y}_{jh})^2 + \sum_{j=1}^{k} N_j (\bar{Y}_{jh} - \bar{Y}_h)^2
$$

and

$$
\sum_{i=1}^{N} (Y_{ir} - \bar{Y}_r)(Y_{is} - \bar{Y}_s) = \sum \sum (Y_{ijr} - \bar{Y}_{jr})(Y_{ijs} - \bar{Y}_{js})
$$

$$
+ \sum_{j=1}^{k} N_j (\bar{Y}_{jr} - \bar{\bar{Y}}_r)(\bar{Y}_{js} - \bar{\bar{Y}}_s),
$$

whence it is clear that $S_{22} = (H + E)/(N - 1)$, where $H$ and $E$ are the between- and within-groups covariance matrices used in the one-way Manova (cf. pp. 168 and 169 of Morrison, 1967).

From the second of Eqs. (5.5) we see that an overall test of the hypothesis of no correlation between any group membership "variable" and any dependent measure is obtained from the largest characteristic root of $S_{22}^{-1} S_{12}' S_{11}^{-1} S_{12}$. However,

$$
S_{12}' = y'x = \begin{bmatrix} T_{11} - N_1 \bar{X}_1 & T_{12} - N_1 \bar{X}_2 & \cdots & T_{1p} - N_1 \bar{X}_p \\ T_{21} - N_2 \bar{X}_1 & T_{22} - N_2 \bar{X}_2 & \cdots & T_{2p} - N_2 \bar{X}_p \\ \vdots & \vdots & \vdots & \vdots \\ T_{k1} - N_k \bar{X}_1 & T_{k2} - N_k \bar{X}_2 & \cdots & T_{kp} - N_k \bar{X}_p \end{bmatrix}
$$

whence

$$
S_{12}' S_{11}^{-1} S_{12} = [1/(N-1)]
$$

$$
\times \begin{bmatrix} \sum_{j=1}^{k} N_j (\bar{X}_{j1} - \bar{\bar{X}}_1)^2 & \sum N_j (\bar{X}_{j1} - \bar{\bar{X}}_1)(\bar{X}_{j2} - \bar{\bar{X}}_2) & \cdots \\ \vdots & \sum N_j (\bar{X}_{j2} - \bar{\bar{X}}_2)^2 & \\ \sum_{j=1}^{k} N_j (\bar{X}_{j1} - \bar{\bar{X}}_1)(\bar{X}_{jp} - \bar{\bar{X}}_p) & \cdots & \sum_{j=1}^{k} N_j (\bar{X}_{jp} - \bar{\bar{X}}_p)^2 \end{bmatrix}
$$

$$
= H/(N-1),
$$

whence the procedures applied to this special case reduce to solving the equations

$$
\frac{1}{N-1}[H - (H+E)\lambda]b = 0, \quad \text{or} \quad [H(H+E)^{-1} - \lambda]b = 0.
$$

For the overall test of the null hypothesis we go directly into the Heck charts with the largest value of $\lambda$. However, as Morrison (1967, p. 166) points out, this is identical to going into the greatest characteristic root tables of Appendix A with the value $\theta = c/(1 + c)$, where $c$ is the largest characteristic root of $HE^{-1}$. This latter procedure is precisely the Manova procedure. To see this, rewrite the Canona equations in terms of $c$ as

$$[\mathbf{H} - (\mathbf{H} + \mathbf{E})\lambda]\mathbf{b} = [\mathbf{H}(1 - \lambda) - \mathbf{E}\lambda];$$

but $\lambda = c/(1 + c)$, whence $1 - \lambda = 1/(1 + c)$, and solving the Canona equations is equivalent to solving

$$\left[ \frac{\mathbf{H}}{1 - c} - \frac{c\mathbf{E}}{1 - c} \right]\mathbf{b} = 0, \qquad \text{or} \qquad (1 - c)^{-1}[\mathbf{H} - c\mathbf{E}]\mathbf{b} = 0,$$

or, after multiplying by $1 - c$,

$$[\mathbf{H} - c\mathbf{E}]\mathbf{b} = 0,$$

which is identical to the set of Manova equations. QED

What meaning is to be attached to the coefficients by which Canona tells us the group membership dummy variables are to be weighted? From the maximization criteria which define Canona, it follows that they give the contrast among the means which accounts for the greatest percentage of the between group variability in the discriminant function—which equals $\bar{D}_j - \bar{\bar{D}}$ for each group $j$, where $\bar{D}_j$ is that group's mean score on the discriminant function.

## 5.3   Likelihood-Ratio Tests of Relationships between Sets of Variables

Just as there are likelihood-ratio tests as competitors to the $\lambda_{\max}$ tests in Manova, so there are likelihood-ratio tests of the overall null hypothesis of no relationship between two sets of variables. Specifically,

$$H_0 : \mathbf{S}_{xy} = 0$$

is tested by comparing

$$\chi^2 = -\left( N - \frac{(p + q)}{2} - \frac{3}{2} \right) \ln\left( \frac{|\mathbf{S}|}{|\mathbf{S}_x| \, |\mathbf{S}_y|} \right)$$

$$= -\left( N - \frac{(p + q)}{2} - \frac{3}{2} \right) \sum_{i=1}^{s} \ln(1 - R_i^2), \tag{5.7}$$

where $R_i$ is the $i$th coefficient of canonical correlation, with the chi-square distribution having $pq$ degrees of freedom, where $\mathbf{S}$ is the $(p + q) \times (p + q)$ covariance matrix of all the variables, ignoring the division into subsets. This test makes good intuitive sense, since the ratio of determinants involved can be rewritten (using the formulas for determinants of partitioned matrices given in Digression 2) as $|\mathbf{S}_y - \mathbf{S}_{xy}' \mathbf{S}_x^{-1} \mathbf{S}_{xy}| / |\mathbf{S}_y|$, that is, the ratio of the "generalized variance" of the set of $X$ variables with the effects of the $Y$ variables "partialed out" to the "generalized variance" of the "raw" $Y$ variables. The concept of generalized variance is discussed in the next chapter. Another interpretation of formulas (5.7), given in the second line, is based on the product of the portions of variance left *un*explained by the first, second, . . . , sth canonical correlations. The big advantage of (5.7) over the gcr test is that (5.7) is readily generalizable to a test of the hypothesis of no relationships between any pair of $k$ sets of variables, as we shall see in the next section.

The last expression in (5.7) makes clear the fact that the likelihood-ratio test of the significance of canonical correlation, like the corresponding tests for Manova, essentially involves a pooled test of *all* sources of linear relationship between the two sets of variables. The likelihood-ratio test therefore provides a test of the overall null hypothesis that $\mathbf{S}_{xy} = \mathbf{0}$, but it must *not* be interpreted (as, unfortunately, many authors do, as was discussed in Section 4.5) as a test of the statistical significance of $R_c^2$.

Nor may the individual components of $-M \ln U$, $-M \ln(l - R_i^2)$, be treated as chi-square variates having $p + q - (2i - 1)$ degrees of freedom, Tatsuoka (1971c) notwithstanding.[1] Nor may the residual after "removing" the effects of the first $i$ variates, $-M \sum_{j=i+1}^{s} (1 - R_i^2)$, be treated as a chi-square variate having $(p - i)(q - i)$ degrees of freedom, despite the fact that this procedure has been almost universally recommended by authors who employ the likelihood-ratio test of the overall null hypothesis in Canona. (It is, for instance, incorporated into the BMD and SPSS programs for Canona as the sole procedure for testing individual coefficients of canonical correlation.) See Section 4.5 of the present Primer, Lancaster (1963), and Harris (1974a, b) for further details on the bases for these proscriptions against partitioned-$U$ tests. Recall, at a minimum, that for $s = 10$ (a not unusually high value for practical applications of Canona), and for a sample size over 200, the partitioned-$U$ test of $R_c^2$ has yielded true Type I error rates of 61% for tests at a nominal $\alpha$ of .05 and 24% for tests at the nominal .01 level.

As a final note, it might be suggested that if we accept the logic of the overall test provided in Eqs. (5.7), then the corresponding use of gcr statistics would require testing the overall $H_0$ of no relationship between the two sets of variables by treating each of the $s$ coefficients of canonical correlation as providing a separate, independent test of the same null hypothesis and pooling them via the familiar (for example, Winer, 1971, p. 49) formula,

$$\chi^2 = -2 \sum_{i=1}^{s} \ln p_i \qquad \text{with} \quad 2s \text{ df,}$$

where $p_i$ is the probability value associated with the test of the significance of $R_i$. This would, however, have the same disadvantage as was pointed out for the maximum-likelihood test in Section 4.5: it would lead us to reject the overall null hypothesis in some situations where no further information could be provided about *how* the two sets of variables were related, since no pair of canonical variates was statistically significant.

---

[1] The simultaneous test procedure (Gabriel, 1969, 1970) for the significance of $R_c^2$ corresponding to the likelihood-ratio-based stp described for Manova in footnote 3 would require that $-[N - (p + q)/2 - 3/2] \ln(1 - R_c^2)$ exceed the critical value of the chi-square distribution with $pq$ degrees of freedom. This is clearly a test procedure which has uniformly lower power than the corresponding gcr test, and thus can be rejected in favor of this latter test.

## 5.4   Generalization and Specialization of Canonical Analysis

A number of variations on the procedures described in this chapter have been proposed. For instance, Horst (1961a,b) has provided computational algorithms for a situation in which $m$ sets of variables are to be related by selecting a linear transformation of the variables within each set such that the sum of the $m(m - 1)/2$ intercorrelations between combined variables is as large as possible. Carroll (1968) has approached Horst's problem by maximizing the largest characteristic root of the matrix of intercorrelations among the $m$ canonical variates. Roy (1957) has discussed the canonical correlation between two sets of *residual* scores (each set having been corrected for the linear effects of the same third matrix). Also Meredith (1964) has proposed that canonical analysis be based on covariance matrices which have been "corrected for" unreliability, in the spirit of factor analytic procedures (cf. Chapter 7). Significance tests for these methods have not been developed, but they may nevertheless be useful descriptively. The reader is referred to the papers just cited for computational procedures, which will not be presented in detail in this Primer. Instead, a maximum-likelihood significance test for the null hypothesis of no relationships between any pair of $m$ sets of variables—which would logically precede either Horst's or Carroll's analyses or pairwise canonical analyses in accordance with the procedures of this chapter—will be presented. Then the consequences of requiring that the canonical variates for the left- and right-hand sets of variables be identical, will be explored.

Wilks (1935) provides a test for the $m$-set case which is a straightforward generalization of the test described in Eq. (5.7). Specifically,

$$H_0 : S_{ij} = 0 \qquad \text{for} \quad i \neq j,$$

where $S_{ij}$ is a submatrix containing the covariances of the variables in set $i$ with those in set $j$, is tested by comparing

$$\chi^2 = -(N - m - \sum p_i) \ln \left( \frac{|S|}{|S_1| \, |S_2| \cdots |S_m|} \right), \qquad (5.8)$$

where $p_i$ is the number of variables in set $i$; $S$ the covariance matrix of all $\sum p_i$ variables; and $S_i$ the $p_i \times p_i$ covariance matrix of the variables in set $i$; with the chi-square distribution having $p_1 p_2 \cdots p_m$ degrees of freedom.

One common situation in which the question of the relationship between two or more sets of variables arises, occurs when an experimental treatment or a battery of tests is administered repeatedly to the same group of subjects. The canonical correlation between any two administrations of the test battery could then be considered a generalized reliability coefficient, and the canonical correlation between any two repetitions of the treatment (that is, any two trials) would afford a measure of the stability of behavior under those conditions. Finally, if different treatments were administered on different occasions, but the set of dependent measures used to assess treatment effects was the same on all occasions, the canonical $R$ between any two such occasions would be a measure of the generalizability of the subjects' behavioral tendencies (predispositions, traits) across experimental conditions. (Note that this is the obverse of the usual interest in establishing *differences* in behavior as a function of conditions.)

In all of these examples, it would seem natural to require that the *same* linear combinations of each set of variables be employed in computing canonical correlations, rather than leaving the computing algorithm free to pick different canonical variates for the two sets. Thus, in the two-set case, we would seek to maximize

$$r^2(\mathbf{a}) = (\mathbf{a}'\mathbf{S}_{xy}\,\mathbf{a})^2/[(\mathbf{a}'\mathbf{S}_x\,\mathbf{a})(\mathbf{a}'\mathbf{S}_y\,\mathbf{a})], \qquad (5.9)$$

rather than the square of the expression in Eq. (5.1). In the $m$-set case, we would adopt Horst's criterion (mentioned above) of maximizing the sum of the $m(m-1)/2$ intercorrelations between pairs of combined variables, but with the added restriction that the same combining coefficients be used within each set. The author (Harris, 1973a) has developed expressions for the special case in which the various $p \times p$ main diagonal submatrices of the population covariance matrices are identical, but not for the more general case.

The solution under the assumption of homogeneity of main-diagonal submatrices requires taking the greatest characteristic root and associated characteristic vector of $(\bar{\mathbf{S}})^{-1}\bar{\mathbf{S}}_{ij}$, where $\bar{\mathbf{S}}$ is the simple average of the $m$ main-diagonal submatrices and $\bar{\mathbf{S}}_{ij}$ is the simple average of the $m(m-1)/2$ above-diagonal submatrices. The sampling distribution of the resulting correlation measure has not been derived.

## DEMONSTRATION PROBLEM
## AND
## SOME REAL DATA EMPLOYING CANONICAL CORRELATION

1.  Given the following data:

| Subject | $X_1$ | $X_2$ | $Y_1$ | $Y_2$ | $Z_1$ | $Z_2$ |
|---------|-------|-------|-------|-------|-------|-------|
| 1 | 1 | 0 | 4 | 2 | 11 | 25 |
| 2 | 1 | 0 | 1 | 3 | 14 | 25 |
| 3 | 1 | 0 | 3 | 5 | 12 | 26 |
| 4 | 1 | 0 | 2 | 2 | 11 | 30 |
| 5 | 0 | 1 | 2 | 2 | 13 | 25 |
| 6 | 0 | 1 | 4 | 5 | 12 | 30 |
| 7 | 0 | 1 | 5 | 4 | 12 | 27 |
| 8 | 0 | 1 | 5 | 5 | 15 | 28 |

(a)  Conduct a canonical analysis on the relationship between variables $Y_{1-2}$ and variables $Z_{1-2}$. Be sure to include in your analysis the two sets of canonical coefficients and the canonical correlations between the two sets of variables. Do not worry about significance tests.

(b)  Compute the two canonical variates for each of the eight subjects. Designate by $Y_i^*$ the left-hand canonical variate for subject $i$, and by $Z_i^*$ the right-hand canonical variate for subject $i$.

(c)   Calculate the correlation between the two canonical variates, that is, compute the Pearson product–moment correlation between $Y*$ and $Z*$. Compare this number with your findings in part (a).

(d)   Do a multiple regression of $Z*$ on $Y_1$ and $Y_2$. Report the betas and the multiple correlation between $Z*$ and $Y_1$ and $Y_2$. Compare these results with those of part (a).

(e)   Do a multiple regression of $Y*$ predicted from $Z_1$ and $Z_2$. Report the regression coefficients and the multiple correlation. Compare these results with those of part (a).

2.   Still working with the data of Problem 1:

(a)   Conduct a canonical analysis of the relationship between variables $X_{1-2}$, on the one hand, and variables $Y_1$, $Y_2$, $Z_1$, and $Z_2$. Report both sets of canonical coefficients, the canonical correlation between the two sets of variables, and the appropriate test statistic ($F$ or greatest root) for testing the overall null hypothesis.

(b)   Treating subjects 1–4 as members of one group and Ss 5–8 as members of a second group, conduct a one-way Manova on the differences among the groups in mean response vectors. Report the discriminant function and the appropriate statistic for testing the overall null hypothesis.

(c)   Label the discriminant function obtained in part (b) as $(c_{Y1}, c_{Y2}, c_{Z1}, c_{Z2})$. Calculate for each subject the single score, $D_i = c_{Y1} Y_{1i} + c_{Y2} Y_{2i} + c_{Z1} Z_{1i} + c_{Z2} Z_{2i}$. Now conduct a $t$-test on the difference in $\mathbf{D}$ between the two groups.

(d)   Using Hotelling's $T^2$, conduct a test of the overall null hypothesis of no difference between the two groups in overall mean response vectors.

(e)   Compare and contrast the results obtained in parts (a)–(d).

3.   As part of a cooperative study of differences between patients classified as clearly paranoid and patients classified as clearly nonparanoid in behavior in a binary prediction experiment and in the Prisoner's Dilemma game, each patient who volunteered for the study went through a psychiatric interview and also took a 400-question "personality inventory" containing, among other things, items from several scales of the MMPI. As a result, we have available both interview (the psychiatrist's ratings of suspiciousness of the study and the overall degree of paranoia) and paper-and-pencil (scores on the Suspicion, Paranoia, and Schizophrenia scales of the MMPI) measures of paranoia. The matrix of intercorrelations among these five measures is reported in the tabulation. Conduct and interpret a canonical analysis of these data.

| | MMPI Scales | | | Interviewer Ratings | |
|---|---|---|---|---|---|
| | Susp | Par | Schiz | Susp' | Par' |
| Susp | 1 | .49[a] | .63[a] | .24 | .12 |
| Par | | 1 | .63[a] | .42[a] | .40[a] |
| Schiz | | | 1 | .34[b] | .14 |
| Susp' | | | | 1 | .54[a] |
| Par' | | | | | 1 |

$N = 44$.

[a] $p < .05$ ⎫ by standard test for significance of
[b] $p < .01$ ⎭   a single correlation coefficient.

4.   How are we to interpret the "extra" sets of canonical coefficients obtained in Problem 3? Note:   You may use any mixture of hand and computer computations you find convenient.

*Answers*

1.  (a)   **Canonical analysis on $Y_1$ and $Y_2$ versus $Z_1$ and $Z_2$**

$$S_{11} = \frac{1}{7}\begin{bmatrix} 15.5 & 8.0 \\ 8.0 & 14.0 \end{bmatrix}, \quad S_{11}^{-1} = \frac{7}{153}\begin{bmatrix} 14.0 & -8 \\ -8 & 15.5 \end{bmatrix}, \quad S_{22} = \frac{1}{7}\begin{bmatrix} 14 & -4 \\ -4 & 32 \end{bmatrix}$$

$$S_{22}^{-1} = \frac{7}{432}\begin{bmatrix} 32 & 4 \\ 4 & 14 \end{bmatrix}, \quad S_{12} = \frac{1}{7}\begin{bmatrix} 0 & 6 \\ 5 & 7 \end{bmatrix}, \quad S_{12}' = \frac{1}{7}\begin{bmatrix} 0 & 5 \\ 6 & 7 \end{bmatrix}$$

$$S_{11}^{-1}S_{12} = \begin{bmatrix} 14 & -8 \\ -8 & 15.5 \end{bmatrix} \frac{1}{153}\begin{bmatrix} 0 & 6 \\ 5 & 7 \end{bmatrix} = \frac{1}{153}\begin{bmatrix} -40 & 28 \\ 77.5 & 60.5 \end{bmatrix}$$

$$S_{22}^{-1}S_{12}' = \begin{bmatrix} 32 & 4 \\ 4 & 14 \end{bmatrix} \frac{1}{432}\begin{bmatrix} 0 & 5 \\ 6 & 7 \end{bmatrix} = \frac{1}{432}\begin{bmatrix} 24 & 188 \\ 84 & 118 \end{bmatrix}$$

$$S_{11}^{-1}S_{12}\,S_{22}^{-1}S_{12}' = \begin{bmatrix} -40 & 28 \\ 77.5 & 60.5 \end{bmatrix} \frac{1}{66{,}096}\begin{bmatrix} 24 & 188 \\ 84 & 118 \end{bmatrix} = \frac{1}{66{,}096}\begin{bmatrix} 1392 & -4216 \\ 6942 & 21{,}709 \end{bmatrix}$$

$$S_{22}^{-1}S_{12}'\,S_{11}^{-1}S_{12} = \begin{bmatrix} 24 & 188 \\ 84 & 118 \end{bmatrix} \frac{1}{66{,}096}\begin{bmatrix} 40 & 28 \\ 77.5 & 60.5 \end{bmatrix} = \frac{1}{66{,}096}\begin{bmatrix} 13{,}610 & 12{,}046 \\ 5{,}785 & 9{,}491 \end{bmatrix}$$

We now have two matrices, either of whose largest characteristic root will give us the square of the maximum canonical correlation. *Always* calculate $\lambda$ for *both* matrices as a check on your calculations. Using the formula for the characteristic roots of a $2 \times 2$ matrix, we obtain

(a)   Using $S_{11}^{-1}S_{12}S_{22}^{-1}S_{12}'$,

$$\lambda = \frac{(a + d) \pm \sqrt{(a - d)^2 + 4bc}}{2}, \quad \text{where the matrix is } \begin{bmatrix} a & b \\ c & d \end{bmatrix},$$

whence

$$\lambda = \frac{23{,}101 \pm \sqrt{295{,}710{,}601}}{2(66{,}096)} = \frac{23{,}101 \pm 17{,}196.2}{2(66{,}096)} = (.30484, .04467).$$

(b)   Using $S_{22}^{-1}S_{12}'S_{11}^{-1}S_{12}$, we obtain

$$\lambda = \frac{23{,}101 \pm \sqrt{295{,}710{,}610}}{2(66{,}096)} = (.30484, .04467).$$

which agrees quite well with the first calculation. The maximum canonical correlation is thus $\sqrt{.30484} = .55212$. (We are only keeping all those decimals for comparison with subsequent calculations.)

Now we want the canonical coefficients. The coefficients for the $Y$ variables are given by the matrix product which begins with the inverse of those variables. Thus

$$[S_{11}^{-1}S_{12} S_{22}^{-1}S'_{12} - I\lambda]a = 0,$$

or

$$\begin{bmatrix} 1392 - 20{,}148.5 & -4216 \\ 6942 & 21{,}709 - 20{,}148.5 \end{bmatrix} \begin{bmatrix} a_{11} \\ a_{12} \end{bmatrix} = 0.$$

Here again we have a check on our calculations. The top and bottom row of the above matrix each define the ratio between $a_{11}$ and $a_{12}$. These ratios should, of course, agree. Thus we obtain

$$a_{11} = -\frac{4216}{18{,}756.5} a_{12} = -.2248a_{12}$$

and

$$a_{11} = -\frac{1560.5}{6942} a_{12} = -.2248a_{12}.$$

Fortunately the two answers agree closely. The coefficients for the right-hand variables are given by

$$\begin{bmatrix} 13{,}610 - 20{,}148.5 & 12{,}046 \\ 5785 & 9491 - 20{,}148.5 \end{bmatrix} \begin{bmatrix} b_{11} \\ b_{12} \end{bmatrix} = 0;$$

$$b_{11} = \frac{12{,}046}{6538.5} b_{12} = 1.8423b_{12} \quad \text{and} \quad b_{11} = \frac{10{,}657.5}{5785} b_{12} = 1.8423b_{12}.$$

We could normalize these vectors to unit length, but there is no need to do so for present purposes, since division of each of two variables by a (possibly different) constant does not affect the correlation between them. Thus we will take

$$\mathbf{a}'_1 = (-.2248,1) \quad \text{and} \quad \mathbf{b}'_1 = (1.8423,1)$$

as the characteristic vectors defining the left- and right-hand canonical variates, respectively.

**(b) and (c)   Calculation of individual canonical variates and correlation between them**

$$\text{Left-hand variate (variable for } Ys) = Y_2 - .2248Y_1$$

$$\text{Right-hand canonical variate} = Z_2 + 1.8423Z_1$$

These calculations are straightforward, but as a means of checking your answers, the two canonical variates for each $S$ are presented in the tabulation.

| Subject | $Y^*$ | $Z^*/10$ |
|---------|--------|----------|
| 1 | 1.1008 | 4.5265 |
| 2 | 2.7752 | 5.0792 |
| 3 | 4.3256 | 4.8108 |
| 4 | 1.5504 | 5.0265 |
| 5 | 1.5504 | 4.8950 |
| 6 | 4.1008 | 5.2108 |
| 7 | 2.8760 | 4.9108 |
| 8 | 3.8760 | 5.5635 |

$$r_{Y*Z*} = \frac{\sum Y*Z* - \frac{(\sum Y*)(\sum Z*)}{N}}{\sqrt{\left(\sum Y*^2 - \frac{(\sum Y*)^2}{N}\right)\left(\sum Z*^2 - \frac{(\sum Z*)^2}{8}\right)}}$$

$$= \frac{112.3265 - (22.1552)(40.0231)/8}{\sqrt{\left(72.5431 - \frac{(22.1552)^2}{8}\right)\left(200.8789 - \frac{(40.0231)^2}{8}\right)}}$$

$$= .55220,$$

which is relatively close to the largest canonical correlation computed in part (a).

**(d)   Multiple regression of $Z*$ on $Y_1$ and $Y_2$**

$\mathbf{B} = \mathbf{S}_{22}^{-1}\mathbf{s}_{12}$,   where $\mathbf{S}_{22}$ is the covariance matrix of the predictor variables ($Y_1$ and $Y_2$) and $\mathbf{s}_{12}$ is the column vector of covariances of $Y_1$ and $Y_2$ with $Z*$. We must therefore calculate these two correlations, giving

$$S_{Y_1 Z*} = \sum Y_1 Z - \frac{(\sum Y_1)(\sum Z*)}{8} = 130.6753 - \frac{(26)(20.0231)}{8} = .6002$$

$$S_{Y_2 Z*} = 141.7023 - \frac{(28)(40.0231)}{8} - 1.6211$$

$$\mathbf{B} = \begin{bmatrix} 14 & -8 \\ -8 & 15.5 \end{bmatrix} \quad \frac{1}{153}\begin{bmatrix} .6002 \\ 1.6211 \end{bmatrix} = \frac{1}{153}\begin{bmatrix} -4.5660 \\ 20.3255 \end{bmatrix}$$

Note that $B_1/B_2 = -.2246$, which is quite close to the $-.2248$ of part (a).

$$\text{Multiple } R = \sqrt{\mathbf{s}_{12}'\mathbf{B}/s_2^2} = \sqrt{.197445/.6478} = .55208,$$

which is quite close to the largest canonical $r$ calculated in part (a).

**(e)   Multiple regression of $Y*$ on $Z_1$ and $Z_2$**

$$S_{Z_1 Y*} = 281.9400 - (100)(22.1552)/8 = 5.0000$$

$$S_{Z_1 Y*} = 603.8416 - (216)(22.1552)/8 = 5.6512$$

$$\mathbf{B} = \begin{bmatrix} 32 & 4 \\ 4 & 14 \end{bmatrix} \quad \frac{1}{432}\begin{bmatrix} 5.000 \\ 5.6512 \end{bmatrix} = \frac{1}{432}\begin{bmatrix} 182.6048 \\ 99.1168 \end{bmatrix}.$$

Note that $B_1/B_2 = 1.8423$, which is quite close to the 1.8423 of part (a).

$$\text{Multiple } R = \sqrt{\frac{\mathbf{s}_{12}'\mathbf{B}}{s_{Y*}^2}} = \sqrt{\frac{3.4101}{11.1865}} = .55212.$$

*Note:*   One of the values of going through a problem such as this by hand is that you then have a "test problem" to aid in checking out your computer program. When these data were analyzed using a locally developed program, the computer agreed to five decimal places with the above quoted figures for $\lambda$ and the canonical correlation. It did *not* agree even to two decimal places, however, with the appropriate values for the canonical coefficients. The clincher came when the program's canonical coefficients were used to compute $Y*$ and $Z*$, yielding a correlation lower than that given by the hand-computed coefficients. (Why is this such a crucial test?) We thus had a problem with this program. This problem turned out to have been caused by a simple mistake in punching the statements of an SSP subroutine used by this program onto IBM cards, and was subsequently corrected. ("SSP" stands for IBM's Scientific Subroutine Package, a very widely distributed package of subroutines useful in "building" programs.)

P.S. The comparisons with the computer program should be done starting with the intercorrelation matrix, since this is what the computer actually does.

Thus, for the current problem,

$$\mathbf{R}_{11} = \begin{bmatrix} 1 & .54308 \\ .54308 & 1 \end{bmatrix}, \quad \mathbf{R}_{22} = \begin{bmatrix} 1.0 & -.18898 \\ -.18898 & 1.0 \end{bmatrix}, \quad \mathbf{R}_{12} = \begin{bmatrix} 0 & .26941 \\ .35714 & .33072 \end{bmatrix}$$

$$\mathbf{R}_{11}^{-1}\mathbf{R}_{12}\,\mathbf{R}_{22}^{-1}\mathbf{R}_{12}' = \frac{1}{6.79884}\begin{bmatrix} .143189 & -.456309 \\ .678647 & 2.233037 \end{bmatrix},$$

whence

$$\lambda_1 = .30488, \qquad \lambda_2 = .04472,$$

and

$$\begin{bmatrix} -.28382 & -.067116 \\ .099818 & .0235638 \end{bmatrix}\begin{bmatrix} a_{11} \\ a_{22} \end{bmatrix} = 0 \Rightarrow a_{11} = -.23647 a_{21}.$$

Upon dividing $a_1$ and $a_2$ by their respective standard deviations, we obtain as *raw-score* canonical coefficients $\mathbf{a}_1 = (-.2247, 1)$, just as we had before.

## 2. (a) Canonical analysis of group membership variables versus others

The problem with applying our usual formulas to the variables as defined is that

$$\mathbf{R}_{11} = \begin{bmatrix} 1 & -1 \\ -1 & 1 \end{bmatrix} \Rightarrow |\mathbf{R}_{11}| = 0,$$

so that $\mathbf{R}_{11}^{-1}$ is undefined. We could avoid the problem by defining the membership variables as *deviation* scores, that is,

$$\mathbf{S}_{11} = \begin{bmatrix} 4 & 0 \\ 0 & 4 \end{bmatrix}, \qquad \text{rather than} \qquad \begin{bmatrix} 2 & -2 \\ -2 & 2 \end{bmatrix}.$$

This demonstrates one reason that group membership variables are referred to as *dummy variables*: the sum of these "deviation scores" is greater than 0! However, a simpler approach is to delete one of the two redundant (because linearly related) variables, whence

$$\mathbf{S}_{11} = s_1^2 = 2; \qquad \mathbf{S}_{11}^{-1} = \tfrac{1}{2};$$

$$\mathbf{S}_{12} = \begin{bmatrix} -3 & -2 & -2 & -2 \end{bmatrix}; \qquad \mathbf{S}_{22} = \begin{bmatrix} 15.5 & 8.0 & 0 & 6.0 \\ & 14.0 & 5.0 & 7.0 \\ & & 14.0 & -4.0 \\ & & & 32.0 \end{bmatrix};$$

$$\mathbf{S}_{22}^{-1} = \frac{1}{10}\begin{bmatrix} .97555 & -.62604 & .21825 & -.01868 \\ & 1.41060 & -.57911 & -.26358 \\ & & .98063 & .20834 \\ & & & .39970 \end{bmatrix}$$

$$\mathbf{S}_{11}^{-1}\mathbf{S}_{12} = \begin{bmatrix} -\tfrac{3}{2} & -1 & -1 & -1 \end{bmatrix}; \qquad \mathbf{S}_{22}^{-1}\mathbf{S}_{12}' = .4\begin{bmatrix} -.51840 \\ .18558 \\ -.46862 \\ -.15822 \end{bmatrix}$$

$$\mathbf{S}_{11}^{-1}\mathbf{S}_{12}\mathbf{S}_{22}^{-1}\mathbf{S}_{12}' = .48754.$$

Clearly the characteristic root of this matrix is its single entry, so that

$$\lambda = R_c^2 = .48754$$

and the characteristic vector for the group membership variables is any arbitrary scalar, and the single entry 1.0 when normalized

$$\mathbf{S}_{22}^{-1}\mathbf{S}_{12}' \, \mathbf{S}_{11}^{-1}\mathbf{S}_{12} = .2 \begin{bmatrix} -.51840 \\ .18558 \\ -.46862 \\ .15822 \end{bmatrix}$$

$$= \begin{bmatrix} [-3 & -2 & -2 & -2] \\ .31104 & .20736 & .20736 & .20736 \\ -.11135 & -.07423 & -.07423 & .07423 \\ .28117 & .18745 & .18745 & .18745 \\ .09493 & .06329 & .06329 & .06329 \end{bmatrix}$$

Since the first column of this matrix is exactly $1\frac{1}{2}$ times each of the other three columns, the matrix is clearly of rank 1, and

$$\lambda = \text{trace} = .31104 - .07423 + .18745 + .06329 = .48755,$$

which checks rather closely with our previous figure. The canonical correlation between the group membership variables and the others is thus given by $\sqrt{.48755} = .6977$.

The parameters for our gcr test of the significance of this correlation are

$$s = 1, \qquad m = (|1 - 4| - 1)/2 = 1, \qquad n = (8 - 5 - 2)/2 = 1/2.$$

Whence

$$F = \frac{3/2}{2} \frac{\lambda}{1 - \lambda} = \frac{3}{4}\left(\frac{.48755}{.51245}\right)$$

$$= .7136 \qquad \text{with} \quad 4 \text{ and } 3 \text{ df,} \quad \text{ns.}$$

The solution for the right-hand canonical variates proceeds as follows:

$$-.17651b_1 + .20736(b_2 + b_3 + b_4) = 0 \Rightarrow b_1 = 1.17478(b_2 + b_3 + b_4)$$

$$-.11135b_1 - .56178b_2 - .07423(b_3 + b_4) = 0 \Rightarrow b_2 = -\frac{.20504}{.69259}(b_3 + b_4)$$

$$= -.29605(b_3 + b_4),$$

whence

$$b_1 = 1.17478(.70365)(b_3 + b_4) = .82699(b_3 + b_4).$$

$$.28117b_1 + .18745b_2 - .30010b_3 + .18745b_4 = 0 \Rightarrow b_3 = \frac{.36448}{.12207}b_4 = 2.96157b_4,$$

whence

$$b_3 + b_4 = 3.96157b_4 \Rightarrow \mathbf{b} = (3.2762, -1.1728, 2.9616, 1).$$

Using the fourth equation as a check,

$$.09493b_1 + .06329(b_2 + b_3) - .42426b_4 = .00004,$$

which is reassuringly close to zero.

**(b)   One-way Manova on $P$ versus $NP$**

$$\mathbf{T} = \begin{bmatrix} 10 & 12 & 48 & 106 \\ 16 & 16 & 52 & 110 \end{bmatrix}; \qquad \mathbf{G}' = [26 \quad 28 \quad 100 \quad 216]$$

$$\mathbf{H} = \frac{\mathbf{T'T}}{4} - \frac{\mathbf{GG'}}{8} = \begin{bmatrix} 4.5 & 3 & 3 & 3 \\ 3 & 2 & 2 & 2 \\ 3 & 2 & 2 & 2 \\ 3 & 2 & 2 & 2 \end{bmatrix}; \quad \mathbf{E} = \begin{bmatrix} 11 & 5 & -3 & 3 \\ 5 & 12 & 3 & 5 \\ -3 & 3 & 12 & -6 \\ 3 & 5 & -6 & 30 \end{bmatrix}.$$

Cheating a bit by using HEINV, we find that the largest characteristic root is given by $\lambda = .95141$, with associated characteristic vector (the discriminant function) given by

$$(.70041, \quad -.25071, \quad .63315, \quad .21377),$$

or

$$D = 3.2765\,Y_1 - 1.1728\,Y_2 + 2.9618Z_1 + Z_2\,.$$

Note that $\lambda/(1 + \lambda) = .48755$, precisely equal to the $R_c^2$ of part (a), and that the discriminant function is almost identical to the right-hand canonical variate of part (a).

The parameters for the gcr test of the significance of the differences between the two groups are the same as for part (a).

**(c)   *t*-Test on discriminant scores**

$$t = \frac{\bar{X}_1 - \bar{X}_2}{\sqrt{(\sum x_1^2 + \sum x_2^2)(\frac{1}{2})}} = \frac{1.6468}{\sqrt{.198646 + 5.502477}}$$

$$= \frac{1.6468}{\sqrt{.475094}} \Rightarrow t^2 = 5.7082,$$

and

$$t = 2.38919 \quad \text{with} \quad 6 \text{ df}, \qquad p > .05$$

| Subject | *D*-14 | Subject | *D*-14 |
|---------|--------|---------|--------|
| 1 | .60912 | 5 | .47460 |
| 2 | .15663 | 6 | 1.55899 |
| 3 | .00350 | 7 | 1.86880 |
| 4 | .27715 | 8 | 3.73131 |

**(d)   Hotelling's $T^2$ on difference between two groups**

$$T^2 = \frac{N_1 N_2}{N_1 + N_2}(\bar{\mathbf{X}}_1 - \bar{\mathbf{X}}_2)'\mathbf{S}^{-1}(\bar{\mathbf{X}}_1 - \bar{\mathbf{X}}_2)$$

$$= \frac{n}{2}\left(\frac{\mathbf{T}_1 - \mathbf{T}_2}{n}\right)' 2(n-1)\mathbf{E}^{-1}\left(\frac{\mathbf{T}_1 - \mathbf{T}_2}{n}\right) = \frac{n-1}{n}(\mathbf{T}_1 - \mathbf{T}_2)'\mathbf{E}^{-1}(\mathbf{T}_1 - \mathbf{T}_2);$$

$$T^2 = \left(1 - \frac{1}{n}\right)(6,4,4,4)\frac{1}{10}\begin{bmatrix} 1.3950 & -.77621 & .59749 & .10936 \\ -.77621 & 1.4644 & -.71486 & -.30942 \\ .59749 & -.71466 & 1.3235 & .32409 \\ .10936 & -.30942 & .32409 & .43878 \end{bmatrix}\begin{bmatrix} 6 \\ 4 \\ 4 \\ 4 \end{bmatrix}$$

$$= \tfrac{3}{4}[36(1.3950) + 48(-.77621 + .59749 + .10936) + 32(-.71486 - .30942 + .32409)$$

$$+ 16(1.4644 + 1.3235 + .43878)]$$

$$= \tfrac{3}{4}(7.611152) = 5.7084,$$

which is quite close to the 5.7082 we calculated as the square of the univariate $t$ for the difference between the two groups in mean "discriminant score."

Our test of significance is based on

$$F = \frac{N_1 + N_2 - p - 1}{(N_1 + N_2 - 2)(p)} T^2 = \frac{3}{6(4)} (5.7084) = \frac{3}{4} (.9514) = .7136 \qquad \text{with} \quad 3 \text{ and } 4 \text{ df.}$$

(Be sure you understand why we do not compare $T^2$ directly with Student's $t$-distribution.)

**(e)   Comparisons**

   (i)   The canonical coefficients for the right-hand variables (the $Y$s and $Z$s) are practically identical to the coefficients of the Manova-derived discriminant function. They are also identical with the $T^2$-derived discriminant function, though you were not required to calculate this last function.

   (ii)   When these coefficients are used to obtain a single combined score for each subject, and a standard univariate $t$-test is conducted on the difference between the two groups on the resulting new variable, the square of this is the same as the value of Hotellings $T^2$ statistic. It is also equal to 6 ($= N - 2$) times the largest characteristic root (in fact, the only nonzero root) obtained in the Manova, and it is further equal to $6\lambda/(1 - \lambda)$, where $\lambda$ is the largest root obtained in Canona with membership variables included.

   (iii)   The significance tests conducted in parts (a), (b), and (d) are identical. The reader may wish to verify that exactly the same results are obtained if the "dummy variable" approach is taken in part (a).

**3.**   To quote from Harris, Wittner, Koppell and Hilf (1970, p. 449):

   An analysis of canonical correlation (Morrison, 1967, Chapter 6) indicated that the maximum possible Pearson $r$ between any linear combination of the MMPI scale scores and any linear combination of the interviewer ratings was .47.[8] The weights by which the standard scores of the original variables must be multiplied to obtain this maximum correlation are $-.06$, 1.30, and $-.05$ for Susp, Pa, and Sc, respectively, and .73 and .68 for Susp' and Pa'.

   [8] Morrison (1967) provides two tests of the statistical significance of canonical $R$: a greatest characteristic root test (p. 210) and a likelihood-ratio test (p. 212). By the first test, a canonical $R$ of .49 is significantly different from zero at the .05 level. By the second test, the canonical $R$ between the original measures is statistically significant, . . . . Both tests must be interpreted very cautiously . . . , since they each assume normally distributed variables.

Note that in this paper the author fell into the very trap mentioned in Section 5.3: interpreting the likelihood-ratio test as testing the statistical significance of the correlation between the first pair of canonical variates.

# 6 | PRINCIPAL COMPONENT ANALYSIS
## Relationships within a Single Set of Variables

The techniques we have discussed so far have had in common their assessment of relationships between two sets of variables: two or more predictors and one outcome variable in multiple regression; one predictor (group membership) variable and two or more outcome variables in Hotelling's $T^2$; two or more group membership variables and two or more outcome variables in Manova; and two or more predictors, two or more outcome variables in canonical correlation. Now we come to a set of techniques (principal component analysis and factor analysis) which "look inside" a single set of variables and attempt to assess the structure of the variables in this set independently of any relationship they may have to variables outside this set. Principal component analysis (PCA) and factor analysis (FA) may in this sense be considered as logical precursors to the statistical tools discussed in Chapters 2–5. However, in another sense they represent a step beyond the concentration of earlier chapters on observable relationships among variables to a concern with relationships between observable variables and unobservable (latent) processes presumed to be generating the observations.

Since there are usually a large number of theories (hypothetical processes) which could describe a given set of data equally well, we might expect that this introduction of latent variables would produce a host of techniques, rather than any single procedure for analyzing data which could be labeled uniquely as "factor analysis." This is in fact the case, and this multiplicity of available models for factor analysis makes it essential that the investigator be prepared to commit himself to some assumptions about the structure of the data in order to select among alternative factor analysis procedures.

The particular procedure discussed in the current chapter—principal component analysis, or PCA—is a "hybrid" technique which, like factor analysis, deals with a single set of variables. Unlike the factors derived from FA, however, principal components are very closely tied to the original variables, with each subject's score on a principal component being a linear combination of his scores on the original variables. Also unlike factor analytic solutions, the particular transformations of the original variables generated by PCA are unique. We shall turn next to a description of the procedures which provide this uniqueness.

## 6.1  Definition of Principal Components

A principal component analysis (PCA) of a set of $m$ original variables generates $m$ new variables, the *principal components*, $PC_1, PC_2, \ldots, PC_m$, with each principal component being a linear combination of the $S$s' scores on the original variables, that is,

$$
\begin{aligned}
PC_1 &= b_{1,1} X_1 + b_{1,2} X_2 + \cdots + b_{1,m} X_m = \mathbf{Xb}_1; \\
PC_2 &= b_{2,1} X_1 + b_{2,2} X_2 + \cdots + b_{2,m} X_m = \mathbf{Xb}_2; \\
&\;\;\vdots \qquad\qquad\qquad\qquad\qquad\qquad\qquad\;\; \vdots \\
PC_m &= b_{m,1} X_1 + b_{m,2} X_2 + \cdots + b_{m,m} X_m = \mathbf{Xb}_m;
\end{aligned}
\tag{6.1}
$$

or, in matrix form, $\mathbf{PC} = \mathbf{XB}$, where each column of $B$ contains the coefficients for one principal component.

The coefficients for $PC_1$ are chosen so as to make its variance as large as possible. The coefficients for $PC_2$ are chosen so as to make the variance of this combined variable as large as possible, subject to the restriction that scores on $PC_1$ and $PC_2$ (whose variance has already been maximized) be uncorrelated. In general, the coefficients for $PC_i$ are chosen so as to make its variance as large as possible, subject to the restrictions that it be uncorrelated with scores on $PC_1$ through $PC_{i-1}$. Actually, an additional restriction, common to all $m$ PCs, is required. Since multiplication of any variable by a constant $c$ produces a new variable having a variance $c^2$ times as large as the variance of the original variable, we need only use arbitrarily large coefficients in order to make $s^2_{PC}$ arbitrarily large. Clearly it is only the relative magnitudes of the coefficients defining a PC that is of importance to us. In order to eliminate the trivial solution, $b_{i,1} = b_{i,2} = \cdots = b_{i,m} = \infty$, we require that the squares of the coefficients involved in any PC sum to unity, that is, $\sum_j b^2_{i,j} = \mathbf{b}'_i \mathbf{b}_i = 1$.

The mean of the scores for our sample of $S$s on any given PC is, of course,

$$
\overline{PC}_i = b_{i,1} \bar{X}_1 + b_{i,2} \bar{X}_2 + \cdots + b_{i,m} \bar{X}_m,
$$

whence, letting $x_{jk} = X_{jk} - \bar{X}_k$,

$$
\begin{aligned}
s_{PC_i} &= [1/(N-1)] \sum (b_{i,1} x_1 + b_{1,2} x_2 + \cdots + b_{i,m} x_m)^2 \\
&= [1/(N-1)] \sum (b^2_{i,1} x^2_1 + b^2_{i,2} x^2_2 + \cdots + b^2_{i,m} x^2_m + 2 b_{i,1} b_{i,2} x_1 x_2 \\
&\qquad + 2 b_{i,1} b_{i,3} x_1 x_3 + \cdots + 2 b_{i,m-1} b_{i,m} x_{m-1} x_m) \\
&= b^2_{i,1} s^2_1 + b^2_{i,2} s^2_2 + \cdots + b^2_{i,m} s^2_m \\
&\qquad + 2 b_{i,1} b_{i,2} s_{12} + \cdots + 2 b_{i,m-1} b_m s_{m-1,m} \\
&= \sum_j b^2_{i,j} s^2_j + \sum_j \sum_k b_{i,j} b_{i,k} s_{jk} \\
&= \mathbf{b}'_i \mathbf{S}_x \mathbf{b}_i,
\end{aligned}
\tag{6.2}
$$

where $S_x$ is the variance–covariance matrix of the $X$s. Using the method of Lagrangian multipliers (Digression 1), we find that, for $b_1$,

$$L = b_1' \, S_x \, b_1 - \lambda(b_1' \, b_1 - 1)$$

and, referring to Section D2.10,

$$dL/db_1 = 2S_x \, b_1 - 2\lambda b_1 = 0 \qquad \text{iff} \qquad [S_x - \lambda I]b_1 = 0,$$

where $0$ is a column vector each of whose $m$ entries is zero.

Thus we have a set of $m$ homogenous equations in $m$ unknowns. In order for this set of equations to have a nontrivial solution, the determinant $|S_x - \lambda I|$ must equal 0. As pointed out in Section D2.12, computing the determinant with $\lambda$ left in as an unknown produces an $m$th degree polynomial in $\lambda$ which, when set equal to zero, constitutes the characteristic equation. Solving this equation for $\lambda$ produces $m$ roots, some of which may be zero if there is a linear dependence among the original variables. Any one of the nonzero roots can be "plugged into" the matrix equation, and the resulting set of equations solved for the coefficients of $b$. Which of these $m$ sets of coefficients is to be taken as $PC_1$?

Recall that each root of the characteristic equation $\lambda_i$ makes the matrix equation $[S_x - \lambda_i I]b_i = 0$ true, whence $b_i' S_x \, b_i = \lambda_i b_i' b_i$. However, we require that $b_i' b_i = 1$, whence $\lambda_i = b_i' S_x \, b_i$. In other words, each root of the characteristic equation is equal to the variance of the combined variable generated by the coefficients of the characteristic vector associated with that root. Since it is this variance which we set out to maximize, the coefficients of the first principal component will be the characteristic vector associated with the *largest* characteristic root. As it turns out, $PC_2$ is computed via the characteristic vector corresponding to the second largest characteristic root, and in general $PC_i$ has the same coefficients as the characteristic vector associated with the $i$th largest characteristic root $\lambda_i$. To show this for the case of $PC_2$, consider the problem of maximizing $b_2' S_x b_2$ subject to the *two* side conditions that $b_2' b_2 = 1$ and that $b_2' b_1 = 0$. This leads to

$$L = b_2' \, S_x \, b_2 - \lambda(b_2' \, b_2 - 1) - \theta(b_2' \, b_1),$$

whence

$$dL/db_2 = 2S_x \, b_2 - 2\lambda b_2 - \theta b_1 = 0 \qquad \text{iff} \qquad 2[S_x - \lambda I]b_2 = \theta b_1;$$

whence

$$2b_1'[S_x - \lambda I]b_2 = \theta b_1' b_1;$$

whence

$$\theta = 2b_1' \, S_x \, b_2 \, .$$

However, from the definition of $b_1$, namely $S_x \, b_1 = \lambda_1 b_1$, we have

$$b_2' \, S_x \, b_1 = \lambda_1 b_2' \, b_1 = 0;$$

whence we know that $\theta = 0$, and $b_2$ must satisfy the equations

$$[S_x - \lambda I]b_2 = 0.$$

Thus $\mathbf{b}_2$ will be one of the characteristic vectors of $\mathbf{S}_x$, and in particular, given our goals, will be the characteristic vector associated with the second largest characteristic root.

As one step in the above derivations, we showed that $\mathbf{b}_2' \mathbf{S}_x \mathbf{b}_1$, which is of course the covariance of $PC_1$ and $PC_2$, is equal to zero. Using the same approach, it can be proved that in general the orthogonality of (lack of correlation among) the characteristic vectors of $\mathbf{S}_x$ is paralleled by the absence of any correlation among the PCs computed from those sets of coefficients.

## 6.2   Interpretation of Principal Components

It has already been demonstrated that the PCs of a set of variables are uncorrelated with each other, and that they are hierarchically ordered in terms of their variances, with the $i$th PC having the $i$th largest variance. The variance of a variable is of course a measure of the extent to which subjects differ in their scores on that variable. It is thus reasonable to interpret $PC_1$ as that linear combination of the original variables which maximally discriminates among our subjects, just as the discriminant function derived in Manova *maximally discriminates* among the various groups. PCA is thus a sort of "internal discriminant analysis."

It is also true that the PCs partition the total variance of the original variables into $m$ additive components. From the general properties of characteristic roots (Section D2.12), we know that

$$\sum_j \lambda_j = \sum_j s_{PC_j}^2 = \text{trace } \mathbf{S}_x = \sum_i s_i^2.$$

It is customary to describe this last property of PCA as indicating that $PC_j$ "accounts for" a certain proportion—$s_{PC_j}^2 / \sum s_i^2$—of the variation among the original variables. It would not, of course, be difficult to construct a set of $m$ variables which also had the property that their variances summed to $\sum s_i^2$, but which had no relationship at all to those original variables. For instance, we could use a random number generator to construct $m$ unit normal distributions, and then multiply each by $\sum s_i^2 / m$. We thus need to provide some additional justification for the phrase, "accounts for," that is, for the assertion that PCA in some way explains, or describes the structure of, the variation among the original variables.

The strongest support for this way of talking about PCs comes from the fact that knowledge of our Ss' scores on the $m$ principal components, together with knowledge of the coefficients defining each PC, would be sufficient to reproduce the Ss' scores on the original variables perfectly. Just as the PCs are defined as linear combinations of the original variables, so the original variables can be defined as linear combinations of the PCs. In fact, the coefficients which must be used to generate $X_j$ in the equation

$$X_j = c_{j,1} PC_1 + c_{j,2} PC_2 + \cdots + c_{j,m} PC_m \qquad (6.3)$$

are simply the weights variable $j$ receives in the various linear compounds which define the PCs, that is, $c_{j,k} = b_{k,j}$ for all $k, j$. Since we can reproduce the score made by each $\mathbf{S}$ on each original variable, we can, *a fortiori*, reproduce any measure or set of measures defined on the original variables—such as, for example, the covariance

matrix or the intercorrelation matrix of the original variables. The latter two "repro-
ductions" do not require detailed knowledge of each individual's score, but only the
values of the characteristic roots and vectors. Specifically, the covariance of two
original variables $X_i$ and $X_j$ is given by the formula

$$s_{ij} = [1/(N - 1)] \sum [(b_{1,i}pc_1 + b_{2,i}pc_2 + \cdots + b_{m,i}pc_m)(b_{1,j}pc_1 + \cdots + b_{m,j}pc_m)]$$

$$= b_{1,i}b_{1,j}\lambda_1 + b_{2,i}b_{2,j}\lambda_2 + \cdots + b_{m,i}b_{m,j}\lambda_m; \tag{6.4}$$

whence

$$\mathbf{S}_x = \mathbf{BDB'},$$

where $\mathbf{B}$ is an $m \times m$ matrix whose $j$th column contains the normalized characteristic
vector associated with $\lambda_j$; $\mathbf{D}$ is a diagonal matrix whose $j$th main diagonal entry is $\lambda_j$;
and $pc_i$ is a deviation score on the $i$th principal component.

To illustrate this reproducibility property, consider a very simple example in
which we seek the principal components of the system of variables consisting of
variables 1 and 3 from Data Set 1 (see p. 31). The covariance matrix of this system is

$$\mathbf{S}_x = \begin{bmatrix} 17.5 & 7.0 \\ 7.0 & 5.0 \end{bmatrix},$$

whence (taking advantage of the computational formulas provided in Section 6.4),

$$\lambda = \tfrac{1}{2}[22.5 \pm \sqrt{12.5^2 + 4(49)}] = 20.635, 1.865;$$

$$c^2 + (\lambda_1 - a)^2 = 49 + 3.135^2 = 58.83;$$

$$\mathbf{b}_1 = (7, 3.135)/7.66 = (.913, .409);$$

and

$$\mathbf{b}_2 = (-7, 15.635)/17.11 = (-.409, .913).$$

The results of a PCA are typically reported in the form of a *factor pattern*, that is, a
matrix $\mathbf{B}$ whose columns contain the normalized coefficients of the principal components (that is, the normalized eigenvectors of $\mathbf{S}_x$). For the present analysis, the
factor pattern would be

|  | $PC_1$ | $PC_2$ |
|---|---|---|
| $X_1$ | .913 | −.409 |
| $X_3$ | .409 | .913 |
| $\lambda_j = s_{pc_j}^2$ | 20.635 | 1.865 |

(It is also common practice to list the characteristic root for each PC at the bottom of
the column containing the coefficients which define that PC.)

The PCs are defined in terms of the original variables as follows:

$$.913X_1 + .409X_3 = PC_1, \qquad -.409X_1 + .913X_3 = PC_2.$$

However, as the form in which these two equations are listed is meant to suggest, we
can as readily consider the equations as a pair of simultaneous equations to be solved

for $X_1$ and $X_3$, whence

$$.370X_1 + .167X_3 = .409PC_1$$
$$-.370X_1 - .835X_3 = .913PC_2 \quad \Rightarrow \quad \begin{array}{l} X_3 = .409PC_1 + .913PC_2 \\ X_1 = .913PC_1 - .409PC_2 . \end{array}$$

Note that the coefficients used in the linear combination of PCs that "generates" or "reproduces" a particular original variable bear a close relationship to (and are in fact identical to) the coefficients in that row of the *factor pattern*.

※ This relationship was stated without proof a few paragraphs earlier. To show that it is correct, consider the equation defining $PC_j$, namely

$$\mathbf{Xb}_j = b_{j,1}X_1 + b_{j,2}X_2 + \cdots + b_{j,m}X_m = PC_j . \tag{6.5}$$

Our assertion is that

$$X_i = b_{1,i}PC_1 + b_{2,i}PC_2 + \cdots + b_{m,i}PC_m . \tag{6.6}$$

If this is correct, then substitution of expressions like (6.6) for each occurrence of an $X_i$ in (6.5) should "satisfy" the equation, that is should produce an expression which equals $PC_j$. Carrying out these substitutions gives

$$b_{j,1}(b_{1,1}PC_1 + b_{2,1}PC_2 + \cdots + b_{j,1}PC_j + \cdots + b_{m,1}PC_m)$$
$$+ b_{j,2}(b_{1,2}PC_1 + b_{2,2}PC_2 + \cdots + b_{j,2}PC_j + \cdots + b_{m,2}PC_m)$$
$$+ b_{j,m}(b_{1,m}PC_1 + b_{2,m}PC_2 + \cdots + b_{j,m}PC_m + \cdots + b_{m,m}PC_m)$$

$$\|$$

$$PC_1(b_{j,1}b_{1,1} + b_{j,2}b_{1,2} + \cdots + b_{j,m}b_{1,m})$$
$$+ PC_2(b_{j,1}b_{2,1} + b_{j,2}b_{2,2} + \cdots + b_{j,m}b_{2,m})$$
$$+ PC_j(b_{j,1}^2 + b_{j,2}^2 + \cdots + b_{j,m}^2)$$
$$+ PC_m(b_{j,1}b_{m,1} + b_{j,2}b_{m,2} + \cdots + b_{j,m}b_{m,m})$$

$$\|$$

$$\sum_{i=1}^{m} \mathbf{b}_j' \mathbf{b}_i PC_i = PC_j .$$

However, we know that the PCs are orthogonal, that is, that $\mathbf{b}_j \mathbf{b}_i = 1$ if $i = j$ but zero otherwise, so that $\sum \mathbf{b}_j' \mathbf{b}_i PC_i = PC_j$, and the assertion is proved.

※ Rather than working with the symbols for the PCs and the $X$s, we could have worked with each $S$'s score on each of these variables and conducted a multiple regression analysis on the relationship between scores on the PCs (treated as predictor variables) and scores on the original variables (each $X_i$ being treated in turn as the outcome variable). Let $\mathbf{x} = [\mathbf{x}_1, \mathbf{x}_2, \ldots, \mathbf{x}_m]$ be the $N \times m$ matrix of deviation scores on the $X$ variables; $\mathbf{pc} = [\mathbf{pc}_1, \mathbf{pc}_2, \ldots, \mathbf{pc}_m]$ be the $N \times m$ matrix of deviation scores on the principal components; and $\mathbf{B} = \mathbf{b}_1, \mathbf{b}_2, \ldots, \mathbf{b}_m$ be the factor pattern, that is, the $m \times m$ matrix of the characteristic vectors of $\mathbf{S}_x$. The multiple regression analysis seeks, for each $X_i$, a vector $\mathbf{c}_i$ such that $\mathbf{pc}\,\mathbf{c}_i = \mathbf{x}_i$. More generally, the results of all $m$ of these regression analyses can be "stored" in the matrix $\mathbf{C} = [\mathbf{c}_1, \mathbf{c}_2, \ldots, \mathbf{c}_m]$. The analysis of the relationship between $\mathbf{c}_i$ and $\mathbf{pc}$ takes the form

$$\mathbf{c}_i = (\mathbf{pc'pc})^{-1}(\mathbf{pc'x}_i);$$

or, more generally,

$$\mathbf{C} = (\mathbf{pc'pc})^{-1}(\mathbf{pc'x}).$$

Thanks to the orthogonality of the *pcs*, $\mathbf{pc'pc} = \mathbf{D}$, a diagonal matrix whose main diagonal entries are the characteristic roots of $\mathbf{S}_x$, so that $(\mathbf{pc'pc})^{-1} = \mathbf{D}^{-1}$ is a diagonal matrix whose main diagonal entries are the reciprocals of the $\lambda_i$. However, by the definition of the *PCs*, $\mathbf{pc} = \mathbf{xB}$, whence $\mathbf{pc'} = \mathbf{B'x'}$, whence

$$\mathbf{pc'x} = \mathbf{B'(x'x)} = (\mathbf{x'xB})',$$

so that

$$\mathbf{C} = \mathbf{D}^{-1}\mathbf{B'(x'x)} = \mathbf{D}^{-1}(\mathbf{x'xB})'.$$

However, the characteristic vectors were obtained as solutions to the equations $[\mathbf{x'x} - \lambda_i\mathbf{I}]\mathbf{b}_i = 0$, whence we know that $\mathbf{x'xb}_i = \lambda_i\mathbf{b}_i$ and therefore $\mathbf{x'xB} = \mathbf{BD}$, whence

$$\mathbf{C} = \mathbf{D}^{-1}\mathbf{D'B'} = (\mathbf{D}^{-1}\mathbf{D})\mathbf{B'} = \mathbf{B'},$$

as was initially asserted.

*Table 6.1*

*Relations between Original Scores and Scores on Principal Components for $X_1$ and $X_3$ of Data Set 1*

| Subject | $x_1$ | $x_3$ | $pc_1$ | $pc_2$ | $.913pc_1 - .409pc_2$ | $.409pc_1 + .913pc_2$ |
|---------|-------|-------|--------|--------|------------------------|------------------------|
| A | 2 | 0 | 1.826 | −.818 | 2.002 | 0.000 |
| B | 4 | 1 | 4.061 | −.723 | 4.003 | 1.001 |
| C | 1 | 3 | 2.140 | 2.330 | 1.001 | 3.003 |
| D | −7 | −3 | −7.618 | .124 | | |
| E | 0 | −1 | −.409 | −.913 | | |
| $\sum x_i^2$ | 70 | 20 | $\sum d_{ij}$ | 82.61 | 7.46 | |
| $s^2$ | 17.5 | 5.0 | | 20.635 | 1.865 | |

To demonstrate this with the PCA of variables $X_1$ and $X_3$ from Data Set 1, Table 6.1 reports each $S$'s scores on $x_1$, $x_3$, $pc_1$, and $pc_2$. From the deviation scores on $PC_1$ and $PC_2$ we compute $(\mathbf{pc})'(\mathbf{pc})$ as

$$\begin{bmatrix} 20.635 & 0 \\ 0 & 1.865 \end{bmatrix},$$

and the matrix of covariances between *PCs* and *Xs* as

$$\begin{bmatrix} 75.362 & 33.744 \\ -3.066 & 6.308 \end{bmatrix},$$

whence

$$\mathbf{C} = \begin{bmatrix} 1/20.635 & 0 \\ 0 & 1/1.865 \end{bmatrix} \mathbf{pc'x} = \begin{bmatrix} .913 & .409 \\ -.411 & .913 \end{bmatrix},$$

as we had predicted. Finally, Table 6.1 also reports the results of applying these definitions of $X_1$ and $X_3$ in terms of the PCs to each $S$'s scores on $pc_1$ and $pc_2$. As can be seen, the original observations are indeed perfectly reproduced—except for round-off error in the computations.

This, then, is one sense in which the PCs "account for" variation among the original variables: the multiple $R$ between the PCs and each original variable is 1.0, and thus knowledge of scores on the PCs is sufficient to exactly reproduce scores on the original variables.

How unique is this contribution? Are there other linear combinations of the original variables which would also have a perfect multiple correlation with each of the original variables? The answer is of course "Yes." One conspicuous example takes as the coefficients for the *j*th linear compound a vector of all zeros except for the *j*th entry. More generally, *any* linearly independent set of vectors would provide a perfect multiple correlation. Thus this property, by itself, does not represent an advantage of principal components over any other set of linear transformations of the original variables.

What, then, makes PCs "better" than other linear compounds as explanations of the data? Are PCs, for instance, more likely to reflect the "true," underlying sources of the variability among the $X$s? A simple hypothetical example shows that this is not the case. Assume that scores on two observable variables, $X_1$ and $X_2$, are generated by a pair of underlying variables, $H_1$ and $H_2$, with $X_1$ being equal to the sum of the underlying processes, and $X_2 = H_1 + 3H_2$. [This implies that $H_1 = (3X_1 - X_2)/2$ and $H_2 = (X_2 - X_1)/2$.] Assume further that $H_1$ and $H_2$ are uncorrelated and that each has a unit normal distribution (and thus a mean and standard deviation of zero and one, respectively). Such a state of affairs would produce an observed covariance matrix of

$$\begin{bmatrix} 2 & 4 \\ 4 & 10 \end{bmatrix}$$

and a factor pattern of

|         | $PC_1$  | $PC_2$ |
|---------|---------|--------|
| $X_1$   | .383    | −.235  |
| $X_2$   | .922    | .974   |
| $s_{pc}^2$ | 11.655 | .345   |

Thus the linear compounds suggested by PCA as explanations of the data are $.383X_1 + .922X_2$ and $.974X_2 - .235X_1$, as contrasted to the true generating variables's relationships to $X_1$ and $X_2$ (in normalized form), $.707X_2 - .707X_1$ and

$.948X_1 - .316X_2$. Looked at from another angle, PCA suggests that $X_1$ and $X_2$ are related to underlying processes by $X_1 = .383PC_1 - .235PC_2$ and $X_2 = .922PC_1 + .974PC_2$, as contrasted to the "true" relationships of $X_1 = H_1 + H_2$ and $X_2 = H_1 + 3H_2$. Thus there is no guarantee that PCA will uncover the true generating processes. The advantages of PCA must lie elsewhere. It is to these advantages we now turn.

## 6.3  Uses of Principal Components

These advantages are (at least) twofold.

### 6.3.1  Uncorrelated Contributions

First, the principal components are *orthogonal* to each other, as the vectors in the trivial (identity) transformation $H_j = X_j$ are not. This is of benefit in two ways. First, the comparisons we make among subjects with respect to their scores on $PC_i$ are uncorrelated with (do not provide the same information as) comparisons based on their scores on $PC_j$. Second, any subsequent analysis of the relationship between the variables in the set on which the PCA was conducted and some other variable or set of variables can use $Ss$' scores on the PCs, rather than their scores on the original variables, without any loss of information. This will have the effect of greatly simplifying the computations involved in these subsequent analyses, as well as providing an automatic "partitioning" of the relationships into those due to correlations between $PC_1$ and the set of outcome variables, those due to correlations between $PC_2$ and the set of outcome variables, etc. For instance, a multiple regression analysis conducted on the relationship between a set of uncorrelated variables (such as $m$ PCs) and some outcome variable yields an $R^2 = \sum r_{iy}^2$, the sum of the squares of the separate Pearson $r$s between $y$ and each of the uncorrelated predictor variables, and regression coefficients each of which is simply equal to $r_{iy}$. Also, if neither the means nor the within-group scores on some set of outcome variables are correlated with each other, the test statistic resulting from a multivariate analysis of variance conducted on these uncorrelated variables will simply equal $(k - 1)/(N - k) \max_i(F_i)$, where $F_i$ is the univariate $F$-ratio obtained from a univariate Anova conducted on the single outcome measure $X_i$. Interpretation of the results of these analyses is also facilitated by the partitioning, in multiple regression, of the overall $R^2$ into $m$ uncorrelated sources of predictability, and in the case of Manova, by the partitioning of the differences among the groups into their differences on each of $m$ uncorrelated variables. This gain in interpretability of results will, however, be illusory if the uncorrelated variables do not themselves have simple substantive interpretations. This raises the *labeling problem*, the very persistent and usually difficult task of finding substantive interpretations of some set of hypothetical latent variables which have been derived through PCA or FA. This usually involves an examination of the *factor structure*, that is, of the correlation between each original variable and each latent variable. Since the partitioning and computational advantages of employing PCs accrue as well to *any* orthogonal set of linear compounds of the original variables, researchers will naturally prefer to employ a set of linear compounds whose factor structure takes on a very simple form. Empirically, the factor structures resulting from PCA tend *not* to have "simple structure," so that many researchers prefer to treat the results of a PCA as merely an intermediate step which provides one of many possible sets of uncorrelated latent variables. These various possible sets of orthogonal linear compounds are all related by a simple set of transformations known as *orthogonal rotations* (to be discussed in more detail in Sections 6.5.1 and 6.5.2), so that PCA can indeed serve as a starting point for the search for simple structure. If such a procedure is followed, however, the researcher automatically loses the second major advantage of PCA, its hierarchical ordering of the latent variables in terms of the percentage of variance for which each accounts.

### 6.3.2   Computational Convenience

The principal components resulting from PCA, unlike the latent variables represented by any other set of linear compounds of the original variables, are designed to account for as high a percentage of the variation among the variables with as few PCs as possible. The $PC_1$ accounts for a larger portion of the variance of the system than any other linear compound (except a simple multiple of itself), while each successive PC accounts for as high a percentage of the total variance as is possible while remaining uncorrelated with all preceding PCs. If, as is often the case, the first few PCs account for some large percentage (80%? 90%?) of the total variance, the researcher may be willing to sacrifice completeness of description (which may well be illusory anyway, since PCA blithely partitions all of the variance despite the fact that a fair portion of the system's total variance may reflect sampling fluctuation and/or relationship to hypothetical processes unique to a given variable) for compactness of description by, essentially, throwing away $S$s' scores on the remaining PCs. Thus even if the PCA factor structure is to be subjected to a series of rotations in search of simple structure, the PCA will prove very useful to the researcher in providing some evidence on the total number of hypothetical, latent variables he needs to include in his final description of the structure of the original set of variables. It must be kept in mind, however, that this reduction in the dimensionality of the system may cost more in terms of the accuracy of subsequent analyses of the data than is immediately apparent from the small percentage of variance left unaccounted for by the "discarded" PCs. The linear compound which maximizes differentiation among the $S$s in terms of their scores on the original variables (that is, $PC_1$) will not in general be the same as the linear compound which correlates most highly with some criterion variable, or which maximally discriminates between various experimental or clinical groups. Thus, for instance, it is possible that $PC_4$ may correlate more highly with some criterion variable than does $PC_1$, so that its deletion from the full set of PCs would reduce $R^2$ by a considerably greater percentage than its share of the total variation among the $X$s.

### 6.3.3 Principal Component Analysis as a Means of Handling Linear Dependence

The computational procedures for multiple regression, $T^2$, Manova, and canonical analysis, all require the inversion of the covariance matrix of one or more sets of variables. When there is a linear relationship among the variables in the set—that is, when the covariance matrix $S$ is of lower rank than the number of variables—$|S| = 0$, and $S^{-1}$ is therefore undefined. In previous chapters we have recommended that such a situation be handled by deleting one or more of the variables involved in this linear dependence until the number of remaining variables equals the rank of $S$. A more elegant approach to this same problem would be to perform a PCA of the variables in each set involved, and then perform the multiple regression or other analysis of between-set relationship on the PCs, the number of which will match the rank of the original covariance matrix. The resultant measure of relationship ($R^2$ or $R_c$) and test of significance ($T$, gcr, or $F$) will be identical, no matter which approach (deletion of variables or preliminary PCA) is taken and no matter which variables from among the subset of linearly dependent variables are eliminated in the deletion procedure. A decision between the two approaches must therefore be made on one of two grounds:

(a)   computational convenience,

or

(b)   interpretability of the resulting regression weights, discriminant function, or canonical coefficients.

So far as computational convenience is concerned, the crucial question is whether several analyses involving this set of variables, or only one, are (is) to be performed. The additional work of computing PCs is worthwhile only if several analyses are to be carried out on these same data, thus leading to savings of computational effort every time the inverse or determinant of $S$ is required in a subsequent analysis.

In terms of interpretability, the crucial point is that *each* of the following sets of variables contains all of the information available in the full set of $p$ measures: the $s$ PCs; any $s$ "factors" resulting from rigid or oblique rotation of the $s$ PCs; the $s$ variables in the reduced set of outcome measures remaining after any $p - s$ of the variables involved in the linear dependence have been deleted. Thus any set of regression coefficients, any discriminant function, or any canonical variate obtained by beginning the analysis with any one of the above sets of $s$ variables can be translated without loss of information into a linear combination of any of the other sets of variables.

By way of illustration, consider the Manova you conducted in the Demonstration Problem in Chapter 4 on differences among the COMP, NMO, COOP, and IND groups in their behavior in the Prisoner's Dilemma game. (Cf. Data Set 3, pp. 68–69.) In that study the four outcome measures—CC, CD, DC, and DD—were constrained to add to 40 for each subject. This linear dependence was handled by deleting DD, leading to a gcr of 1.3775 and a first discriminant function of $D_1 = .832\text{CC} - .159\text{CD} + .532\text{DC}$. However, since $\text{CC} + \text{CD} + \text{DC} = 40 - \text{DD}$, an infinite number of equally valid discriminant functions, each leading to exactly the same gcr, can be obtained from the expression, $D_1 = (.832 - c)\text{CC} - (c + .159)\text{CD} + (.532 - c)\text{DC} + c\text{DD}$, where $c$ is any constant whatever. In particular, if $c = .832$, we obtain a discriminant function in which CC receives no *direct* weight in the discriminant function, all of its contributions having been "channeled through" or "absorbed into" the other three variables. As the reader can readily confirm, this is precisely the discriminant function we would have obtained had we applied the Manova procedure after deleting CC, rather than DD. Similarly, the expressions which result by taking $c = -.159$ or $c = .532$ are those which would have followed deletion of CD or DC, respectively. Thus there is no basis for the advantage sometimes claimed for preliminary PCA over deletion of variables, namely that PCA yields an expression which involves *all* of the original variables, since such an expression is readily obtained from the deletion procedure simply by taking into account the nature of the linear relationship among the measures.

It might be argued that PCA provides a means of discovering the fact of linear dependence and of specifying the nature of the dependence. This is true, since a perfect linear relationship among the $p$ variables will result in only $s < p$ nonzero eigenvalues of $S$, and the rows of the "factor pattern" resulting from PCA provide $p$ equations relating the $p$ variables perfectly to only $s$ unknowns, which set of equations can be solved in various ways to display the linear dependence among the original measures. However, the comments in Chapter 3 on the inadvisability of using the determinant of $S$ as a test for linear dependence apply to PCA, too. It is highly unlikely that random sampling from any multivariate population will yield a sample covariance matrix having a determinant of zero or fewer than $p$ nonzero characteristic roots, *unless* a linear dependence has been "built into" the set of variables by the researcher (or by his supplier of tests). Even in the latter case, roundoff errors or relatively small mistakes in transcribing or keypunching the data will lead to a nonzero determinant or to a full complement of $p$ PCs. However, this state of affairs *is* more likely to be detected upon examination of the PCA, since most of us have much stronger intuitive feeling for "proportion of variance accounted for" figures (for example, the last eigenvalue divided by the sum of all the characteristic roots) than for "near-zero" determinants—which may be $10^8$ or larger for "near-miss" covariance matrices.

At any rate, a preliminary PCA on the $E$ matrix from the Demonstration Problem in Chapter 4 yields

$$\text{PC}_1 = .757\text{CC} - .013\text{CD} - .099\text{DC} - .646\text{DD};$$

$$\text{PC}_2 = .398\text{CC} - .334\text{CD} - .635\text{DC} + .571\text{DD};$$

and

$$\text{PC}_3 = .134\text{CC} - .799\text{CD} + .580\text{DC} + .085\text{DD}.$$

(A $\text{PC}_4$ was computed by the HEINV program, but it accounted for only .0002% of the total variance.)

Note that we chose to perform the initial PCA on $E$, rather than on $H$ or on the total covariance matrix, ignoring groups $H + E$. Where the separation of the subjects into groups is, as in the present case, on the basis of some set of experimental manipulations, we would not wish to run the risk of confusing ways in which the variables are alike in their responsiveness to experimental treatments—as reflected in the **H**-matrix—with their tendency to covary within each group as a consequence of common genetic or developmental roots—as reflected in **E**. If, however, the groups are established on the basis of some organismic variable such as sex, occupational status, or broad age groupings, whether PCA is to be applied to **E** or to **H** + **E** will depend on the population to which the experimenter wishes to generalize his results. If he wishes his analysis to be relevant to subpopulations each of which is homogenous with respect to the variable used to assign *S*s to groups, the PCA should be applied to **E**. If instead he wishes it to be relevant to a more general population in which the current "independent" variable is allowed to vary naturally, **H** + **E** should be analyzed. None of these decisions will have any effect on the ultimate outcome of the Manova, since either set of PCs contains all of the information contained in the original variables, so there can be no difference in results. However, performing the PCA on **E** will yield the greatest computational savings in the subsequent Manova.

At any rate, we compute $\mathbf{H}_{pc}$ for the three principal components via the formula, $\mathbf{H}_{pc} = \mathbf{L'HL}$, where **L** is the factor pattern for the PCA (each column representing one of the PCs) and **H** is the original $4 \times 4$ hypothesis matrix. Of course $\mathbf{E}_{pc}$ is a diagonal matrix having the variances of the PCs on the main diagonal. The result is a gcr of 1.3775, as had been the case for the original analysis deleting DD, and a first discriminant function of

$$D_1 = .725PC_1 + .959PC_2 + .686PC_3$$

$$= .665CC - .577CD + .289DC - .377DD$$

$$= 1.042CC - .200CD + .666DC$$

in nonnormalized form. After normalizing so as to achieve a sum of squared coefficients equal to unity, we have

$$D_1 = .832CC - .159CD + .532DC,$$

as before. Similarly, we could convert our original discriminant function in terms of CC, CD, and DC into one involving $PC_1$, $PC_2$, and $PC_3$ by substituting in expressions for each original variable as a function of the three PCs.

Given the tedium of computing $\mathbf{H}_{pc}$, and the complete equivalence of the gcr and discriminant function produced by the deleted-variable and initial-PCA approaches to handling linear dependence, the initial-PCA approach would seem ill advised *unless*:

(a)   the researcher is interested in the results of the PCA in their own right, and would thus perform the PCA anyway;
(b)   the researcher plans several other analyses of the same set of data; or
(c)   the researcher expects the PCs or some rotation thereof to have such a clear and compelling substantive interpretation that the discriminant function expressed as a function of the PCs is more meaningful to him than any of the discriminant functions based on the original variables.

### 6.3.4  Single-Symbol Formulas for Simple Cases of PCA

We now present formulas which can be used to obtain principal components of $2 \times 2$, $3 \times 3$, and uncorrelated covariance matrices. These formulas were derived via procedures discussed in Digression 2.

### (a)  $2 \times 2$ Covariance Matrix

Where the covariance matrix is given by

$$\begin{bmatrix} a & c \\ c & b \end{bmatrix},$$

the characteristic equation is given by

$$\lambda^2 - (a + b)\lambda + (ab - c^2) = 0;$$

whence

$$\lambda = \left[ \frac{(a + b) + \sqrt{(a - b)^2 + 4c^2}}{2}, \frac{(a + b) - \sqrt{(a - b)^2 + 4c^2}}{2} \right],$$

and the characteristic vectors are $(-c/(a - \lambda), 1)$ in nonnormalized form, whence

$$\mathbf{b}_1' = (c, \lambda_1 - a)/\sqrt{c^2 + (\lambda_1 - a)^2},$$
$$\mathbf{b}_2' = (-c, a - \lambda_2)/\sqrt{c^2 + (a - \lambda_2)^2}$$

in normalized form.

If $a = b$, then the characteristic roots are $a + c$ and $a - c$, with corresponding characteristic vectors in normalized form of $(\sqrt{2}, \sqrt{2}) = (.707, .707)$ and $(\sqrt{2}, -\sqrt{2}) = (.707, -.707)$. Which root and vector correspond to the first principal component depends on the sign of $c$. If $c$ is positive, the sum of the two variables is the largest principal component; if $c$ is negative, the difference accounts for a greater portion of the variance.

### (b)  $3 \times 3$ Covariance Matrix

The covariance matrix

$$\begin{bmatrix} a & b & c \\ b & d & e \\ c & e & f \end{bmatrix}$$

yields the characteristic equation

$$\lambda^3 - (a + d + f)\lambda^2 + (ad + df + af - b^2 - c^2 - e^2)\lambda$$
$$- (adf - ae^2 - fb^2 - dc^2 + 2bce) = 0,$$

which is rather difficult to develop further algebraically. See Digression 3 for procedures for solving cubic equations.

If, however, $a = d = f$, the characteristic equation becomes

$$(\lambda - a)^3 - (b^2 + c^2 + e^2)(\lambda - a) - 2bce = 0.$$

The "discriminant" of this equation is $b^2 c^2 e^2 - (b^2 + c^2 + e^2)^3/27$, which is always negative. (Can you see why?) Thus a trigonometric solution is called for, namely,

$$(\lambda - a) = K \cos(\phi/3), \; K \cos(120° + \phi/3), \; K \cos(240° + \phi/3),$$

where

$$\cos \phi = \sqrt{27} bce/(b^2 + c^2 + e^2)^{3/2}$$

and

$$K = 2\sqrt{(b^2 + c^2 + e^2)/3} = \sqrt{b^2 + c^2 + e^2}/.86603.$$

Table D3.1 provides a convenient listing of the trigonometric functions involved in this solution.

If any one of the covariances (say, $b$) is equal to 0, then $\cos \phi = 0$, whence $\phi = 90°$, $\phi/3 = 30°$, and $(\lambda - a) = (.86603K, -.86603K, 0)$, whence

$$\lambda = a + \sqrt{c^2 + e^2}, a - \sqrt{c^2 + e^2}, a.$$

The characteristic vector corresponding to the largest characteristic root is then given by

$$[c/\sqrt{2c^2 + 2e^2}, e/\sqrt{2c^2 + 2e^2}, \sqrt{2}]$$

in normalized form. The other two principal components are given by

$$[-c/\sqrt{2c^2 + 2e^2}, -e/\sqrt{2c^2 + 2e^2}, .707] \quad \text{and} \quad [e/\sqrt{c^2 + e^2}, -c/\sqrt{c^2 + e^2}, 0].$$

In the equicovariance (or equicorrelation) case, where $a = d = f$ and $b = c = e$, we have $\lambda = a + 2b, a - b, a - b$, and the corresponding characteristic vectors are $(3^{-1/2}, 3^{-1/2}, 3^{-1/2}) = (.86603, .86603, .86603)$ and any two vectors orthogonal to (uncorrelated with) the first, for example, $((\frac{2}{6})^{1/2}, -(\frac{1}{6})^{1/2}, -(\frac{1}{6})^{1/2})$ and $(0, 2^{-1/2}, -2^{-1/2})$.

If you wish to employ the iterative procedure (cf. Section 4.6) for finding the characteristic vectors in the $3 \times 3$ case, it may be helpful to note that

$$\begin{bmatrix} a & b & c \\ b & d & e \\ c & e & f \end{bmatrix}^2 \quad \begin{bmatrix} a^2 + b^2 + c^2 & (a + d)b + ce & (a + f)c + be \\ & b^2 + d^2 + e^2 & (d + f)e + bc \\ (\text{symmetric}) & & c^2 + e^2 + f^2 \end{bmatrix},$$

that is, each main diagonal entry in the squared matrix is the sum of the squares of the entries in the corresponding row (or column) of the original matrix. The off-diagonal entry in a given row or column is calculated by adding the product of the other two off-diagonal entries in the original matrix to the product of the entry in the same position in the original matrix times the sum of the diagonal entry in the same row and the diagonal entry in the same column. Note that the square of a matrix with equal main diagonal entries will *not* in general have equal main diagonal entries.

## (c)  Uncorrelated $m \times m$ Matrix

The uncorrelated matrix

has characteristic roots equal to $a_{11}, a_{22}, \ldots, a_{mm}$.

The characteristic vector corresponding to the root $\lambda_j = a_{jj}$ has coefficients which are all zero except for the $j$th position, which equals unity.

## (d)  Equicovariance Case for $m \times m$ Matrix

If every main diagonal entry is equal to a common value $s^2$, and every off-diagonal entry is equal to a common value $s^2 r$, then the gcr, $\lambda_1 = s^2[1 + (m - 1)r]$, and the corresponding normalized characteristic vector is $(m^{-1/2}, m^{-1/2}, m^{-1/2}, \ldots, m^{-1/2})$. Each remaining root is equal to $s^2(1 - r)$, with vectors equal to any set which are orthogonal to the first characteristic vector.

### 6.3.5  Examples of PCA

***Example 6.1***  *Experimental games*  Flint, in his UNM doctoral dissertation study (1970), had each subject play each of of four different kinds of experimental games: Prisoner's Dilemma (PD), Chicken, Battle of Sexes (BS), and Apology. Half of the subjects (the fixed $S$s) had the same "partner" in all four games, while the other $S$s (the mixed $S$s) had a different partner for each game. Crossed with this mixed–fixed partner variable was the payoff procedure: either one mil per point earned (regular payoff) or one of seven graduated "prizes" (50¢, $1, $1.50, $2.00, $2.50, $3.00, or $3.50), depending on $S$'s ranking in terms of total points earned across all four games, relative to other $S$s occupying the same role (row player or column player)—the ranked payoff procedure. To summarize the four experimental conditions:

> FREG:    fixed partner, regular payoff procedure,
> FRANK:   fixed partner, ranked payoff procedure,
> MREG:    mixed partners, regular payoff procedure,
> MRANK:   mixed partners, ranked payoff procedure.

The main focus of the research was on the consistency of individual differences in behavior across games, and the amount of attenuation in between-game correlations produced by "mixing" partners. The analysis thus concentrated on canonical correlation techniques (cf. Chapter 5), but it also seemed worthwhile to explore the structure of the several available indices of behavior. These outcome variables included, for each game played by a pair, CC (the number of mutually cooperative responses), DD (mutually competitive responses), CD (the number of trials on which this player received less than his partner), and DC (the number of times $S$ received more than his partner), as well as $\rho_0$ the correlation between the two players' choices. A PCA was conducted separately on the correlation matrix for game 1 (Battle of

Sexes) under FREG conditions, and game 2 (Prisoner's Dilemma) under FREG conditions. The results are given in Table 6.2. In both cases, the fifth characteristic root was essentially zero—as it ought to be, since there is a linear relationship among CC, CD, DC, and DD, whence these four variables contribute only three linearly independent variables. (Listed in Table 6.2, incidentally, is the factor *structure* resulting from each analysis—the correlations between original variables and PCs—not the factor pattern, which includes the actual coefficients defining the PCs.)

*Table 6.2*

|  | FREG, Battle of Sexes | | | | FREG, Prisoner's Dilemma | | | |
|  | $PC_1$ | $PC_2$ | $PC_3$ | $PC_4$ | $PC_1$ | $PC_2$ | $PC_3$ | $PC_4$ |
|---|---|---|---|---|---|---|---|---|
| CC | .627 | .347 | .641 | −.274 | .985 | −.147 | −.089 | .016 |
| CD | −.602 | .751 | −.125 | .242 | −.697 | .346 | −.583 | −.236 |
| DC | −.699 | −.647 | .298 | .044 | −.766 | −.328 | .451 | −.320 |
| DD | .815 | −.163 | −.546 | .105 | −.846 | .310 | .212 | .378 |
| $\rho_0$ | .780 | −.109 | .226 | .573 | .550 | .791 | .375 | −.191 |
| $\lambda_j$ | 2.515 | 1.141 | .866 | .477 | 3.062 | .864 | .737 | .337 |

As can be seen from examination of the two tables, the major differences between game 1 and game 2 with respect to the first principal component of the relationship among these five variables are (a) the fact that $PC_1$ accounts for a larger percentage of the variance in Prisoner's Dilemma (61% vs 50%), and (b) the fact that DD correlates positively with $PC_1$ in Battle of Sexes but negatively in Prisoner's Dilemma. The problem of significance tests for such comparisons will be discussed later in this chapter. On a purely descriptive level, however, these two PCAs are fairly consistent with a game-theoretic analysis of the logical properties of the two games.

In BS, the only way for both Ss to achieve equal and satisfactory outcomes is for the pair to establish a pattern of alternation between CD and DC outcomes. Thus $PC_1$ for BS seems to reflect (negatively) the pair's overall understanding of the game, since pairs low on $PC_1$ would have the desirable characteristic of few CCs or DDs and many nonmatching choices (CD and DC outcomes). It is also true in BS that on any single CD or DC trial the payoffs to the two players are unequal (though better for each than the payoffs for CC or DD), so that CD − DC (which is essentially $PC_2$) represents S's willingness to accept (or inability to avoid) lower payoffs than his partner, and might be considered a "dominance" dimension. Essentially $PC_3$ is the difference between CC and DD, which, since the payoffs for a CC outcome are higher than those for a DD outcome, might differentiate between pairs who could not seem to achieve alternation but who at least had the "sense" to avoid the disastrous DD outcome versus those who lack even this much ability in social interaction. Finally, $\rho_0$ is increased by lock-ins on CC or DD, and decreased by CD or DC outcomes (it is the Pearson r between the two subjects' choices on each trial), so "successful" pairs should have a low value of $\rho_0$ and thus a low score on $PC_4$.

In PD, the CC outcome provides each player with his second best outcome, and DD with his second worst. However, S's best outcome is available only through DC, which entails partner's getting his worst outcome. On the other hand, CD gives $S$ his worst outcome and $S$'s partner his best outcome. The D response is thus a dominant response (better against whichever choice your partner makes) though C is better "in the long run" *if* you are altruistic and if your partner is trustworthy. It thus seems reasonable that $PC_1$ for this game would essentially be a comparison of CC versus any other outcome. Primarily $PC_2$ seems to reflect $\rho_0$, which will be high for any pair of $S$s who "lock in" on either CCs or DDs, while $PC_3$ represents DC − CD, which has the same dominance interpretation as in BS. There is no ready interpretation of $PC_4$ (which, after all, accounts for only about 7.0% of the total variance). More importantly, all of these interpretations of the PCs are clouded by the high amount of "noise" reflected by the nonzero loadings of each variable on each PC.

A second analysis on Flint's data treated all 20 variables (CC, CD, DC, DD, and $\rho_0$ for each of four kinds of game) as a single outcome vector. The analysis was based on the 20 × 20 intercorrelation matrix, with the correlations being computed on the *deviation* of each score from the mean on that variable for that experimental condition, so as to eliminate possible effects of differences attributable to the experimenter's manipulations. The results are given in Table 6.3. Interpreting this PCA is rather rough. The first PC is clearly a general "factor" which does not have an obvious pattern *except* that, with the exception of DD (which has all positive loadings on $PC_1$) the loadings for game 1 and 4 variables have the same signs, with the signs for game 2 and game 3 variables being just the opposite of the 1–4 signs. Thus one

*Table 6.3*

*PCA on 20-Variable Outcome Vector*

| Game | | $PC_1$ | $PC_2$ | $PC_3$ | $PC_4$ | $PC_5$ | $PC_6$ | $PC_7$ |
|---|---|---|---|---|---|---|---|---|
| Battle | CC | .101 | −.103 | .282 | −.440 | −.395 | .612 | −.040 |
| of sexes | CD | −.544 | .466 | −.304 | −.229 | −.246 | −.186 | .315 |
| | DC | −.099 | −.224 | .133 | .720 | −.236 | .378 | −.247 |
| | DD | .638 | −.250 | .080 | −.232 | .312 | −.445 | −.092 |
| | | .503 | .076 | .343 | −.401 | .534 | .001 | .051 |
| Prisoner's | CC | −.634 | −.156 | .665 | −.168 | −.174 | −.028 | −.149 |
| dilemma | CD | .387 | −.067 | −.464 | .012 | .109 | .460 | .113 |
| | DC | .462 | −.169 | −.418 | .138 | .104 | −.006 | −.102 |
| | DD | .474 | .340 | −.499 | .164 | .140 | −.197 | .203 |
| | | −.294 | .113 | .582 | −.114 | .057 | −.314 | .004 |
| Chicken | CC | −.586 | −.656 | −.338 | −.194 | −.014 | −.075 | .043 |
| | CD | .022 | .783 | .025 | −.214 | −.026 | .092 | .135 |
| | DC | .503 | .305 | .414 | .360 | −.058 | .234 | .118 |
| | | .604 | .057 | .276 | .272 | .110 | −.181 | −.343 |
| | | −.258 | −.625 | −.362 | −.042 | .312 | −.079 | −.118 |
| Apology | CC | .679 | −.255 | .064 | −.016 | −.443 | −.228 | −.036 |
| | CD | −.312 | −.175 | .247 | .494 | .421 | −.017 | .590 |
| | DC | −.400 | .506 | −.313 | −.248 | .129 | .125 | −.571 |
| | DD | .234 | −.396 | .131 | −.499 | −.320 | .269 | .274 |
| | | .665 | −.144 | .016 | −.343 | −.482 | .059 | .078 |

possible interpretation of $PC_1$ is that it is a contrast between behavior in games 1 and 4 (BS and Apology) and behavior in games 2 and 3 (PD and Chicken). This makes sense, since alternation is an attractive strategy in games 1 and 4, but not in games 2 and 3. Very tentatively it might further be suggested that the next four components tap behavior in games 3, 2, 1, and 4, respectively, since 59.6, 59.0, 48.3, and 46.1% of the contribution of the given component comes through variables reflecting behavior in that game. Beyond that, we are at a loss. The first five components account for only about 64% of the total variance, so clearly there is a great deal of variability left unexplained.

***Example 6.2***   *Attitudes toward cheating*   A questionnaire seeking to ascertain their attitudes toward cheating was sent to a random sample of UNM students. The even-numbered questions between 12 and 22 asked how often the respondent engaged in various practices (copying another student's assignments; copying someone's paper during a test; using crib notes during a test; plagiarizing papers; handing in someone else's research paper with his consent; stealing someone else's paper to hand in as your own), while the odd-numbered questions 13–23 asked how often the "typical UNM student" engaged in each of these same activities. Responses

*Table 6.4*

*PCA on Questions 12–23 of Cheating Questionnaire*

| Question (activity) | Correlation of resp to question with | | | | Sum of squared correlations with $PC_1$–$PC_4$ |
|---|---|---|---|---|---|
| | $PC_1$ | $PC_2$ | $PC_3$ | $PC_4$ | |
| 12  (Copy assignment?) | .321 | −.693 | −.161 | .008 | .609 |
| 14  (Copy test paper?) | .400 | −.354 | .502 | .467 | .754 |
| 16  (Crib notes during test) | .161 | −.522 | .535 | −.025 | .585 |
| 18  (Plagiarize?) | .332 | −.689 | −.155 | −.377 | .750 |
| 20  (Use other's paper?) | .342 | −.338 | −.481 | .177 | .495 |
| 22  (Stolen other's paper?) | .191 | −.142 | −.524 | .615 | .710 |
| 13 | .635 | .171 | .135 | .035 | .452 |
| 15 | .788 | .068 | .294 | .201 | .753 |
| 17  How often typical | .730 | .264 | .272 | −.203 | .718 |
| 19  UNM student | .702 | −.100 | −.296 | −.455 | .797 |
| 21  does these things | .679 | .410 | −.181 | −.060 | .666 |
| 23 | .624 | .411 | −.168 | .144 | .608 |
| Eigenvalue | 3.455 | 1.936 | 1.420 | 1.085 | 7.896 |
| % Variance accounted for | 28.8 | 16.2 | 11.8 | 9.1 | 69.8 |

to these twelve questions were subjected to a PCA based on the intercorrelation matrix, with the resulting factor structure listed partially in Table 6.4. This table reports only those PCs whose associated characteristic roots were greater than 1.0 (the average value for all 12). The remaining characteristic roots were:

Root:  .831  .731  .630  .527  .451  .380  .325  .229
%:    7.1   5.9   5.2   4.4   3.8   3.0   2.6   2.0

Thus the characteristic roots after the third show a very slow decline. We would certainly expect a distinction between the even-numbered variables (how often the respondent cheats in various ways) and the odd-numbered questions to emerge from any factor analysis of these scores, and this will be one of the criteria for judging adequacy of PCA as compared with other solutions. There is a hint of this distinction in this "component structure" in that the loadings on the first, general component are uniformly higher for odd-numbered than for even-numbered variables, and in that, with one exception, loadings on component number two are positive for all odd-numbered questions but negative for all even-numbered questions. The remaining two PCs are more difficult to interpret. Making the arbitrary decision to concentrate on loadings of .5 or better leads to a description of $PC_3$ as representing the difference between responses to questions 14 and 16 versus question 22 (and possibly question 20), which might be considered a difference between "public" and "private" cheating. Essentially $PC_4$ taps only responses to question 22.

## 6.4 Significance Tests for Principal Components

So far we have discussed PCs on a purely descriptive level. We have thus required no distinction between population and sample PCs and no assumptions about the nature of the distributions of the various variables. However, in using the PCs as an aid in describing the "structure" underlying a system of variables, or in "explaining" the covariance among these variables, we are usually hopeful that this description or explanation will be relevant to the *population* of observations from which our particular *sample* of observations was drawn. The number of inferential tests which have been developed for PCs is rather small.

Further, far fewer significance tests are available for PCs derived from a *correlation* matrix **R** than for PCs derived from a *covariance* matrix **S**. At first glance this may seem strange, since **R** is simply a special case of **S**, where the covariances are computed on standard scores rather than raw scores. Since the investigator is free to begin his PCA with whatever transformation(s) of his data seem most meaningful to him—raw scores, logarithms, squares, cube roots—it seems somewhat churlish to tell him that if he happens to choose transformation to standard scores he cannot use many of the significance tests which are otherwise available to him. However, there *is* a major difference between using, say, a logarithmic transformation on the data to take into account multiplicative relationships among variables, and converting all scores to standard scores.

Let $X_i$ be a single observation in a sample of $N$ observations from some population, $\bar{X}$ and $s$ being the sample mean and standard deviation of the $N$ observations. The $\log(X_i)$ involves the single variable $X_i$, while $z_i = (X_i - \bar{X})/s$ involves the *three* variables $X_i$, $\bar{X}$, and $s$. Large values of $\log(X_i)$ can arise only through large values of $X_i$, while large values of $z_i$ can arise if $X_i$ is large, $\bar{X}$ is small, or $s$ is small. Another way of phrasing this difference is to point out that we could specify, in advance of collecting any data, what value of $\log(X_i)$ would replace each value of $X_i$ which might be observed in the data, while we have no idea what value of $X_i$ will lead to a transformed score of $z_i$ until we have collected *all* of the data and computed $s$.

Somewhat more formally:

If $X_i$ is normally distributed, then so is $aX_i + b$, where $a$ and $b$ are preselected constants; and the distribution of $x_i^2$ has the same shape as a chi-square distribution with one degree of freedom. However, $z_i = x_i/s$ has a much more complex distribution which was derived by Cramer (1946), who also showed that a nonlinear transformation of $z_i$ has the $t$-distribution. The distribution of correlation-based PCs *is* therefore more complex than the distribution of covariance-based PCs. However, when no test appropriate to some situation is available for PCs obtained from PCA of an intercorrelation matrix, but is for covariance-matrix-derived PCs, the latter may be used as a rough guide under the rationale that you could have simply taken standard scores as your basic measures.

A second caution which must be noted with respect to the significance tests discussed below: they are all based on the assumption of large (say, $N - m > 30$) samples. Small-sample confidence intervals and $p$-values can be expected to be larger than these given here, though how much larger is not generally known. The following summary of available tests is patterned closely after that provided by Morrison (1967, Section 7.7).

## 6.4.1  Sampling Properties of Covariance-Based PCs

1.  $L_i$ (the $i$th *sample* characteristic root) is distributed independently of $\mathbf{b}_i$ (the $i$th sample characteristic vector).

2.  $\sqrt{df}\,(L_i - \lambda_i)$ is distributed as a normal distribution with mean zero and variance $2\lambda_i^2$, and is distributed independently of the other characteristic roots.

3.  $H_0 : \lambda_{q+1} = \lambda_{q+2} = \cdots = \lambda_{q+r}$, that is, the hypothesis that the $r$ intermediate characteristic roots are truly identical in terms of population values, can be tested by comparing

$$-df \sum_{q+1}^{q+r} \ln(L_j) + (df)r \ln[(\textstyle\sum L_j)/r]$$

with the desired critical value of the chi-square distribution having $\frac{1}{2}r(r+1) - 1$ degrees of freedom. The term "df" here and in test (2) above refers to the number of degrees of freedom going into each estimate of an $s_i$, and will usually be equal to the number of observations $-1$. Where, however, deviations scores are taken about each of several group means in obtaining a "pooled" covariance matrix which is independent of differences in group means on the different variables, $df = N - k$, where $N$ is the number of subjects and $k$ is the number of groups; "ln" stands for the natural logarithm, that is, the logarithm to the base $e = 2.78\ldots$. This test, developed by Anderson (1963), makes good intuitive sense, since it can be rewritten as

$$\chi^2 = (df)\ln(\bar{L}/\textstyle\prod_j L_j),$$

where $\bar{L}$ is the mean of the $r$ roots being tested for equality and $\prod_j L_j$ is the product of these roots. If the sample values are all equal, then $L_j = \bar{L}$ for all the relevant $j$, whence $\bar{L}^r$ identically equals $\prod L_j$. The farther the $L_j$s depart from equality, the greater will be the ratio between the $r$th power of their mean and their product. (The logarithm of 1 is, of course, zero.)

4.   The *confidence interval* about a given $\lambda_i$ is given by Anderson (1963) as

$$\frac{1_i}{1 + z_{\alpha/2}\sqrt{2/\text{df}}} < \lambda_i < \frac{1_i}{1 - z_{\alpha/2}\sqrt{2/\text{df}}}$$

where $z_{\alpha/2}$ is the $(1 - \alpha/2)$ percentile of the normal distribution and the confidence interval constructed is at the $(1 - \alpha)$-level of confidence. Thus, for instance, the 95% confidence interval for the first PC derived from a covariance matrix each of whose elements had 72 degrees of freedom would be between $.755L_1$ and $1.488$ $[= 1/(1 - 1.96\sqrt{2/72})]$ times the observed value of the first PC.

5.   $H_0 : \boldsymbol{\beta}_i = \boldsymbol{\beta}_{i0}$ ; that is, the null hypothesis that the characteristic vector associated with the distinct root $L_i$ is equal to a specific " null hypothesis " vector, is tested by comparing

$$\text{df}(L_i\boldsymbol{\beta}'_{i0}\, \mathbf{S}^{-1}\boldsymbol{\beta}_{i0} + (1/L_i)\boldsymbol{\beta}'_{i0}\, \mathbf{S}\boldsymbol{\beta}_{i0} - 2)$$

with the chi-square distribution having $m - 1$ degrees of freedom. This test makes good intuitive sense, since it is essentially equal to the number of degrees of freedom going into each covariance estimate, multiplied by $[w + (1/w) - 2]$, where $w$ is the ratio between the variance of the hypothesized linear combination of the original variables—$\boldsymbol{\beta}'_{i0}\, \mathbf{S}\boldsymbol{\beta}_{i0}$—and the variance of $\text{PC}_i$—$L_i$. If the two are equal, then $w = 1$ and $w + (1/w) = 2$. Any inequality, however, produces a value of $w + (1/w) > 2$. (For instance, $\frac{1}{2} + 2 = 2\frac{1}{2}$; $3 + \frac{1}{3} = 3.33$; etc.) This is a very useful test in determining whether the substantive interpretation of a particular principal component is viable, or whether the discrepancy between the coefficients which would support your interpretation perfectly and those actually obtained in the PCA cannot reasonably be attributed to chance fluctuation. For instance, $\text{PC}_1$ obtained from the PCA of the data for game 1, FREG condition (Example 6.1) was interpreted as "essentially" equal to $(CC + DD) - (CD + DC)$, which implies coefficients of .5 for CC and DD and $-.5$ for CD and DC. The test described in this paragraph can be used to quantify the "essentially" by assigning a specific value to the probability that the differences between these "ideal" coefficients and the ones actually obtained are due solely to "chance" fluctuation.

### 6.4.2   Sampling Properties of Correlation-Based PCs

1.   $H_0$ : last $m - 1$ characteristic roots are equal, which is equivalent to

$$H_0 : \rho_{ij} = \rho \qquad \text{for all} \quad i \neq j,$$

can be tested by comparing

$$(\text{df}/\hat{\lambda}^2)\left[\sum_i \sum_j (r_{ij} - \bar{r})^2 - \hat{\mu}\sum_{k-1}(\bar{r}_k - \bar{r})^2\right]$$

with the chi-square distribution having $\frac{1}{2}(m - 1)(m - 2)$ degrees of freedom; where

$$\bar{r}_k = \left(\sum_{i \neq k}^{m} r_{ik}\right)/(m - 1)$$

is the mean of the correlations in row $k$ of $\mathbf{R}$;

$$\bar{r} = \left(\sum_{i>j}\sum r_{ij}\right)/[m(m-1)/2]$$

is the mean of all the correlations in $\mathbf{R}$;

$$\hat{\lambda} = 1 - \bar{r}; \quad \text{and} \quad \hat{\mu} = (m-1)^2(1-\hat{\lambda}^2)/[m-(m-2)\hat{\lambda}^2].$$

This test developed by Lawley (1963) essentially involves a comparison of the variability of row means with the variability among all the correlation coefficients, ignoring which variables are involved.

  2.  $H_0$: first $p$ principal components provide a perfect description of the population factor structure, the remaining PCs having population values of zero, is tested by comparing

$$\chi^2 = (\text{df} - 2m/3 - p/3 - 5/6)\ln(|\hat{\mathbf{R}}|/|\mathbf{R}|) \qquad (6.7)$$

with the chi-square distribution having $\frac{1}{2}[(m-p)^2 - m - p]$ degrees of freedom, where $\hat{\mathbf{R}}$ is the correlation matrix "reproduced" from the loadings on the first **p** PCs only, and $\mathbf{R}$ is the observed correlation matrix. The ratio between $|\hat{\mathbf{R}}|$ and $|\mathbf{R}|$ will of course approach 1 (and the natural logarithm of this ratio, zero) as the goodness of fit of the reproduced correlations to the observed correlations increases. The natural logarithm of the ratio of the two determinants can be approximated for very large samples by the sum of the squares of the discrepancies between observed and reproduced correlations. This test was actually developed (by Lawley, 1940, and Bartlett, 1954) for testing the adequacy of the fit of the maximum-likelihood factor solution involving $p$ common factors, and significant findings must therefore be interpreted cautiously, since both minres and maximum-likelihood factor solutions (Chapter 7) provide better fits than selection of the same number of PCs from a PCA. A nonsignificant result, however, definitely indicates that including additional PCs is unnecessary to obtain a fit which is "perfect" within the limits of sampling fluctuation.

  3.  $H_0$: $\rho_{ij} = 0$ for all $i, j$; that is, there is no structure to be explained, was tested in Chapter 2.

## 6.5  Rotation of Principal Components

  As was pointed out earlier, the uniqueness of the coefficients derived in PCA is achieved by requiring a descending order of importance (percentage of variance accounted for) among the PCs. This will generally be quite useful in providing evidence as to how many latent variables need be assumed in order to account for a sizeable percentage of the variance in the system of original variables. Indeed, Kaiser (1960) has argued that a whole host of criteria involving both statistical and practical considerations suggest the number of PCs having associated characteristic roots greater than $\sum s_i^2/m$ ($= 1$ when the PCA is performed on $\mathbf{R}$) as the best single criterion for the number of factors to be assumed in *any* analysis of structure, whether

PCA or one of the multitude of factor analysis models we will discuss in Chapter 7. However, this hierarchical structure may (and usually does) correspond quite poorly to the researcher's preconceptions (theoretical commitments?) as to the nature of the latent variables which might be producing the intercorrelations among the observable variables. He might, for instance, feel that all latent variables should be of roughly equal importance, or he might feel that each original variable should be "produced" by a relatively small number of latent "determinants." Once the variance-maximization criteria of PCA are abandoned, an infinitude of alternative sets of uncorrelated linear combinations of the original variables become available to the researcher. Fortunately, each of these alternative "factorizations" is related to the others (including PCA) by a set of operations known equivalently as "orthogonal transformations" or "rigid rotations." To provide a concrete framework within which to develop definitions of and formulas for these techniques, let us consider an earlier example in somewhat greater detail.

**Example 6.3**  *Known generating variables*    Assume that our observable variables, $X_1$ and $X_2$, are generated as:

$$X_1 = H_1 + H_2, \qquad X_2 = H_1 + 3H_2,$$

where $r_{H_1 H_2} = 0$, and each latent variable has a unit normal distribution.
    Then

$$s_1^2 = \sum (h_1 + h_2)^2 = 2; \qquad s_{12} = \sum (h_1 + h_2)(h_1 + 3h_2) = 1 + 3 = 4;$$

and

$$s_2^2 = \sum (h_1 + 3h_2)^2 = 10; \qquad \text{that is,} \quad \mathbf{S} = \begin{bmatrix} 2 & 4 \\ 4 & 10 \end{bmatrix},$$

whence, applying PCA to $\mathbf{S}$, we obtain

$$\lambda = \tfrac{1}{2}[12 \pm \sqrt{8^2 + 4(4^2)}] = 6 \pm 5.655 = 11.656, .344,$$

whence

|        | \multicolumn{2}{c}{S-Derived Factor Pattern} | \multicolumn{2}{c}{S-Derived Factor Structure[a]} |
|--------|---------|---------|---------|---------|
|        | $PC_1$  | $PC_2$  | $PC_1$  | $PC_2$  |
| $X_1$  | .3828   | −.9239  | .9239   | −.3828  |
| $X_2$  | .9239   | .3828   | .9974   | .0721   |
| $s_{pc}^2$ | 11.656 | .344   | 1.848   | .152    |

[a] $r_{X_i PC_j} = b_{ij} \sqrt{\lambda_j} / s_i$.

Further,

$$\mathbf{R} = \begin{bmatrix} 1 & \sqrt{.8} \\ \sqrt{.8} & 1 \end{bmatrix} = \begin{bmatrix} 1 & .894 \\ .894 & 1 \end{bmatrix},$$

whence, via PCA, we obtain

|  | R-Derived Factor Pattern | | R-Derived Factor Structure | |
|---|---|---|---|---|
|  | $PC_1^*$ | $PC_2^*$ | $PC_1^*$ | $PC_2^*$ |
| $X_1$ | .707 | −.707 | .973 | .231 |
| $X_2$ | .707 | .707 | .973 | −.231 |
| $\sum r_{X_i PC_j}^2$ | 1.893 | .107 | 1.893 | .107 |

In addition to these two PCA-derived factor structures, we have the "true" structure, which we determine as

$$h_1 + h_2 = x_1, \qquad h_2 = (x_2 - x_1)/2,$$

$$h_1 + 3h_2 = x_2, \qquad h_1 = x_1 - h_2 = (3x_1 - x_2)/2,$$

$$r_{x_1 h_1} = \sum x_1 h_1 / \sqrt{\sum x_1^2 \cdot \sum h_1^2} = (3s_1^2 - s_{12})/\sqrt{s_1^2(9s_1^2 - 6s_{12} + s_2^2)} = 1/\sqrt{2}$$

$$= .707, \qquad \text{etc.}$$

whence

|  | "True" Pattern | | "True" Structure | |
|---|---|---|---|---|
|  | $H_1$ | $H_2$ | $H_1$ | $H_2$ |
| $X_1$ | .9486 | −.707 | .707 | .707 |
| $X_2$ | −.3162 | .707 | .3162 | .9486 |
| $\sum r_{X_i H_j}^2$ | | | .6 | 1.4 |

## 6.5.1   Basic Formulas for Rotation

Each of the three factor structures can be plotted on a graph having the latent variables as axes (dimensions) (Figs. 6.1). However, examination of these figures shows (allowing for some numerical inaccuracies) that the relationship between variables $X_1$ and $X_2$ remains constant, with the cosine of the angle between them being equal to $r_{12} = .894$. The only differences among the three factor structures lie in the orientation of the two references axes. We thus ought to be able to find a means of expressing the effects of these "rotations." In the general case of an observed variable plotted as a function of two latent variables $L_1$ and $L_2$ and then redescribed after rotation through an angle $\theta$ in terms of $L_1^*$ and $L_2^*$, we have Fig. 6.2. (Note that in plotting a variable, the length of the vector connecting it to the origin is equal to its "communality," that is, the sum of the squares of its loadings on the factors of the factor structure is equal to 1 when all PCs are retained.) By consideration of the triangle formed by points 1, 2, and 3 and the 234 triangle, we see that $b_1^* = \overline{12} + \overline{45} = \cos \theta \, b_1 + \sin \theta \, b_2$ (where $\overline{ij}$ is the length of the line segment connecting points

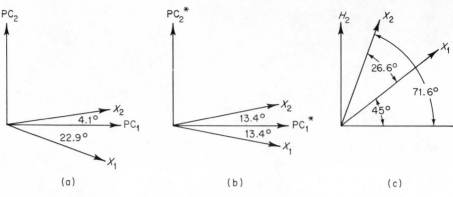

**Figure 6.1.** Factor structures, Example 6.3.

$i$ and $j$); and $b_2^* = \overline{34} - \overline{23} = \cos \theta\, b_2 - \sin \theta\, b_1$. Put in matrix form, this becomes

$$[b_1^* \quad b_2^*] = [b_1 \quad b_2] \begin{bmatrix} \cos \theta & -\sin \theta \\ \sin \theta & \cos \theta \end{bmatrix} = [b_1 \quad b_2]\mathbf{T}.$$

The reader may readily verify that the second two factor structures can indeed be obtained from our PCA of the covariance matrix by rotations of $-9.5°$ and $67.9°$, respectively, that is, by transformation

$$\begin{bmatrix} .973 & .226 \\ -.226 & .973 \end{bmatrix} \quad \text{and} \quad \begin{bmatrix} .384 & .923 \\ -.923 & .384 \end{bmatrix}.$$

We could of course pick any other value of $\theta$ which gives us a "desirable" factor structure. For example, we might wish to have a zero loading of one variable on one of our hypothetical variables and zero elsewhere. This would be accomplished by a $\theta$

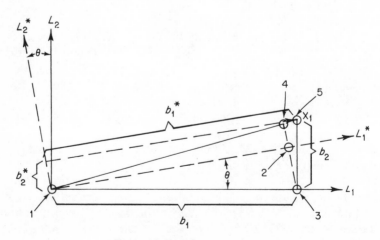

**Figure 6.2.** Rotation, general case.

of $-22.9$ and a transformation matrix $\mathbf{T}$ of

$$\begin{bmatrix} .9206 & .3884 \\ -.3884 & .9206 \end{bmatrix},$$

leading to a factor structure of

| | "Simple" Factor Structure | |
|---|---|---|
| | $H_1^*$ | $H_2^*$ |
| $X_1$ | 1.000 | .000 |
| $X_2$ | .891 | .454 |
| $\sum r_{X_i H_j}^2$ | 1.794 | .206 |

Note, incidentally, that in general $\mathbf{TT}' = \mathbf{T'T} = \mathbf{I}$, which is the defining property of an *orthogonal matrix*.

When we move to three or more dimensions, it should be fairly clear that we can rotate any pair of axes while holding all other axes fixed, thus changing the loadings of the original variables only on that particular pair of latent variables. The matrix expression of this process of pairwise rotation of reference axes is

$$\mathbf{T} = \mathbf{T}_1 \cdot \mathbf{T}_2 \cdot \mathbf{T}_3 \cdot \cdots \cdot \mathbf{T}_r,$$

where each orthogonal matrix on the right is simply an $m \times m$ identity matrix with the exception of the entries

$$t_{ii} = \cos \theta_{ij}, \qquad t_{ji} = \sin \theta_{ij}, \qquad t_{ij} = -\sin \theta_{ij}, \qquad \text{and} \qquad t_{jj} = \cos \theta_{ij},$$

where $\theta_{ij}$ is the angle through which axes $i$ and $j$ are rotated; and $\mathbf{T}$ is the single transformation matrix which, when premultiplied by the original factor structure $\mathbf{P}$, yields the new factor structure $\mathbf{P}^*$ which is the result of the $r$ [usually equal to $p(p-1)/2$, where $p$ is the number of PCs or factors retained] pairwise rotations of axes. If in fact some of the original PCs are "ignored" and thus kept fixed throughout the various rotations, the rows and columns corresponding to these "dropped" dimensions can be deleted, making $\mathbf{T}$ and each component $\mathbf{T}_i$ only $p \times p$ $(p < m)$. For any large number of "retained" PCs, the process of rotating to obtain a factor structure which is easy to interpret is quite time consuming, since $p(p-1)/2$ pairs of axes must be considered. The researcher pursuing the "intuitive-graphical" approach to rotation will find an extensive table of trigonometric functions, augmented by a slide rule with trig scales, extremely helpful.

### 6.5.2   Objective Criteria for Rotation

This section draws heavily on Harman's (1967, Chapter 14) summary of analytic methods. The indeterminacy we have seen in the positioning of our reference axes, and thus in the configuration of "loadings" (correlations) of the observable variables on the latent variables, has been (probably needlessly) a sore point with factor analysts for many years. To a mathematician there is no indeterminacy, since the configuration of vectors representing the relationships among the variables is the same, regardless of how we choose to orient the axes we use in obtaining a numerical expression of this geometric configuration. However, most psychologists are at least as interested in describing the axes (dimensions) themselves as in describing the relations among the original variables. Concern over the issue of rotation of axes was probably heightened by the extravagant claims of early factor analysts that factor analysis yields information about *the* "true" sources of intercorrelations among the observable variables, that is, about the true, fundamental dimensions underlying responses in some situation or class of situations. A plethora of versions of the truth is rather damaging to such claims; it was therefore important to these pioneers in the field to develop principles for selecting just one of the infinite number of possible solutions as *the* solution. We know now, of course, that these claims were based on a fundamental misconception about the nature of scientific explanations. PCA and FA, like any other techniques for the construction of models for the explanation of empirical data, can at best establish the plausibility of one of many alternative explanations of a given set of data. However, there is no harm in choosing the positions of our reference axes in such a way as to make the task of providing substantive interpretations of the hypothetical variables as simple as possible, so long as we recognize that factor structures having "simple structure" are no more (nor less) valid than any of the other factor structures which are obtainable by rotation of our initial PCA. No consideration of simple structure will, for instance, generate the "true" structure underlying Example 6.3. (We might, however, postulate—as Einstein has and as is done implicitly whenever we use maximum-likelihood statistical techniques—that Nature does not play tricks on us and that all "true" natural laws are simple ones. This would then justify putting greater faith in "simple structure" solutions than in alternative factor structures.)

There are (not necessarily unreasonably or even unfortunately) many kinds of simplicity. In order to make the search for simple structure an objective process, rather than a subjective art, it is necessary to specify exactly what criteria for and/or measures of simplicity are to be employed. The first person to attempt an explicit definition of "simple structure" was Thurstone (1947). He phrased his criteria (as have all subsequent researchers) in terms of the loadings withinn the $m \times p$ rotated factor structure, as follows:

(1)   Each row of the factor structure should contain at least one zero. (This sets as a minimum criterion of simplicity of description of the variables the criterion that not all of the hypothetical variables should be required to describe a given original variable.)

(2)   Each column of the factor structure should contain at least $p$ zeros. (This sets a minimum condition on the simplicity of our description of a given latent variable in terms of the original variables.)

(3)   Every pair of columns should contain several original variables whose loadings vanish in one column but not in the other. (This criterion aids in distinguishing between the two hypothetical variables in terms of their relationship to the original variables.)

(4)   If the number of factors (or retained PCs) is four or more, every pair of columns of the factor structure should contain a large number of responses with zero loadings in *both* columns. (This is useful in distinguishing this particular pair of latent variables from the other latent variables.)

(5)   For every pair of columns, only a small number of original variables should have nonzero loadings in both columns. [This is essentially a rephrasing of condition (4).]

There are two major drawbacks to Thurstone's criteria:

(a)   they can almost never be satisfied by any set of real data, and
(b)   no objective measures are provided of how far we are from satisfying the criteria.

Thurstone's criteria have, however, provided a general framework for other authors' search for numerically specifiable criteria for simple structure.

Around 1953 or 1954, several authors, starting from slightly different conceptualizations of simple structure, arrived independently at the *quartimax* criterion. To see how these apparently different criteria lead to precisely the same results, we need to consider the fact that the *communality* of each original variable (the portion of its variance it shares with other variables in the system, which is equal to the sum of the squares of that variable's loadings on the various hypothetical variables as long as these latent variables are uncorrelated) remains constant as the reference axes are rotated. If the rotation is from an initial PCA, the communalities will of course be 1.0 for all variables, but latent variables generated by factor analysis, or factor structures based on only some of the PCs from a PCA, yield communalities less than 1.0. In a graphical plot of a factor structure, the communality of a variable is mirrored by the length of its vector, which is clearly unaffected by the orientation of the reference axes. Formally, then

$$h_j^2 = \sum_i L_{ji}^2 = \text{constant} \qquad \text{for each variable } X_j, \qquad (6.8)$$

where $L_{ji}$ is the correlation of variable $j$ with latent variable $i$, that is, is the entry in the $j$th row, $i$th column of the factor structure. However, if this is constant across rotations, then so is the sample variance of the communalities, namely

$$\sum (h_j^2 - \overline{h_j^2})^2 = \sum \sum (h_j^2)^2 - (\sum h_j^2)^2/p = \sum \sum L_{ji}^4 + 2 \sum_{\substack{u=1 \\ u>v}}^{p} \sum_{v=1}^{p} (\sum L_{ju}^2 L_{jv}^2) - p(\overline{h_j^2})^2$$

$$= Q + CP - \text{constant}; \qquad (6.9)$$

that is, the variance (actually, sum of squared deviations from the mean communality) of the communalities is decomposable, in a manner reminiscent of the decomposition of total variance in Anova into within- and among-group sums of squares, into a term $Q$ which is the sum of the fourth powers of the loadings of the original variables on the latent variables and a term $CP$ which is the sum, over all pairs of columns of the factor structure, of the sums of cross-products of the loadings in each pair of columns. Since the sum of $Q$ and $CP$ is a constant, maximization of $Q$ (which is what was proposed by Ferguson, 1954; Newhaus and Wrigley, 1954; and Saunders, 1953) is equivalent to minimization of CP (which was proposed by Carroll, 1953). Any rotation scheme which attempts to minimize *CP* or maximize *Q* is known as a *quartimax* rotation.

Maximization of *Q* can be seen as desirable from several different viewpoints. In the first place, $Q = \sum \sum L_{ji}^2/mp$ (the subtractive term being a constant) is equal to the variance of all *mp* squared loadings, and will be large to the extent that all loadings are close either to zero or to one, which certainly tends to simplify the interpretative problem. Secondly, $Q - \sum_j (\sum_i L_{ji}^2)^2/p$ (the subtractive term again being a constant) is equal to the sum of the variances of the squared loadings within each row, which will be large to the extent that each variable has a near-unity loading on one or two latent variables and near-zero loadings on all other latent variables. Finally, *Q* is also linearly related to the kurtosis of the distribution of squared loadings, which will again be high to the extent that there is a sharp differentiation between high and low loadings within the factor structure. (Kurtosis is, for a unimodal distribution, a measure of its "peakedness," that is, the tendency for observations to cluster both tightly around the mean and broadly in the tails.)

Minimization of *CP* can be seen to be desirable on two closely related grounds. First, the description of a given variable in terms of a pair of latent variables (that is, as plotted against only these two latent variables) approaches maximum simplicity when one of the two axes passes directly through (is colinear with) the vector defining the variable, since this produces a zero loading on the other latent variable. This situation would be mirrored numerically by a zero product of the two loadings, and the magnitude of the product of the squared loadings would therefore be an indication of how close we were to colinearity of the vector defining the original variable and one of the reference axes. Therefore *CP*, which is simply the sum of all such cross-products in the factor structure, is an overall measure of how close we have come to the ideal situation in which each "variable vector" is colinear with one of the *p* reference axes. Also *CP* is the sum across all pairs of columns of the sums of cross-products of loadings within pairs of columns. Each such pairwise sum of cross-products is linearly related to the covariance of the loadings in the two columns involved. Thus *CP* is an overall measure of the redundancy of our descriptions of the latent variables in terms of the original variables. To the extent that *CP* approaches zero, we can readily differentiate between pairs of latent variables by examining the pattern of their loadings on the original variables.

Despite the multiplicity of desirable properties of quartimax rotations, however, the quartimax criterion currently runs far behind the popularity of the *varimax* criterion for rotation. The reason for this is most clearly seen through consideration of the definition of *Q* in terms of the variances of the loadings within *rows* of the matrix. Maximizing *Q* is thus perfectly compatible with a factor structure in which a large portion of the variance is accounted for by a *general factor* on which each and every variable has a very high loading—which is precisely the feature of the factor structure produced by PCA which most researchers find unacceptable and which led to consideration of the possibility of rotating this initial solution in the first place. Kaiser (1956, 1958) proposed maximizing instead the variance of the loadings within *columns* of the matrix. This of course precludes the emergence of a strong general factor, and tends to produce a factor structure in which each latent variable contributes roughly the same amount of variance. Specifically, the *raw varimax* criterion is

$$V^* = (1/m^2) \sum_i^p \left[ m \sum_j^m L_{ji} - \left( \sum_j^m L_{ji}^2 \right)^2 \right]. \tag{6.10}$$

Note that the sum of the squared loadings within a given *column* of the matrix is *not* constant across rotations, and the sum across all columns of these column sums of squared loadings is *not* the same as the sum of the squared communalities. This criterion is called the *raw varimax criterion* because Kaiser has proposed a correction to each loading in the factor structure before carrying out the rotation. The raw varimax procedure tends to give equal weight to variables having very low communalities and those having near-unity communalities. Kaiser finds this undesirable, and corrects for this tendency by dividing each loading within a given row of the factor structure by the square root of the communality of that variable. After the rotation process has been completed the effects of the "Kaiser normalization" are removed by multiplying each loading in the rotated structure by the square root of the communality of the variable described in that row. A varimax rotation employing Kaiser normalization of the loadings is referred to as *normal varimax* or simply *varimax* rotation.

Computationally, varimax rotation is simplified by the fact that the angle of rotation for any given pair of reference axes which produces the maximum increase in $V$ (the normalized varimax criterion) can be solved for explicitly, rather than requiring trial-and-error search. Specifically, for the latent variables $r$ and $s$, the optimal angle of rotation $\phi_{rs}$ is given by

$$\tan(4\phi) = \frac{4[m \sum d_{rs} c_{rs} - (\sum c_{rs})(\sum d_{rs})]}{m \sum d_{rs}^2 - (\sum d_{rs})^2 - 4m \sum c_{rs}^2 + 4(\sum c_{rs})^2}, \tag{6.11}$$

where

$$d_{rs} = v_{jr}^2 - v_{js}^2$$

is the difference between the squared normalized loadings of variable $j$ on factors $r$ and $s$;

$$c_{rs} = v_{jr} v_{js}$$

is the product of the loadings of variable $j$ on the two factors; and

$$v_{ji} = L_{ji} / \sqrt{h_j^2}.$$

The right-hand side of Eq. (6.11) seems to be related to the correlation between $d_{rs}$ and $2c_{rs}$, but the author has been unable to move from this astute observation to a convincing intuitive rationale for the reasonableness of (6.11). (Note that each summation is over the $m$ variables from $i = 1$ through $i = m$.) Because of the cyclic nature of $\tan \phi$, there will be two choices of $4\phi$ which satisfy (6.11). Which to select is determined by the sign of the numerator and of the denominator of (6.11) as indicated in Table 6.5. (Table 6.5 is derived from a consideration of the second derivative of $V$, the normalized varimax criterion, with respect to $\phi$.) A "large iteration" of the varimax rotation scheme requires that latent variable (factor) 1 be rotated with factor 2; the new factor 1 is then rotated with original factor 3; etc. until all $p(p-1)/2$ pairs of factors have been rotated. Kaiser has shown that each large iteration increases the value of $V$, and that $V$ has an upper bound of $(p-1)/p$, so that the rotation scheme eventually converges. In practice, rotations are continued until the difference between $V$ for the large iteration just completed and the value of $V$ after the preceding large iteration falls below some criterion, or until the rotation angles recommended for the next large iteration all fall below some minimum value.

Kaiser has presented evidence that normalized varimax rotation satisfies the very desirable criterion of "factorial invariance," that is, the independence of the definition of a set of latent variables in terms of a battery of original variables from the "contaminating" effects of the particular other variables included in the set on which the initial factor solution is obtained.

More specifically, Kaiser proves mathematically that where the vectors representing the original variables fall into two colinear clusters (that is, two factors provide sufficient explanation of the variation among the original variables), the angle of rotation satisfying (6.11) is independent of the number of variables. While he was unable to provide a mathematical proof beyond the two-factor case, he does provide a rather convinving empirical demonstration of the invariance property for a case with four factors and 24 variables, which he adds in several stages, beginning with six variables.

Incidentally, an expression similar to (6.11) is available for quartimax rotation, namely

$$\tan(4\phi_q) = 4\sum c_{rs}d_{rs}/(\sum(d_{rs}^2 - 4c_{rs}^2),\tag{6.12}$$

where the subscript on $\phi$ serves as a reminder that this is the angle of rotation for the quartimax rotation scheme, and where the appropriate quadrant for $4\phi$ is given by Table 6.5.

Finally, Lawley and Maxwell (1963) have outlined least-squares methods for rotating to factor structures which come as close as possible to an *a priori* pattern of ones and zeros. The researcher having strong theoretical bases for predicting the factor structure would do well to consult the work of Lawley and Maxwell (1963), the practical significance of whose work has been greatly increased by Joreskög's (1967, 1969, 1971) development of efficient computer programs for such "confirmatory factor analysis."

*Table 6.5*

*Quadrant within Which $4\phi$ Must Fall as Function of Signs of Numerator and Denominator of Expression (6.11)*

|  |  | Sign of numerator | |
|---|---|---|---|
|  |  | + | − |
| Sign of | + | $0° \le 4\phi \le 90°$ | $270° \le 4\phi \le 360°$ |
| denominator | − | $90° \le 4\phi \le 180°$ | $180° \le 4\phi \le 270°$ |

### 6.5.3   Examples of Rotated PCs

To illustrate the very simplest case of rotated principal components, we shall apply formulas (6.11) and (6.12) to the "known-generator" case, Example 6.3. With just two hypothetical variables, these formulas produce the desired solution in just one step, no iteration being required. Since it does not matter which of the several structures generated by rotation of the initial PC solution we use, we will choose to work with the intuition-based "simple structure" solution, since the first row (loadings of $X_1$) is so simple. Table 6.6 gives the required computations. Based on

*Table 6.6*

*Intermediate Calculations for Quartimax and Varimax Rotation*

| $j$ | $2L_{j1}L_{j2} = 2c_j$ | $L_{j1}^2 - L_{j2}^2 = d_j$ | $4c_j^2$ | $d_j^2$ | $2c_jd_j$ | $d_j^2 - 4c_j^2$ |
|---|---|---|---|---|---|---|
| 1 | 0 | 1 | 0 | 1 | 0 | 1 |
| 2 | .8090 | .5878 | .6545 | .3455 | .4755 | −.3090 |
| Sum | .8090 | 1.5878 | .6545 | 1.3455 | .4755 | .6910 |

these intermediate computations, we have as the angle of rotation for a quartimax solution,

$$\tan(4\phi) = 2(.4755)/.6910 = 1.3763,$$

whence

$$4\phi_q = 53.96° \quad \text{and} \quad \theta_q = 13.49°,$$

whence

$$\sin\theta_q = .23345 \quad \text{and} \quad \cos\theta_q = .97237,$$

and

$$\mathbf{T} = \begin{bmatrix} .972 & -.233 \\ .233 & .972 \end{bmatrix},$$

whence we have

|  | Quartimax factor structure | |
|---|---|---|
|  | $H_1$ | $H_2$ |
| $X_1$ | .972 | −.233 |
| $X_2$ | .972 | .233 |
| $\sum r_{X_jH_j}^2$ | 1.890 | .109 |

This is identical (except for computational error in the third digit) to the correlation-derived (though not to the covariance-derived) PCA, as can be confirmed by applying formula (6.12) to that factor structure. This might lead us to suspect that the PCA of a two-variable system will always have optimally simple structure in the quartimax sense if it is based on the intercorrelation matrix.

Applying formula (6.11), we obtain

$$\tan(4\phi) = 2[2(.4755) - (1.5878)(.8090)]/[2(1.3455) - (1.5878)^2 - 2(.6545) + (.8090)^2]$$
$$= -2(.33353)/(-.48463) = 1.3764,$$

whence

$$4\phi = -126.04° \quad \text{and} \quad \phi = -31.51°, \quad \sin\phi = -.52250, \quad \cos\phi = .85264;$$

whence

$$T = \begin{bmatrix} .853 & .5225 \\ -.5225 & .853 \end{bmatrix},$$

whence we have

|  | Varimax factor structure | |
|---|---|---|
|  | $H_1$ | $H_2$ |
| $X_1$ | .8526 | .5225 |
| $X_2$ | .5225 | .8526 |
| $\sum r^2_{X_i H_j}$ | 1.0000 | 1.00000 |

The varimax rotation has certainly produced level contribution of factors, though the resulting structure seems in this case to be less interpretable than the quartimax (or PCA) solution. In terms of the varimax simplicity criterion, the varimax structure produces a value of

$$[(.727 - .5)^2 + (.273 - .5)^2] \times 2 = .206,$$

as compared with values of .0207, .0000, and .0424 for the S-derived PCA, the R-derived PCA, and the intuitive simple solution, respectively.

As an example of rotation of a larger factor structure, $PC_1$–$PC_4$ (those having eigenvalues greater than 1.0) from the PCA of the cheating questionnaire (Example 6.2) were subjected to a varimax rotation, yielding the rotated structure of Table 6.7. For comparison purposes, the entire, twelve-component factor structure was also subjected to varimax rotation, producing as the first four factors, the structure described in Table 6.8.

Considering first the results of rotating only the four most important PCs, we find that varimax rotation has accentuated the difference between one's own cheating and perception of others' cheating. Now the first and most important (26% of total variance) factor involves only others's cheating, the even-numbered questions having near-zero correlations with factor 1. The other factors are not as easy to label. In

*Table 6.7*

*Varimax Rotation of $PC_1$–$PC_4$,*
*Cheating Questionnaire*

| Question | $PC_1$ | $PC_2$ | $PC_3$ | $PC_4$ |
|---|---|---|---|---|
| 12 | −.047 | −.661 | .299 | .282 |
| 14 | .181 | −.017 | .835 | .157 |
| 16 | −.075 | −.292 | .661 | −.238 |
| 18 | −.011 | −.856 | .132 | −.004 |
| 20 | .111 | −.421 | −.019 | .552 |
| 22 | .039 | −.014 | .014 | .841 |
| 13 | .643 | −.048 | .187 | .023 |
| 15 | .727 | −.047 | .463 | .089 |
| 17 | .792 | −.096 | .159 | −.238 |
| 19 | .588 | −.643 | −.193 | −.007 |
| 21 | .785 | −.032 | −.184 | .120 |
| 23 | .725 | .096 | −.098 | .251 |
| $\sum r^2$ | 3.112 | 1.871 | 1.597 | 1.315 |
| % Total | 25.9 | 15.6 | 13.3 | 10.9 |

general, perceptions of others's cheating have quite low loadings on factors 3 and 4. Another way of saying this is that reports of one's own cheating behavior are more "factorially complex" than perceptions of others's cheating. At any rate, based on examination of the variables having loadings greater than .5 (in absolute value) on a given factor, factor 2 (own copying of assignments, own plagiarizing of another's paper, others's plagiarizing) seems to reflect abstention from cheating on assignments (except that the low loading of variable 13 does not fit this description); factor 3

*Table 6.8*

*Varimax Rotation of All Twelve PCs,*
*Cheating Questionnaire*

| Question | $PC_1$ | $PC_2$ | $PC_3$ | $PC_4$ |
|---|---|---|---|---|
| 12 | −.049 | −.224 | .903 | .085 |
| 14 | .031 | −.068 | .956 | .020 |
| 16 | −.036 | −.102 | .166 | −.020 |
| 18 | −.031 | −.935 | .073 | .032 |
| 20 | .070 | −.089 | .050 | .113 |
| 22 | .045 | −.028 | .018 | .984 |
| 13 | .123 | −.005 | .077 | .009 |
| 15 | .120 | −.025 | .283 | .017 |
| 17 | .217 | −.038 | .060 | −.056 |
| 19 | .237 | −.252 | −.037 | .015 |
| 21 | .889 | .037 | .039 | .059 |
| 23 | .228 | .038 | .021 | .114 |
| $\sum r^2$ | 0.876 | 1.017 | 1.051 | 1.010 |
| % Total | 8.2 | 8.5 | 8.8 | 8.4 |

seems to reflect own cheating on tests (though the .46 loading for variable 15 suggests that we might be able to include others's cheating on exams in the factor), and factor 4 seems to reflect one's own frequency of particularly flagrant cheating (handing in another's paper, and especially stealing someone else's paper).

An initial glance at the results of rotating all 12 PCs (Table 6.8) as compared with rotation of only the first four PCs (Table 6.7), suggests that the two are quite different; however, closer inspection reveals that each variable which has the highest loading on that factor in one matrix also has the highest loading on that factor in the other matrix. Further, the rank order among the variables having high loadings is quite consistent across the two structures. In general, the pattern which results from rotating the full twelve-component structure is quite similar to the pattern which results from rotating only the first four components, except that in the former the highest loading within a given column is greatly accentuated at the "expense" of other loadings in the column. The extent of this accentuation is indicated by the fact that within each column of the full twelve-factor matrix (only the first four columns of which are reproduced in Table 6.9) there is one loading which is at least three times as large in absolute magnitude as the next largest loading, the smallest of these "largest loadings" being .825. There is a strong tendency for rotation of a complete component structure to lead to a "matching up" of factors with original variables, as indicated in Table 6.9, a list of largest loadings. This tendency will of course obscure more subtle patterns such as those offered as interpretations of the results displayed in Table 6.7. Note, too, the extremely level contributions of the 12 factors. In fact, the factor we suspect to be the most meaningful (factor 1) has the third lowest sum of squared loadings (total contribution to variance accounted for) of the 12 factors. Probably, too, if we wished to use this structure to estimate the communalities of the variables, we should simply sum the loadings for a given variable *without* its loading on the factor on which it has its highest loading (which is really just measuring itself). These are going to be rather low communalities. Such are the results of overfactoring.

Note that we could have, had we chosen to do so, listed all 12 hypothetical

Table 6.9

Large Loadings for Cheating Questionnaire

| Factor: | 1 | 2 | 3 | 4 | 5 | 6 | 7 | 8 | 9 | 10 | 11 | 12 |
|---|---|---|---|---|---|---|---|---|---|---|---|---|
| Largest loading on factor: | .889 | .935 | .956 | .984 | −.970 | −.945 | −.969 | .944 | −825 | .919 | .876 | .885 |
| Next largest loading: | .237 | −.252 | .283 | .114 | −.141 | −.251 | −.173 | .231 | −.221 | .230 | .234 | .213 |
| Variable having highest loading: | 21 | 18 | 14 | 22 | 20 | 13 | 16 | 12 | 15 | 23 | 19 | 17 |
| Sum of (loading)$^2$: | .988 | 1.017 | 1.051 | 1.010 | 1.019 | 1.061 | 1.014 | 1.004 | .846 | 1.040 | .949 | 1.000 |

variables (four rotated PCs and eight unrotated PCs) in a single factor structure after we had rotated the first four PCs. Applying rotations to the first four PCs only has not in any way affected the relationship between the original variables and the last eight PCs, and we can at any time "restore" perfect reproducibility of the original intercorrelation matrix by adding the unrotated PCs to our "prediction" equations. The difference between rotation of the full PCA factor structure and rotation of only the "important" PCs is that the latter procedure must attempt to simplify the descriptions of all $m$ PCs, while the former can concentrate on simplifying the descriptions of those $p$ PCs which account for most of the variance and are thus most worth describing.

### 6.5.4 Individual Scores on Rotated PCs

In Section 6.2 we showed, through multiple regression analysis of the relationship between PCs and original variables, that scores on PCs would be perfectly "predicted" from scores on the original variables, and that the coefficients of the regression equations for computing scores on $PC_i$ were simply the entries in column $i$ of the factor pattern (or the entries in that column of the factor structure multiplied by $s_j/\lambda_i$). The situation after rotation of the PCs is not quite so simple. It is still true that the multiple correlation between each rotated PC and the original variables is 1.0. (The reverse is *not* true. A given original variable will generally *not* be perfectly reproducible from scores on the rotated PCs. How can this be? Remember that we will generally *not* rotate all of the PCs.) However, the entries of the factor loading matrix (that is, of the factor structure) may be quite different from the weights given the different original variables in the formulas relating rotated PCs to original variables. Harking back to our known-generator problem, for instance, we note that all four correlations between an original variable and a latent variable are positive, while the formulas which would be used to compute $h_1$ and $h_2$ from scores on $x_1$ and $x_2$ are $h_1 = 3x_1 - x_2$ and $h_2 = x_2 - x_1$. This discrepancy should come as no surprise after our experience with multiple regression and the great difficulty in predicting the regression a variable will receive on the basis of its simple correlation with the outcome variable.

In the known-generator problem we started with the factor pattern (whose columns give the coefficients to be used in computing each latent variable from scores on the original variables) and constructed the factor structure (whose entries are latent variable–original variable correlations) from this information. Generally, of course, we shall have to work in the opposite direction. It would therefore be helpful to have available a set of formulas for computing the factor pattern for rotated PCs from the factor structure.

One approach to developing such formulas is to consider the original variables to be a set of predictor variables from which we wish to "predict" scores on one (and eventually all) of the rotated PCs. We saw in Chapter 2 that all the information needed to develop a regression equation for predicting standard scores on $Y$ from $z$-scores on a set of $X$s were the intercorrelations among the predictors and between each predictor and $Y$. In our present case, the predictor–predicted correlations for predicting a particular rotated PC are simply the loadings in the corresponding column of the rotated factor structure.

Thus, for instance, the regression coefficients for predicting $z$-scores on $H_1$ (the first of the "true" generating variables) from Example 6.3 are obtained by computing

$$\mathbf{b}_{h_1} = \mathbf{R}^{-1}\mathbf{r}_{x,h_1} = \begin{bmatrix} 1 & \sqrt{.8} \\ \sqrt{.8} & 1 \end{bmatrix}^{-1} \begin{bmatrix} .707 \\ .3162 \end{bmatrix} = \frac{1}{.2}\begin{bmatrix} 1 & -.894 \\ -.894 & 1 \end{bmatrix} \begin{bmatrix} .707 \\ .3162 \end{bmatrix} = \begin{bmatrix} 2.121 \\ -1.580 \end{bmatrix}.$$

For $H_2$, we have

$$\mathbf{b}_{h_2} = \frac{1}{.2}\begin{bmatrix} 1 & -.894 \\ -.894 & 1 \end{bmatrix} \begin{bmatrix} .707 \\ .9486 \end{bmatrix} = \begin{bmatrix} -.707 \\ 1.580 \end{bmatrix}.$$

Since the procedures for computing $\mathbf{b}_{h_1}$ and $\mathbf{b}_{h_2}$ have in common premultiplication by $\mathbf{R}^{-1}$, we could have solved for the two sets of coefficients somewhat more compactly by writing

$$\mathbf{B}_h = [\mathbf{b}_{h_1}\ \mathbf{b}_{h_2}] = \mathbf{R}^{-1}\mathbf{F} = \mathbf{R}^{-1}[\mathbf{r}_{x,h_1}\ \mathbf{r}_{x,h_2}], \tag{6.13}$$

that is, simply by premultiplying the (rotated) factor structure by $\mathbf{R}^{-1}$.

Note that the coefficients we obtain are those employed in the $z$-score regression equation, that is,

$$z_{h_1} = 2.121z_1 - 1.580z_2 = (2.121/\sqrt{2})X_1 - (1.580/\sqrt{10})X_2$$
$$= 1.5X_1 - .5X_2;$$

and

$$z_{h_2} = -.707z_1 + 1.580z_2 = -.5X_1 + .5X_{2'}.$$

Since we generated $X_1$ and $X_2$ in the first place on the basis of their relationships to two unit normally distributed hypothetical variables, we know that these are also the raw-score formulas for $H_1$ and $H_2$. More generally, we would have to multiply both sides of the expression relating $h_i$ to the $X$s by $s_{h_i}$ to obtain an expression for $H_i$, the raw score on hypothetical variable $i$; that is,

$$\mathbf{H}_i = s_{h_i}\mathbf{X}\mathbf{b}_{h_i},$$

where $\mathbf{H}_i$ is a column vector of scores on hypothetical variable (rotated PC) $i$. However, we *cannot* ordinarily determine $s_{h_i}$—nor, fortunately, do we need to know the absolute magnitudes of subjects' raw scores on the hypothetical variables.

Note, too, that

$$R^2_{h_1 \cdot x} = .707(2.125) + .3162(-1.580) = 1.000,$$

and

$$R^2_{h_2 \cdot \mathbf{x}} = -.707(.707) + .949(1.58) = .999,$$

as we might have expected from the fact that there was an exact relationship between PCs and $X$s before rotation. We shall discover when we discuss factor analysis that this same basic multiple regression technique can be used to *estimate* factor scores on the basis of the factor structure and $\mathbf{R}$, but the resulting $R^2$ values will be *less* than unity.

The multiple regression approach to deriving scores on rotated PCs is a straight-forward application of what is hopefully a very familiar technique. However, it also requires the inversion of the full $p \times p$ matrix of intercorrelations among the original variables, thus negating to a large degree one of the purported advantages of PCs, their computational convenience. Given the perfect relationship between rotated PCs and original variables, we might suspect that there would be some computa-tionally simpler method of computing scores on the rotated PCs. In particular, since "rotation" simply involves a set of linear transformations of the intercorrelations between PCs and $X$s, we might expect rotation to have the effect of carrying out a similar set of transformations on the PCs themselves. As it turns out (and as we shall shortly prove), this is almost correct. Rigid rotation of a PCA-derived factor structure—that is, postmultiplication of $\mathbf{F}$ by $\mathbf{T} = \mathbf{T}_1 \cdot \mathbf{T}_2 \cdot \cdots \cdot \mathbf{T}_r$—has the effect of postmultiplying $\mathbf{Z}_{pc}$—the $N \times m$ matrix of standard scores on the $m$ PCs—by $\mathbf{T}$ as well, and thus of postmultiplying $\mathbf{Z}_x$—the matrix of standard scores on $X$—by $\mathbf{B}_z \cdot \mathbf{T}$, where $\mathbf{B}_z$ is the standard-score factor pattern from the original PCA. In other words,

$$\mathbf{Z}_h = \mathbf{Z}_{pc} \cdot \mathbf{T} = \mathbf{Z}_x \cdot \mathbf{B}_z \cdot \mathbf{T}; \tag{6.14}$$

where the matrices are as defined above. Thus, returning to Example 6.3 (the known-generator problem), we have, for scores on the "true" hypothetical variables (whose correlations with the original variables are obtained by rotating the S-based PC axes $-67.5°$),

$$\mathbf{Z}_h = \mathbf{Z}_{pc} \cdot \begin{bmatrix} .3827 & .924 \\ -.924 & .3827 \end{bmatrix} = \mathbf{Z}_x \cdot \begin{bmatrix} .1586 & -2.2277 \\ .8558 & 2.0639 \end{bmatrix} \begin{bmatrix} .3827 & .9239 \\ -.9239 & .3827 \end{bmatrix}$$

$$= \mathbf{Z}_x \cdot \begin{bmatrix} 2.119 & -.706 \\ -1.579 & 1.581 \end{bmatrix} = \mathbf{X} \cdot \begin{bmatrix} 1.498 & -499 \\ -.499 & .500 \end{bmatrix}.$$

Note that the entries of the standard-score factor pattern are obtained by multiplying each entry in the S-derived factor pattern by $s_i/\sqrt{\lambda_j}$. Be sure you understand why.

※ Now, the proof—which can, of course, be skipped by trusting readers. We shall prove Eq. (6.14) by showing that its application

(a)   yields $z$-scores on the rotated PCs which are uncorrelated, and
(b)   yields correlations between the original variables and the PCs which are identical to the correlations contained in the rotated factor structure.

The proof shall actually be given only for a single pair of PCs (indeed, only the first two, since the only property of the PCs essential to the proof is their zero intercorrelations). Since part of the proof establishes that the scores on the rotated PCs are uncorrelated, too, the proof can easily be extended to the scores obtained after rotating a second pair of axes—including possibly one of the new axes obtained from the just-rotated pair of PCs—etc. First, consider the correlation between $h_1$ and $h_2$, the rotated axes corresponding to the first two PCs. Applying (6.14), we have

$$z_{h_1} = \cos\theta\, z_{pc_1} + \sin\theta\, z_{pc_2};$$

and

$$z_{h_2} = -\sin\theta\, z_{pc_1} + \cos\theta\, z_{pc_2}.$$

Thus

$$
\begin{aligned}
r_{h_1,\,h_2} &= \sum z_{h_1 h_2}/(N-1)\\
&= [1/(N-1)]\sum[(\cos\theta\, z_{pc_1} + \sin\theta\, z_{pc_2})(-\sin\theta\, z_{pc_1} + \cos\theta\, z_{pc_2})]\\
&= [1/(N-1)]\sum[-\sin\theta\cos\theta\, z_{pc_1}^2 + (\cos^2\theta - \sin^2\theta)z_{pc_1}z_{pc_2}\\
&\quad + \sin\theta\cos\theta\, z_{pc_2}^2].
\end{aligned}
$$

However,

$$\sum z_{pc_1}^2 = \sum z_{pc_2}^2 = N-1, \quad \text{and} \quad \sum z_{pc_1}z_{pc_2} = (N-1)r_{pc_1,pc_2},$$

so that we have

$$r_{h_1,h_2} = (\cos^2\theta - \sin^2\theta)r_{pc_1,pc_2} = 0,$$

since we know that any two PCs are uncorrelated.

Next, consider the correlation between, say, $X_1$ and $h_1$:

$$
\begin{aligned}
r_{x_1,\,h_1} &= \sum z_1 z_{h_1}/(N-1) = [1/(N-1)]\sum z_1(\cos\theta\, z_{pc_1} + \sin\theta\, z_{pc_2})\\
&= [1/(N-1)][\cos\theta\sum z_1 z_{pc_1} + \sin\theta\sum z_1 z_{pc_2}]\\
&= \cos\theta\, r_{x_1,\,pc_1} + \sin\theta\, r_{x_1,\,pc_2}.
\end{aligned}
$$

This final expression is exactly the expression for the loading of $X_1$ on the first rotated PC, that is, the entry in the first row, first column of the rotated factor structure. The extension to other $X$s and the second rotated PC should be obvious. It should also be clear from the above why it is *not* true that rotation through an angle $\theta$ leads to the same transformation being applied to the *raw score* factor pattern. For example, $\sum pc_1^2 \neq \sum pc_2^2$, so that we would have had a nonzero term, $\sin\theta\cos\theta(\sum pc_2^2 - \sum pc_1^2)$, in the numerator of the expression for $r_{h_1,h_2}$.

## DEMONSTRATION PROBLEM

A.   Do the following "by hand." (Use of a desk calculator is all right, as is doing part C first.)

1.   Conduct a PCA on the *covariance* matrix of the predictor variables employed in the Demonstration Problem in Chapter 2. Report the factor pattern and the factor structure separately.

**2.** Compute each $S$'s score on each of the PCs computed in Problem 1. Now compute the sample variance on scores on $PC_1$, $PC_2$, and $PC_3$. To what previously computed statistics do these variances correspond?

**3.** Using the $Ss$' scores on the PCs, compute all intercorrelations among the PCs and between the PCs and the original variables. To what do these intercorrelations correspond?

**4.** "Interpret" the PCs obtained in Problem 1. For each interpretation (for example, "$PC_2$ is simply the sum of $X_1$, $X_2$, and $X_3$") specify the "ideal" characteristic vector implied by that interpretation, and conduct a significance test to determine whether the actually obtained characteristic vector can reasonably be interpreted as being a random fluctuation from that ideal characteristic vector.

**5.** Test the hypothesis that the second and third characteristic vectors are identical except for sampling fluctuation.

**6.** Use loadings on $PC_1$ alone to "reproduce" the covariance matrix. Test the null hypothesis that the discrepancies between observed and "reproduced" covariances are due solely to random fluctuation.

**7.** Repeat Problem 6 on the basis of loadings on $PC_1$ and $PC_2$.

**8.** Repeat Problem 6 on the basis of loadings on $PC_1$, $PC_2$, and $PC_3$.

**9.** Why must the significance tests in Problems 4–8 be interpreted with extreme caution?

**10.** Conduct a multiple regression analysis (MRA) of the relationship between the outcome variable and scores on $PC_1$, $PC_2$, and $PC_3$. How does this compare with the results of the MRA you performed before on the relationship between $Y$ and $X_1 - X_3$?

**11.** Conduct a MRA of the relationship between the outcome variable and scores on $PC_1$ and $PC_2$. How does this compare with the results of Problem 10 and with the results of the MRA you performed before on the relationship between $Y$ and $X_1 - X_3$?

**12.** Conduct a MRA of the relationship between $Y$ and $PC_1$. How does this compare with Problems 10 and 11?

**13.** Use graphical methods to rotate the factor structure obtained in Problem 1 to what you consider the most easily interpretable structure.

**14.** Do a varimax rotation of $PC_1$, $PC_2$, and $PC_3$. Compare with Problem 13.

**15.** Do a varimax rotation of $PC_1$ and $PC_2$. Compare with Problems 13 and 14.

**16.** Compute a score for each $S$ on each of the three hypothetical variables obtained in Problem 13. Compute the variance of each hypothetical variable. Could you have predicted these variances in advance?

**17.** Compute a score for each $S$ on each of the two hypothetical variables obtained in Problem 15. Compute the variances. Could you have predicted them in advance?

**B.** Repeat the analyses of part A, beginning with the intercorrelation matrix.

**C.** Perform as many of the analyses in parts A and B as possible using available computer programs.

## Answers

**A.**

**1.** Note that the first covariance-derived principal component accounts for 64% of the variation in *raw scores*, but only 50% of the variation in *standard scores*.

| | Factor pattern | | | | Factor structure | | |
|---|---|---|---|---|---|---|---|
| | $PC_1$ | $PC_2$ | $PC_3$ | | $PC_1$ | $PC_2$ | $PC_3$ |
| $X_1$ | −.2643 | −.3175 | .9107 | | −.7006 | −.6208 | .3521 |
| $X_2$ | .1211 | .9259 | .3579 | | .1741 | .9819 | .0750 |
| $X_3$ | .9568 | −.2049 | .2062 | | .9873 | −.1560 | .0310 |
| $\lambda_j$ | 17.567 | 9.559 | .3737 | $\sum L^2$ | 1.4959 | 1.3739 | .1306 |
| % accounted | .639 | .348 | .0136 | % variance accounted for | .499 | .458 | .0435 |

**2.** On computing the variance of each column of numbers, we find that (except for roundoff errors) the variance of principal component *i* is equal to the eigenvalue associated with that principal component.

| | Score on | | |
|---|---|---|---|
| Subject | $PC_1$ | $PC_2$ | $PC_3$ |
| A | .6911 | .4629 | 6.9503 |
| B | −.5949 | 3.9679 | 6.7009 |
| C | 7.7962 | −1.2623 | 6.2144 |
| D | 3.0550 | −2.3211 | 7.7638 |
| E | 8.7871 | 4.6106 | 7.4511 |

**3.** The correlations between PCs are all zero, while the correlations between *X*s and PCs are given by the entries in the factor structure.

| | $X_1$ | $X_2$ | $X_3$ | $PC_1$ | $PC_2$ | $PC_3$ |
|---|---|---|---|---|---|---|
| $X_1$ | 1 | −.70502 | −.58387 | −.7005 | −.6298 | .3521 |
| $X_2$ | | 1 | .02111 | .1741 | .9819 | .0750 |
| $X_3$ | | | 1 | .9873 | −.1560 | .0310 |
| $PC_1$ | | | | 1 | 0 | 0 |
| $PC_2$ | | | | | 1 | 0 |
| $PC_3$ | | | | | | 1 |

**4.** First, taking the extremely simple hypotheses that $\mathbf{B}'_1 = (0\ 0\ 1)$, $\mathbf{B}'_2 = (0\ 1\ 0)$, and $\mathbf{B}'_3 = (1\ 0\ 1)$, we have

$$\chi^2 = -4[16.5/17.567 + 17.567(171/1004) - 2]$$

$= 7.725$     with   4 df,               ns           for the first $H_0$;

$$\chi^2 = -4[8.5/9.559 + 9.559(435/1004) - 2]$$

$= 12.123$     with   4 df,             $p \cong .05$       for the second $H_0$;

and

$$\chi^2 = -4[2.5/.3737 + .3737(2243/1004) - 2]$$

$= 22.099$     with   4 df,             $p < .001$      for the third $H_0$.

Since the major discrepancy occurs in our description of the $PC_3$, we might be tempted to modify only that description and retest for significance. However, this is not really legitimate, since we know that PCs are uncorrelated and we thus cannot change our description of $PC_3$ without also changing our predictions for $PC_1$ and $PC_2$ so as to keep them orthogonal to $PC_3$. This suggests that we should really consider ourselves as constructing a hypothesized factor pattern. That then suggests that we could simply see what covariance matrix is implied by our hypothesized factor pattern, and test that against the observed covariance matrix. The significance test described in Section 6.4.2, however, applies to correlation matrices, not covariance matrices (though we recommended simply substituting $\mathbf{S}$ and $\hat{\mathbf{S}}$ for $\mathbf{R}$ and $\hat{\mathbf{R}}$ when you wanted to test that sort of hypothesis). The correlation matrix implied by our initial descriptions is simply a $3 \times 3$ diagonal matrix, that is, corresponds to the prediction that all off-diagonal correlations are equal to zero in the population. However, we have a test for that, namely

$$\chi^2 = (\text{complicated constant}) \times \ln(1/|\mathbf{R}|) = -1.5 \ln(.1789)$$

$$= 2.58 \quad \text{with} \quad 3 \text{ df,} \quad \text{ns.}$$

This seems inconsistent with the results of our tests of the separate predictions until we realize that those predictions involve not only predictions about interrelationships among the original variables but also hypotheses about the *variances* of the original variables. If we test the hypothesis, for instance, that the variance of variable 1 is truly .3737 in the population (which is what our hypothesized factor pattern implies), then we obtain

$$\chi^2 = 4(2.5)/(.3737) = 26.76 \quad \text{with} \quad 4 \text{ df,} \quad p \ll .01.$$

Of course, if the assumption of no nonzero intercorrelations were true, then the population value of the third characteristic root would be 2.5, not .3737, and our "reproduced" value of $s_1^2$ would be identically equal to its observed value.

At any rate, trying out different sets of hypothesized PCs leads finally to the conclusion that we must use coefficients which reflect the relative magnitudes of the actual PC coefficients if we are to avoid significant chi-square values especially for $PC_3$. Remember, however, that these tests are all large-sample approximations and probably grossly underestimate the true probabilities of our hypotheses.

**5.** $q = 1, r = 2; H_0 : \lambda_2 = \lambda_3$.

$$\chi^2 = -4[\ln(9.559) + \ln(.3737)] + 4(2) \ln[(9.559 + .3737)/2]$$

$$= 4 \ln[4.966^2/(9.559(.3737))] = 4(1.932)$$

$$= 7.728 \quad \text{with} \quad 2 \text{ df,} \quad p < .05.$$

Thus we can reject the hypothesis that $PC_2$ and $PC_3$ have identical eigenvalues in the population.

6.  The reproduced covariance matrix is $PDP'$, where $P$ in this case is merely a column vector consisting of the coefficients of $PC_1$. Thus

$$\hat{S} = \begin{bmatrix} -.2643 \\ .1211 \\ .9568 \end{bmatrix} \begin{bmatrix} 17.567 \end{bmatrix} \begin{bmatrix} -.2643 & .1211 & .9568 \end{bmatrix} \begin{bmatrix} 1.227 & -.562 & -4.442 \\ & .258 & 2.036 \\ & & 16.082 \end{bmatrix}$$

Using the test in Eq. (6.7), p. 177 with $|\hat{R}|/|R|$ replaced with $|\hat{S}|/|S|$ whence

$$\chi^2 = -(5 - 3/3 - 2/3 - 11/6)\ln(.00013/16.5) \simeq \infty.$$

It becomes a little clearer why the hypothesis that one PC is sufficient can be so soundly rejected when we consider the predicted correlation matrix, namely

$$\hat{R} = \begin{bmatrix} 1 & -.999 & -1 \\ & 1 & .999 \\ & & 1 \end{bmatrix}.$$

*Any* deviation from a correlation of unity is of course highly significant, since unity correlations are by definition perfect, and any deviation from perfection permits rejection of the null hypothesis.

7.  Based on $PC_1$ and $PC_2$:

$$\hat{S} = \begin{bmatrix} -.2643 & -.3175 \\ .1211 & .9259 \\ .9569 & -.2049 \end{bmatrix} \begin{bmatrix} 17.567 & \\ & 9.559 \end{bmatrix} \begin{bmatrix} -.2643 & .1211 & .9569 \\ -.3175 & .9259 & -.2049 \end{bmatrix} \begin{bmatrix} 2.191 & -3.372 & -3.821 \\ & 8.452 & .222 \\ & & 16.482 \end{bmatrix};$$

$$\hat{R} = \begin{bmatrix} 1 & -.783 & -.635 \\ & 1 & .019 \\ & & 1 \end{bmatrix}$$

$$\chi^2 = -1.5 \ln(|\hat{R}|/|R|) = 5.87 \quad \text{with} \quad (9 + 3)/2 = 6 \text{ df}, \quad \text{ns}.$$

8.  Perfect reproduction, $\chi^2 = 0$.

9.  The very small sample size.

10.  In terms of $z$-scores, $z_y = .87053 z_{pc_1} + 44984 z_{pc_2} + .12319 z_{pc_3}$;

$$R^2 = \sum r^2_{y,pc_i} = .97536, \quad \text{whence} \quad R = .9876.$$

This matches perfectly (well, almost perfectly) the results from the MRA using the original scores. We can get back to the regression equation based on original scores by substituting into the above regression equation the definitions of $z_{pc_i}$ in terms of the original scores.

11.  $S = \begin{bmatrix} 17.567 & 0 \\ 0 & 9.560 \end{bmatrix}$; $b = \begin{bmatrix} .210 \\ .147 \end{bmatrix}$; $b^* = \begin{bmatrix} .871 \\ .450 \end{bmatrix}$.

$$R^2 = .961, \quad \text{whence} \quad R = .980.$$

**12.** $R^2 = .758$; $R = .871$.

**13.** The three pairwise plots are shown in Figs. 6.3. Try $\theta_{12} = -6.6°$, $\theta_{13} = +2.8°$, $\theta_{23} = +1.8°$.

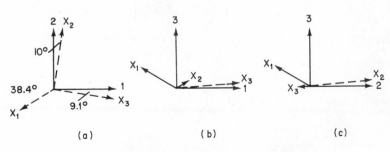

*Figure 6.3.* Rotation of factor structure from Problem 1.

Thus

$$\mathbf{T}_{12} = \begin{bmatrix} .993 & .115 & 0 \\ -.115 & .993 & 0 \\ 0 & 0 & 1 \end{bmatrix}; \quad \mathbf{T}_{13} = \begin{bmatrix} .999 & 0 & -.049 \\ 0 & 1 & 0 \\ .049 & 0 & .999 \end{bmatrix}; \quad \mathbf{T}_{23} = \begin{bmatrix} 1 & 0 & 0 \\ 0 & 1.0 & -.031 \\ 0 & .031 & 1.0 \end{bmatrix};$$

$$\mathbf{T} = \mathbf{T}_{12} \cdot \mathbf{T}_{13} \cdot \mathbf{T}_{23} = \begin{bmatrix} .992 & .113 & -.053 \\ -.115 & .993 & -.025 \\ .049 & .031 & .999 \end{bmatrix};$$

whence the rotated structure is

$$\mathbf{FT} = \begin{bmatrix} -.606 & -.685 & .404 \\ .063 & .997 & .041 \\ .999 & -.042 & -.017 \end{bmatrix}.$$

The varimax criterion (sum of variances of squared loadings within each column) is .8938 for this rotated structure, as compared with .9049 before rotation!

**14.** Since each variable's communality is 1.000, Kaiser normalization is superfluous (since it requires that we divide the loadings within each row by the square root of the communality of that variable). The calculations for the first large iteration went as follows:

For $PC_1$ and $PC_2$:

| $d = L_1^2 - L_2^2$ | $c = L_1 L_2$ |
|---|---|
| .10576 | .43532 |
| −.93404 | .17087 |
| .94983 | −.15397 |

$$\tan(4\theta_{12}) = \frac{4[3(-.26141) - (.12065)(.45222)]}{3(1.78580) - (.12065)^2 - 4(3)(.24241) + 4(.4522)^2}$$

$$= -3.355/3.252 = -1.03173,$$

whence

$$\theta_{12} = -11.475°,$$

whence

$$\mathbf{T}_{12} = \begin{bmatrix} .98 & .199 & 0 \\ -.199 & .98 & 0 \\ 0 & 0 & 1 \end{bmatrix}$$

and the new structure is

| $H_1$ | $H_2$ | $PC_3$ |
|---|---|---|
| −.5635 | −.7480 | .352 |
| −.0248 | .9970 | .075 |
| .9983 | .0434 | .031 |

For $H_1$ vs $PC_3$:

| $d = L_1^2 - L_2^2$ | $c = L_1 L_2$ |
|---|---|
| .19363 | −.19835 |
| −.00501 | −.00186 |
| .99564 | .03095 |

$$\tan(4\theta_{1*3}) = \frac{4[3(-.00758) - (1.18426)(-.16926)]}{3(1.02882) - (1.18426)^2 - 12(.04030) + 4(.16926)^2}$$

$$= .71084/1.31499 = .54057,$$

whence

$$\theta_{1*3} = 7°5.91',$$

whence

$$\mathbf{T} = \begin{bmatrix} .99233 & 0 & -.12358 \\ 0 & 1 & 0 \\ .12358 & 0 & .99233 \end{bmatrix}$$

and the new structure is

| $H_1'$ | $H_2$ | $H_3$ |
|---|---|---|
| −.5157 | −.748 | .4189 |
| −.0153 | .997 | .0775 |
| .9945 | .0434 | −.0926 |

For $H_2$ vs $H_3$:

$$\tan(4\theta_{2*3*}) = .78368/.48735 = 1.60804; \qquad \theta = 14°31.9';$$

$$\mathbf{T} = \begin{bmatrix} 1 & 0 & 0 \\ 0 & .96801 & -.24091 \\ 0 & .25091 & .96801 \end{bmatrix}$$

and the new structure is

| $H'_1$ | $H'_2$ | $H'_3$ |
|--------|--------|--------|
| $-.516$ | $-.619$ | .593 |
| $-.015$ | .985 | $-.175$ |
| .995 | .019 | $-.101$ |

After this first large iteration, the varimax criterion is now 1.078 (versus .905 initially). Since it ultimately converges to 1.083 (we cheated by looking at the computer's work), it hardly seems worth continuing the process by hand. The final varimax-rotated structure is

| $H''_1$ | $H''_2$ | $H''_3$ |
|---------|---------|---------|
| $-.582$ | $-.460$ | .671 |
| $-.025$ | .974 | $-.223$ |
| .980 | .002 | $-.197$ |

**15.** Varimax rotation of $PC_1$ and $PC_2$ was the first step of Problem 14. This produces a varimax criterion of .916, as compared with .905 for the original structure and .875 for the intuitive solution. However, this assumes either that we do not wish to employ Kaiser normalization, or that we wish to apply the varimax criterion to the entire factor structure (retain all three PCs), even though rotating only two of the PCs. If we retain only the two PCs and *do* use Kaiser normalization, we obtain

$$\tan(4\theta) = -2.0532/1.0880 = -1.887; \qquad \theta = -15°31',$$

and the new structure is

| $H_1$ | $H_2$ |
|-------|-------|
| $-.509$ | $-.786$ |
| $-.095$ | .993 |
| .993 | .114 |

**16.** Regression approach:

$$\mathbf{B}_h = \mathbf{R}^{-1}\mathbf{F} = \begin{bmatrix} 5.635 & 3.911 & 3.209 \\ 3.911 & 3.715 & 2.206 \\ 3.209 & 2.206 & 2.828 \end{bmatrix} \begin{bmatrix} .041 & -.095 & 2.38 \\ .068 & .932 & 1.69 \\ 1.019 & -.118 & 1.34 \end{bmatrix}$$

$$\begin{bmatrix} -.606 & -.685 & .404 \\ .063 & .997 & .041 \\ .999 & -.042 & -.017 \end{bmatrix}$$

"Direct" approach:

$$\mathbf{Z}_h = \mathbf{Z}_{pc} \cdot \mathbf{T} = \mathbf{Z}_x \cdot \begin{bmatrix} -.09971 & -.16237 & 2.35550 \\ .08424 & .87311 & 1.70691 \\ .92729 & -.26920 & 1.37015 \end{bmatrix} \begin{bmatrix} .035 & -.099 & 2.35 \\ .067 & .929 & 1.68 \\ 1.018 & -.120 & 1.37 \end{bmatrix}$$

$$\begin{bmatrix} .992 & .113 & -.053 \\ -.115 & .993 & -.025 \\ .049 & .031 & .999 \end{bmatrix}$$

The variance of each $z_{h_i}$ is 1.0. The variance of raw scores on the hypothetical variables (rotated PCs) is impossible to determine.

**17.**

$$\mathbf{R}^{-1}\mathbf{F} = \mathbf{k}^{-1} \cdot \begin{bmatrix} -.509 & -.786 \\ -.095 & .993 \\ .993 & .114 \end{bmatrix} = \begin{bmatrix} -.053 & -.180 \\ -.153 & .866 \\ .965 & -.009 \end{bmatrix} = \mathbf{B}_h \cdot$$

$$\mathbf{Z}_{pc} \cdot \mathbf{T} = \begin{bmatrix} -.09971 & -.16237 \\ .08424 & .87311 \\ .92729 & -.26920 \end{bmatrix} \begin{bmatrix} .96355 & +.26752 \\ -.26752 & .96355 \end{bmatrix} \begin{bmatrix} -.053 & -.183 \\ -.152 & .864 \\ .966 & -.011 \end{bmatrix}.$$

Note that the multiple regression method requires (even after $\mathbf{R}^{-1}$ has been found) a total of 18 multiplications, while the "direct" approach requires only 12. This discrepancy equals $p^2 m - pm^2 = pm(p - m)$, where $p$ is the number of original variables, and $m$ is the number of rotated PCs.

**B.  Computations based on correlation matrix**

**1.**

| | Factor pattern | | | | Factor structure | | |
|---|---|---|---|---|---|---|---|
| | PC$_1$ | PC$_2$ | PC$_3$ | | PC$_1$ | PC$_2$ | PC$_3$ |
| | .705 | .004 | .702 | | .9757 | .0043 | .2191 |
| | -.546 | -.637 | .535 | | -.7574 | -.6311 | .1674 |
| | -.456 | .770 | .440 | | -.6326 | .7622 | .1374 |
| Eigenvalue | 1.927 | .979 | .094 | $\sum L^2$ | 1.927 | .979 | .094 |
| % accounted | 64.2 | 32.6 | 3.1 | | | | |

**2.**  Note that the scores on the PCs are *not* z-scores, even though they are linear combination of true z-scores. Once again the variance of each column of scores is equal to the eigenvalue associated with that PC.

| | Score on | | |
|---|---|---|---|
| Subject | PC$_1$ | PC$_2$ | PC$_3$ |
| A | .9684 | -.3480 | -.0660 |
| B | .1861 | -1.3900 | -.1760 |
| C | -.5190 | 1.1930 | -.3840 |
| D | 1.4510 | .6612 | .3399 |
| E | -2.0900 | -.1210 | .2856 |

**3.**  The correlations between PCs are still zero. The correlations between original scores and PCs are the entries in the **R**-derived factor structure.

**4.** The obvious set of hypothesized PCs is

|  | $PC_1$ | $PC_2$ | $PC_3$ |
|---|---|---|---|
| $\mathbf{P}_0 =$ | 2/6 | 0 | 1/3 |
|  | −1/6 | −.707 | 1/3 |
|  | −1/6 | .707 | 1/3 |

that is, the hypotheses that $PC_1$ is the difference between $X_1$ and the other two original variables; that $PC_2$ is the difference between $X_3$ and $X_2$; and that $PC_3$ is the simple sum of the three original variables. These hypotheses lead to predicted variances of 1.929, .979, and .151, respectively, and to values of $\boldsymbol{\beta}'_0 \mathbf{S}^{-1}\boldsymbol{\beta}_0$ (actually $\beta_0 \mathbf{R}^{-1}\beta_0$ in the present case, of course) of .905, 1.075, and 10.10. The end result is a chi-square value for each test of 2.976, .204, and 2.208, all with four degrees of freedom and none statistically significant. Using $\mathbf{P}_0$ to reproduce the intercorrelation matrix yields an $\hat{\mathbf{R}}$ of

$$\begin{bmatrix} 1 & -.592 & -.592 \\ & 1 & -.070 \\ & & 1 \end{bmatrix},$$

to which we apply our overall significance test, obtaining

$$\chi^2 = 1.5 \ln(.246/.179) = .481 \quad \text{with} \quad 3 \text{ df, ns.}$$

**5.** $\chi^2 = 4 \ln[.5365^2/((.979)(.094))] = 4 \ln(3.74) = 4.58 \quad \text{with} \quad 2 \text{ df, ns.}$

**6.** Same as Problem A6.

**7.** Using $PC_1$ and $PC_2$ yields

$$\hat{\mathbf{R}} = \begin{bmatrix} 1 & -.737 & -.613 \\ & 1 & -.002 \\ & & 1 \end{bmatrix};$$

whence

$$\chi^2 = -1.5 \ln(.079/.179) = 3.26 \quad \text{with} \quad 6 \text{ df, ns.}$$

**8.** Perfect reproduction.

**9.** Small sample sizes and use of **R**-matrix.

**10.** The correlations between the PCs and $Y$ are −.9266, .2257, and .2594, respectively. These are also the $z$-score coefficients for predicting $z_y$ from $z$-scores on the PCs.

$$R^2 = .9266^2 + .2257^2 + .2594^2 = .9678; \quad R = .988.$$

These values of $R^2$ and $R$ match both Problem A10 and the results of the Demonstration Problem in Chapter 2. The coefficients for the regression equation are different here, but comparability is restored when the definitions of the PCs in terms of the original variables are substituted.

**11.** Keeping only $PC_1$ and $PC_2$, we simply drop the term corresponding to $PC_3$ from our regression equation, thereby decreasing $R^2$ to .901 and $R$ to .950. This is a greater loss than from dropping the third covariance-derived PC, but less than dropping the third original variable.

**12.** $R^2 = .850$, $R = .922$. We get more "mileage" out of the first **R**-derived PC (in this particular example, though *not* as a general rule) than out of the first **S**-derived PC or out of the first original variable.

**13–15.** Since the **R**-derived factor structure is the same as the covariance-derived factor structure, except for a rigid rotation, any objective rotation scheme (such as varimax) will produce the same final rotated factor structure regardless of whether we start with **R** or **S**. The same should be true of graphical-intuitive solutions (at least approximately) except in so far as one or the other starting point gives different insights into the possibilities for simple structure. As a check on your work, the results of the first large iteration in the varimax rotation scheme are given below.

| $\theta_{12} = +30°10'$ | | | $\theta_{2*\,3} = 12°12'$ | | | $\theta_{1*\,3*} = 45°$ | | |
|---|---|---|---|---|---|---|---|---|
| $PC_1^*$ | $PC_2^*$ | $PC_3$ | $PC_1^*$ | $PC_2^*$ | $PC_3^*$ | $PC_1^{**}$ | $PC_2^{**}$ | $PC_3^{**}$ |
| .846 | −.487 | .219 | .846 | −.429 | .317 | .822 | −.429 | −.374 |
| −.972 | −.165 | .167 | −.972 | −.126 | .199 | −.547 | −.126 | .828 |
| −.164 | .977 | .137 | −.164 | .984 | −.072 | −.167 | .984 | .065 |

**16, 17.** By the same reasoning as in the answer to Problems 13–15, we would expect the expressions for the relationships between original variables and $z$-scores on hypothetical variables to be the same as if we had begun with the covariance matrix—we do, after all, "wind up" with the same factor structure. However, being too lazy to perform the rotations by hand, the author was lead to recognize a problem which is apt to confront a user whose computer program does not compute factor scores as an option. Namely, if such a user wishes to employ the "direct approach" to computing scores on rotated PCs, he will not in general know the angles of rotation used by the program in satisfying, say, a varimax criterion, and thus will not know **T**. However, we have

$$\mathbf{F}^* = \mathbf{FT},$$

where $\mathbf{F}^*$ is the rotated factor structure; whence

$$\mathbf{F'F}^* = \mathbf{F'FT};$$

whence

$$\mathbf{T} = (\mathbf{F'F})^{-1}(\mathbf{F'F}^*),$$

so that we can compute **T** knowing only the original and rotated factor structures. This may sound like scant comfort, since it involves inverting a matrix. However, $\mathbf{F'F}$ is only $p \times p$ *and* is a diagonal matrix (this latter property is generally *not* true of $\mathbf{F}^{*\prime}\mathbf{F}^*$), so that its inverse is simply a diagonal matrix whose main diagonal entries are the reciprocals of the characteristic roots of **S** or of **R**. To illustrate this on the graphically rotated PCs:

$$\mathbf{F'F}^* = \begin{bmatrix} -1.271 & -1.397 & .374 \\ .719 & -.664 & -.037 \\ .015 & .011 & .093 \end{bmatrix}; \quad \mathbf{T} = \begin{bmatrix} -.660 & -.725 & .194 \\ .735 & -.678 & -.038 \\ .160 & .117 & .990 \end{bmatrix};$$

$$\mathbf{Z}_h = \mathbf{Z}_x \begin{bmatrix} .508 & .994 & 2.290 \\ -.393 & -.644 & 1.745 \\ -.329 & .778 & 1.435 \end{bmatrix} \mathbf{T} = \mathbf{Z}_x \begin{bmatrix} .034 & -.102 & 2.37 \\ .066 & .927 & 1.68 \\ 1.018 & -.121 & 1.33 \end{bmatrix},$$

as was obtained in Problem A16.

# 7 | FACTOR ANALYSIS
## The Search for Structure

Several times in the previous chapter we mentioned the possibility of retaining only the first few PCs—those accounting for a major percentage of the variation in the original system of variables—as a more economical description of that system. The PCs having low associated eigenvalues are likely to be describing error variance, anyway, or to be representing influences (causal factors?) which affect only one or a very few of the variables in the system. However, the basic PCA model leaves no room for error variance or for nonshared (specific) variance. Models which *do* provide explicitly for a separation of shared and unique variance lead to the statistical techniques known collectively as *factor analysis* (FA).

Excellent texts on computational procedures and uses of factor analysis are available (cf. especially Harman, 1967, and Mulaik, 1972). This chapter will therefore confine itself to a brief survey of the various types of factor analytic solution, together with an attempt to add a bit of insight into the bases for preferring FA to PCA or vice versa.

### 7.1 The Model

$$z_{ij} = f_{j1}h_{i1} + f_{j2}h_{i2} + \cdots + f_{jp}h_{ip} + d_j u_{ij} = \sum_{k=1}^{p} f_{jk}h_{ik} + d_j u_{ij} \qquad (7.1)$$

where $z_{ij}$ is subject $i$'s z-score on original variable $j$; $h_{ik}$ is subject $i$'s z-score on hypothetical variable $k$; $u_{ij}$ is subject $i$'s z-score on the *unique* (nonshared) variable associated with original variable $j$; and $f_{jk}$ and $d_j$ are coefficients which must usually be estimated from the data. Written in matrix form this becomes

$$\mathbf{Z} = \begin{bmatrix} z_{11} & z_{12} & \cdots & z_{1m} \\ z_{21} & z_{22} & \cdots & z_{2m} \\ \vdots & \vdots & \vdots & \vdots \\ z_{N1} & z_{N2} & \cdots & z_{Nm} \end{bmatrix} = [\mathbf{z}_1 \quad \mathbf{z}_2 \cdots \mathbf{z}_m] = [\mathbf{H} \vdots \mathbf{U}]\begin{bmatrix} \mathbf{F} \\ \text{-----} \\ \mathbf{D} \end{bmatrix}; \qquad (7.2)$$

where [A | B] represents a combined matrix obtained by appending the columns of the matrix **B** immediately to the right of the columns of the matrix **A**; **F** (the common-factor pattern) is an $m \times p$ matrix giving the common-factor coefficients $f_{jk}$; **D** is a diagonal matrix whose $j$th main diagonal entry is $d_j$; $H = [h_1 \ h_2 \ \cdots \ h_p]$ is an $N \times p$ matrix giving each **S**'s score on the $p$ common factors; and **U** is an $N \times m$ matrix giving each **S**'s score on the $m$ unique factors. It is assumed that the unique factors are uncorrelated with each other.

At first glance this would appear to be a much less parsimonious model than the PCA model, since it involves $m + p$ hypothetical (latent) variables, rather than just $m$. However, the sole purpose of the $m$ unique factors is to permit elimination of unique variance from the description of the relationships among the original variables, for which description only the $p$ (usually $\ll m$) common factors are employed. We shall thus need expressions for the common-factor portion of the model, namely

$$\mathbf{Z}_c = \begin{bmatrix} \hat{z}_{11} & \hat{z}_{21} & \cdots & \hat{z}_{m1} \\ \hat{z}_{12} & \hat{z}_{22} & \cdots & \hat{z}_{m2} \\ \vdots & \vdots & \vdots & \vdots \\ \hat{z}_{1N} & \hat{z}_{2N} & \cdots & \hat{z}_{mN} \end{bmatrix} = \mathbf{HF}; \qquad (7.3)$$

where the carets emphasize the fact that the entries of $\mathbf{Z}_c$ are only approximations to or estimates of the entries in **Z**.

Note that nothing thus far has been said about the intercorrelations among the latent variables. In PCA the latent variables were always uncorrelated. In some forms of FA, however, this is no longer true. In the case of correlated factors, we must provide the additional information contained in the *interfactor correlation matrix*,

$$\mathbf{R}_h = [r_{h_i h_j}]. \qquad (7.4)$$

This information is often provided implicitly in the factor structure, $\mathbf{S} = \mathbf{FR}_h$, which gives the correlations between original variables and factors.

This relationship implies that **S** and **F** are identical in the case of orthogonal factors, in which case $\mathbf{R}_h = \mathbf{I}$. That this is true can be seen by considering the multiple regression equation for "predicting" $z$-scores on $X_i$ from $z$-scores on a set of uncorrelated predictors, namely

$$\hat{z}_i = r_{x_i h_1} z_{h_1} + r_{x_i h_2} z_{h_2} + \cdots + r_{x_i h_p} z_{h_p},$$

which, by comparison with (7.2), shows the equivalence between **F** and **S**.

Note that **F** in the present context is *not* the same kind of matrix as the factor pattern matrices discussed in Chapter 6. It is *not*, in particular, equal to $\mathbf{B}_z$, the standard-score factor pattern used in Eq. (6.14). In PCA, the emphasis has traditionally been on describing the PCs in terms of the original variables, so that $\mathbf{B}_z$ refers to the coefficients which generate $z$-scores on the PCs from $z$-scores on the original variables. In FA, on the other hand, the emphasis has traditionally been on describing the original variables in terms of the hypothetical variables (we shall argue later that this emphasis should be reversed), so that **F** refers to the coefficients used to

"generate" $z$-scores on the $X$s from (unknown and, as we shall see, usually unknow-
able) $z$-scores on the hypothetical variables. PCA factor patterns might then be
referred to as $X$-to-$H$ patterns, while FA factor patterns are $H$-to-$X$ patterns.

Actually, the reproduced (estimated) scores of the subjects on the original variables are seldom obtained. Instead, the emphasis in FA is traditionally on reproducing the intercorrelations among the original variables. Of particular relevance to this
goal are the following matrix equations:

$$R = [1/(N-1)]ZZ' = R_r + D'D = \begin{bmatrix} h_1^2 & r_{12} & \cdots & r_{1m} \\ r_{12} & h_2^2 & \cdots & r_{2m} \\ \vdots & \vdots & \vdots & \vdots \\ r_{1m} & r_{2m} & \cdots & h_m^2 \end{bmatrix} + \begin{bmatrix} d_1^2 & & O \\ & d_2^2 & \\ O & & \ddots \\ & & & d_m^2 \end{bmatrix}; \quad (7.5)$$

$$R_r = FR_h F' = SF' = FS' \quad (= FF' \text{ for orthogonal factors}).$$

The goal of FA is to select values for the communalities $h_j^2$ and the pattern
coefficients $f_{jk}$ which will do the best job of reproducing the off-diagonal elements of
$R$.

Many of the concepts and computational techniques required in factor analysis
were introduced in the chapter on PCA. The additional considerations which arise in
FA are essentially three in number:

(i)   the problem of estimating communalities;
(ii)  the goal of repoducing $R$ as opposed to maximizing percentage of variance
accounted for; and
(iii) the difficulties of describing nonorthogonal factors.

These differences have some important consequences for the usefulness of FA as
compared with PCA. For one thing, the introduction of communalities adds yet
another dimension along which factor "solutions" may differ. Whereas any factori-
zation of the complete $S$- or $R$-matrix is related to all other factorizations by a rigid
rotation, this is true of factorizations of $R_r$ only if they are based on identical communalities. For another, the relationship between original scores and factor scores is
no longer a perfect one. The fact that only part of the variance of a given variable is
accounted for by our system of common factors introduces uncertainty into our
reproduction of $r_{ij}$. Factor scores must therefore be obtained through multiple regression methods, with the multiple $R$ being less than unity. However, the "direct
approach" to computing scores on rotated PC (cf. Section 6.5.4) can be generalized
to computing *estimates* of scores on rotated factors. Our original estimates are

$$H = ZR^{-1}S;$$

our estimates of scores on rotated factors would therefore be

$$H^* = ZR^{-1}S^* = ZR^{-1}(ST) = (ZR^{-1}S)T = HT,$$

where $T$ is the transformation matrix which accomplishes the rotation of the original
factor structure $S$ to the new factor structure $S^*$.

## 7.2   Communalities

As pointed out in Digression 2, the rank of a matrix is closely related to the dimensionality of the system of variables involved in that matrix. If the rank of the matrix $\mathbf{A}$ is $p$, then $p$ columns can be found such that any column of $\mathbf{A}$ can be written as a linear combination of these columns. Because of the linear relationships between variances of variables and variances of linear combinations of these variables, this implies that if a correlation matrix is of rank $p$, any of the original variables can be expressed as a linear combination of $p$ latent variables. This implication is further supported by examination of the Laplace expansion of the determinantal equation used to compute the characteristic roots of $\mathbf{R}$. If the rank of $\mathbf{R}$ is $p$, then there will be exactly $p$ nonzero and $m - p$ zero characteristic roots.

Now, it is extremely unusual—except in the case where the researcher has deliberately included "derived" measures defined as linear combinations of other variables in the system—for a sample covariance or correlation matrix to have anything other than full rank $m$. However, if the main diagonal entries are adjusted to values other than unity, it may be possible to bring the values of the determinant of $\mathbf{R}_c$—and more generally the values of all $(p + 1) \times (p + 1)$ or larger minor determinants of $\mathbf{R}_c$—very close to zero. It could then be argued—and is so argued by factor analysts—that the apparent full rank of $\mathbf{R}$ was an artifact of including in the analysis error variance and specific variance, neither of which can have anything to do with the relationships *among* different variables in the system. These relationships can thus be described in terms of $p$ (hopefully $\ll m$) latent "generating" variables.

(Note that $\mathbf{R}_c$ is the correlation matrix with communalities—values $h_i^2$ equal to the portion of the total variance of $X_i$ which is shared with (held in common with) the other variables—substituted for the original 1.0 entries on the main diagonal. In other words, $\mathbf{R}_c = \mathbf{R} - \mathbf{D}(1 - h_i^2)$, where $\mathbf{D}(a_i)$ is a diagonal matrix whose $i$th main diagonal entry is $a_i$.)

There are three general procedures for estimating communalities:

(i)   through a direct theoretical solution of the conditions necessary to achieve minimum possible rank;

(ii)   through an iterative process in which a set of communalities is assumed, a FA of the reduced matrix is carried out and used to compute a new set of communalities which are substituted into $\mathbf{R}_r$ for a second solution, etc., until the communalities which are assumed before a given FA is carried out and those computed on the basis of the resulting factor pattern are consistent;

(iii)   through various empirical approximations.

### 7.2.1   Theoretical Solution

To obtain a theoretical solution for the communalities needed to achieve rank $p$, algebraic symbols for the communalities are inserted in the main diagonal of $\mathbf{R}$ and all determinants of minors of order $(p + 1) \times (p + 1)$ are set equal to zero. Some of these equations will involve the unknown communalities and can be used to solve for them, while others will involve only off-diagonal elements and will thus establish conditions on the observed correlation coefficients which must be satisfied in order that rank $p$ be obtained. Actually, this latter class of conditions can always be obtained through the requirement that the various available solutions for a given communality produce the same value of $h_j$. A number of authors (for example, Harman, 1967) have, by consideration of the number of determinants of a given order existing within a matrix of particular size, established the minimum rank which can be achieved by adjustment of the communalities, irrespective of the values of the off-diagonal elements, and the number of linearly independent conditions on these off-diagonal elements which are needed to achieve a yet lower rank. Thus, for instance, Harman (1967, p. 72) states that "three variables can always be described in terms of one common factor," while 54 linearly independent conditions on the off-diagonal entries of a $12 \times 12$ correlation matrix must be satisfied in order to achieve rank 1. To see how this works, consider the three-variable case. If all $2 \times 2$ minors are to vanish, then we must have

$$c_1 c_2 - r_{12} = 0; \qquad c_1 c_3 = r_{13}; \qquad \text{and} \qquad c_2 c_3 = r_{23};$$

whence

$$c_3/c_2 = r_{13}/r_{12} \qquad \text{and} \qquad c_2/c_1 = r_{23}/r_{13};$$

whence

$$c_1 = |r_{12} r_{13}/r_{23}|; \qquad c_2 = |r_{12} r_{23}/r_{13}|; \qquad c_3 = |r_{13} r_{23}/r_{12}|.$$

Applying these results to our PCA and MRA Demonstration Problem data, we find that $c_1 = (.705)(.524)/.021$, which equals approximately 23! However, of course communalities greater than 1.0 cannot exist, so we must conclude that it is *not* possible to achieve rank 1 for these data in a meaningful way. Instead, we can ask only that the rank be 2, for which it is necessary that

$$|\mathbf{R}_c| = c_1 c_2 c_3 + 2r_{12} r_{13} r_{23} - c_1 r_{23}^2 - c_2 r_{13}^2 - c_3 r_{12}^2 = 0.$$

For these data, $r_{12} r_{13} r_{23}$ and $r_{23}^2$ are both very small, so that an approximate solution (really, a family of solutions) is provided by the equation

$$c_1 c_2 c_3 = c_2 r_{13}^2 + c_3 r_{12}^2 = .497 c_2 + .341 c_3,$$

whence

$$c_1 = .497/c_3 + .341/c_2.$$

Thus, for instance, $c_2 = c_3 = 1$ and $c_1 = .838$ would serve as one satisfactory set of communalities, while $c_1 = c_2 = 1$ and $c_3 = .755$, or $c_1 = c_2 = c_3 = .916$ would also yield a reduced correlation matrix of rank 2. Cases where communalities producing a desired rank $p$ can be found, but only if one or more communalities are greater than unity, are called *Heywood* cases, and indicate either that more than $p$ factors are needed to describe the system or that nonlinear relationships among the variables exist. At any rate, it seems clear to the author that the conditions listed by Harman as necessary in order that a rank of 1 be attainable, namely that

$$\text{rank 1 attainable} \Leftrightarrow r_{js} r_{jt}/r_{st} = h_j^2 = \text{constant} \qquad \text{for all} \quad s, t \neq j \qquad (7.6)$$

should be supplemented by the requirement that $0 < h_j^2 < 1$. A similar set of conditions for attaining rank 2 have been developed, namely

$$h_j^2 = \frac{r_{aj} \begin{vmatrix} r_{jb} & r_{jd} \\ r_{cb} & r_{cd} \end{vmatrix} - r_{cj} \begin{vmatrix} r_{jb} & r_{jd} \\ r_{ab} & r_{ad} \end{vmatrix}}{\begin{vmatrix} r_{ab} & r_{ad} \\ r_{cb} & r_{cd} \end{vmatrix}} = \text{constant} \qquad (7.7)$$

for all $a$, $b$, $c$, $d$ different from each other and $\neq j$; again subject to the restriction that this constant value lie between zero and unity. Actually, these conditions need not be met exactly to make it reasonable to assume a rank of 1 or 2, since these will be subject to a certain amount of error variability. If the various values of $h_j^2$ are fairly close to each other, they may simply be averaged together and the analysis continued. Spearman and Holzinger (1925) have provided formulas for the sampling error of "tetrad differences" of the form $r_{jt} r_{su} - r_{ju} r_{st}$, which must all be zero within sampling error if (7.6) is to hold. Alternately, one may use as a rough guide the requirement that the standard deviation of the various estimates of a particular communality be of the same order of magnitude as $\sqrt{N-3}$, the standard deviation of a sampling distribution of correlation coefficients drawn from a population in which the true correlation is zero. Ultimately, of course, the decision as to whether a particular assumption about the rank of **R** is tenable should rest on the significance of the difference between the observed and reproduced correlations resulting from use of those communality estimates in a factor analysis.

The expressions for $4 \times 4$ and higher-order determinants are sufficiently complex that no one has as yet worked out closed-form expressions for the conditions necessary to attain rank 3 or higher. Instead, it is recommended that minres or maximum-likelihood procedures, which require that the number of factors be specified and then produce appropriate communalities as a by-product of the analysis, be employed where a rank higher than 2 is suspected.

### 7.2.2   Empirical Approximations

A very wide variety of rough-and-ready approximations to communalities have been suggested for use in those cases (for example, principal factor analysis) where communalities must be specified before the analysis can begin. It can be shown, for instance, that

$$R_{j \cdot c}^2 \leq h_j^2 \leq r_{jj};$$

where $R_{j \cdot c}^2$ is the square of the multiple correlation between variable $j$ and the $m - 1$ other variables; and where $r_{jj}$ is the reliability with which variable $j$ is measured. Thus the squared multiple correlation can be used to set a lower bound on $h_j^2$ while the reliability of the variable sets an upper bound on $h_j^2$. This $R_{j \cdot c}^2$ is readily obtainable as 1.0 minus the reciprocal of the $j$th main diagonal entry of $\mathbf{R}^{-1}$. However, $r_{jj}$ is itself a theoretical construct which can only be approximated by, for example, the correlation between two successive measurements of the same group on that variable. This test–retest reliability could be deflated by real changes in the subjects between the two measurements or it might be inflated artificially by memory effects.

Other approximations—$\max_j \sum r_{ij}^2$, $\max_{i,k}(r_{ij}r_{jk}/r_{ik})$, sum of the squared loadings on those PCs having eigenvalues greater than 1.0—have been suggested but none has been shown to be superior to the others in generating communalities which reduce the rank of $\mathbf{R}$ as much as possible.

### 7.2.3   Iterative Procedure

Under this procedure, a rank $p$ is assumed and an initial guess as to communalities is made. A principal factor analysis of the reduced correlation matrix is conducted (that is, the characteristic roots and vectors are obtained), and the sum of the squared loadings of each variable on the first $p$ factors is obtained. These are taken as the new communality estimates and are "plugged into" the main diagonal of $\mathbf{R}$. A PFA is conducted on this new $\mathbf{R}_r$, and the process continued until each communality inserted into $\mathbf{R}_r$ matches the sum of the squared loadings of that variable on the first $p$ factors derived from a PFA of $\mathbf{R}_r$. Wrigley (1958) found that this method converges for all the matrices he tried, with the convergence being fastest when $R_{j.c}^2$ was taken as the initial estimate of $h_j^2$ and especially poor when unities were used as initial estimates.

## 7.3   Factor Analysis Procedures Requiring Communality Estimates

### 7.3.1   Principal Factor Analysis

Principal factor analysis (PFA) differs computationally from principal component analysis (PCA) only in that the main diagonal entries of $\mathbf{R}$ are replaced by communalities. (PFA can be carried out on the covariance matrix, in which case the $j$th main diagonal entry would be $h_j^2 s_j^2$. However, factor analysts almost invariably take $\mathbf{R}$, rather than $\mathbf{S}$, as their starting point.) The first principal factor $(PF_1)$ accounts for as large a percentage of the *common* variance as possible, with its associated eigenvalue being equal to the sum of the squared loadings. By way of illustration, we can conduct a PFA of the PCA Demonstration Problem data (Chapter 6). First, taking that set of values for the communalities for which $\sum h_j^2$ is as large as possible while ensuring that $\mathbf{R}_c$ has rank 2, we find that

$$\mathbf{R}_c = \begin{bmatrix} .821 & -.705 & -.584 \\ -.705 & 1.000 & .021 \\ -.584 & .021 & 1.000 \end{bmatrix};$$

whence

$$|\mathbf{R}_c| = 0, \qquad S_2 = 1.803, \qquad \text{and} \qquad S_1 = 2.821;$$

whence our three eigenvalues are 1.8415, .9795, and 0.0, with corresponding eigenvectors

$$
\begin{bmatrix} -.669 \\ .570 \\ .475 \end{bmatrix}, \qquad
\begin{bmatrix} -.005 \\ -.639 \\ .769 \end{bmatrix}, \qquad \text{and} \qquad
\begin{bmatrix} .745 \\ .516 \\ .426 \end{bmatrix};
$$

whence the factor structure and pattern are

|  | $PF_1$ | $PF_2$ | $PF_3$ |
|---|---|---|---|
| $X_1$ | $-.906$ | $-.006$ | |
| $X_2$ | $.775$ | $-.632$ | |
| $X_3$ | $.648$ | $.761$ | |
| $\sum r^2_{X_i PF_j}$ | $1.842$ | $.980$ | $0$ |

## 7.3.2  Triangular Decomposition

From the point of view of most factor analysts, the primary value of principal factor analysis is that it provides a convenient way of obtaining an initial solution which can subsequently be rotated to obtain a more compelling and/or more easily interpreted set of loadings. Any other procedure which factored the initial matrix into orthogonal hypothetical variables capable of "reproducing" $\mathbf{R}_r$ would do as well as PFA's eigenvector approach for this purpose. One such alternative procedure for which computational algorithms are readily available is the method of *triangular decomposition*. This procedure, like PFA, yields as a first factor a general factor on which all original variables have nonnegligible loadings. The second factor, however, has a zero correlation with the first original variable; the third factor has zero correlations with the first two original variables; etc. In other words, the factor structure is a truncated lower triangular matrix—lower triangular because all above-diagonal loadings are precisely zero and truncated because, unless $\mathbf{R}$ rather than $\mathbf{R}_r$ is factored, there will be fewer columns than original variables. The method is also known as the *square root* method of factoring. To see why, consider the $3 \times 3$ case. We seek a matrix $\mathbf{T}$ such that

$$
\mathbf{TT'} = 
\begin{bmatrix} a & 0 & 0 \\ b & c & 0 \\ d & e & f \end{bmatrix}
\begin{bmatrix} a & b & d \\ 0 & c & e \\ 0 & 0 & f \end{bmatrix}
=
\begin{bmatrix} h_1^2 & r_{12} & r_{13} \\ r_{12} & h_2^2 & r_{23} \\ r_{13} & r_{23} & h_3^2 \end{bmatrix};
$$

whence

$$a = \sqrt{h_1^2}; \qquad b = r_{12}/a; \qquad d = r_{13}/a;$$

$$c = \sqrt{h_2^2 - b^2}; \qquad e = (r_{23} - bd)/c; \qquad f = \sqrt{h_3^2 - d^2 - e^2}. \tag{7.8}$$

As is readily apparent, the square roots of the main diagonal elements, corrected for the effects of preceding factors, play a crucial role in these formulas. Usually row and column interchanges (equivalent to relabeling of the rows and columns of $\mathbf{R}$) are used to ensure that the factors come out in decreasing order of importance. For instance, loadings on the first factor are proportionate to the correlations of the other variables with the variable listed in the first row and column of $\mathbf{R}_r$, so that it would seem natural to put in that position the variable $j$ for which the sum of the squared intercorrelations with other variables $\sum r_{ij}^2$ is largest. See Harman (1967) or Mulaik (1972) for computational details.

Triangular decomposition (TDA) would clearly be a desirable factor analytic procedure if the researcher's theory led him to predict a hierarchical factor structure like the one produced by TDA. However, most researchers employing factor analysis find such a structure unacceptable for much the same reasons PFA is not popular as a final solution. Since the TDA factor pattern for a given set of communalities is obtainable via a rigid rotation of the PFA factor pattern, the only remaining basis for choosing between them is computational convenience. The wide availability of canned computer programs for PFA gives the edge to that technique unless a particular user must resort to desk calculator procedures. The triangular decomposition formulas are more easily adapted to calculator operations than are PFA computations. However, an even simpler initial factoring procedure for such "hand" calculations is provided by centroid analysis, which will be discussed in the next section.

Before moving on to centroid analysis, however, let us illustrate TDA on the data from the PCA Demonstration Problem (Chapter 6). We could of course use the explicit formulas derived in (7.8) for the $3 \times 3$ case, but it may be more informative to consider instead a more general computational algorithm, namely

$$\mathbf{f}_n = \mathbf{r}_n / \sqrt{{}_n h_n^2}; \qquad \mathbf{R}_{n+1} = \mathbf{R}_n - \mathbf{f}_n \mathbf{f}_n', \qquad \text{for} \quad n = 1, 2, \ldots, q \leq p; \qquad (7.9)$$

where $\mathbf{f}_n$ is the $n$th column of the $z$-score factor pattern; $\mathbf{r}_n$ is the $n$th column of $\mathbf{R}_n$; $\mathbf{R}_n$ is the residual correlation matrix after the effects of the first $n - 1$ factors have been partialed out; and ${}_n h_n^2$ is the communality of the $n$th residual variate of $\mathbf{R}_n$, that is, the $n$th main diagonal entry of $\mathbf{R}_n$. Applying this iterative process to $\mathbf{R}$, the matrix with "communalities" of unity, gives

$$\mathbf{f}_1 = \frac{1}{\sqrt{1}} \begin{bmatrix} 1 \\ -.705 \\ -.584 \end{bmatrix};$$

$$\mathbf{R}_2 = \begin{bmatrix} 1.0 & -.705 & -.584 \\ -.705 & 1.0 & .021 \\ -.584 & .021 & 1.0 \end{bmatrix} - \begin{bmatrix} 1.0 & -.705 & -.584 \\ -.705 & .497 & .412 \\ -.584 & .412 & .341 \end{bmatrix} = \begin{bmatrix} 0 & 0 & 0 \\ 0 & .503 & -.391 \\ 0 & -.391 & .659 \end{bmatrix};$$

$$\mathbf{f}_2 = \frac{1}{\sqrt{.503}} \begin{bmatrix} 0 \\ .503 \\ -.391 \end{bmatrix} = \begin{bmatrix} 0 \\ .710 \\ -.551 \end{bmatrix}; \qquad \mathbf{R}_3 = \begin{bmatrix} 0 & 0 & 0 \\ 0 & 0 & 0 \\ 0 & 0 & .355 \end{bmatrix};$$

whence

|  |  | $f_1$ | $f_2$ | $f_3$ | $\sum h_i^2$ |
|---|---|---|---|---|---|
| $\mathbf{F} =$ | $X_1$ | 1.0 | 0 | 0 | 1.000 |
|  | $X_2$ | $-.705$ | .710 | 0 | 1.000 |
|  | $X_3$ | $-.584$ | $-.551$ | .596 | 1.000 |
|  | $\sum f_{jk}^2$ | 1.838 | .807 | .356 |  |

On the other hand, starting with rank 2 communalities gives

$$\mathbf{f}_1 = \frac{1}{\sqrt{.821}} \begin{bmatrix} .821 \\ -.705 \\ -.584 \end{bmatrix} = \begin{bmatrix} .906 \\ -.778 \\ -.645 \end{bmatrix};$$

$$\mathbf{R}_2 = \begin{bmatrix} 0 & 0 & 0 \\ 0 & .395 & -.480 \\ 0 & -.480 & .584 \end{bmatrix}; \quad \mathbf{f}_2 = \begin{bmatrix} 0 \\ .628 \\ -.765 \end{bmatrix};$$

|  | $f_1$ | $f_2$ | $\sum h_i^2$ |
|---|---|---|---|
| $\mathbf{F} =$ | .906 | 0 | .821 |
|  | $-.778$[a] | .628 | 1.0 |
|  | $-.645$ | $-.765$ | 1.0 |
| $\sum f^2$ | 1.842 | .980 |  |

### 7.3.3   Centroid Analysis

The centroid method was developed by Thurstone (1947) as a computationally simpler approximation to PFA. Since canned computer programs for PFA are now readily available, there is little need for centroid analysis unless the researcher finds himself away from his computer. If all entries of **R** are positive, the first centroid factor is simply the unweighted sum of all the original variables. The $n$th centroid factor is the unweighted sum of the residual variates, that is, of the measures corrected for the effects of the first $n - 1$ centroid factors. To illustrate, let us compute the centroid factors of **R** for the PCA Demonstration Problem (Chapter 6). To simplify matters, we *reflect* $X_1$, that is, we perform the analysis on

$$\mathbf{R}_1' = \begin{array}{c} \\ -X_1 \\ X_2 \\ X_3 \end{array} \begin{array}{c} -X_1 \quad X_2 \quad X_3 \\ \begin{bmatrix} 1.0 & .705 & .584 \\ .705 & 1.0 & .021 \\ .584 & .021 & 1.0 \end{bmatrix} \end{array}; \quad \sigma_{f_1}^2 = [1 \cdots 1]\mathbf{R}\begin{bmatrix} 1 \\ \vdots \\ 1 \end{bmatrix} = 5.620;$$

$$\mathbf{f}_1 = \left(\frac{1}{5.62}\right)\mathbf{R}\begin{bmatrix} 1 \\ 1 \\ 1 \end{bmatrix} = \frac{1}{2.370}\begin{bmatrix} 2.289 \\ 1.726 \\ 1.605 \end{bmatrix} = \begin{bmatrix} .965 \\ .728 \\ .677 \end{bmatrix};$$

$$\mathbf{R}_2' = \begin{bmatrix} 1.0 & .705 & .584 \\ .705 & 1.0 & .021 \\ .584 & .021 & 1.0 \end{bmatrix} - \begin{bmatrix} .931 & .701 & .653 \\ .701 & .530 & .493 \\ .653 & .493 & .458 \end{bmatrix} = \begin{bmatrix} .069 & .004 & -.069 \\ .004 & .470 & -.472 \\ -.069 & -.472 & .542 \end{bmatrix}.$$

Reflecting variable 3 this time, we obtain

$$
\mathbf{R}_2'' =
\begin{array}{c}
\phantom{-X_1} \\
\begin{array}{ccc}
-X_1 & X_2 & -X_3
\end{array}
\end{array}
$$

$$
\mathbf{R}_2'' =
\begin{matrix}
-X_1 \\ X_2 \\ -X_3
\end{matrix}
\begin{bmatrix}
.069 & .004 & .069 \\
.004 & .470 & .472 \\
.069 & .472 & .542
\end{bmatrix} ; \qquad
\sigma_{f_2}^2 = 2.171; \qquad
\mathbf{f}_2 =
\begin{bmatrix}
.096 \\ .641 \\ .735
\end{bmatrix} ;
$$

$$
\mathbf{R}_3'' =
\begin{bmatrix}
.060 & -.058 & -.001 \\
-.058 & .059 & .002 \\
-.001 & .002 & .002
\end{bmatrix} ; \qquad
\mathbf{R}_3''' =
\begin{matrix}
X_1 \\ X_2 \\ -X_3
\end{matrix}
\begin{bmatrix}
X_1 & X_2 & -X_3 \\
.060 & .058 & .001 \\
.058 & .059 & .002 \\
.001 & .002 & .002
\end{bmatrix} ;
$$

$$
\sigma_{f_3}^2 = .243; \qquad
\mathbf{f}_3 =
\begin{bmatrix}
.242 \\ .242 \\ .010
\end{bmatrix} .
$$

Now, expressing our loadings in terms of the original variables, we obtain

|       |         | $f_1$   | $f_2$   | $f_3$   | $\sum h_i^2$ |
|-------|---------|---------|---------|---------|--------------|
|       | $X_1$   | −.965   | −.096   | .242    | .999         |
| $F =$ | $X_2$   | .728    | .641    | .242    | .998         |
|       | $X_3$   | .677    | −.735   | −.010   | .998         |
|       | $\sum r^2$ | 1.919 | .960    | .116    |              |

## 7.4   Methods Requiring Estimate of Number of Factors

The primary goal of factor analysis is usually thought to be to reproduce as accurately as possible the original intercorrelation matrix from a small number of hypothetical variables to which the original variables are linearly related. The factor analytic methods discussed so far have not explicitly incorporated this goal into the factoring process; the goal is represented only in the process of selecting communalities, and even here the emphasis is on the approximate rank of the reduced correlation matrix, with no clear cut specification of how much error in reproduction of **R** is to be tolerated. Each of two as yet relatively unused factor analytic procedures establishes a measure of goodness of fit of the observed to the reproduced correlation matrix and then selects factor loadings in such a way as to optimize this measure. The minimum-residuals, or *minres*, method seeks to minimize the sum of the squared discrepancies between observed and reproduced correlations, while the *maximum-likelihood* approach finds the factor structure which maximizes the likelihood of the observed correlation matrix occurring, each likelihood being conditional on that particular proposed structure's being the population structure.

Each of these methods has the added advantage of providing a natural base for a significance test of the adequacy of the factor solution. An additional advantage of the maximum-likelihood method is that it readily generalizes to *confirmatory* factor analysis, that is, to situations in which E specifies in advance of performing the analysis certain patterns which the final factor structure should fit. This theoretical commitment is usually expressed in the form of certain loadings which are preset to zero or to unity, with the maximum-likelihood method being used to estimate the remainder of the loadings and then to test one or more of the null hypotheses that:

(a)   the *a priori* reproduced correlation matrix $\mathbf{R}_{ap}$, implied by the factor loadings obtained under the specified side conditions, is the population matrix (If this hypothesis is true, then the differences between the off-diagonal elements of $\mathbf{R}$ and of $\mathbf{R}_{ap}$ are all due to sampling fluctuations.);

(b)   the *post hoc* factors derived with no restrictions on the loadings (other than that they lie between zero and unity, inclusive) imply a correlation matrix which is in fact the population correlation matrix;

(c)   the *a priori* restrictions on the factor structure are correct in the population.

Therefore, no significant improvement in goodness of fit will be observed when factor loadings are estimated after dropping these restrictions. Thus a test of $H_0$: "*a priori* restrictions hold" against $H_1$: "no *a priori* restrictions" will not show significance.

The test used for each of these hypotheses follows the same form as Eq. (6.7), with the numerator determinant being the determinant of the correlation matrix reproduced under the most restrictive assumptions in each case. See the work of Mulaik (1972, pp. 381–382) for details.

The disadvantages of minres and maximum-likelihood factor analysis are that

(a)   each requires prior commitment as to the number of factors which underlie responses on the original variables, and

(b)   the computational procedures for each are extremely complex, even with the aid of a computer.

The first disadvantage is actually no more serious than the problem of knowing when to stop adding variables to a prediction equation in stepwise multiple regression, since the significance test provided by (6.7) can be applied repeatedly for 1, 2, 3, ... factors, stopping when hypothesis (b) fails to be rejected. The problem, as in stepwise multiple regression, is that the Type I error rate for the statistical test in (6.7) applies to *each* test, and not to the overall probability of unnecessarily adding a factor in a sequential process such as just described. However, Joreskög (1962, cited by Mulaik, 1972) points out that this latter probability is actually less than or equal to the significance level of each separate test.

On the other hand, the computational problem is a very serious one. Until quite recently, only iterative procedures requiring large amounts of computer time and not carrying any guarantee of convergence to a final solution were available. However, Joreskög (1967, 1971) published procedures for maximum-likelihood factor analysis (which procedures are straightforwardly adaptable to minres factor analysis) which are known to converge and which do this much faster than any of the previously available techniques. Joreskög has made his computer program available to anyone sending him or Educational Testing Service (Princeton, New Jersey) a magnetic tape, but his procedures have not yet been incorporated into any widely available package of canned computer programs, and Joreskög's use of subroutines available at his computer center as part of the operating system make the process of conversion to the user's local computer nontrivial. (John Cotton reports in a personal communication that the University of California at Santa Barbara experienced no conversion problems.)

Neither of these procedures would ordinarily be used unless at least four variables were being analyzed, since a $3 \times 3$ correlation matrix can always (except for Heywood cases) be perfectly reproduced by a single factor. Moreover, where a theoretical solution for communalities producing perfect fit with a specified number of factors can be obtained, there is again no need for minres or maximum-likelihood procedures. In fact, it can be shown that *a PFA performed on a reduced correlation matrix having the same communalities as those produced by a minres or maximum-likelihood solution yields a factor structure which is identical (except possibly for an orthogonal rotation) to that minres or maximum-likelihood solution.* Thus minres and/or maximum-likelihood solutions are apt to be sought only when fairly large numbers of original variables are involved and therefore only in situations where desk-calculator computation is impractical and use of a computer is essential. Morrison (1967) and Harman (1967) each provide outlines of the pre-Jöreskog algorithm for maximum-likelihood solutions on a computer, while Mulaik (1972) and of course Jöreskog (1969, 1970, 1973) provide a similar outline of the latest and most efficient technique.

## 7.5   Other Approaches to Factor Analysis

All of the approaches mentioned so far can be considered members of the PFA "family" of procedures. TDA can be accomplished either directly or via orthogonal rotation of a PFA solution; centroid analysis is merely a computationally convenient approximation to PFA; and minres and maximum-likelihood solutions are PFA applied to reduced correlation matrices having "optimal" communalities inserted. An approach to factor analysis which is *not* simply a special case of or orthogonal rotation of a PFA, but which follows the basic model in (7.1) and (7.2), is *multiple group* factor analysis. This method differs from the preceding ones in at least two ways: (a) It requires specification of both the number of factors and the communalities. (b) It yields oblique factors as an initial solution. (All of the other methods we have examined yield orthogonal factors as the initial solution, though these may subsequently be subjected to oblique rotation.) The basic procedure is somewhat reminiscent of centroid analysis in that composite variables, each equal to the unweighted sum of some subset of the original variables, form the basis of the analysis. The researcher specifies, either on theoretical grounds or via more formal methods of cluster analysis, which variables "go together" to form a particular cluster of measures of essentially the same underlying concept. Each factor is then simply the unweighted sum of all the variables in a particular cluster, with the resultant factor *structure* simply being a matrix of correlations of original variables with these composite variables. It is of course highly unlikely that these composite variables will be uncorrelated with each other. As with any other oblique solution the researcher may wish to submit $\mathbf{R}_h$ (the matrix of correlations among the factors) to PCA in order to generate a pair of orthogonal reference axes.

A wide variety of other approaches to the analysis of within-set variability have been taken which do *not* fit (7.1) and (7.2). Among the more important of these are *cluster analysis* (Tryon and Bailey, 1970) and *image analysis* (Guttman, 1953). Cluster analysis, instead of assuming the same linear relationship between observed and hypothetical variables for all subjects, seeks out different profiles of scores on the various measures which are typical of different *clusters* of individuals. This is often closer to the researcher's actual aims than is the continuous ordering of all subjects on the same underlying dimensions. ("Subjects" and "individuals" of course refer here to whatever sampling units the researcher employs.) Image analysis differs from PFA and related approaches primarily in that it permits the possibility of correlations among the various unique factors, restricting the common factors to explanation of that portion of the variance of each variable which is directly predictable via linear relationships from knowledge of the subjects' scores on the other measures, that is, to the variable's *image* as "reflected" in the other variables in the set. This amounts to taking the population value of $R_{i \cdot c}^2$ (the squared multiple correlation between variable $i$ and the other variables) as the *exact* value of variable $i$'s communality, rather than as a lower bound on $h_i^2$.

Mulaik (1972, Chapter 8) and Pruzek (1973), in a more complete form, have pointed out several interesting relationships among image analysis, maximum-likelihood factor analysis, and *weighted components analysis*, in which each original variable is rescaled by multiplying it by the best available estimate of the square root of its error variance before applying PCA to the matrix of variances and covariances of the rescaled data. Equivalently, we may apply PCA to the matrix $\mathbf{E}^{-1}\mathbf{R}\mathbf{E}^{-1}$, where $\mathbf{E}$ is a diagonal matrix whose $i$th diagonal entry is the error variance of variable $i$. Image analysis is interpretable as weighted component analysis in which $1 - R_{i \cdot c}^2$ is taken as the estimate of the error variance of variable $i$. ($R_{i \cdot c}^2$ is the squared multiple correlation between variable $i$ and the other variables.) Maximum-likelihood FA is also interpretable as a special case of weighted component analysis. Another very important point made by Mulaik and Pruzek is that image analysis, unlike other approaches to FA, yields completely *determinate* factor scores, that is, factors each of which have a multiple correlation of unity with scores on the original variables.

Also worthy of mention again here is the laudable movement toward use of *confirmatory* (as opposed to purely exploratory) factor analysis (cf. the preceding section).

### 7.6   Factor Loadings versus Factor Scores

By far the most common procedure for interpreting (naming) the factors resulting from a principal components analysis or a factor analysis is to single out for each factor those variables having the highest loadings (in absolute value) on that factor. The highly positive loadings then help to define one end of the underlying dimension, while the highly negative loadings (if any) define the opposite end. The author suggests that this practice should be supplanted in most cases by examination instead of the linear equation which relates subjects' scores on that factor to their scores on the original variables. For principal components analysis these two procedures yield identical results, since the PCA factor pattern can be read either rowwise to obtain the definitions of the variables in terms of the principal components or columnwise to define the principal components in terms of scores on the original variables. As soon, however, as communalities are introduced or the pattern is rotated, this equivalence between the variable–factor correlations (that is, the loadings) and the factor–score coefficients (the coefficients in the linear equation relating factors to scores on original variables) is lost. For purposes of the present discussion it will be assumed that the factor–score coefficients are obtained via multiple regression techniques, which leads to a multiple $R$ of 1.0 in the case of unity communalities but less than unity as the communalities depart from unity. This makes clear the close connection between the problem of interpreting or naming factors and the problem of interpreting a multiple regression equation. This latter problem is closely related to the issue of how to assess the importance of a single variable's contribution to such a regression equation. Reviewing briefly some of our "findings" about these two issues from Chapter 2:

(a)   The correlation between a single predictor and the outcome measure is a very poor "predictor" of how high a weight that variable will receive in the multiple regression equation. A variable having a near-zero correlation with $Y$ by itself may have one of the largest regression weights in the multiple regression equation, while the variable which is the best single predictor of $Y$ may receive a near-zero weight in the regression equation.

(b)   As (a) makes clear, a given variable's contribution to the total predictability achieved by the set of variables is highly dependent on context, that is, on what other measures are included in the predictor battery.

(c)   There are a number of possible measures of a particular predictor variable's "importance." One of these is simply the absolute value of that variable's standardized ($z$-score) regression coefficient $b_i$, since large values of $b_i$ indicate that subjects' scores on that predictor variable will have a big influence on their scores on the linear combination of variables which defines their predicted scores on $Y$. Another measure is $b_i r_{iy}$, the product of the variable's regression weight and its correlation with $Y$. This has the advantage that the sum of this measure over all predictor variables exactly equals $R^2$, so that using this measure "partitions" $R^2$ into additive components—assuming that each product is positive. The product can also be rewritten as

$$b_i r_{iy} = b_i^2 + \sum_{i>j} b_i b_j r_{ij},$$

which makes clear the dependence of variable $i$'s contribution on its correlations with the other variables. A third measure is $b_i^2 (1 - R_{i.c}^2)$, which turns out to be equal to the decrease in $R^2$ which would result from dropping variable $i$ from the prediction equation. A final measure is the correlation between variable $i$ and the linear composite constituting $Y$, which turns out to be equal to $r_{iy}/R^2$, that is, proportionate to the simple correlation between that variable and $Y$.

This last mentioned relationship has been used by some authors (for example, Cooley and Lohnes, 1971) to argue that regression coefficients should be ignored in assessing a variable's contribution to the prediction of $Y$. This argument draws heavily on the analogy to traditional practice in interpreting the results of a factor analysis. However, the facts cited in points (a) and (b) seem to the author to demonstrate that this argument should be directly reversed. The uselessness of $r_{iy}$ as a "predictor" of $b_i$ in the multiple regression context strongly suggests that we should put the major interpretative burden in factor-naming, not on the loadings (which are analogous to $r_{iy}$), but on the factor–score coefficients (analogous to the $b_i$s)—or possibly on the loading-coefficient products (analogous to $b_i r_{iy}$) or on the corrected squared factor–score coefficients, analogous to $b_i^2(1 - R_{i \cdot c}^2)$. (Recall that $R_{i \cdot c}^2$, the squared multiple correlation between variable $i$ and the other variables, can be readily computed as 1.0 minus the reciprocal of the $i$th main diagonal entry of $\mathbf{R}^{-1}$.)

It is true that if variable $i$ has a high loading on factor $j$, subjects scoring high on variable $i$ would tend to score high on factor $j$. However, this is relatively useless information if it turns out that the factor–score coefficient corresponding to variable $i$ is close to zero, since this tells us that the greatest predictability of a subject's position on factor $j$ is achieved with a linear equation which essentially ignores variable $i$. Had we been interested in hypothetical variables only as they relate to single variables, we would have had no need for factor analysis in the first place. Further, the factor score coefficients come much closer to an operational definition of the factor than do the loadings.

This emphasis on factor–score coefficients as providing the most direct interpretation of the factors was one of Guttman's principal motives for developing image analysis. Guttman (1955, 1956 as cited by Mulaik, 1972) pointed out that if the multiple correlation between a factor and the original variables drops as low as .707 $(= \sqrt{.5})$, it is possible to find two sets of factor scores which are uncorrelated with each other but which are both consistent with the factor loadings! He further showed that PFA leads to determinate factor scores if $h_j^2 = R_{j \cdot c}^2$—which is of course the basic assumption of image analysis. (Mulaik, 1972, states this as a *necessary* condition for determinate factor scores, which it clearly is not, since $h_j^2 = 1$ also leads to determinate factor, or principal component, scores.)

One implication of turning to an emphasis on factor–score coefficients is that if our goal is to describe the nature of the variables underlying the covariance among our observed variables, then our criterion for rotation should be, not the simplicity of the factor structure, but the simplicity of $\mathbf{B}_z$, the $m \times p$ matrix of regression coefficients predicting $z$-scores on the factors from $z$-scores on the original variables. The author has been unable to unearth any systematic treatment of such rotation techniques in the literature.

## 7.7   Relative Merits of Principal Component Analysis versus Factor Analysis

Before launching into a general discussion, let us return to our PFA of the PCA Demonstration Problem (Chapter 6) data for some examples. We already have the $\mathbf{R}$-derived PCA from this Demonstration Problem, and the factor structure from a PFA based on communalities $h_1^2 = .821$, $h_2^2 = h_3^2 = 1.0$. On first attempting a PFA based on equal communalities, it was discovered that solution of the theoretical expression for the rank 2 communalities in that case could *not* ignore the $2r_{12}r_{13}r_{23}$ and $r_{23}^2$ terms. Solving the complete cubic equation, $c^3 + .01738 - .83840c = 0$ (cf. Digression 3) led to the value $c = .905$ (rather than the .916 previously estimated). The factor structure, the Kaiser-normalized factor structure (which is what is submitted to rotation in most applications of varimax rotation), and the $z_x$-to-$z_f$ factor pattern for each of these three analyses are compared in Tables 7.1–7.3. Note that all three factor structures are highly similar, with the major effect of lowering variable 1's communality $(PFA_1)$ having been to reduce its loadings on the hypothetical variables, with a slight additional tendency to raise the loadings of variables 2 and 3. Both of these effects can be seen as arising from the requirement that the sum of the squared loadings in a given row must equal the communality for that variable.

*Table 7.1*

Comparison of Factor Structures for PCA versus Two PFAs of Same Data

|  | PCA ($PC_1$ and $PC_2$) | | | $PFA_1$ | | | $PFA_2$ | | |
|---|---|---|---|---|---|---|---|---|---|
|  | $PC_1$ | $PC_2$ | $h_j^2$ | $f_1$ | $f_2$ | $h_j^2$ | $f_1$ | $f_2$ | $h_j^2$ |
| $z_1$ | .976 | .004 | .952 | −.906 | .004 | .821 | .951 | .004 | .905 |
| $z_2$ | −.757 | −.631 | .972 | .775 | −.632 | 1.000 | −.738 | −.600 | .905 |
| $z_3$ | −.633 | .762 | .981 | .648 | .762 | 1.000 | −.617 | .724 | .905 |
| $\sum r_{xf}^2$ | 1.926 | .979 | 2.905 | 1.842 | .979 | 2.821 | 1.831 | .884 | 2.715 |

*Table 7.2*

Comparison of Kaiser-Normalized Factor Structures

|  | PCA | | | $PFA_1$ | | | $PFA_2$ | | |
|---|---|---|---|---|---|---|---|---|---|
|  | $PC_1$ | $PC_2$ | $h_j^2$ | $f_1$ | $f_2$ | $h_j^2$ | $f_1$ | $f_2$ | $h_j^2$ |
| $z_1$ | 1.000 | .004 | 1.000 | −1.000 | .005 | 1.000 | 1.000 | .004 | 1.000 |
| $z_2$ | −.768 | −.640 | 1.000 | .775 | −.632 | 1.000 | −.768 | −.640 | 1.000 |
| $z_3$ | −.639 | .769 | 1.000 | .648 | .762 | 1.000 | −.639 | .769 | 1.000 |
| $\sum r_{xf}^2$ | 1.998 | 1.001 | 3.000 | 2.021 | .980 | 3.000 | 1.998 | 1.001 | 3.000 |

*Table 7.3*

Comparison of Factor–Score Coefficients

|  | PCA | | $PFA_1$ | | $PFA_2$ | |
|---|---|---|---|---|---|---|
|  | $PC_1$ | $PC_2$ | $f_1$ | $f_2$ | $f_1$ | $f_2$ |
| $z_1$ | .705 | .004 | .000 | .000 | .494 | .004 |
| $z_2$ | −.546 | −.637 | .762 | .648 | −.383 | −.612 |
| $z_3$ | −.456 | .770 | .632 | .776 | −.320 | .740 |
| $R_{f \cdot x}^2$ | 1.000 | 1.000 | 1.000 | 1.000 | .948 | .903 |

   The similarity in the factor structures for the PCA and $PFA_2$ is even greater. Each column of the $PFA_2$ factor structure is directly proportionate to the corresponding column of PCA, the "reduction factor" being .975 ($= \sqrt{1 - .095/1.926}$) and .950 ($= \sqrt{1 - .095/.979}$) for loadings on factor 1 and factor 2, respectively. Thus the pattern of the loadings is identical for the two analyses, and we gain nothing by our insertion of communalities except perhaps a feeling of having virtuously "chastised" ourselves for having error variance in our data. As the equivalences in parentheses in the sentence before last suggest, this close match between PCA and equal-communalities PFA is no freak coincidence. In fact, *for any equal-communalities PFA, the jth column of the factor structure is equal to the jth column of the unity-communalities* (R-*based PCA*) *factor structure multiplied by* $\sqrt{1 - (1 - c)/\lambda_j}$ *, where c is the common communality and* $\lambda_j$ *is the eigenvalue associated with the jth PC.*

※To see this, consider the general problem of performing an equal-communalities PFA. We seek the eigenvalues and eigenvectors of

$$\mathbf{R}_c = \mathbf{R} - \mathbf{D}(1 - h_j^2) = \mathbf{R} - (1 - c)\mathbf{I}$$

in the equal-communalities case. Thus we seek solutions to the matrix equation

$$[\mathbf{R}_c - \lambda\mathbf{I}]\mathbf{a} = \mathbf{0};$$

that is, to

$$[\mathbf{R} - (1 - c)\mathbf{I} - \lambda\mathbf{I}]\mathbf{a} = \mathbf{0};$$

whence to

$$[\mathbf{R} - (\lambda + (1 - c))\mathbf{I}]\mathbf{a} = \mathbf{0};$$

that is, to

$$[\mathbf{R} - \lambda^*\mathbf{I}]\mathbf{a} = \mathbf{0}, \qquad \text{where} \quad \lambda^* = \lambda + (1 - c).$$

However, $\lambda_j^*$ is the jth characteristic root (eigenvalue) of **R**. Clearly, then, the characteristic vectors of $\mathbf{R}_c$ are the same as those of **R**. However, when we multiply by $\sqrt{\lambda_j}$ to obtain the loadings on factor $j$, we multiply by

$$\sqrt{\lambda^* - (1 - c)} = \sqrt{\lambda^*} \cdot \sqrt{1 - (1 - c)/\lambda^*}. \quad \text{QED}$$

   Next, note that Kaiser normalization makes the structures even more similar, tending to "undo" the initial effects of inserting communalities. In fact, the Kaiser-normalized factor structure for the equal-communalities PFA is *identical* to the Kaiser-normalized factor structure for the **R**-derived PCA of the same data! Thus what is currently the most popular procedure for analytic rotation, the procedure which Kaiser claims satisfies the long-sought goal of factor invariance, essentially amounts to ignoring any differences among the variables in communality. Of course, we "shrink" the loadings back to nonunity communalities after completing the rotation—but that is like saying that we are to append a note reminding ourselves that not all of the variance for a given variable is shared variance.

   Table 7.3 compares the relationship between factor scores and original variables for each of the three analyses. As was pointed out in Section 7.1, the major cost of

PFA as compared to PCA (ignoring the difference in computational effort) is the resulting loss of contact between original variables and hypothetical variables, as reflected in the drop in $R^2_{f \cdot x}$ (the squared multiple correlation between each factor and the original variables) from unity to lower values when nonunity communalities are inserted. As Table 7.3 makes clear, the magnitude of this drop is generally considerably smaller than the "average" percentage reduction in communality. In fact, for the present data, lowering the communality of variable 1 to .8209 had *no* discernible effect on $R^2_{f \cdot x}$. (This is due to the fact that the communality selected for variable 1 is identically equal to its squared multiple $R$ with the other two variables, which Guttman (1956) showed to be a sufficient condition for determinate factor scores.) In the equal-communalities case, the reduction in $R^2_{f \cdot x}$ is directly proportional to the decrease (from unity) in communality and inversely proportional to the magnitude of the characteristic root associated with that particular factor, that is, $R^2_{f \cdot x} = 1 - (1 - c)/\lambda_j$, where $c$ is the common communality and $\lambda_j$ is the $j$th characteristic root of $\mathbf{R}$, the "full" correlation matrix. Thus, for those PCs which are retained (usually those having eigenvalues greater than unity), the reduction in $R^2_{f \cdot x}$ is less than the reduction in $h^2_j$. It is, moreover, concentrated in the less important factors, thus being consistent with our intuitive feeling that the low-variance PCs are more likely to be describing nonshared variance than are the first few PCs. (Note that we did not say "*confirming* our intuition …," since it is far from clear in what sense inserting communalities "removes" nonshared variance.) This relatively small effect on $R^2_{f \cdot x}$ of inserting communalities is typical of PFA. For instance, Harman (1967, p. 355) gives an example of a PFA of an eight-variable correlation matrix in which two factors were retained and in which the mean communality was .671, but $R^2_{1 \cdot x} = .963$ and $R^2_{2 \cdot x} = .885$. Harman also, incidentally, provides a general formula for the magnitude of the squared multiple correlation between factor $j$ and the original variables, namely

$$R^2_{f_j \cdot x} = 1 - \mathbf{r}'_{x,f_j} \mathbf{R}^{-1} \mathbf{D}(1 - h^2_j) \mathbf{r}_{x,f_j}. \tag{7.10}$$

However, *any* loss of contact between the original variables and the hypothetical variable is to be regretted if it does not gain us anything in terms of the simplicity of our description of the factor loadings or of the $x$-to-$f$ factor pattern—which it certainly does *not* in the equal-communalities case. It therefore seems compelling to the author that *PCA is greatly preferable to PFA unless strong evidence exists for differences among the original variables in communalities.*

Table 7.3 does suggest that two quite similar factor structures may yield $x$-to-$f$ factor patterns which differ greatly in their interpretability. For instance,

$$PC_1 = .705z_1 - .546z_2 - .456z_3,$$

while $f_1$, the first factor from $PFA_1$, is given by

$$PFA_1 = .762z_2 + .632z_3.$$

In other words, $f_1$ does not involve $x_1$ at all, even though $x_1$ has a loading of $-.906$ on $f_1$. This reinforces the emphasis of Section 7.6 on interpreting factor–score coefficients, rather than factor loadings. Note, however, that the expressions for $PC_1$

and $f_1$ are not as discrepant as they appear at first, since we have

$$z_1 \simeq -.906f_1 = -.690z_2 - .573z_3\,,$$

whence

$$PC_1 = .705z_1 - .546z_2 - .456z_3 \simeq -1.032z_2 - .860z_3 = -1.20f_1.$$

Nevertheless, the task of *interpreting* $f_1$ as initially expressed will probably be considerably simpler than that of interpreting $PC_1$, even though $Ss$' scores on the two will be highly similar. In general, then, inserting communalities *may* lead to more readily interpretable factors if communalities are unequal.

This is, however, a considerably weaker claim for the virtues of FA than is implied by the usual statement that FA "removes" nonshared variance which PCA leaves. The altogether too tempting inference that the relationships revealed by FA are more "reliable" than those contained in a PCA of the same data, since error variance has been "removed" along with all the other nonshared variance, must of course be strongly resisted. In fact, the opposite is probably closer to the truth.

Mulaik (1972), working from classic reliability theory, has shown that weighted component analysis (cf. Section 7.5) in which the weighting factor for each variable is equal to its true variance (the variance of true scores on that measure, which is equal to its observed variance minus error of measurement) minimizes the proportion of error variance to be found in the first (or first two or first three ...) PCs. This procedure is equivalent to performing a PFA with reliabilities in the main diagonal of **R**. However, the uniqueness of a variable (1 minus its communality) is equal to the sum of its error variance and its specificity (the latter being its reliable, but nonshared variance, that is, that variance attributable to factors which are not common to any of the other variables). It seems quite likely that in most situations where FA is employed, the variables would be nearly homogeneous with respect to their reliabilities, but heterogeneous with respect to their communalities. In such cases, if accurate communality estimates are obtained, relatively more of the total error variance in the system will be retained in the first $p$ principal factors than in the first $p$ PCs. (Of course, factors which are as nearly errorfree as possible would be obtained by inserting reliabilities, rather than either unity or communalities in the main diagonal of **R**. However, obtaining accurate estimates of reliabilities is at least as difficult as determining communalities.) A systematic survey of the relative homogeneity of communalities and reliabilities in "typical" situations would help to resolve this issue. In the meantime, glib references to FA's "removal" of nonshared variance are apt to be misleading.

# 8 | THE FOREST REVISITED

There are at least four issues which cut across the specific materials of Chapters 2–7, but which require some acquaintance with the material in those chapters for meaningful discussion. The present chapter discusses: (a) the applicability of the multivariate techniques discussed so far to data having "weaker" than interval scale properties; (b) the robustness of multivariate significance tests in the presence of violations of the assumptions (especially multivariate normality) used to derive them; (c) how nonlinear relationships among or between variables can be detected, and how to handle nonlinearity when it is detected; and (d) the utility of the very general techniques provided for testing the multivariate general linear hypothesis as an alternative to the more specialized procedures described in Chapters 2–7. Finally, some suggestions for further study are offered.

## 8.1    Scales of Measurement and Multivariate Statistics

As was mentioned briefly in Chapter 1, a number of psychologists, most notably S. S. Stevens (e.g., 1951, 1968), have taken the position that most of the most common and most powerful statistical procedures—including all of those we have discussed in this Primer—are meaningless unless the researcher can establish that his data were generated through measurement processes having at least interval scale properties. For all other data (that is, almost all data gathered by psychologists, sociologists, or political scientists), other statistical procedures designed specifically for ordinal or nominal data must be employed—if they are available. Such nominal- and ordinal-scale versions of multivariate statistical procedures are not in general available, except in the limited sense that the entries in a correlation matrix can be Spearman rank order correlations or phi coefficients, instead of Pearson $r$s, and any of the multivariate procedures discussed in this Primer can take the correlation matrix as its starting point. This substitution of nominal or ordinal coefficients for the "interval-scale" Pearson $r$ does not eliminate the problem that, from the point of view of Steven's adherents, the optimization criteria (for example, percent of variance accounted for, ratio of among-group to within-group variability), as well as the fundamental operation of computing various linear combinations of scores on the original variables, are meaningless. Thus if Stevens's strictures are adhered to, multivariate statistical procedures of any sort would be available for and applicable to only a very small percentage of the data collected by social and behavioral scientists.

That the author does *not* accept Stevens' position on the relationship between strength of measurement and " permissible " statistical procedures should be evident from the kinds of data used as examples throughout this Primer: level of agreement with a questionnaire item, as measured on a five-point scale having attached verbal labels; dichotomous variables such as sex converted to 0–1 numeric codes; and the qualitative differences among the treatments administered $k$ experimental groups converted to $k-1$ group membership 0–1 variables. The most fundamental reason for this willingness to apply multivariate statistical techniques to such data, despite the warnings of Stevens and his associates, is the fact that the validity of statistical conclusions depends only on whether the numbers to which they are applied meet the *distributional* assumptions (usually multivariate normality and homogeneity of covariance matrices from group to group) used to derive them, and *not* on the scaling procedures used to obtain the numbers. In other words, statistical tests are " blind " as to how the numbers " fed " them were generated. Moreover, we have such strong mathematical and empirical evidence of the robustness of statistical procedures under violation of normality or homogeneity of variance assumptions (cf. Section 8.2) that the burden of proof must be presumed to be on the shoulders of those who claim that a particular set of data can be analyzed only through " nonparametric " (a better phrase would be " distribution free ") statistical techniques.

The various alternative measures of correlation which have been proposed for " non-interval-scale " data provide especially convincing demonstrations of the low returns obtained by fretting over whether data have " truly " interval scale properties. Most introductory statistics texts provide special computational formulas for Spearman's $\rho$ (or $r_s$) for use when two ordinal variables are to be related; the point biserial correlation coefficient $r_{pb}$ and the biserial correlation $r_b$ for the relationship between a dichotomous, 0–1 variable and an interval-scale measure; and the phi coefficient $\phi$ or the tetrachoric correlation $r_{tet}$ when two dichotomous variables are to be related. (Cf. Glass and Stanley, 1970, for computational formulas.) These various measures of relationship fall into two groups.

The special formulas for $r_s$, $r_{pb}$, and $\phi$ turn out to yield precisely the same numerical values as would be obtained by blindly applying the formula for Pearson's product–moment coefficient of correlation $r$ to the ranked or dichotomized data. (If the original data were not in the form of ranks or 0–1 measures, $r$ computed on the original data will generally differ from $r_s$, $r_{pb}$, or $\phi$.) Moreover, the significance levels for $r_s$, $r_{pb}$, and $\phi$, computed so as to take into account the special nature of the sets of data being related, are so close to the corresponding significance levels for Pearson's $r$ as to be practically indistinguishable. Table 8.1 lists these significance levels for a few values of $N-2$, the degrees of freedom on which the sample estimate of the corresponding population correlation is based. This table provides, by the way, an excellent example of the robustness of Pearson's $r$ under violations of the assumption of normality. A set of ranks has a rectangular distribution, and a set of 0–1 scores has a binomial distribution, and yet ignoring the departures from normality introduces very little error in the resultant significance test. Thus the only reason for using the special computational formula for $r_s$, $r_{pb}$, or $\phi$, rather than simply using the Pearson $r$ formula, is computational convenience. For most problems in which multivariate statistical procedures are required, a computer program will be used for computing correlations, so that computational convenience is actually on the side of Pearson's $r$, since most computer programs employ $r$ rather than $r_s$, $r_{pb}$, or $\phi$. One can only pity the not entirely apocryphal doctoral student whose advisor sends him in search of a computer program which can compute $r_s$, $r_{pb}$, or $\phi$, since his data have "only" ordinal or dichotomous properties. The computational formula for $r_s$ also illustrates the complete bankruptcy of the argument sometimes made by Stevens' followers that level of measurement determines the kinds of arithmetic operations that can be carried out on the data. For instance, addition and subtraction are said to be meaningless for ordinal data and multiplication and division of pairs of interval-scale numbers is *verboten*. Following these proscriptions would successfully eliminate $r_s$, since it relies on squared differences between the rankings.

*Table 8.1*

Representative   Critical   Values   for   Measures   of
Association[a]

| df = N − 2 | r | $r_s$ | $r_{pb}$ | $\phi$ |
|:---:|:---:|:---:|:---:|:---:|
| 5 | .754 | .786 | .754 | .740 |
| 10 | .576 | .591 | .576 | .566 |
| 15 | .482 | .490 | .482 | .475 |
| 20 | .423 | .428 | .423 | .418 |
| 25 | .381 | .385 | .381 | .378 |

[a] All critical values are for $\alpha = .05$.

The remaining two commonly used correlation measures—the biserial correlation $r_b$ and the tetrachoric coefficient of correlation $r_{tet}$—are "wishful thinking" measures. Each is the correlation measure which would have been obtained if the subjects' "true" scores on what the researcher assumes are the underlying bivariate normal dimensions had been correlated, that is, if these true scores had not been "distorted" from their "truly" normal distribution by floor effects, ceiling effects, the crudity of the available (pass–fail) scale, or a discrepancy between fact and assumption. Use of these measures is based on the untestable assumption that true relationships always yield normally distributed variables.

There is a third, much less commonly used, set of measures of association which *do* involve more than a renaming of or wishful adjustment of Pearson's $r$. Kendall's tau $\tau$ is, like $r_s$, a measure of the association between two sets of ranks. However, $\tau$ has the straightforward property (not shared by $r_s$) of being equal to the difference between the probability that two randomly selected subjects will be ordered identically on both variables, and the probability that their relative positions on one variable will be the reverse of their relative positions on the other variable. Along similar lines, Goodman and Kruskal (1954) have developed measures of association for two-way contingency tables (bivariate frequency distributions for nominal data) which are equal to the gain in probability of correctly predicting a randomly selected subject's category on the second variable which results from knowing his category on the first variable. Thus $\tau$ and the Goodman–Kruskal measures have probabilistic interpretations that are quite different from the "shared variance" interpretation on which Pearson's $r$, $r_{pb}$, $r_s$, $\phi$, $r_b$, and $r_{tet}$ are based. The researcher may find these probabilistic interpretations more congenial in some situations and may therefore prefer $\tau$ or the Goodman–Kruskal measures. Such a preference carries with it, however, the penalities of computational inconvenience and inability to integrate the resulting measure into multivariate statistical procedures.

It is hoped that the above discussion of alternatives to Pearson's $r$ has convinced the reader that searching out statistical procedures and measures of relationship especially designed for nominal or ordinal data is usually a waste of time. This is *not* to say, however, that the researcher may simply ignore the level of measurement provided by his data. It is indeed crucial for the investigator to take this factor into account in considering the kinds of *theoretical statements* and generalizations he makes on the basis of his significance tests. The basic principle is this: Consider the range of possible transformations, linear or otherwise, which you could apply to your data without changing any important properties or losing any of the useful information "stored" in your numbers. Then restrict your generalizations and theoretical statements to those whose truth or falsehood would not be affected by the most extreme such permissible transformation. This is, as the reader who has encountered Stevens's classification of strength of measurement into nominal, ordinal, interval, and ratio will recognize, precisely the logic he employs in proscribing, for example, arithmetic means as measures of central tendency for data which have only ordinal properties. Data with only ordinal scale properties may be subjected to any monotonic (order-preserving) transformation without affecting any important properties. Therefore any two sets of ordinal data for which it is true that $\bar{X}_1 > \bar{X}_2$, but that the largest single observation falls in group 2 *or* the smallest single observation falls in group 1 can be converted to a set of transformed data in which $\bar{X}'_1 < \bar{X}'_2$ simply by making the largest observation many orders of magnitude larger than any other observation (this of course does not affect its rank in the distribution) or by making the smallest observation many orders of magnitude smaller than any other observation. For instance

| Group | Original data | $\bar{X}$ | Transformed data | $\bar{X}'$ |
|-------|---------------|-----------|------------------|------------|
| 1 | 1, 7, 3, 2, 2 | 3.0 | 1, 70, 3, 2, 2 | 15.6 |
| 2 | 4, 6, 6, 3, 6 | 5.0 | 4, 6, 6, 3, 6 | 5.0 |

It is perfectly legitimate to compute the two sample means for either the original or the transformed data and to use "interval scale" statistical tests such as the $t$-test to determine whether the two sets of numbers could have arisen through random sampling from populations of numbers having identical means. However, any generalization from the statistical tests to theoretical statements about the true effects of whatever *conceptual* independent variable differentiates the two groups, on the *conceptual* dependent variable $E$ was attempting to measure is meaningless (not verifiable) if indeed transformations like the one we applied ("multiply the largest value by ten and leave the others unchanged") are permissible—as they must be if the original data have *only* interval-scale properties, since in that case *any* monotonic transformation leaves all important properties unchanged. However, most researchers (and most readers of research reports) would find this sort of transformation completely *un*acceptable, would feel that it does lose (more strongly, distort) meaningful information contained in the data. For instance, assume the original numbers represented each subject's stated level of agreement (on a scale ranging from 0 to 10) with the proposition that "Men should be barred from any job for which there is a qualified woman candidate until the sex ratio for persons holding such jobs reaches unity." Probably no psychologist would claim that the difference between subjects responding "6" and "7" in antimasculinism (assuming this to be the conceptual dimension that the question was designed to tap) is the same as the difference in antimasculinism between subjects who responded "2" and "3." Thus the measurement process would readily be admitted not to guarantee interval-scale measurement. On the other hand, the difference in antimasculinism represented by the difference in responses of "6" and "7" is almost certainly nowhere near ten times as great as the conceptual difference between responses of "2" and "3." In other words, the measurement process—as is quite commonly the case—provides *some* information about relative magnitudes of differences between individuals on the property or dimension being measured, but not *complete* or fully reliable information. Perhaps such measurement processes should be labeled as having "reasonably ordinal" properties, since only *reasonable* monotonic transformations of the resulting numbers are permissible. The conditions for reasonableness of a transformation cannot be easily summarized mathematically, but they probably include at least the following two conditions:

(a)  The transformation must be *a priori*, that is, specified (or at least specifiable) before the actual data obtained in the study are examined.

(b)  The transformation must preserve the ordering not only of all pairs of observations actually obtained in the study, but also of all pairs of observations which *might* have been obtained.

These two conditions are most easily met by continuous functions which can be expressed in closed form, such as square root, logarithmic, and arc sine transformations—and indeed empirical sampling studies have shown that such "reasonable" or "well-behaved" transformations of data have miniscule effects on the results of normal-curve-based statistical tests.

On the other hand, if the researcher has committed himself to constructing a theoretical framework or a set of empirical operations which depends *only* on ordinal-scale measurement of his concepts, then he must confine himself to statements which are meaningful under *any* monotonic transformation, not just reasonable ones. For example, many sociologists have concerned themselves with the issue of the *shape* of various societies's class structure. Does a bar graph of the number of persons in that society falling into the upper, middle, and lower classes have a pyramidal, rectangular, or inverted pyramidal shape? These same sociologists often claim that any given indicator of social status (for example, income) provides a continuous, ordinal measure of true status, with the dividing line between the classes being a matter of arbitrary decision on the researcher's part. However, if this is true, then the question of shape is a meaningless one, since the percentage of subjects falling into the various classes (and thus the shape of the frequency distribution) can be manipulated at will simply by moving the (arbitrary) dividing lines. If a pyramidal shape is desired, put the dividing line between the upper and middle classes at an annual income of $250,000 and the lower–middle dividing line at $15,000; for an inverted pyramid, put the lines at $2000 and $50; etc. Statistical tests such as chi-square goodness-of-fit tests can certainly be used to indicate whether any given resultant frequency distribution differs significantly from a rectangular one, but under the researcher's assumptions about strength of measurement, such tests tell us more about his preconceptions about a particular society than about any empirical property of that society. The point is that the issue of strength of measurement comes into play *not* in determining what statistical tests are "permissible," nor in deciding whether the results of such tests are valid, but rather in deciding what sorts of meaningful (that is, verifiable) theoretical statements about relationships among the conceptual variables being tapped by $E$'s measurements can be made. (The class structure example just cited also illustrates that Stevens' classification of levels of measurement as nominal, ordinal, interval, or ratio provides not even an ordinal scale of "level of measurement," since there is no question in the researcher's mind that ordinal measurement of social status is possible, but considerable doubt as to whether nominal measurement is possible! See Brown's (1965) *Social Psychology* text for a full discussion of the problem of providing nonarbitrary definitions of class boundaries.)

There are some multivariate statistical procedures which are more likely than others to provide answers to questions which are not theoretically meaningful even for "reasonably ordinal" data. The most striking example is provided by *profile analysis* of data undergoing $T^2$ or Manova analysis. If the researcher would find it reasonable or acceptable to change the unit and/or origin of *individual variables* within the set of outcome measures, the shape of the grand mean profile and parallelism or lack thereof of the various group profiles become largely a matter of the arbitrary choices of unit and origin for each of the scales, and profile analysis, while valid statistically, would have little theoretical value for the researcher. Reasonable monotonic transformations applied to *every* variable in the vector of outcome measures would not, however, seriously affect the results of a profile analysis. The reader familiar with univariate Anova will recognize the parallels between the present discussion of profile analysis and the issue of the interpretability of interaction effects. The general conclusion for both Anova and profile analysis is that interactions which involve an actual *reversal* of the direction of a difference (called *crossover* interactions) are unaffected by monotonic transformations, while interactions which do not involve a crossover but merely a change in the absolute magnitude of a difference may be made to "disappear" by judicious selection of a transformation.

Two additional areas in which $E$ needs to give especially careful thought to whether the results of his multivariate statistical tests may be greatly affected by arbitrary scaling decisions are:

(a)  attempting to determine the presence or absence of significant nonlinearity in the relationships among variables (cf. Section 8.3);

(b)  assessing the relative importance of the individual variables in a set as contributors to multiple $R$, $R_c$, $T^2$, the gcr obtained from a Manova, or a principal component on the basis of *raw-score* regression, discriminant function, or characteristic vector coefficients.

## 8.2  Effects of Violations of Distributional Assumptions in Multivariate Analysis

All of the significance tests outlined in this Primer were derived under at least the assumption of multivariate normality—that is, that the observed data vectors are independent random samples from a population in which any linear combination of the variables in the data vector is normally distributed. In addition, $T^2$ and Manova tests involve the assumption that the populations from which the different groups' data were sampled all have identical covariance matrices. These two assumptions are almost certainly *not* valid for any real set of data—and yet they are *nearly* valid for many sets of data. Moreover, the fact that a particular assumption was used in deriving a test does *not* mean that violation of that assumption invalidates the test, since the test may be quite robust under (that is, insensitive to) violations of the assumptions used to derive it. We have extremely strong evidence—both mathematical and empirical—that the univariate tests of which the tests in this Primer are direct generalizations and which were derived under the assumptions of (univariate) normality and homogeneity of variance are in fact extremely robust under violation of those assumptions (cf., for example, Boneau, 1960; Winer, 1971; Lindquist, 1953; Donaldson, 1968; and Norris and Hjelm, 1961). The major exceptions to this statement occur for very small and unequal sample sizes and/or one-tailed tests. The author strongly feels that one-tailed statistical tests are almost never appropriate in a research situation (and in very few applied situations) since they require that the researcher be willing to respond to a fantastically large effect in the direction opposite to his prediction in exactly the same way he would to a miniscule difference among his experimental groups. (However see Part 4 of the book by Steger, 1971, for opposing views on this issue.) We are therefore left with only the unequal-$n$ restriction on our general statement about the robustness of normal-curve-based univariate statistical tests. As a general guideline, normal-curve-based tests on a single correlation coefficient can be considered valid for almost any unimodal $X$ and $Y$ population for any number of observations greater than about 10; while normal-curve-based $F$- or $t$-tests can be considered valid for even $U$-shaped population distributions so long as two-tailed tests are used; the ratio between the largest and smallest sample variance is no greater than about 20 to 1; the ratio between the largest and smallest sample size is no greater than about 4; and the total degrees of freedom for the error term $(s_c^2$ or $MS_w)$ is 10 or more. When the violations of assumptions are close to these bounds, a test at the nominal 5% level might actually have close to a 10% probability of yielding a false rejection of $H_0$. For even grosser violations of the assumptions, transformation of the data so as to meet the assumptions more closely (for example, converting the observations to ranks based on all the observations in all groups, which procedure is referred to as Kruskal–Wallis Anova) or use of a specialized test procedure (for example, Welch's $t^*$-test for the difference between two sample means arising from populations with grossly different variances) may be called for. The book by Winer (1971) is a good reference for such transformations and tests.

Unfortunately, the situation is not quite so clearcut for the multivariate tests described in this Primer. It is easy to present an intuitively compelling argument that tests based on Roy's union–intersection principle (for example, the significance tests provided for multiple regression and gcr tests for $T^2$, Manova, and canonical analysis) should display the same kind of robustness as the univariate tests. In each case the multivariate test involves computing the largest possible value of a univariate test statistic $(r, t, \text{ or } F)$ which could result from applying the univariate statistical procedure to any linear combination of the several variables, and the relevant distribution is thus the distribution of the maximum value of such optimized univariate statistics. If the input to this maximum-value distribution (that is, the distribution of the univariate statistic) is valid despite violation of a particular assumption, then so should be the resulting multivariate distribution. Put another way, we know that the univariate statistic computed on any *a priori* combination of the variables would follow the normal-curve-based univariate distribution quite well despite violations of the assumptions, so any nonrobustness must arise solely in the adjustments needed to correct for the fact that our linear combinations was deliberately picked to provide as large a $t$ (or $r$ or $F$) as possible—an unlikely source of nonrobustness. However, there is many a slip between intuitive conviction and mathematical or empirical proof, and the latter two are in somewhat short supply where the issue of robustness of multivariate significance tests is concerned. What *is* known includes the following:

(a)  Significance tests on the overall significance of a multiple correlation coefficient and on individual regression coefficients are unaffected by the use of *cluster sampling*, a departure from simple random samplings in which a random sample of large units (for example, census tracts or mental hospitals) is selected from the population and then simple random sampling (or additional layers of cluster sampling) is carried out within each macro unit (Frankel, 1971). However, Cotton (1967) points out that most experimental research employs volunteer (or at least self-selected) subjects from an available pool, which constitutes neither random sampling nor cluster sampling, so that the most useful studies of robustness would be studies of the "behavior" of various statistics under conditions of random *assignment* to treatments of a very nonrandom sample of subjects.

(b)  A multivariate generalization of the central limit theorem assures us that for sufficiently large sample sizes vectors of sample means have a multivariate normal distribution (Ito, 1969). The catch comes in trying to specify how large "sufficiently large" is.

(c)  The true significance levels for Hotelling's $T^2$ match the nominal significance levels, despite very nonhomogenous covariance matrices, so long as $N_1 = N_2$. Large discrepancies in both sample sizes and covariance matrices can lead to correspondingly large discrepancies between the true and nominal significance levels (Ito and Schull, 1964).

(d)   The power of the test of homogeneity of covariance matrices reported in Section 3.6.1 appears to be low when nominal and actual significance levels of the overall test in Manova are quite close, and high when the true significance level is considerably higher than the nominal significance level (Korin, 1972). Thus the multivariate test of homogeneity of covariance matrices must be taken somewhat more seriously than Bartlett's test in the univariate situation, which latter test has high power to detect departures from homogeneity which are too small to have any appreciable effect on the overall $F$-test.

(e)   Nonparametric alternatives to the tests described in this Primer are under development for use when the data show evidence of gross violation of multivariate normality or homogeneity of covariance matrices (cf. Krishnaiah, 1969).

The issue of robustness of multivariate tests is the focus of a great deal of current effort by mathematical statisticians, and large strides can be expected within the next few years. Most indications are that multivariate techniques will prove as robust as their univariate counterparts.

## 8.3   Nonlinear Relationships in Multivariate Statistics

All of the statistical procedures discussed in this Primer analyze only the *linear* relationships among the variables within a set or between the variables of different sets. By this we mean that the general formula relating one variable to another variable or to a set of other variables involves only the variables themselves (no logarithms, squares, square roots, etc.), connected only by + or − signs (no multiplicative combinations). The exception to this is that independent, group-membership variables are treated as though they were nominal variables, with no attempt to specify the form of their functional relationship to other variables—except through specific contrasts (*à la* Scheffé) used to specify the source of significant overall effects. Thus the multivariate procedures where the question of linearity arises are:

(a)   those involving combinations of continuous predictor variables, such as the predicted score in multiple regression, the canonical variate in canonical analysis, and the factor score in PCA or factor analysis; and

(b)   those involving combinations of outcome variables, such as the discriminant function in $T^2$ or Manova, the canonical variate in canonical analysis, and the principal component in PCA.

(Note that the canonical variate is mentioned under both categories, as is appropriate to the symmetry of canonical analysis, and that considering the hypothetical variables—principal components, factors—in PCA and FA as the independent variables is the traditional, but not the necessary, way of looking at these analyses.) Of these various multivariate procedures, multiple regression is the one whose users are most apt to plan specifically for, test for, and/or adjust for nonlinearity, though there is certainly just as great a likelihood of nonlinear relationships obscuring the results of the other procedures.

With respect to linearity, we are in much the same situation as with normality and with interval scale measurement: almost no variables have a truly linear relationship over their full range, and yet a great many have a nearly linear relationship over the ranges apt to be included in a given study. Thus, for instance, Weber's law (that any two physical stimuli having a given *ratio* of physical energies will yield the same *difference* in resulting subjective sensations) is known to break down for very low or very high levels of physical energy, but is nevertheless a very useful empirical law over the broad middle range of physical intensity. Moreover, most kinds of nonlinear relationships that can be anticipated on theoretical grounds can be incorporated into a linear model by use of appropriate transformations.

To illustrate this last point, consider a study by Steele and Tedeschi (1967) in which they attempted to develop an index based on the payoffs used in an experimental game which would be predictive of the overall percentage of cooperative choices $C$ made in that game by a pair or pairs of subjects. They had each of 42 subject pairs play a different two-person, two-choice game, the payoffs for the various games being generated randomly. Four payoffs, labeled as T, R, P, and S, were used in each of the 42 games. On theoretical grounds, the authors expected the best index to be a ratio of some two of the following measures: T, R, P, S, T − R, T − P, T − S, R − P, R − S, P − S, or a square root or logarithmic transformation of such a ratio. This unfortunately yields 135 "candidates" for best index. The authors selected among these (and 75 other possible indices generated unsystematically) by correlating them all with $C$, then using the 15 having the highest correlation with $C$ as predictor variables in a multiple regression analysis, and finally comparing the squared regression weights each of these 15 attained in the best predictor equation. Rather than detailing the many ways in which these procedures betray a lack of understanding of the goals of multiple regression analysis, let us consider a more efficient approach to the authors's initial question. The stipulation that the index should involve the ratio of two of the ten measures specified above can be formulated somewhat more generally as

$$C = \prod_{i=1}^{10} c_i^{\beta_i}, \tag{8.1}$$

where $c_1 = T$; $c_2 = R$; ...; $c_{10} = P − S$; and $\beta_i$ is an exponent associated with term $i$. Thus, for instance, the index $(T − R)/(T − S)$ is obtained from (8.1) by taking $\beta_5 = 1$, $\beta_7 = −1$, and all other $\beta_i = 0$. This looks very far from a linear relationship. However, taking logarithms of both sides of (8.1) gives

$$\log C = \sum_{i=1}^{10} \beta_i \log c_i. \tag{8.2}$$

Thus we can apply standard linear multiple regression procedures to (8.2), letting $Y = \log C$ and $X_i = \log c_i$. The $b_i$ which optimally fit (8.2) will provide optimal exponents for (8.1) as well. If we now ask for the optimal weights under the assumption that the ideal index is logarithmically related to the ideal ratio, that is, that

$$C = \log\left(\prod_{i=1}^{10} c_i \beta_i\right) = \sum_{i=1}^{10} \beta_i \log c_i,$$

we see at once that standard multiple regression analysis again applies with the same predictor variables as before, but this time with "raw" cooperation percentage as the predicted variable. If, finally, we assume that the ideal index is the square of a ratio of differences, we need only take $\log(C^2) = 2 \log C$ as our predicted variable. The transformations suggested here may of course lead to distributions sufficiently non-normal to be a problem in significance testing.

This example is not meant to imply that all situations involving nonlinearity where the precise nature of the relevant functions can be spelled out on *a priori* grounds can be reduced by appropriate transformations to linear form. For instance, had we wanted to find the best function of the payoffs to use as a predictor of a whole set of measures of the outcomes of a game (for example, CC, CD, and DC as defined in Data Set 3, p. 68), where the function of the payoffs is to be multiplicative while only linear combinations of the outcome measures are to be considered, we could *not* reduce the problem to the usual (linear) canonical correlation procedures. However, most *a priori* specifications can be reduced to linear combinations of transformed variables. If not, there is little which can be done beyond taking the linear formulas (or polynomial expressions—see below) as approximations to the true combining rule.

Where no *a priori* specification of the nature of any nonlinearity can be made, several options are open to the investigator, including:

(a)   proceeding with the usual linear analysis on grounds either of convenience or of the ease of interpreting linear combinations, recognizing that measures of relationship (multiple $R$, canonical $R$, $T^2$, gcr in Manova, or the percent of variance accounted for by $PC_1$) may be lower than if transformed variables or additional, nonlinear terms were used;

(b)   performing an initial linear analysis, followed by a test for nonlinearity and a modified analysis if and only if the test reveals statistically reliable nonlinearity;

(c)   performing an analysis employing a polynomial model (squares, cubes, etc., *and* cross-products of the original variables), which is highly likely to provide a close approximation to *any* function, followed by deletion of terms in the model which do not make a statistically significant contribution;

(d)   converting the continuous variables into nominal, group-membership variables and performing factorial Anova instead of multiple regression, or Manova instead of canonical analysis.

Approach (a) needs no further explanation, except to point out that it is a particularly compelling choice in the case of PCA or FA, where the "predictor" variables are unobservable anyway. Procedure (b) is an extension to other multivariate techniques of the "step-up" variety of stepwise multiple regression procedures such as are employed by the BMDO2R program (cf. Digression 4), where the variables to be added are squares, cubes, and cross-products of the original variables. Formal significance tests for such stepwise procedures are presented in Table 2.2, p. 48. Similar tests for $T^2$, Manova, and Canona are less well known, though Dempster (1969) applies stepwise procedures to $T^2$ and cites the work of Pothoff and Roy (1964) for the application to Manova. For all these techniques the most intuitively meaningful measure of the improvement provided by addition of one or more terms is the increase in the square of the correlation between the augmented set of variables and the other variable or set of variables in the situation. For multiple regression and canonical analysis this measure is clearly multiple $R^2$ and $R_c^2$, respectively. Since $T^2$ and Manova are (as was shown in Chapter 4) special cases of Canona, the corresponding measure for these techniques is the change in $\theta = \lambda/(1 + \lambda)$, where $\lambda$ is the gcr of $\mathbf{E}^{-1}\mathbf{H} = SS_{a,D}/SS_{w,D}$ (the "stripped" $F$-ratio computed on the discriminant function) in the general case of Manova and $= T^2/(N_1 + N_2 - 2)$ in the special case of Hotelling's $T^2$. For each of these two techniques, $\theta$ represents the square of the canonical correlation between the group membership variables and the outcome measures, and thus provides a simple, readily interpretable measure of the importance of the additional terms.

Such formal procedures may well be preceded by "eyeball tests" of the curvilinearity of relationships. It was pointed out in Chapter 2 (cf. especially the Demonstration Problem) that the best visual check of the linearity assumption in multiple regression is provided by a plot of $Y$ versus $\hat{Y}$ or versus $Y - \hat{Y}$ (that is, against the predicted or the residual scores on $Y$). The primary basis for this conclusion is the fact that curvilinearity of the relationship between $Y$ and a given predictor variable $X_i$ does *not* necessarily imply curvilinearity in the relationship between $Y$ and $\hat{Y}$—which is all that is necessary in multiple regression analysis. By similar arguments it can be seen that a plot of $C_x$ versus $C_y$, where $C_x$ and $C_y$ are the canonical variates of the left- and right-hand sets of variables, is the best overall visual check for curvilinearity in canonical analysis.

Approach (c) is practical only if the investigator has available a very large number of observations, since the total number of variables involved if $k$ original variables, their squares, their cubes, and the cross-products of all these variables are included is $3k + 3k(3k - 1)/2$—which increases quite rapidly with $k$ and has the potential for exhausting the degrees of freedom for the analysis quite rapidly.

Approach (d) has the major problem of being rather sensitive to the particular cutting points established for each variable's set of categories, as well as yielding some very small subgroups of subjects having particular combinations of scores on the various measures, unless a large amount of data is available.

### 8.4   The Multivariate General Linear Hypothesis

All of the significance tests described in Chapters 2–5—including specific comparisons explaining statistically significant overall measures—can be represented as tests of a single, very general hypothesis: the *multivariate general linear hypothesis* (mgl hypothesis). The mgl hypothesis can be stated as

$$\mathbf{C\beta M} = \mathbf{0}, \tag{8.3}$$

where

$$\boldsymbol{\beta} = \begin{bmatrix} \beta_{1,1} & \beta_{1,2} & \cdots & \beta_{1,p} \\ \beta_{2,1} & \beta_{2,2} & \cdots & \beta_{2,p} \\ \vdots & \vdots & \vdots & \vdots \\ \beta_{q,1} & \beta_{q,2} & \cdots & \beta_{q,p} \end{bmatrix}$$

is a $q \times p$ matrix giving the parameters relating the $q$ predictor variables (usually group-membership variables) to the $p$ outcome measures via the multivariate general linear model (mgl model),

$$\mathbf{Y} = \mathbf{X\beta} + \mathbf{\varepsilon}, \tag{8.4}$$

where $\mathbf{Y}$ is an $N \times p$ matrix of scores of the $N$ subjects on the $p$ outcome measures, and is of rank $t$; $\mathbf{X}$ is an $N \times q$ matrix of scores on the predictor variables, is usually referred to as the design matrix, and is of rank $r$; $\mathbf{\varepsilon}$ is an $N \times p$ matrix of residual scores, that is, of errors of prediction, which, for purposes of significance testing, are assumed to be sampled from a multivariate normal population having mean vector $(0, 0, \ldots, 0)$ and covariance matrix $\mathbf{\Sigma}$ of full rank $p$. Thus $\mathbf{\beta}$ is essentially a matrix of population regression coefficients.

Referring back to Eq. (8.3), $\mathbf{C}$ is a $g \times q$ ($g \le r$, the rank of $\mathbf{X}$) matrix whose $j$th row specifies a linear combination of the $q$ parameters which relate the predictor variables to the $j$th outcome measure, while $\mathbf{M}$ is a $p \times u$ matrix ($u \le t$, the rank of $\mathbf{Y}$) whose $i$th column specifies a linear combination of the $p$ parameters which relate the outcome measures to the $i$th predictor variable. Thus, for instance, taking

$$\mathbf{C} = \begin{bmatrix} 1 & 0 & 0 & -1 \\ 0 & 1 & 0 & -1 \\ 0 & 0 & 1 & -1 \end{bmatrix} \quad \text{and} \quad \mathbf{M} = \begin{bmatrix} 1 & 0 & 0 & \cdots & 0 \\ 0 & 1 & 0 & \cdots & 0 \\ 0 & 0 & 1 & \cdots & 0 \\ \vdots & \vdots & \vdots & \ddots & \vdots \\ 0 & 0 & 0 & \cdots & 1 \end{bmatrix}, \text{ with } \mathbf{X} = \begin{bmatrix} 1 & 0 & 0 & 1 \\ 1 & 0 & 0 & 1 \\ \vdots & \vdots & \vdots & \vdots \\ 1 & 0 & 0 & 1 \\ 0 & 1 & 0 & 1 \\ \vdots & \vdots & \vdots & \vdots \\ 0 & 1 & 0 & 1 \\ 0 & 0 & 1 & 1 \\ \vdots & \vdots & \vdots & \vdots \\ 0 & 0 & 1 & 1 \end{bmatrix}$$

(thus setting $g = 3$, $q = 4$, $u = p$), yields a test of the hypothesis that

$$\begin{bmatrix} \beta_{1,1} - \beta_{4,1} \\ \beta_{2,1} - \beta_{4,1} \\ \beta_{3,1} - \beta_{4,1} \end{bmatrix} = \begin{bmatrix} \beta_{1,2} - \beta_{4,2} \\ \beta_{2,2} - \beta_{4,2} \\ \beta_{3,2} - \beta_{4,2} \end{bmatrix} = \cdots = \begin{bmatrix} \beta_{1,p} - \beta_{4,p} \\ \beta_{2,p} - \beta_{4,p} \\ \beta_{3,p} - \beta_{4,p} \end{bmatrix} = \begin{bmatrix} 0 \\ 0 \\ 0 \end{bmatrix},$$

where $\hat{\beta}_{i,j} - \hat{\beta}_{4,j}$ (the least-squares estimates of the corresponding population parameters) will be equal to $\bar{Y}_{i,j} - \bar{Y}_j$, and $\beta_1$ to $\beta_3$ thus correspond to $\mu_1$ to $\mu_3$; $\beta_4$ to $\bar{\mu} = (\mu_1 + \mu_2 + \mu_3)/3$. In other words, these choices of matrices yield the overall null hypothesis for a three-group, $p$-outcome-measure Manova. (Note that the direct equivalent of Eq. (8.3) is a statement of equality between a $3 \times p$ matrix of differences between $\beta$s whose columns are the column vectors listed above and a $3 \times p$ matrix consisting entirely of zeros.) Similarly, taking $\mathbf{C}$ as before but letting

$$\mathbf{M} = \begin{bmatrix} 2 \\ -1 \\ -1 \\ 0 \\ \vdots \\ 0 \end{bmatrix}$$

yields the null hypothesis that none of the three groups has a population mean difference other than zero between $Y_1$ and the average of $Y_2$ and $Y_3$. On the other hand, letting $\mathbf{M}$ be an identity matrix while

$$\mathbf{C} = [1 \quad -1 \quad 0 \quad 0]$$

yields the null hypothesis that groups 1 and 2 do not differ in their population mean outcome vectors.

If the rows of $\mathbf{C}$ meet an *estimability condition* (which is always fulfilled by any *contrast* among the columns of $\boldsymbol{\beta}$, that is, by any set of coefficients which sum to 0), then the null hypothesis (8.3) is tested by computing the greatest characteristic root (gcr) of $\mathbf{E}^{-1}\mathbf{H}$, where

$$\mathbf{E} = \mathbf{M'Y'}[\mathbf{I} - \mathbf{X}_1(\mathbf{X}_1\mathbf{X}_1)^{-1}\mathbf{X}_1']\mathbf{XM}; \tag{8.5a}$$

and

$$\mathbf{H} = \mathbf{M'Y'X}_1(\mathbf{X}_1'\mathbf{X}_1)^{-1}\mathbf{C}_1'[\mathbf{C}_1(\mathbf{X}_1'\mathbf{X}_1)^{-1}\mathbf{C}_1']\mathbf{C}_1(\mathbf{X}_1'\mathbf{X}_1)^{-1}\mathbf{X}_1'\mathbf{YM}, \tag{8.5b}$$

where $\mathbf{X}_1$ is an $N \times r$ submatrix of $\mathbf{X}$ which has the same rank as $\mathbf{X}$. (Such a submatrix is called a *basis* for $\mathbf{X}$, since the subjects' scores on the remaining $q - r$ independent variables can be determined via linear equations from their scores on these first $r$ measures.) This gcr is then compared to the appropriate critical value in the gcr tables of Appendix A with degree-of-freedom parameters:

$$s = \min(g, u); \qquad m = (|g - u| - 1)/2; \qquad n = (N - r - u - 1)/2. \tag{8.6}$$

This is a very broad model which subsumes overall tests, profile analysis, specific comparisons, and a wide range of other tests. One approach to teaching multivariate statistics would be to begin with the mgl hypothesis and present all other significance tests as special cases thereof. This Primer has eschewed that approach for both didactic and practical reasons. On didactic grounds, the multivariate general linear hypothesis is simply too abstract to be readily digestible until the user has built up a backlog of more specific procedures. Having developed some understanding of and experience with multiple regression, $T^2$, Manova, and canonical analysis, the user can be impressed with and perhaps put to use the elegance and generality of the mgl model; without such a background, the main reaction is apt to be one of awe, fear, and/or bewilderment. On the practical level, formulas (8.5) are much more complex than the expressions we had to deal with in Chapters 2–5. Equations (8.5) simply do not lend themselves to "hand" exploration of even the simplest cases. Even where a computer program for the mgl model is available, the very generality of the model means that the user must devote considerable labor to selecting and punching cards for the $\mathbf{C}$, $\mathbf{M}$, and design matrices for even the simplest hypotheses, and this procedure must be repeated for each hypothesis to be tested. Where a canned program for a specific procedure exists, the user will generally find it much more convenient to use it rather than resorting to a mgl hypothesis program. Finn's (1973) MULTIVARIANCE program is reported to be much more convenient than the BMDX73 program in this respect.

On the other hand, a canned program for the mgl hypothesis can prove useful where the "special procedure" program either is not available or requires that some condition which is *not* true of the researcher's data (for example, equal cell sizes in Manova, as required by BMDX69) be met. (Many of these cases can, however, be handled with much less card preparation than for the mgl hypothesis by using a canonical correlation program in which one set of variables are group-membership variables.) For the researcher who has need of the mgl model, Morrison (1967, Chapter 5) provides listings of the **C**, **M**, and design matrices for many of the most often used statistical procedures. Pruzek (1971), in a paper which incorporates many of the suggestions for "reform" of statistics which were discussed in Section 1.0, illustrates the utility of the mgl model as a unifying tool in discussing methods and issues in multivariate data analysis.

## 8.5  Where to Go from Here

The primary advice to the novice is: get Thee to some data. By far the best way to consolidate your understanding of the techniques discussed in this Primer is to apply them to data that are important to you. As suggested in Chapter 1, statistical under-standing resides in the fingertips, be they punching cards for a computer run or operating a desk calculator. Secondary advice, once you have had some experience with the techniques discussed in this Primer, is to use reading of other texts and of journal articles to extend your knowledge along one or more of three lines: greater mathematical sophistication, greater variety of techniques, and acquaintance with opposing points of view.

More sophisticated treatments of multivariate statistics are available in most other texts in this area, especially Morrison (1967) and Anderson (1958). However, the reader who encountered matrix algebra for the first time in this Primer will probably find these two texts much more comprehensible if he first strengthens his knowledge of matrix algebra by reading, for example, Horst (1963) or Searle (1966)—or by taking a course on linear algebra or finite math, or an introductory mathematical statistics course.

The Morrison and Anderson texts are more comprehensive than the Primer in their coverage of statistical techniques, as are the books by Press (1972) and Overall and Klett (1972). A book which covers very nearly the same areas as the present Primer, which is quite readable, and which emphasizes geometric interpretations much more than we have in this book is Tatsuoka's (1972b) excellent text. At the time of this writing, books on multivariate statistics by R. Darrell Bock and Neal Timm were in press and were reported to be well worth investigation.

The reader will probably wish to supplement the rather cursory treatment of factor analysis in the current Primer with a text dealing solely with FA, such as Harman (1967), Mulaik (1972), or Comrey (1973).

For specific procedures not covered in the present text, Tryon and Bailey (1970) should be consulted on cluster analysis, and Tatsuoka has published highly readable booklets on discriminant analysis (1971a,b) and on multivariate significance tests of group differences (1972).

Finally, to those readers who have made it this far: It is the author's hope that he has managed to show the utility of and the basic simplicity of multivariate statistics, while providing enough mathematical and computing skills to enable you actually to compute and interpret such statistics.

# 1 | DIGRESSION

## Finding Maxima and Minima of Polynominals

A problem which crops up time after time in statistics is that of finding those values of $b_0$, $b_1, \ldots, b_m$ which cause a criterion function

$$C = b_0 + a_1 b_1 + a_2 b_2 + \cdots + a_m b_m + c_1 b_1^2 + c_2 b_2^2 + \cdots + c_m b_m^2$$

to take on its largest (or smallest) possible value. The $b$s here are unknowns, while the $a$s and the $c$s are known constants (usually sums or sums of squares of data values). This is a problem for which the tools of differential calculus are ideally suited. We shall of course barely tap the wealth of techniques available in that branch of mathematics.

### D1.1   Derivatives and Slopes

The problem of maximizing (or minimizing) $C$ is approached by setting the *derivatives* of $C$ with respect to $b_0$, $b_1$, $b_2$, $\ldots$, $b_m$ equal to zero and then solving the resultant set of $m + 1$ equations in $m$ unknowns. This approach makes good intuitive sense, since the derivative of any function $f(x)$ with respect to $x$—$df(x)/dx$—is essentially a measure of how rapidly $f(x)$ increases as $x$ increases, that is, of the *rate of change* of $f(x)$, or the number of units of change produced in $f(x)$ by a unit change in $x$. If the function is a continuous one which takes on its maximum value at $x = x^*$, then $f(x)$ must decrease on either side of $x^*$, that is, for any change up or down in the value of $x$. Thus for values of $x$ slightly less than $x^*$, the derivative of $f(x)$ must be positive; for values of $x$ slightly greater than $x^*$, the derivative of $f(x)$ must be negative; and at $x^*$ the derivative must temporarily equal zero. (A similar argument establishes that $df/dx$ must be equal to zero at the function's minimum, too. Higher-order derivatives can be used to tell whether a maximum or a minimum has been located, but the context will usually make this clear, anyway, and for purposes of this text, you may assume that whichever extremum—maximum or minimum—is sought by the derivative procedure has been found.) The only "trick," then, is to be able to compute the derivatives of $C$. This requires application of a very few, intuitively sensible, rules.

***Rule 1*** *The derivative of a constant is zero.* A constant is any expression which does not include within it the variable with respect to which the derivative is being taken. Thus, for instance, $d(a_1 b_1)/db_2$ is equal to zero, since $a_1 b_1$ is not modified in any way by (has a rate of change of zero with respect to) changes in $b_2$.

**Rule 2**   $d(cx)/dx = c$.   A plot of $cx$ against $x$ will of course be a straight line whose slope (increase in vertical direction divided by increase in horizontal direction for any two points on the line) is equal to $c$. Put another way, if $y$ is equal to a constant times $x$, each unit of increase in $x$ is magnified (or shrunk, if $c < 1$) by $c$ in its effect on $y$, with $c$ units of change in $y$ for every one unit change in $x$.

**Rule 3**   $d(cx^2)/dx = 2cx$.   This is the most "difficult" of the rules, since $f(x)$ plotted against $x$ no longer forms a straight line, but a "positively accelerated" curve, for which the rate of change in $f(x)$ is itself dependent on the value of $x$. For values of $x$ close to zero, small changes in $x$ produce very small changes in $x^2$ (and thus in $cx^2$), while the same increment applied to a large value of $x$ makes a relatively large change in $x^2$. For instance, $(.2)^2 - (.1)^2$ is only $.03$, but $(25.1)^2 - (25.0)^2 = 5.01$. Note, too, that the ratio of the change in $x^2$ to the change in $x$ is $.03/.10 = 2(.15)$ in the first case and $5.01/.1 = 2(25.05)$ in the second case—exactly equal to twice the mean of the values $x$ takes on during the change. The derivative of a function for a given value of the differentiating variable (the variable with respect to which the derivative is taken, that is, the variable whose effects on some function we wish to assess) is essentially the *instantaneous* rate of change or, put geometrically, the slope of a straight line tangent to the curve representing the function at a point directly above the specified value of the differentiating variable. As can be seen from a plot of $f(x) = x^2$ against $x$, this slope is close to zero when $x$ is zero, but rapidly increases as $x$ increases.

**Rule 4**   *The derivative of a sum is equal to the sum of the derivatives, that is,*

$$d[f(x) + g(x) + \cdots]/dx = df(x)/dx + dg(x)/dx + \cdots,$$

or

$$d[\textstyle\sum f_j(x)]/dx = \sum [df_j(x)/dx].$$

This rule makes excellent intuitive sense, since it essentially asserts that net rate of change is equal to the sum of the separate rates; for example, the total rate at which a reservoir into which each of three streams brings 2000 gallons per minute and out of which the city takes 4000 gallons per minute, while evaporation loss is 1000 gallons per minute, is $2000 + 2000 + 2000 - 4000 - 1000 = 1000$ gallons per minute. As you might suspect from the frequency with which summation signs occur in statistics, this is an oft-used rule.

**Application of Rules 1–4**   As an example of the application of these rules, consider a situation (for example, univariate regression) in which we want to select values of $b_0$ and $b_1$ which will make $E = \sum (Y - b_0 - b_1 X)^2$ as small as possible. First, in order to get the criterion function into a form in which Rules 1–4 can be applied, we must carry out the squaring operation, yielding

$$E = \sum (Y^2 + b_0^2 + b_1^2 X^2 - 2b_0 Y - 2b_1 XY + 2b_0 b_1 X).$$

We next apply Rule 4, to give

$$dE/db_0 = \sum d(Y^2 + b_0^2 + \cdots + 2b_0 b_1 X)/db_0$$

$$= \sum [d(Y^2)/db_0 + d(b_0^2)/db_0 + \cdots + d(2b_0 b_1 X)/db_0]$$

and

$$dE/db_1 = \sum [d(Y^2)/db_1 + d(b_0^2)/db_1 + \cdots + d(2b_0 b_1 X)/db_1].$$

Now we can apply Rules 1–3 to the separate derivatives. Thus, using Rule 1:

$$d(Y^2)/db_0 = d(b_1^2 X^2)/db_0 = d(-2b_1 XY)/db_0$$

$$= d(Y^2)/db_1 = d(b_0^2)/db_1 = d(-2b_0 Y)/db_1 = 0.$$

Note that, for instance, $b_1^2 X^2$, even though it involves the unknown value $b_1$, is just a constant when differentiating with respect to $b_0$, since changes in $b_0$ have no effect whatever on $b_1^2 X^2$.

Using Rule 2 tells us that

$$d(-2b_0 Y)/db_0 = -2Y; \qquad d(2b_0 b_1 X)/db_0 = 2b_1 X;$$

$$d(-2b_1 XY)/db_1 = -2XY; \qquad d(2b_0 b_1 X)/db_1 = 2b_0 X.$$

Using Rule 3 tells us that

$$d(b_0^2)/db_0 = 2b_0 \qquad \text{and} \qquad d(b_1^2 X^2) = 2b_1 X^2.$$

The end result is that

$$dE/db_0 = \sum (2b_0 - 2Y + 2b_1 X) = 2Nb_0 - 2\sum Y + 2b_1 \sum X;$$

and

$$dE/db_1 = \sum (2b_1 X^2 - 2XY + 2b_0 X) = 2b_1 \sum X^2 - 2\sum XY + 2b_0 \sum X.$$

Finally, to find the values of $b_0$ and $b_1$ which maximize $E$, we must ask what pair of values will cause these two derivatives to simultaneously equal zero. Thus we must solve the system of two simultaneous linear equations,

$$Nb_0 + (\sum X)b_1 - \sum Y = 0; \qquad \text{and} \qquad (\sum X)b_0 + (\sum X^2)b_1 - \sum XY = 0.$$

The solution which results (see page 36 of Chapter 2 for details) is

$$b_1 = \sum xy / \sum x^2 \qquad \text{and} \qquad b_0 = \bar{Y} - b_1 \bar{X}.$$

## D1.2   Optimization Subject to Constraints

There are many situations where we wish to find optimum values of some parameter or set of parameters, but where the criterion we are interested in is not strong enough to define completely the "optimal" set of weights. For instance, we might be interested in finding a pair of weights, $b_1$ and $b_2$, which make the correlation between $Y$ and $W = b_1 X_1 + b_2 X_2$ as large as possible. Clearly there is no single pair which will meet this criterion, since the correlation coefficient is unaffected by linear transformations of either or both sets of variables, so that, for instance, $b_1 = 2$ and $b_2 = 1$ will always yield precisely the same correlation of $W$ with $Y$ as $b_1 = 10$ and $b_2 = 5$. Thus we must limit the sets of weights our optimization procedure is allowed to choose among by putting a side condition (or conditions) on the search "request." This is easily incorporated into our use of derivatives to find maxima or minima through the use of *Lagrangian multipliers*.

In order to use Lagrangian multipliers, we must first express our constraint(s) in the form $g_i(b_1, b_2, \ldots, b_m) = 0$, where $g_i$ is some function of the unknowns whose "optimal" values we wish to identify. We then define

$$L = f(b_1, b_2, \ldots, b_m) - \lambda_1 g_1(b_1, b_2, \ldots, b_m) - \lambda_2 g_2(b_1, \ldots, b_m) - \cdots - \lambda_k g_k(b_1, \ldots, b_m),$$

where we have $k$ different side conditions to be satisfied by the optimization process. We then take the derivatives of $L$ with respect to $b_1, b_2, \ldots, b_m$ and set these to zero. These equations, together with the equations provided by the side conditions ($g_1 = 0; g_2 = 0;$ etc.) enable us to solve both for the $\lambda$s and for the $b$s. As an example, suppose we wished to choose values for $b_1$ and $b_2$ which would make $(b_1 x + b_2 y)^2$ as large as possible, subject to the constraint that

$b_1^2 + b_2^2 = 1$. Here

$$L = b_1^2 x^2 + 2b_1 b_2 xy + b_2^2 y^2 - \lambda(b_1^2 + b_2^2 - 1);$$

$$dL/db_1 = 2b_1(x^2 - \lambda) + 2b_2 xy; \qquad dL/db_2 = 2b_1 xy + 2b_2(y^2 - \lambda).$$

Setting derivatives equal to zero and adding the side condition gives

$$b_1(x^2 - \lambda) + b_2 xy = 0; \qquad b_1 xy + b_2(y^2 - \lambda) = 0; \qquad b_1^2 + b_2^2 = 1.$$

The first two equations each yield a value of $b_1/b_2$, namely

$$b_1/b_2 = -xy/(x^2 - \lambda) = -(y^2 - \lambda)/xy;$$

whence $(x^2 - \lambda)(y^2 - \lambda) = (xy)^2$; whence $\lambda$ must either equal 0 or $x^2 + y^2$. Using the value $\lambda = 0$ yields $b_1 = \pm (y/x)b_2$, which, together with the side condition, tells us that

$$b_1 = \mp y/\sqrt{x^2 + y^2} \qquad \text{and} \qquad b_2 = x/\sqrt{x^2 + y^2}.$$

Whence $(b_1 x + b_2 y)^2 = 0$, clearly its minimum value. Using the other value of $\lambda$ gives the same expressions for the $b$s, except that they now both have the same sign, and

$$(b_1 x + b_2 y)^2 = (2xy)^2/(x^2 + y^2) = 4/(1/x^2 + 1/y^2),$$

the maximum we sought.

# 2 | DIGRESSION
## Matrix Algebra

As pointed out in Chapter 2, matrix algebra proves extremely useful in multivariate statistics for at least three reasons:

(1) Matrix notation provides a highly compact summary of expressions and equations which, if a separate symbol for each variable and each constant involved were employed, would be enormously bulky.

(2) The compactness and simplicity of matrix notation greatly facilitates memorization of expressions and equations in matrix form.

(3) The underlying similarity of the operations involved in applying a multivariate technique to a set of 2, 15, or 192 variables is much easier to see when these operations are expressed in matrix notation. Further, the similarity between univariate and multivariate techniques is much more easily seen from matrix expressions than from single-symbol formulas. These two functions combine to lessen considerably the difficulty of deriving solutions to multivariate problems.

### D2.1 Basic Notation

In order to reap the first two benefits it is necessary to be sufficiently familiar with the conventions of matrix notation to be able to translate expressions from matrix form to single-symbol form and vice versa. Admittedly an adequate substitute for this skill in most situations is the ability to match matrix operations with corresponding computer algorithms as described in writeups of available library ("canned") programs, but unless the user of such computer programs is able to convert matrix expressions to single-symbol expressions in at least simple, small-matrix cases, he will be forever at the mercy of "bugs" in the programs he uses. In order to reap the third benefit, it is necessary as well to learn the conventions for manipulating matrix expressions and equations, which conventions are highly similar to the rules of single-symbol algebra in many respects but crucially different in others. The following sections are designed to familiarize the reader with those aspects of matrix notation and matrix manipulation which are necessary in order to follow the derivations outlined in this Primer. The coverage of matrix algebra is by no means comprehensive, there being many techniques other than those covered here, and many alternative interpretations and applications of the techniques that are presented. The reader who wishes to delve further into matrix algebra will find study of the works of Horst (1963) or Searle (1966) rewarding.

$$\mathbf{a} = \begin{bmatrix} a_1 \\ a_2 \\ a_3 \\ \vdots \\ a_m \end{bmatrix}$$

is a *column vector*, that is, a vertical arrangement of $m$ *elements*. A single number or symbol, that is, a *scalar*, can be considered to be a column vector—or a row vector—containing a single element, that is, for which $m = 1$.

$$\mathbf{a}' = \begin{bmatrix} a_1 & a_2 & a_3 & a_4 & \cdots & a_m \end{bmatrix}$$

is a *row vector*, a horizontal arrangement of $m$ elements.

$$\mathbf{A} = [a_{ij}] = \begin{bmatrix} \mathbf{a}_1 & \mathbf{a}_2 & \mathbf{a}_3 & \cdots & \mathbf{a}_p \end{bmatrix} = \begin{bmatrix} \mathbf{a}'_1 \\ \mathbf{a}'_2 \\ \mathbf{a}'_3 \\ \vdots \\ \mathbf{a}'_m \end{bmatrix} = \begin{bmatrix} a_{11} & a_{12} & a_{13} & \cdots & a_{1p} \\ a_{21} & a_{22} & a_{23} & \cdots & a_{2p} \\ a_{31} & a_{32} & a_{33} & \cdots & a_{3p} \\ \vdots & \vdots & \vdots & \cdots & \vdots \\ a_{m1} & a_{m2} & a_{m3} & \cdots & a_{mp} \end{bmatrix}$$

is an $m \times p$ matrix consisting of $m$ rows and $p$ columns of elements $a_{ij}$ being the element located at the intersection of the $i$th row and the $j$th column of the matrix. The quantities $m$ and $p$ are called the *dimensions* of $\mathbf{A}$.

$$\mathbf{A}' = [a_{ji}] = \begin{bmatrix} a_{11} & a_{21} & a_{31} & \cdots & a_{m1} \\ a_{12} & a_{22} & a_{32} & \cdots & a_{m2} \\ a_{13} & a_{23} & a_{33} & \cdots & a_{m3} \\ \vdots & \vdots & \vdots & \cdots & \vdots \\ a_{1p} & a_{2p} & a_{3p} & \cdots & a_{mp} \end{bmatrix}$$

is the transpose of $\mathbf{A}$, with the rows of $\mathbf{A}'$ being the columns of $\mathbf{A}$. If $\mathbf{A}' = \mathbf{A}$, the matrix is said to be *symmetric*. Thus a matrix is symmetric if and only if $a_{ij} = a_{ji}$ for every $i, j$ between 1 and $m, p$. As implied by the above definitions, boldface capital letters refer to matrices, while boldface lowercase letters refer to vectors (a vector being, of course, a matrix one of whose dimensions is unity). The principal exception to this general rule occurs in discussing sets of scores on predictor or outcome variables. Thus $\mathbf{X}$ will refer to a data matrix whose entries are raw scores (each row being the data for a single subject, the columns representing the different variables), $X_{ij}$ being subject $i$'s score on predictor variable $j$, while $\mathbf{x}$ will refer to a data matrix whose entries are *deviation* scores, with $x_{ij}$ being subject $i$'s deviation score, $(X_{ij} - \bar{X}_j)$, on predictor variable $j$. A similar distinction holds between $\mathbf{Y}$ and $\mathbf{y}$.

### D2.2   Linear Combinations of Matrices

The sum of two matrices, $\mathbf{A} + \mathbf{B}$, is defined if and only if (iff) the two matrices have identical dimensions, that is, the same number of rows and columns. In that case,

$$\mathbf{A} + \mathbf{B} = [a_{ij}] + [b_{ij}] = [a_{ij} + b_{ij}].$$

In words, the sum of the two matrices is a new matrix $\mathbf{C}$ each of whose elements $c_{ij}$ is the sum of the corresponding elements, $a_{ij}$, and $b_{ij}$, of the matrices being added together.

The product of a scalar times a matrix is simply a new matrix each of whose entries is equal to the corresponding entry of the matrix being multiplied times the scalar. That is,

$$c\mathbf{A} = \mathbf{A}c = c[a_{ij}] = [ca_{ij}].$$

These two definitions can be combined to define a *linear combination* of two matrices, $c_1 \mathbf{A} + c_2 \mathbf{B}$, as follows:

$$c_1 \cdot \mathbf{A} + c_2 \, \mathbf{B} = c_1[a_{ij}] + c_2[b_{ij}] = [c_1 a_{ij} + c_2 b_{ij}].$$

Of course, $\mathbf{A}$ and $\mathbf{B}$ must have identical dimensions. This notation can be useful in summarizing the process of constructing new variables which are linear combinations of original variables. For instance, if $\mathbf{X}_0$ represents an $N \times p$ matrix which records the scores of $N$ subjects on $p$ baseline measures of performance, and $\mathbf{X}_1$ represents the scores of the same $N$ Ss on the same $p$ performance measures based on an observation period twice as long as the baseline period, $\mathbf{X}_1 - 2\mathbf{X}_0$ represents a matrix of change scores which takes into account the difference in the length of the two observation periods.

## D2.3 Multiplication of Matrices

$\mathbf{a}'\mathbf{b}$, the scalar product of two vectors, is defined only if the two vectors have the same number of elements, and is in that case equal to the *scalar* (single element, or one-dimensional vector, or $1 \times 1$ matrix) $\sum a_j b_j$. Note that if $\mathbf{a}$ and $\mathbf{b}$ are both column vectors, $\mathbf{ab}'$ and $\mathbf{ba}'$ are *not* scalars, but $m \times m$ matrices, with $\mathbf{ab}' = [a_i b_j]$, a square (generally nonsymmetric) matrix whose $(i, j)$th element is the product of the $i$th element of $\mathbf{a}$ with the $j$th element of $\mathbf{b}$ and $\mathbf{ba}' = [b_i a_j] = (\mathbf{ab}')$. This will become clearer when matrix products are defined below. This is our first example of a difference between matrix algebra and single-symbol algebra. In single-symbol algebra, the order of the symbols in a chain of multiplications is unimportant, that is, $x \cdot y \cdot z \cdots = x \cdot z \cdot y \cdots = y \cdot x \cdot z \cdots$, etc. In matrix algebra, however, very careful attention must be paid to the order of multiplication of matrices, since $\mathbf{AB}$ is not in general equal to $\mathbf{BA}$—frequently only one of these two products is even defined. In more formal terms, single-symbol multiplication is commutative; matrix multiplication is not.

$$\mathbf{Ab} = \begin{bmatrix} \mathbf{a}'_1 \\ \mathbf{a}'_2 \\ \mathbf{a}'_3 \\ \vdots \\ \mathbf{a}'_p \end{bmatrix} \times \begin{bmatrix} b_1 \\ b_2 \\ \vdots \\ b_m \end{bmatrix} = \begin{bmatrix} \sum a_{1j} b_j \\ \sum a_{2j} b_j \\ \sum a_{3j} b_j \\ \vdots \\ \sum a_{pj} b_j \end{bmatrix}.$$

Note that for the product $\mathbf{Ab}$ to be defined, the number of elements in $\mathbf{b}$ must equal the number of elements in each row of $\mathbf{A}$. Similarly,

$$\mathbf{b}'\mathbf{A} = [b_1 b_2 b_3 \cdots b_m] \times [\mathbf{a}_1 \mathbf{a}_2 \cdots \mathbf{a}_p] = \left[ \sum_{i=1}^{m} a_{i1} b_i \ \sum a_{i2} b_i \ \sum a_{i3} b_i \cdots \sum a_{ip} b_i \right] = (\mathbf{A}'\mathbf{b})'.$$

Finally, the product of two matrices $\mathbf{AB}$ is defined if and only if the two matrices are *conformable*, that is, if and only if the number of columns of $\mathbf{A}$ (the premultiplying matrix) is equal to the number of rows of $\mathbf{B}$ (the postmultiplying matrix). If this condition is satisfied, then

$$\mathbf{AB} = \begin{bmatrix} \mathbf{a}'_1 \\ \mathbf{a}'_2 \\ \vdots \\ \mathbf{a}'_m \end{bmatrix} \overset{[\mathbf{b}_1 \quad \mathbf{b}_2 \quad \cdots \quad \mathbf{b}_p]}{} = \begin{bmatrix} \mathbf{a}'_1 \mathbf{b}_1 & \mathbf{a}'_1 \mathbf{b}_2 & \cdots & \mathbf{a}'_1 \mathbf{b}_p \\ \mathbf{a}'_2 \mathbf{b}_1 & \mathbf{a}'_2 \mathbf{b}_2 & \cdots & \mathbf{a}'_2 \mathbf{b}_p \\ \vdots & & \cdots & \vdots \\ \mathbf{a}'_m \mathbf{b}_1 & \mathbf{a}'_m \mathbf{b}_2 & \cdots & \mathbf{a}'_m \mathbf{b}_p \end{bmatrix} = \begin{bmatrix} \overset{r}{\underset{k=1}{\sum}} a_{1k} b_{k1} & \sum a_{1k} b_{k2} & \cdots & \sum a_{1k} b_{kp} \\ \sum a_{2k} b_{k1} & \sum a_{2k} b_{k2} & \cdots & \sum a_{2k} b_{kp} \\ \vdots & \vdots & \cdots & \vdots \\ \sum a_{mk} b_{k1} & \sum a_{mk} b_{k2} & \cdots & \sum a_{mk} b_{kp} \end{bmatrix}.$$

number of columns in $\mathbf{A}$ and number of rows in $\mathbf{B}$

In other words, the entry in the $i$th row, $j$th column of the product matrix is equal to the sum of the cross-products of the elements in row $i$ of $\mathbf{A}$ with the elements in column $j$ of $\mathbf{B}$.

Until you become familiar with matrix multiplication, you may find it helpful to write the postmultiplying matrix directly above and to the right of the premultiplying matrix and draw lines through the rows of **A** and the columns of **B** as reminders of which row and column contribute to each entry in the product matrix where the row and column lines just mentioned meet. For instance,

$$
\begin{array}{c}
\begin{bmatrix} 5 & 3 & 2 \\ 6 & 1 & 1 \\ 1 & 4 & 7 \\ 2 & 0 & 6 \end{bmatrix} \\[4pt]
\begin{bmatrix} 1 & -3 & -5 & -1 \\ 7 & -1 & -6 & -4 \end{bmatrix}
\begin{bmatrix} 30 & -26 & -46 \\ -55 & -46 & -81 \end{bmatrix}
\end{array}
$$

Note that in general $\mathbf{AB} \neq \mathbf{BA}$. Often (as in the above example), only one of the two orders of multiplication will even be defined.

Various products of matrices arise repeatedly in statistics. For instance, the sum of the squares of a set of numbers or symbols can be conveniently represented as

$$\mathbf{a'a} = \sum a_i^2.$$

In particular, if $\mathbf{x}_j$ is a column vector containing $N$ subjects' deviation scores on variable $X_j$, that is, if

$$\mathbf{x}_j' = [(X_{1,j} - \bar{X}_j), (X_{2,j} - \bar{X}_j), \dots (X_{N,j} - \bar{X}_j)] = [x_{1,j} x_{2,j} \cdots x_{N,j}],$$

then

$$\mathbf{x}_j' \mathbf{x}_j = \sum_{i=1}^{N} x_{i,j}^2 \quad \text{and} \quad \mathbf{x}_j' \mathbf{x}_j/(N-1) = s_j^2,$$

the sample variance of variable $X_j$ and an unbiased estimate of the variance $\sigma_j^2$ of the population from which these $N$ observations were randomly sampled. In the case of a single variable, the matrix notation for the computation of a sample variance is not notably more compact than the corresponding single-symbol notation. However, recall that the product of two matrices **AB** records the results of postmultiplying each row of **A** by each column of **B**. Thus if we let **x** be a data *matrix* each of whose $N$ rows reports one of our subject's deviation scores on each of **p** variables, then $\mathbf{x'x}$ is a $p \times p$ matrix whose $(i, j)$th element is $\sum_{u=1}^{N} x_{u,i} x_{u,j}$, the sum of the cross-products of the $N$ Ss' scores on variables $i$ and $j$. Further, $\mathbf{x'x}/(N-1) = \mathbf{S}$ is the $p \times p$ *covariance* matrix whose $(i, j)$th element is the sample covariance, $s_{ij} = \sum x_i x_j/(N-1)$, of variables $i$ and $j$, the $j$th main diagonal element being $s_{jj} = s_j^2$, the sample variance of variable $j$. We can also record the process of computing a covariance matrix very compactly in terms of matrix manipulations as

$$\mathbf{S} = (\mathbf{X'X} - \bar{\mathbf{X}}\bar{\mathbf{X}}'N)/(N-1),$$

where **X** is the $N \times p$ raw-score data matrix and $\bar{\mathbf{X}}$ is a $p$-element column vector whose $i$th entry is $\bar{X}_i$, the mean of variable $i$.

Another very useful application of matrix products is to the manipulation of linear combinations (linear compounds) of the original variables in a set of data. If we define subject $i$'s score on the new variable $W$ as

$$W_i = w_1 X_{i,1} + w_2 X_{i,2} + \cdots + w_p X_{i,p},$$

we can represent the set of $N$ scores on the new variable $W$ as the column vector $\mathbf{Xw}$ which results when the $N \times p$ data matrix of scores on the original variables is postmultiplied by the

$p$-element column vector $\mathbf{w}$ whose entries are the weights, $w_1, w_2, \ldots, w_p$, which define the linear compound. More importantly, and as the reader will wish to verify for himself, the variance of this linear compound $s_w^2$ can be expressed as a function of the variances and covariances of the original variables by the relationship

$$s_w^2 = \mathbf{w}'\mathbf{S}_x\,\mathbf{w} = w_1^2 s_1^2 + \cdots + w_p^2 s_p^2 + 2w_1 w_2 s_{12} + 2w_1 w_2 s_{13} + \cdots + 2w_{p-1} w_p s_{p-1,p}$$
$$= [1/(N-1)] \sum (w_1 x_{i,1} + w_2 x_{i,2} + \cdots + w_p x_{i,p})^2$$
$$= (\mathbf{w}'\mathbf{x}'\mathbf{x}\mathbf{w})/(N-1).$$

Further the covariance of the linear compound $\mathbf{Xw}$ with the linear compound $\mathbf{Yb}$ is given by

$$s_{W,\,B} = \mathbf{w}'\mathbf{S}_{xy}\mathbf{b} = \mathbf{b}'\mathbf{S}'_{xy}\mathbf{w}$$
$$= \sum (w_1 x_1 + w_2 x_2 + \cdots + w_p x_p)(b_1 y_1 + b_2 y_2 + \cdots + b_m y_m)/(N-1),$$

where

$$\mathbf{S}_{xy} = [\mathbf{s}_{xy_1}\mathbf{s}_{xy_2} \cdots \mathbf{s}_{xy_m}]$$

is a $p \times m$ matrix whose $(i, j)$th entry is the covariance of the $i$th $X$ variable with the $j$th $Y$ variable.

## D2.4   Permissible Manipulations

Expressions and equations involving matrices can be manipulated much as can scalar (single-symbol) expressions, *except* that no matrix operation corresponding simply to division by a scalar exists (we must instead multiply by the inverse of the matrix, where that exists), and care must be taken to distinguish between different orders of multiplication. The following rules may be helpful in solving equations involving matrices:

***Rule 1***   $\mathbf{A} = \mathbf{B}$ iff $a_{ij} = b_{ij}$ for all $i, j$. In other words, two matrices are equal if and only if their elements are identical.

***Rule 2***   The order of additions and subtractions is immaterial. Thus

$$\mathbf{A} + \mathbf{B} - \mathbf{C} = \mathbf{A} - \mathbf{C} + \mathbf{B} = \mathbf{B} + \mathbf{A} - \mathbf{C} = \mathbf{B} - \mathbf{C} + \mathbf{A}, \quad \text{etc.}$$

Note, however, that the sum or difference of two matrices exists if and only if they are of identical dimensions, that is, have the same number of rows and the same number of columns. The sum or difference of two matrices is simply a matrix each of whose elements equals the sum or difference of the corresponding elements of the original matrices, that is, $\mathbf{A} + \mathbf{B} = \mathbf{C}$ iff $c_{ij} = a_{ij} + b_{ij}$ for all $i, j$.

***Rule 3***

$$\mathbf{A}(\mathbf{B} + \mathbf{C}) = \mathbf{AB} + \mathbf{AC}; \qquad \mathbf{A}(\mathbf{BC}) = (\mathbf{AB})(\mathbf{C}); \qquad (\mathbf{AB})' = \mathbf{B}'\mathbf{A}';$$

but $\mathbf{AB} \neq \mathbf{BA}$ in general though there are specific choices of pairs of matrices for which the order of multiplication is immaterial (for example, a matrix and its inverse).

***Rule 4***   The validity of an equation is unaffected if both sides are pre- or postmultiplied by the same matrix. In other words, if $\mathbf{A} = \mathbf{B}$, then $\mathbf{AC} = \mathbf{BC}$ and $\mathbf{DA} = \mathbf{DB}$.

## D2.5   Computation of Determinants

Probably the most common application of matrices is in summarizing a set of equations which are to be solved for the values of $m$ unknowns. In other words, we have $\mathbf{Ax} = \mathbf{c}$, where $\mathbf{x}$ is a column vector of $m$ unknown values; $\mathbf{A}$ is a $k \times m$ matrix each of whose $k$ rows gives the coefficients by which the unknowns are multiplied in the left-hand side of an equation involving the unknowns, and $\mathbf{c}$ is a column vector of $k$ constants to which the various expressions on the left-hand side of the equations are known to be equal. Actually, it is fairly easy to show that it is sufficient to consider only systems of equations in which $\mathbf{A}$ is a square matrix and $\mathbf{c}$ and $\mathbf{x}$ are of the same dimensionality. If we have fewer than $m$ equations in the $m$ unknowns (and thus fewer rows than $m$ in $\mathbf{A}$), we obviously will not be able to solve for all of the unknowns solely in terms of the constants. If we have more than $m$ equations, the system is "over-determined" and will in fact be internally inconsistent unless it is true that $(k - m)$ of the equations can be written as linear combinations of the other $m$ equations. Having reduced the system to $m$ equations in $m$ unknowns we will now be faced with one of two cases: $\mathbf{c} = \mathbf{0}$ (the null vector), in which case we refer to this as a *homogenous* system, or $\mathbf{c} \neq \mathbf{0}$ (the nonhomogeneous case). If $\mathbf{c} = \mathbf{0}$, then we always have $\mathbf{x} = \mathbf{0}$ (that is, $x_1 = x_2 = \cdots = x_m = 0$) as one solution which satisfies the set of equations. If there is to be a "nontrivial" solution as well, it must be possible to express any given equation as a linear combination of the other equations. If, on the other hand, $\mathbf{c} \neq \mathbf{0}$, then a solution to the set of equations is possible only if no equation is linearly dependent on the others.

As implied by the last two sentences, an equation is *linearly dependent* on a set of $n$ other equations if and only if $n$ nonzero constants can be found such that adding together those multiples of the $n$ equations leads to an equation identical to the dependent equation. Similarly, a set of $n$ column or row vectors, $\mathbf{x}_1, \mathbf{x}_2, \ldots, \mathbf{x}_n$, are linearly dependent on each other if a set of nonzero constants $k_1, k_2, \ldots, k_n$ can be found such that $\sum c_k \mathbf{x}_k = \mathbf{0}$.

It would thus be extremely helpful in solving systems of equations if there existed a simple procedure for determining, from the matrix $\mathbf{A}$ of coefficients alone, whether or not the $m$ equations are linearly independent. It is for this purpose that *determinants* were developed. The determinant of a square matrix can be defined in any one of several ways, as follows:

**Definition 1**   (*Fundamental definition*)   $|\mathbf{A}|$, the determinant of $\mathbf{A}$, is the sum of all possible products of $m$ elements of $\mathbf{A}$ in which each row and each column contributes exactly one element, with each of these products being multiplied by $(-1)^i$, where $i$ is the number of inversions of the subscripts on the elements from the "natural" ordering in which row and column subscripts match. To illustrate this definition, consider its application to a $3 \times 3$ matrix

$$\begin{bmatrix} a_{11} & a_{12} & a_{13} \\ a_{21} & a_{22} & a_{23} \\ a_{31} & a_{32} & a_{33} \end{bmatrix}.$$

The possible products in which each of the row subscripts occurs only once and each of the column subscripts occurs only once are $a_{11}a_{22}a_{33}$, $a_{11}a_{23}a_{32}$, $a_{12}a_{23}a_{31}$, $a_{12}a_{21}a_{33}$, $a_{13}a_{21}a_{32}$, and $a_{13}a_{22}a_{31}$. We have written these products in "row order," such that the row subscripts form the sequence "123." The number of inversions for each product—the value of $i$ in $(-1)^i$—is therefore equal to the number of pairwise interchanges necessary to bring the sequence of *column* subscripts back to the "natural" 123 order. Thus $i = 0$ for $a_{11}a_{22}a_{33}$;

  1 for $a_{11}a_{23}a_{32}$   ("132 becomes "123" when 3 and 2 are interchanged);

  1 for $a_{12}a_{21}a_{33}$   (interchanging 2 and 1 in "213" restores the natural order);

  2 for $a_{12}a_{23}a_{31}$   ("231" becomes "213" and then "123" in successive interchanges);

etc. The final result is that

$$|\mathbf{A}| = a_{11}a_{22}a_{33} - a_{11}a_{23}a_{32} - a_{12}a_{21}a_{33} + a_{12}a_{23}a_{31} + a_{13}a_{21}a_{32} - a_{13}a_{22}a_{31}$$

$$= a_{11}(a_{22}a_{33} - a_{23}a_{32}) - a_{12}(a_{21}a_{33} - a_{23}a_{31}) + a_{13}(a_{21}a_{32} - a_{22}a_{31})$$

$$= -a_{21}(a_{12}a_{33} - a_{32}a_{13}) + a_{22}(a_{11}a_{33} - a_{31}a_{13}) - a_{23}(a_{11}a_{32} - a_{12}a_{31})$$

$$= a_{11}(a_{22}a_{33} - a_{23}a_{32}) - a_{21}(a_{12}a_{33} - a_{13}a_{32}) + a_{31}(a_{12}a_{23} - a_{22}a_{13}); \quad \text{etc.}$$

**Definition 2**    (*Expansion by cofactors*)    Note that $|\mathbf{A}|$ for the 3 × 3 case can be written in any of the following equivalent forms:

$$|\mathbf{A}| = \quad a_{11}\begin{vmatrix} a_{22} & a_{23} \\ a_{32} & a_{33} \end{vmatrix} - a_{12}\begin{vmatrix} a_{21} & a_{23} \\ a_{31} & a_{33} \end{vmatrix} + a_{13}\begin{vmatrix} a_{21} & a_{22} \\ a_{31} & a_{32} \end{vmatrix}$$

$$= -a_{21}\begin{vmatrix} a_{12} & a_{13} \\ a_{32} & a_{33} \end{vmatrix} + a_{22}\begin{vmatrix} a_{11} & a_{13} \\ a_{31} & a_{32} \end{vmatrix} - a_{23}\begin{vmatrix} a_{11} & a_{12} \\ a_{31} & a_{32} \end{vmatrix}$$

$$= \quad a_{11}\begin{vmatrix} a_{22} & a_{23} \\ a_{32} & a_{33} \end{vmatrix} - a_{21}\begin{vmatrix} a_{12} & a_{13} \\ a_{32} & a_{33} \end{vmatrix} + a_{31}\begin{vmatrix} a_{12} & a_{13} \\ a_{22} & a_{23} \end{vmatrix}; \quad \text{etc.}$$

In each case we have each of the elements of a single row or a single column of the matrix multiplied by $(-1)^{i+j}$, where $i$ and $j$ are the subscripts of that element, multiplied again by the determinant of the 2 × 2 submatrix which results if row $i$ and column $j$ are deleted from the original matrix (since, by the fundamental definition, elements from row $i$ and column $j$ are no longer eligible for inclusion in any product which already includes $a_{ij}$). This approach to determining $|\mathbf{A}|$ can be shown to hold for the general case, that is,

$$|\mathbf{A}| = \sum_{j=1}^{m} a_{ij}(-1)^{i+j}M_{ij} \qquad \text{for   any choice of } i$$

$$= \sum_{i=1}^{m} a_{ij}(-1)^{i+j}M_{ij} \qquad \text{for   any choice of } j,$$

where the *minor* of $a_{ij}$, $M_{ij}$, equals $|\mathbf{A}_{(ij)}|$ and $\mathbf{A}_{(ij)}$ is the $(m-1) \times (m-1)$ submatrix which results when row $i$ and column $j$ are deleted from $\mathbf{A}$. The product $(-1)^{i+j}M_{ij}$ is known as the *cofactor* of $a_{ij} = C_{ij}$, so that we can also write this definition of $|\mathbf{A}|$ as

$$|\mathbf{A}| = \sum_{j} a_{ij}C_{ij} = \sum_{i} a_{ij}C_{ij}$$

for any choice of a fixed $i$ or a fixed $j$. For a large original matrix, this formula must be applied to each $(m-1) \times (m-1)$ minor to reduce it to an expression involving $(m-2) \times (m-2)$ determinants, each of these in turn being reduced by reapplication of this " cofactor definition" to an expression involving determinants of $(m-3) \times (m-3)$ submatrices, etc.

## D2.6    Properties of Determinants

These two definitions are sufficient to permit us to examine some of the properties of determinants. These properties mirror quite closely (as they were designed to do) the properties of sets of simultaneous linear equations, in that $|\mathbf{A}|$ is unaffected by the sorts of operations which leave the validity of a set of equations unaffected and $|\mathbf{A}| = 0$ if there is linear dependence among the equations of the corresponding system of equations. In particular:

(a)   If two rows or two columns of **A** are identical, $|\mathbf{A}| = 0$. This is certainly a desirable property, since if two of the $m$ equations in our system are identical, they are perfectly redundant and we thus have an insufficient amount of information to solve for all $m$ unknowns. Similarly, if the coefficients multiplying a particular pair of variables are identical in every equation, then $x_i$ and $x_j$ always enter into each equation as a unit, and there is therefore no way to discriminate their *relative* contributions to the values of **c**.

(b)   The value of $|\mathbf{A}|$ is unchanged by addition to any row of **A** of a constant times any other row of **A**, or by addition to any column of **A** of a constant times any other column of **A**. This property mirrors the standard technique of adding or subtracting equations (rows of coefficients) to or from each other in order to eliminate variables (bring some coefficients to zero).

(c)   If any row or any column of **A** consists entirely of zeros, $|\mathbf{A}| = 0$. This follows directly from the cofactor definition of the determinant, and reflects the fact that in any system of equations, if one of the equations "vanishes" completely after a series of additions to or subtractions from that equation of other equations, the eliminated equation was redundant and the system lacks sufficient information to solve for all $x_j$.

From these properties, we obtain yet another definition (more properly, method of computing) of the determinant, namely:

**Definition 3**   (*Row or column operations*)   $|\mathbf{A}|$ can be computed by using row or column operations (adding a constant times one row or column to another row or column) to reduce **A** to *triangular* form (a form in which only above-diagonal entries are nonzero), since the determinant of a triangular matrix is simply equal to the product of the diagonal elements. (All other products must necessarily involve one or more below-diagonal elements and thus equal zero.)

We can illustrate the triangularization process just described by applying it to the general symmetric $3 \times 3$ matrix, as follows:

$$
\begin{vmatrix} a & b & c \\ b & d & e \\ c & e & f \end{vmatrix} = \begin{vmatrix} a & b & c \\ b - \dfrac{b}{a}a & d - \dfrac{b^2}{a} & e - \dfrac{bc}{a} \\ c - \dfrac{c}{a}a & e - \dfrac{bc}{a} & f - \dfrac{c^2}{a} \end{vmatrix} = \begin{vmatrix} a & b & c \\ 0 & \dfrac{ad - b^2}{a} & \dfrac{ae - bc}{a} \\ 0 & \dfrac{ae - bc}{a} & \dfrac{af - c^2}{a} \end{vmatrix}
$$

$$
= \begin{vmatrix} a & b & c \\ 0 & \dfrac{ad - b^2}{a} & \dfrac{ae - bc}{a} \\ 0 & \dfrac{ae - bc}{a} - \left(\dfrac{ae - bc}{ad - b^2}\right)\left(\dfrac{ad - b^2}{a}\right) & \dfrac{af - c^2}{a} - \dfrac{(ae - bc)^2}{a(ad - b^2)} \end{vmatrix}
$$

$$
= \begin{vmatrix} a & b \\ 0 & \dfrac{ad - b^2}{a} \\ 0 & 0 \end{vmatrix} \quad \dfrac{(af - c^2)(ad - b^2) - (ae - bc)^2}{a(ad - b^2)}
$$

$$
= [(af - c^2)(ad - b^2) - (ae - bc)^2]/a
$$

$$
= [a^2df = ac^2d = ab^2f + b^2c^2 - a^2e^2 + 2abce - b^2c^2]/a
$$

$$
= adf + 2bce - ae^2 - dc^2 - fb^2,
$$

which matches (except for a different lettering system) the formula given in Section D2.8.

Applied to a specific numerical example (namely, the determinant of **A** for the single-sample $T^2$ example of Section 3.1), we have

$$\begin{vmatrix} 5196.5 & -853.1 & -656.7 \\ -853.1 & 739.7 & -34.7 \\ -656.7 & -34.7 & 882.7 \end{vmatrix} = \begin{vmatrix} 5196.5 & -853.1 & -656.7 \\ 0 & 599.65 & -142.51 \\ 0 & -142.51 & 799.71 \end{vmatrix}$$

$$\begin{vmatrix} 5196.5 & -853.1 & -656.7 \\ 0 & 599.65 & -142.51 \\ 0 & 0 & 765.84 \end{vmatrix} = (5196.5)(599.65)(765.84) = 2.3864 \times 10^9,$$

in perfect agreement with the result cited in Section 3.1, which result was obtained by the cofactor definition of determinants.

## D2.7   Inverses of Matrices

Just as it is desirable to determine whether or not a set of $m$ equations in $m$ unknowns is linearly independent (as is accomplished by asking whether the matrix of coefficients has a nonzero determinant), so it is desirable to find a matrix which, when multiplied by both sides of the matrix equation, $\mathbf{Ax} = \mathbf{c}$, will "isolate" $\mathbf{x}$ on the left-hand side. Such a matrix **B** must have the property that $\mathbf{BA} = \mathbf{I}$, whence, by analogy to scalar arithmetic, such a matrix is known as the *inverse* of **A**, $\mathbf{A}^{-1}$. A given matrix may not have an inverse, just as a given set of equations may not have a solution in the sense that each unknown is expressed solely as a function of the $c_i$ (and just as division of a scalar by zero is undefined). In particular, $\mathbf{A}^{-1}$ exists if and only if **A** is square and $|\mathbf{A}| \neq 0$. The properties of inverses show a (pre-planned) similarity to the properties of sets of linear equations. In particular:

(1)   The inverse of a diagonal matrix is a diagonal matrix whose elements are the reciprocals of the corresponding elements of the original matrix, that is,

$$\begin{bmatrix} a_{11} & & \bigcirc \\ & a_{22} & \\ & & \ddots \\ \bigcirc & & a_{mm} \end{bmatrix}^{-1} = \begin{bmatrix} 1/a_{11} & & \bigcirc \\ & 1/a_{22} & \\ & & \ddots \\ \bigcirc & & 1/a_{mm} \end{bmatrix}.$$

This is appropriate, since we begin with a set of equations of the form $a_{ii} x_i = c_i$, whence of course $x_i = c_i/a_{ii}$.

(2)   If row operations as defined in properties (b), (c), and Definition 3 of Section D2.6 are applied to the combined, $m \times 2m$ matrix $\mathbf{K} = [\mathbf{AI}]$ so as to produce an identity matrix in the first $m$ columns of **K**, the last $m$ columns will contain the inverse of **A**. Working on **A** and **I** simultaneously is equivalent to applying all operations to both sides of the equation. Put another way, conjoining an $m \times m$ identity matrix to **A** before beginning our row operations serves to "record" the operations involved in reducing the left-hand side to an identity matrix, that is, in developing equations in each of which only a single unknown appears on the left.

(3)   Finally, a more formal solution to the problem of finding $\mathbf{A}^{-1}$ is offered, namely: If $|\mathbf{A}| \neq 0$, then

$$\mathbf{A}^{-1} = \frac{1}{|\mathbf{A}|} \begin{bmatrix} C_{11} & C_{21} & \cdots & C_{m1} \\ C_{12} & C_{22} & \cdots & C_{m2} \\ \vdots & \vdots & \cdots & \vdots \\ C_{1m} & C_{2m} & \cdots & C_{mm} \end{bmatrix}.$$

That this formula will work can be seen by considering the nature of the entries of the product matrix, $\mathbf{P} = \mathbf{C}\mathbf{A}/D$, where $\mathbf{C}$ is the matrix of cofactors and $D$ is the determinant of $\mathbf{A}$. Clearly each main diagonal entry will have the form

$$P_{jj} = (1/D)\sum_k a_{kj}C_{kj} = (1/D)D = 1,$$

while each off-diagonal entry will be of the form

$$P_{kl} = (1/D)\sum_j a_{lj}C_{kj} = 0.$$

That $P_{kl} = 0$ follows from the fact that the summation $\sum a_{lj}C_{kj}$ can be seen, by the cofactor definition of the determinant, to be equal to the determinant of a matrix identical to $\mathbf{A}$ except that row $k$ is replaced by row $l$. Such a matrix will have two identical rows and will thus have a determinant of zero, by property (a) of Section D2.6.

### D2.8   Some Handy Formulas for Inverses in Simple Cases

The following formulas are useful in solving multivariate problems involving only a few variables, either for didactic purposes or in the course of "checking out" a computer program.

I.   2 × 2 Matrices

$$\begin{bmatrix} a & b \\ c & d \end{bmatrix}^{-1} = \frac{1}{D}\begin{bmatrix} d & -b \\ -c & a \end{bmatrix}, \qquad \text{where} \quad D = ad - bc;$$

$$\begin{bmatrix} w & x \\ x & y \end{bmatrix}^{-1} = \frac{1}{wy - x^2}\begin{bmatrix} y & -x \\ -x & w \end{bmatrix}.$$

II.   3 × 3 Matrices

$$\begin{bmatrix} a & b & c \\ d & e & f \\ g & h & i \end{bmatrix}^{-1} = \frac{1}{D}\begin{bmatrix} ei - fh & ch - bi & bf - ce \\ fg - di & ai - cg & cd - af \\ dh - eg & bg - ah & ae - bd \end{bmatrix},$$

where $D = a(ei - fh) + b(fg - di) + c(dh - eg)$.

$$\begin{bmatrix} u & v & w \\ v & x & y \\ w & y & z \end{bmatrix}^{-1} = \frac{1}{(uxz + 2vwy - uy^2 - xw^2 - zv^2)}\begin{bmatrix} xz - y^2 & wy - vz & vy - wx \\ wy - vz & uz - w^2 & vw - uy \\ vy - wx & vw - uy & ux - v^2 \end{bmatrix};$$

$$\begin{bmatrix} a & b & b \\ b & a & b \\ b & b & a \end{bmatrix}^{-1} = \frac{1}{a(a + b) - 2b^2}\begin{bmatrix} a + b & -b & -b \\ -b & a + b & -b \\ -b & -b & a + b \end{bmatrix}.$$

III.   Diagonal Matrices

$$\begin{bmatrix} a & 0 & 0 & \cdots & 0 \\ 0 & b & 0 & \cdots & 0 \\ 0 & 0 & c & \cdots & 0 \\ \vdots & \vdots & \vdots & \vdots & \vdots \\ 0 & 0 & 0 & \cdots & z \end{bmatrix}^{-1} = \begin{bmatrix} 1/a & 0 & 0 & \cdots & 0 \\ 0 & 1/b & 0 & \cdots & 0 \\ 0 & 0 & 1/c & \cdots & 0 \\ \vdots & \vdots & \vdots & \vdots & \vdots \\ 0 & 0 & 0 & \cdots & 1/z \end{bmatrix}.$$

## D2.9   Rank

We have shown that if $D = |A| = 0$, then one or more rows or columns of the matrix **A** are linearly dependent on (equal to a weighted sum of) the remaining rows or columns of the matrix. It is often helpful to know how many independent rows or columns there are in a given matrix. This number is called the *rank* of the matrix, and can be determined by examining *submatrices* of **A**. If an $r \times r$ submatrix of **A** (called a "minor" and obtained from **A** by deleting the row and column in which one or more elements appear) can be found whose determinant equals zero, then the rank of **A** is at most $r - 1$. The rank of a matrix can, of course, also be determined (sometimes quite conveniently) by adding or subtracting multiples of various rows or columns to other rows or columns in order to create one or more rows (columns) consisting entirely of zeros.

If **A** is symmetric, then the rank of the matrix can be determined by concentrating on the *leading principal minor determinants*, that is, the determinants of the $1 \times 1$, $2 \times 2$, $3 \times 3$, ..., $m \times m$ submatrices each of which consists of the first $k$ rows and columns of the matrix. If the first $r$ of these all have positive determinants while the $(r + 1) \times (r + 1)$ and larger leading principal minors all yield zero determinants, then **A** is of rank $r$. Moreover, the quadratic form **b'Ab** corresponding to **A** is said to be "positive semidefinite" and can be shown to be greater than or equal to zero for any choice of coefficients for **b**. This guarantees, for instance, that the variance of any linear combination of the variables involved in a covariance matrix **a'Sa** will always be positive regardless of our choice of combining rule.

## D2.10   Matrix Calculus

It is extremely common in matrix algebra that we have to find the maximum or minimum of an expression $f(b_1, b_2, \ldots, b_m)$, which involves several different coefficients. We seek this extremum by differentiating $f(b_1, b_2, \ldots, b_m)$ with respect to each of the $b_j$ and then setting these derivatives equal to zero. Thus we have the set of equations,

$$df(b_1, b_2, \ldots, b_m)/db_1 = 0;$$

$$df(b_1, b_2, \ldots, b_m)/db_2 = 0;$$

$$\vdots$$

$$df(b_1, b_2, \ldots, b_m)/db_m = 0.$$

A more compact notation would be useful in such a situation. It seems natural to rewrite the above set of equations as

$$df(\mathbf{b})/d\mathbf{b} = \mathbf{0},$$

and this convention has been adopted. It must be kept in mind that $df(\mathbf{b})/d\mathbf{b}$ is nothing more than a column vector of the separate derivatives of $f(\mathbf{b})$, that is,

$$df(\mathbf{b})/d\mathbf{b} = \begin{bmatrix} df(\mathbf{b})/db_1 \\ df(\mathbf{b})/db_2 \\ \vdots \\ df(\mathbf{b})/db_m \end{bmatrix} = \begin{bmatrix} df(b_1, b_2, \ldots, b_m)/db_1 \\ df(b_1, b_2, \ldots, b_m)/db_2 \\ \vdots \\ df(b_1, b_2, \ldots, b_m)/db_m \end{bmatrix}.$$

Having adopted this notation, we find that it produces some very considerable economies of derivation in that we can develop some simple rules for working within that notation instead of translating $df(\mathbf{b})/d\mathbf{b}$ back into the more cumbersome notation. Dwyer and MacPhail (1948) provide a systematic listing of many of these "matrix derivatives." We shall present and justify a few of the ones we shall use most often.

(1) If $f(\mathbf{b})$ is a constant for all choices of $b_1, b_2, b_3$, etc., then

$$df(\mathbf{b})/d\mathbf{b} = d(c)/d\mathbf{b} = \mathbf{0}.$$

(2) If $f(\mathbf{b}) = \mathbf{a}'\mathbf{b} = a_1 b_1 + a_2 b_2 + \cdots + a_m b_m = \mathbf{b}'\mathbf{a}$, whence $df/db_i = a_i$; then

$$df(\mathbf{b})/d\mathbf{b} = d(\mathbf{a}'\mathbf{b})/d\mathbf{b} = \mathbf{a} = d(\mathbf{b}'\mathbf{a})/d\mathbf{b}.$$

(3) If $f(\mathbf{b}) = \mathbf{b}'\mathbf{A}\mathbf{b}$, where $\mathbf{A}$ is symmetric, then

$$df(\mathbf{b})/d\mathbf{b} = d(\mathbf{b}'\mathbf{A}\mathbf{b})/d\mathbf{b} = d(\sum \sum a_{ij}b_i b_j)/d\mathbf{b} = d(\sum a_{ii}b_i^2 + 2\sum_{i>j} \sum a_{ij}b_i b_j)/d\mathbf{b}.$$

Clearly

$$d(\mathbf{b}'\mathbf{A}\mathbf{b})/db_h = 2a_{hh}b_h + 2\sum_{\substack{j \\ j \neq h}} a_{hj}b_j = 2\mathbf{a}_h'\mathbf{b} = 2\mathbf{b}'\mathbf{a}_h ,$$

since only those terms in $\sum \sum a_{ij}b_i b_j$ which involve $b_h$ and thus $a_{hj}$ or $a_{jh}$ (which are identical, since this is a symmetric matrix) will have nonzero derivatives with respect to $b_h$. Thus

$$d(\mathbf{b}'\mathbf{A}\mathbf{b})/d\mathbf{b} = 2\begin{bmatrix} \mathbf{a}_1'\mathbf{b} \\ \mathbf{a}_2'\mathbf{b} \\ \vdots \\ \mathbf{a}_m'\mathbf{b} \end{bmatrix} = 2\mathbf{A}\mathbf{b} = 2\begin{bmatrix} \mathbf{b}'\mathbf{a}_1 \\ \mathbf{b}'\mathbf{a}_2 \\ \vdots \\ \mathbf{b}'\mathbf{a}_m \end{bmatrix}.$$

## D2.11  Partitioned Matrices

One common need is computation of the determinant and/or inverse of a matrix which involves two distinct sets of variables

$$\mathbf{A} = \begin{bmatrix} \mathbf{A}_{11} & \mathbf{A}_{12} \\ \mathbf{A}_{21} & \mathbf{A}_{22} \end{bmatrix}.$$

It would be convenient to have an expression for $\mathbf{A}^{-1}$ in terms of $\mathbf{A}_{11}, \mathbf{A}_{12}, \mathbf{A}_{21},$ and $\mathbf{A}_{22}$. Let us ask for a matrix

$$\mathbf{B} = \begin{bmatrix} \mathbf{B}_{11} & \mathbf{B}_{12} \\ \mathbf{B}_{21} & \mathbf{B}_{22} \end{bmatrix}$$

such that $\mathbf{B}\mathbf{A} = \mathbf{I} = \mathbf{A}\mathbf{B}$. We can satisfy these conditions only if, considering the entries in the first row of $\mathbf{B}\mathbf{A}$,

$$\mathbf{B}_{11}\mathbf{A}_{11} + \mathbf{B}_{12}\mathbf{A}_{21} = \mathbf{I}, \qquad \text{and} \qquad \mathbf{B}_{11}\mathbf{A}_{12} + \mathbf{B}_{12}\mathbf{A}_{22} = \mathbf{0};$$

whence

$$\mathbf{B}_{12} = -\mathbf{B}_{11}\mathbf{A}_{12}\mathbf{A}_{22};$$

and

$$\mathbf{B}_{11}\mathbf{A}_{11} - \mathbf{B}_{11}\mathbf{A}_{12}\mathbf{A}_{22}^{-1}\mathbf{A}_{21} = \mathbf{I}, \qquad \text{or} \qquad \mathbf{B}_{11}(\mathbf{A}_{11} - \mathbf{A}_{12}\mathbf{A}_{22}^{-1}\mathbf{A}_{21}) = \mathbf{I};$$

whence

$$\mathbf{B}_{11} = (\mathbf{A}_{11} - \mathbf{A}_{12}\,\mathbf{A}_{22}^{-1}\mathbf{A}_{21})^{-1}.$$

Now, substituting the expression for $\mathbf{B}_{11}$ into that for $\mathbf{B}_{12}$, we obtain

$$\mathbf{B}_{12} = -\mathbf{B}_{11}\,\mathbf{A}_{12}\,\mathbf{A}_{22} = -(\mathbf{A}_{11} - \mathbf{A}_{12}\,\mathbf{A}_{22}^{-1}\mathbf{A}_{21})^{-1}\mathbf{A}_{12}\,\mathbf{A}_{22}^{-1}.$$

Now, using the second row of $\mathbf{BA}$, we find that

$$\mathbf{B}_{21}\,\mathbf{A}_{11} + \mathbf{B}_{22}\,\mathbf{A}_{21} = \mathbf{0}, \qquad \text{and} \qquad \mathbf{B}_{21}\,\mathbf{A}_{12} + \mathbf{B}_{22}\,\mathbf{A}_{22} = \mathbf{I};$$

whence

$$\mathbf{B}_{22}\,\mathbf{A}_{22} = \mathbf{I} - \mathbf{B}_{21}\,\mathbf{A}_{12}\,.$$

However,

$$\mathbf{B}_{21} = \mathbf{B}_{12}' = -\mathbf{A}_{22}^{-1}\mathbf{A}_{12}'(\mathbf{A}_{11} - \mathbf{A}_{12}\,\mathbf{A}_{22}^{-1}\mathbf{A}_{21})^{-1},$$

so that

$$\mathbf{B}_{22}\,\mathbf{A}_{22} = \mathbf{I} + \mathbf{A}_{22}^{-1}\mathbf{A}_{12}'(\mathbf{A}_{11} - \mathbf{A}_{12}\,\mathbf{A}_{22}^{-1}\mathbf{A}_{21})^{-1}\mathbf{A}_{12}\,.$$

Postmultiplication by $\mathbf{A}_{22}^{-1}$ thus gives

$$\mathbf{B}_{22} = \mathbf{A}_{22}^{-1} + \mathbf{A}_{22}^{-1}\mathbf{A}_{21}(\mathbf{A}_{11} - \mathbf{A}_{12}\,\mathbf{A}_{22}^{-1}\mathbf{A}_{21})^{-1}\mathbf{A}_{12}\,\mathbf{A}_{22}^{-1}.$$

Alternately, we can seek an expression in which the roles of $\mathbf{A}_{22}$ and $\mathbf{A}_{11}$ are interchanged. Beginning our analysis with the second row of the product matrix this time, we obtain

$$\mathbf{B}_{21} = -\mathbf{B}_{22}\,\mathbf{A}_{21}\,\mathbf{A}_{11}^{-1}; \qquad \text{whence} \quad \mathbf{B}_{22} = (\mathbf{A}_{22} - \mathbf{A}_{21}\,\mathbf{A}_{11}^{-1}\mathbf{A}_{12})^{-1}; \quad \text{etc.}$$

Summarizing all these manipulations, our final expressions for $\mathbf{A}^{-1}$ are

$$\mathbf{A}^{-1} = \begin{bmatrix} (\mathbf{A}_{11} - \mathbf{A}_{12}\,\mathbf{A}_{22}^{-1}\mathbf{A}_{21})^{-1} & -(\mathbf{A}_{11} - \mathbf{A}_{12}\,\mathbf{A}_{22}^{-1}\mathbf{A}_{21})^{-1}\mathbf{A}_{12}\,\mathbf{A}_{22}^{-1} \\ -\mathbf{A}_{22}^{-1}\mathbf{A}_{21}(\mathbf{A}_{11} - \mathbf{A}_{12}\,\mathbf{A}_{22}^{-1}\mathbf{A}_{21})^{-1} & \mathbf{A}_{22}^{-1} + \mathbf{A}_{22}^{-1}\mathbf{A}_{21}(\mathbf{A}_{11} - \mathbf{A}_{12}\,\mathbf{A}_{22}^{-1}\mathbf{A}_{21})^{-1}\mathbf{A}_{12}\,\mathbf{A}_{22}^{-1} \end{bmatrix}$$

$$= \begin{bmatrix} \mathbf{A}_{11}^{-1} + \mathbf{A}_{11}^{-1}\mathbf{A}_{12}(\mathbf{A}_{22} - \mathbf{A}_{21}\mathbf{A}_{11}^{-1}\mathbf{A}_{12})^{-1}\mathbf{A}_{21}\mathbf{A}_{11}^{-1} & -\mathbf{A}_{11}^{-1}\mathbf{A}_{12}(\mathbf{A}_{22} - \mathbf{A}_{21}\mathbf{A}_{11}^{-1}\mathbf{A}_{12})^{-1} \\ -(\mathbf{A}_{22} - \mathbf{A}_{21}\mathbf{A}_{11}^{-1}\mathbf{A}_{12})^{-1}\mathbf{A}_{21}\mathbf{A}_{11}^{-1} & (\mathbf{A}_{22} - \mathbf{A}_{21}\mathbf{A}_{11}^{-1}\mathbf{A}_{12})^{-1} \end{bmatrix}.$$

We could now proceed to expand one or both of the above expressions by cofactors to compute the determinant of $\mathbf{A}$, but it is much easier to reduce $\mathbf{A}$ to triangular form, as follows:

$$|\mathbf{A}| = \begin{bmatrix} \mathbf{A}_{11} & \mathbf{A}_{12} \\ 0 & \mathbf{A}_{22} - \mathbf{D}\mathbf{A}_{12} \end{bmatrix},$$

where $\mathbf{D}\mathbf{A}_{11} = \mathbf{A}_{21}$, whence $\mathbf{D} = \mathbf{A}_{21}\,\mathbf{A}_{11}^{-1}$, whence

$$|\mathbf{A}| = |\mathbf{A}_{11}| \cdot |\mathbf{A}_{22} - \mathbf{A}_{21}\,\mathbf{A}_{11}^{-1}\mathbf{A}_{12}|.$$

Alternately, using the same procedure but getting the zero in the first row,

$$|\mathbf{A}| = |\mathbf{A}_{22}| \cdot |\mathbf{A}_{11} - \mathbf{A}_{12}\,\mathbf{A}_{22}^{-1}\mathbf{A}_{21}|.$$

As an example, let $\mathbf{S}$ be the $(m+1) \times (m+1)$ covariance matrix involving $X_1, X_2, \ldots, X_m$ and $Y$. Now $\mathbf{A}_{22} = s_y^2$; $\mathbf{A}_{12} = \mathbf{s}_{xy}$; and $\mathbf{A}_{11} = \mathbf{S}_x$, the covariance matrix of the $X$s alone. Using the expressions developed above, and letting $\mathbf{C} = \mathbf{A}^{-1}$ be partitioned in the same way as $\mathbf{A}$, we find that

$$\mathbf{C}_{22} \ (\text{which is really a scalar}) = (\mathbf{A}_{22} - \mathbf{A}_{21}\,\mathbf{A}_{11}^{-1}\mathbf{A}_{12})^{-1} = (s_y^2 - \mathbf{s}_{xy}'\,\mathbf{S}_x^{-1}\mathbf{s}_{xy})^{-1}$$

$$= 1/[s_y^2(1 - R_{y \cdot x}^2)];$$

and

$$C_{12} = -S_x^{-1}s_{xy}[s_y^2(1 - R_{y \cdot x}^2)]^{-1} = -\mathbf{b}[s_y^2(1 - R_{y \cdot x}^2)]^{-1};$$

whence

$$\mathbf{b} = -C_{12}\,s_y^2(1 - R_{y \cdot x}^2) = -C_{12}/C_{22}; \qquad \text{and} \qquad R_{y \cdot x}^2 = 1 - 1/(s_y^2 \cdot C_{22}).$$

(We have, of course, made use of the formulas for $\mathbf{b}$, the column vector of regression coefficients for predicting $Y$ from the $X$s, and $R_{y \cdot x}^2$, the squared coefficient of multiple correlation, that were developed in Chapter 2.)

In other words, the column vector of regression coefficients is given by the last column of $\mathbf{C}$ (or the last row, since $\mathbf{C}$, like $\mathbf{A}$, is symmetric), where each entry has been normalized by dividing by the lower right-hand entry in $\mathbf{C}$. However, $\mathbf{A}$ is just a covariance matrix with an "extra" variable $Y$ added on, so that we could just as well consider it to be an $\mathbf{S}_x$ matrix which involves $m + 1$ variables, rather than $m$. We could then let $Y$ be relabeled as $X_{m+1}$. Moreover, we could, through relabeling, bring any of the $X$ variables into the last row and column of $\mathbf{S}_x$ and then apply the procedures outlined above. Thus the only role that positioning $Y$ in the last row and column of $\mathbf{A}$ played was to simplify the derivation somewhat, and it should be clear that *any* given row $i$ of the inverse of a covariance matrix is closely related to the regression of variable $i$ on all the other variables involved in that covariance matrix, with

$$R_{i \cdot c}^2 = 1 - 1/(c_{ii}s_i^2),$$

where $\mathbf{c}$ represents the set of all variables other than $i$, and $c_{ij}$ is the $ij$th entry in $\mathbf{S}^{-1}$; and the coefficients in the regression equation which best predicts $i$ from the other variables are simply the entries in the $i$th row of $\mathbf{S}^{-1}$, divided by $c_{ii}$.

## D2.12 Characteristic Roots and Vectors

In the course of finding "optimal" linear combinations of variables, we shall often have recourse to the method of Lagrangian multipliers (cf. Digression 1). When applied to matrix expressions, for example, the variance or covariance of various linear combinations, this will often result in equations of the form

$$\mathbf{Cb} - \lambda\mathbf{Db} = \mathbf{0};$$

whence

$$\mathbf{D}^{-1}\mathbf{Cb} - \lambda\mathbf{Ib} = \mathbf{0},$$

whence

$$[\mathbf{D}^{-1}\mathbf{C} - \lambda\mathbf{I}]\mathbf{b} = \mathbf{0}.$$

It would therefore be extremely handy to have a systematic way of finding solutions to a matrix equation of the form $[\mathbf{A} - \lambda\mathbf{I}]\mathbf{b} = \mathbf{0}$. We know from our earlier discussion of systems of linear equations that this system has a nontrivial solution if and only if $|\mathbf{A} - \lambda\mathbf{I}| = 0$. Expanding this determinant with $\lambda$ left in as an unknown produces a $p$th-order polynomial in $\lambda$ (the *characteristic equation*) which can be solved to yield, in general, $p$ roots (assuming $\mathbf{A}$ to be a $p \times p$ matrix) $\lambda_i$, $i = 1, 2, \ldots, p$. The $i$th *characteristic root* can then be "plugged into" the original matrix equation and the resulting homogenous system of $p$ equations in $p$ unknowns solved for $\mathbf{b}_i$, the *characteristic vector* associated with that characteristic root $\lambda_i$. (An alternative and quite common terminology labels $\lambda_i$ and $b_i$ as the $i$th *eigenvalue* and *eigenvector*, respectively.) By way of illustration, the characteristic roots and vectors of the correlation matrix of the predictor variables of Data Set 1 will be obtained. The determinantal equation is

$$\begin{vmatrix} 1.0 - \lambda & .134 & .748 \\ .134 & 1.0 - \lambda & .650 \\ .748 & .650 & 1.0 - \lambda \end{vmatrix} = 0;$$

whence

$$0 = (1.0 - \lambda)[(1.0 - \lambda)^2 - (.650)^2] - .134[.134(1.0 - \lambda) - .650(.748)]$$

$$+ .748[(.134)(.650) - (1.0 - \lambda)(.748)]$$

$$= (1.0 - \lambda)^3 - (.650^2 + .134^2 + .748^2)(1.0 - \lambda) + 2.0(.134)(.650)(.748)$$

$$= (1.0 - \lambda)^3 - (.9999)(1.0 - \lambda) + .1303.$$

Referring to Digression 3 (which summarizes procedures for solving cubic equations), we see that this equation is already of the form $x^3 + ax + b = 0$, whence

$$\frac{b^2}{4} + \frac{a^3}{27} = \frac{.1303^2}{4} - \frac{.9999^3}{27} = -.033,$$

whence a trigonometric solution is recommended:

$$\cos \phi = -2.59808(.1303) = -.3335 \Rightarrow \phi = 109.5°$$

and

$$1 - \lambda = \frac{1}{\sqrt{3}}(.8039, \quad -.9171, \quad .1132) = .927, \quad -1.058, \quad .131,$$

whence

$$\lambda = (2.058, \quad .869, \quad .073)$$

Alternately, we could have got to this point via the *Laplace expansion*, namely

$$|\mathbf{A} - \lambda\mathbf{I}| = (-\lambda)^p + S_1(-\lambda)^{p-1} + S_2(-\lambda)^{p-2} + \cdots + S_{p-1}\lambda + |\mathbf{A}|,$$

where $S_i = \sum$ (principal minor determinants of size $i \times i$), and a "principal minor determinant" is the determinant of a square submatrix of $\mathbf{A}$ each of whose main diagonal elements is a main diagonal element of $\mathbf{A}$.

Applying the Laplace expansion to our example gives

$$S_1 = 1.0 + 1.0 + 1.0 = 3.0;$$

$$S_2 = \begin{vmatrix} 1 & .134 \\ .134 & 1 \end{vmatrix} + \begin{vmatrix} 1 & .650 \\ .650 & 1 \end{vmatrix} + \begin{vmatrix} 1 & .748 \\ .748 & 1 \end{vmatrix}$$

$$= 3 - .650^2 - .134^2 - .748^2$$

$$S_3 = |\mathbf{R}| = 1 + 2(.134)(.650)(.748) - (.650^2 + .134^2 + .748^2)$$

$$= 1 + .1303 - .9999 = .1304.$$

At this point we turn to Digression 3, finding there that we have a general cubic equation with $y = -\lambda$, $p = 3$, $q = 2.001$, and $r = .1304$. Thus the substitution $-\lambda = x - 3/3$ or $\lambda = 1 - x$ yields

$$|\mathbf{R} - \lambda\mathbf{I}| = (-\lambda)^3 + 3\lambda^2 - 2.001\lambda + .1304.$$

Letting $\lambda = 1 - x$ yields $x^3 - .9999x + .1343 = 0$, which is exactly the equation we obtained via the direct route.

Note that the product and sum of the characteristic roots are equal, respectively, to the determinant and trace (sum of the main diagonal entries) of the matrix: $(.073)(.869) \times (2.058) = .130$ and $.073 + .869 + 2.058 = 3.000$. These two properties are true of the set of characteristic roots derived from any matrix, and are useful as checks on your computations. Characteristic roots have a number of other interesting and useful properties, including:

(1)   The characteristic roots of a positive definite matrix (such as the covariance matrix of a set of variables none of which is linearly dependent on the others) are all positive.

(2)   If an $n \times n$ symmetric matrix is positive semidefinite of rank $r$, it contains exactly $r$ positive roots and $n - r$ zero roots.

(3)   The characteristic roots of a diagonal matrix are the diagonal elements themselves.

With the values of the characteristic roots in hand, we now proceed to compute the characteristic vector associated with each root. For instance, to obtain the vector corresponding to our largest root 2.058 we must solve the following set of equations:

$$-1.058b_1 + .134b_2 + .748b_3 = 0,$$

$$.134b_1 - 1.058b_2 + .650b_3 = 0,$$

$$.748b_1 + .650b_2 - 1.058b_3 = 0.$$

Subtracting the second equation from the first gives $1.192(b_2 - b_1) + .098b_3 = 0$, whence $b_1 = .082b_3 + b_2$. Plugging this into the second equation gives $b_2 = (.661/.924)b_3 = .715b_3$, whence $b_1 = .797b_3$. Since this is a homogenous set of equations, the $b$s are specified only up to a scale factor. Using $b_3$ as our unit of measurement gives $\mathbf{b}_1' = (.797, .715, 1)$ as a solution to these equations and $k(.797, .715, 1)$ as the general solution, where $k$ is any constant. It is quite common to require that $\mathbf{b}_i' \mathbf{b}_i = \sum b_i^2$ be equal to unity, and the particular vector which satisfies that side condition is referred to as the *normalized* characteristic vector. In order to "normalize" our vector, we seek a value of $k$ such that

$$(k\mathbf{b}_1')(k\mathbf{b}_1) = k^2 \mathbf{b}_1 \, \mathbf{b} = 1;$$

whence

$$k = 1/\sqrt{\sum b_i^2} = 1/\sqrt{.797^2 + .715^2 + 1} = 1/1.465 = .683.$$

In other words, the *normalized* characteristic vector corresponding to $\lambda_i$ is obtained by dividing any vector satisfying the defining equations by the square root of the sum of the squares of the coefficients in that vector.

Characteristic roots of *symmetric* matrices have three very important properties (cf. Morrison, 1967, p. 63):

(1)   The characteristic vectors associated with any two unequal characteristic roots are orthogonal, that is, $\mathbf{b}_i' \mathbf{b}_j = 0$ if $\lambda_i \neq \lambda_j$.

(2)   If we define a matrix $\mathbf{P}$ whose columns are the *normalized* characteristic vectors of $\mathbf{A}$, then $\mathbf{P'AP} = \mathbf{D}$, where $\mathbf{D}$ is a diagonal matrix (all entries zero except on the main diagonal) whose diagonal entries are the characteristic roots of $\mathbf{A}$. Thus the characteristic vectors can be used to "diagonalize" $\mathbf{A}$.

(3)   If we transform the variables $x_i$ involved in the quadratic form $\mathbf{x'Ax}$ to a new set of variables $y_i$ in which each $y$-variable is a linear combination of the $x$-variables, the coefficients of that transformation being equal to the coefficients of the $i$th characteristic vector of $\mathbf{A}$ (that is, $y_i = \mathbf{x'b}_i$, where $\mathbf{b}_i$ is the $i$th characteristic vector of $\mathbf{A}$), then

$$\mathbf{x'Ax} = \lambda_1 y_1^2 + \lambda_2 y_2^2 + \cdots + \lambda_p y_p^2.$$

# 3 | DIGRESSION
## Solution of Cubic Equations

It frequently happens that the characteristic equation of a $3 \times 3$ matrix whose characteristic roots we seek (cf. Digression 2) will involve cubes of $\lambda$. It is therefore desirable to have a procedure for solving cubic equations, so that we can obtain analytic solutions for the characteristic roots in at least the $3 \times 3$ case.

Given a cubic equation of the form: $y^3 + py^2 + qy + r = 0$,

(1) Eliminate the $y^2$ term through the substitution, $y = x - p/3$, whence we have

$$x^3 + ax + b = 0,$$

where

$$a = q - p^2/3 \quad \text{and} \quad b = (2p^3 - 9pq + 27r)/27.$$

(2) Calculate the "discriminant," $d = b^2/4 + a^3/27$.

  (i) If $d$ is positive, there will be one real root and two imaginary roots, namely

$$x = A + B, \quad -\tfrac{1}{2}(A + B) + \tfrac{1}{2}(A - B) - 3, \quad \text{and} \quad -\tfrac{1}{2}(A + B) - \tfrac{1}{2}(A - B) - 3,$$

where

$$A = (-b/2 + d)^{1/3}, \quad B = (-b/2 - d)^{1/3}.$$

  (ii) If $d = 0$, there will be three real roots, of which two at least are equal, namely,

$$x = 2(-b/2)^{1/3}, \quad (b/2)^{1/3}, \quad (b/2)^{1/3}.$$

  (iii) If $d$ is negative (by far the most common case where characteristic equations are involved), there will be three real and unequal roots. In this case, use a trigonometric substitution, namely

$$\cos \phi = -b(27/4)^{1/2}(-a^3)^{1/2} = -2.59808b/(-a)^{3/2},$$

whence, having found $\phi$ from a table of cosines,

$$x = 2(-a/3)^{1/2} \cos(\phi/3), \quad 2(-a/3)^{1/2} \cos(\phi/3 + 120°), \quad 2(-a/3)^{1/2} \cos(\phi/3 + 240°)$$
$$= [(-a)^{1/2}/.86603][\cos(\phi/3), -\cos(60° - \phi/3), \sin(\phi/3 - 30°)].$$

The process of evaluating these trigonometric functions can be speeded somewhat by use of Table D3.1. (For finer accuracy, a more extensive table of the trigonometric functions must be consulted.)

*Table D3.1*

*Cosine Functions of Use in Solving Cubic Equations[a]*

| $w =$ | $\phi$ | | $c_1 =$ | $c_2 =$ | $c_3 =$ |
|---|---|---|---|---|---|
| $\cos \phi$ | Deg | Rad | $\cos(\phi/3)$ | $\cos(\phi/3 + 120°)$ | $\cos(\phi/3 + 240°)$ |
| 1.000 | 0.0 | .0000 | 1.0000 | −.5000 | −.5000 |
| .998 | 3.6 | .0633 | .9998 | −.5181 | −.4816 |
| .996 | 5.1 | .0895 | .9996 | −.5256 | −.4740 |
| .994 | 6.3 | .1096 | .9993 | −.5313 | −.4680 |
| .992 | 7.25 | .1266 | .9991 | −.5361 | −.4630 |
| .990 | 8.1 | .1415 | .9989 | −.5403 | −.4586 |
| .988 | 8.9 | .1551 | .9987 | −.5441 | −.4546 |
| .986 | 9.6 | .1675 | .9984 | −.5476 | −.4509 |
| .984 | 10.3 | .1791 | .9982 | −.5508 | −.4474 |
| .982 | 10.9 | .1900 | .9980 | −.5538 | −.4442 |
| .980 | 11.5 | .2003 | .9978 | −.5567 | −.4411 |
| .978 | 12.0 | .2102 | .9976 | −.5594 | −.4382 |
| .976 | 12.6 | .2195 | .9973 | −.5620 | −.4354 |
| .974 | 13.1 | .2285 | .9971 | −.5645 | −.4326 |
| .972 | 13.6 | .2372 | .9969 | −.5668 | −.4300 |
| .970 | 14.1 | .2456 | .9965 | −.5691 | −.4275 |
| .968 | 14.5 | .2537 | .9964 | −.5714 | −.4251 |
| .966 | 15.0 | .2615 | .9962 | −.5735 | −.4227 |
| .964 | 15.4 | .2691 | .9960 | −.5756 | −.4204 |
| .962 | 15.8 | .2766 | .9958 | −.5776 | −.4182 |
| .960 | 16.3 | .2838 | .9955 | −.5796 | −.4160 |
| .958 | 16.7 | .2908 | .9953 | −.5815 | −.4138 |
| .956 | 17.1 | .2978 | .9951 | −.5834 | −.4117 |
| .954 | 17.4 | .3045 | .9948 | −.5852 | −.4097 |
| .952 | 17.8 | .3111 | .9946 | −.5870 | −.4077 |
| .95 | 18.2 | .3176 | .9944 | −.5887 | −.4057 |
| .94 | 19.9 | .3482 | .9933 | −.5969 | −.3964 |
| .93 | 21.6 | .3764 | .9921 | −.6044 | −.3877 |
| .92 | 23.1 | .4027 | .9910 | −.6114 | −.3796 |
| .91 | 24.5 | .4275 | .9899 | −.6179 | −.3719 |
| .90 | 25.8 | .4510 | .9887 | −.6241 | −.3646 |
| .89 | 27.1 | .4735 | .9876 | −.6299 | −.3577 |
| .88 | 28.4 | .4949 | .9864 | −.6354 | −.3510 |
| .87 | 29.5 | .5156 | .9853 | −.6407 | −.3445 |
| .86 | 30.7 | .5355 | .9841 | −.6458 | −.3383 |
| .85 | 31.8 | .5548 | .9830 | −.6507 | −.3322 |
| .84 | 32.9 | .5735 | .9818 | −.6554 | −.3263 |
| .83 | 33.9 | .5917 | .9806 | −.6600 | −.3206 |
| .82 | 34.9 | .6094 | .9794 | −.6644 | −.3150 |
| .81 | 35.9 | .6266 | .9783 | −.6687 | −.3096 |

*Table D3.1* (continued)

| $w = \cos\phi$ | $\phi$ Deg | $\phi$ Rad | $c_1 = \cos(\phi/3)$ | $c_2 = \cos(\phi/3 + 120°)$ | $c_3 = \cos(\phi/3 + 240°)$ |
|---|---|---|---|---|---|
| .80 | 36.9 | .6435 | .9771 | −.6729 | −.3042 |
| .79 | 37.8 | .6600 | .9759 | −.6769 | −.2990 |
| .78 | 38.7 | .6761 | .9747 | −.6809 | −.2938 |
| .77 | 39.6 | .6920 | .9735 | −.6847 | −.2888 |
| .76 | 40.5 | .7075 | .9723 | −.6885 | −.2838 |
| .75 | 41.4 | .7227 | .9711 | −.6922 | −.2789 |
| .74 | 42.3 | .7377 | .9699 | −.6958 | −.2741 |
| .73 | 43.1 | .7525 | .9687 | −.6993 | −.2694 |
| .72 | 43.9 | .7670 | .9675 | −.7028 | −.2647 |
| .71 | 44.8 | .7813 | .9663 | −.7061 | −.2601 |
| .70 | 45.6 | .7954 | .9651 | −.7095 | −.2556 |
| .69 | 46.4 | .8093 | .9638 | −.7127 | −.2511 |
| .68 | 47.2 | .8230 | .9626 | −.7159 | −.2467 |
| .67 | 47.9 | .8366 | .9614 | −.7191 | −.2423 |
| .66 | 48.7 | .8500 | .9601 | −.7222 | −.2380 |
| .65 | 49.5 | .8632 | .9589 | −.7252 | −.2337 |
| .64 | 50.2 | .8763 | .9576 | −.7282 | −.2294 |
| .63 | 50.9 | .8892 | .9564 | −.7312 | −.2252 |
| .62 | 51.7 | .9020 | .9551 | −.7341 | −.2211 |
| .61 | 52.4 | .9147 | .9539 | −.7369 | −.2170 |
| .60 | 53.1 | .9273 | .9526 | −.7398 | −.2129 |
| .59 | 53.8 | .9397 | .9513 | −.7425 | −.2088 |
| .58 | 54.5 | .9521 | .9501 | −.7453 | −.2048 |
| .57 | 55.2 | .9643 | .9488 | −.7480 | −.2008 |
| .56 | 55.9 | .9764 | .9475 | −.7507 | −.1968 |
| .55 | 56.6 | .9884 | .9462 | −.7533 | −.1929 |
| .54 | 57.3 | 1.0004 | .9449 | −.7559 | −.1890 |
| .53 | 58.0 | 1.0122 | .9436 | −.7585 | −.1851 |
| .52 | 58.7 | 1.0240 | .9423 | −.7610 | −.1813 |
| .51 | 59.3 | 1.0356 | .9410 | −.7636 | −.1774 |
| .50 | 60.0 | 1.0472 | .9397 | −.7660 | −.1736 |
| .49 | 60.7 | 1.0587 | .9384 | −.7685 | −.1699 |
| .48 | 61.3 | 1.0701 | .9370 | −.7709 | −.1661 |
| .47 | 62.0 | 1.0815 | .9357 | −.7734 | −.1624 |
| .46 | 62.6 | 1.0928 | .9344 | −.7757 | −.1587 |
| .45 | 63.3 | 1.1040 | .9330 | −.7781 | −.1550 |
| .44 | 63.9 | 1.1152 | .9317 | −.7804 | −.1513 |
| .43 | 64.5 | 1.1263 | .9304 | −.7827 | −.1476 |
| .42 | 65.2 | 1.1374 | .9290 | −.7850 | −.1440 |
| .41 | 65.8 | 1.1483 | .9276 | −.7873 | −.1404 |

*Table D3.1* (continued)

| $w = \cos\phi$ | $\phi$ | | $c_1 = \cos(\phi/3)$ | $c_2 = \cos(\phi/3 + 120°)$ | $c_3 = \cos(\phi/3 + 240°)$ |
|---|---|---|---|---|---|
| | Deg | Rad | | | |
| .40 | 66.4 | 1.1593 | .9263 | −.7895 | −.1367 |
| .39 | 67.0 | 1.1702 | .9249 | −.7917 | −.1332 |
| .38 | 67.7 | 1.1810 | .9235 | −.7939 | −.1296 |
| .37 | 68.3 | 1.1918 | .9221 | −.7961 | −.1260 |
| .36 | 68.9 | 1.2025 | .9207 | −.7983 | −.1224 |
| .35 | 69.5 | 1.2132 | .9193 | −.8004 | −.1189 |
| .34 | 70.1 | 1.2239 | .9179 | −.8026 | −.1154 |
| .33 | 70.7 | 1.2345 | .9165 | −.8047 | −.1119 |
| .32 | 71.3 | 1.2451 | .9151 | −.8067 | −.1084 |
| .31 | 71.9 | 1.2556 | .9137 | −.8088 | −.1049 |
| .30 | 72.5 | 1.2661 | .9123 | −.8109 | −.1014 |
| .29 | 73.1 | 1.2766 | .9108 | −.8129 | −.0979 |
| .28 | 73.7 | 1.2870 | .9094 | −.8149 | −.0945 |
| .27 | 74.3 | 1.2974 | .9079 | −.8169 | −.0910 |
| .26 | 74.9 | 1.3078 | .9065 | −.8189 | −.0876 |
| .25 | 75.5 | 1.3181 | .9050 | −.8209 | −.0841 |
| .24 | 76.1 | 1.3284 | .9036 | −.8228 | −.0807 |
| .23 | 76.7 | 1.3387 | .9021 | −.8248 | −.0773 |
| .22 | 77.3 | 1.3490 | .9006 | −.8267 | −.0739 |
| .21 | 77.9 | 1.3592 | .8991 | −.8286 | −.0705 |
| .20 | 78.5 | 1.3694 | .8976 | −.8305 | −.0671 |
| .19 | 79.0 | 1.3796 | .8961 | −.8324 | −.0637 |
| .18 | 79.6 | 1.3898 | .8946 | −.8343 | −.0603 |
| .17 | 80.2 | 1.4000 | .8931 | −.8362 | −.0569 |
| .16 | 80.8 | 1.4101 | .8916 | −.8380 | −.0535 |
| .15 | 81.4 | 1.4202 | .8900 | −.8398 | −.0502 |
| .14 | 82.0 | 1.4303 | .8885 | −.8417 | −.0468 |
| .13 | 82.5 | 1.4404 | .8869 | −.8435 | −.0434 |
| .12 | 83.1 | 1.4505 | .8854 | −.8453 | −.0401 |
| .11 | 83.7 | 1.4606 | .8838 | −.8471 | −.0367 |
| .10 | 84.3 | 1.4706 | .8822 | −.8489 | −.0334 |
| .09 | 84.8 | 1.4807 | .8806 | −.8506 | −.0300 |
| .08 | 85.4 | 1.4907 | .8791 | −.8524 | −.0267 |
| .07 | 86.0 | 1.5007 | .8775 | −.8541 | −.0234 |
| .06 | 86.6 | 1.5108 | .8759 | −.8558 | −.0200 |
| .050 | 87.1 | 1.5208 | .8742 | −.8576 | −.0167 |
| .048 | 87.2 | 1.5228 | .8739 | −.8579 | −.0160 |
| .046 | 87.4 | 1.5248 | .8736 | −.8582 | −.0153 |
| .044 | 87.5 | 1.5268 | .8733 | −.8586 | −.0147 |
| .042 | 87.6 | 1.5288 | .8729 | −.8589 | −.0140 |

*Table D3.1* (continued)

| $w =$ cos $\phi$ | $\phi$ Deg | $\phi$ Rad | $c_1 =$ cos$(\phi/3)$ | $c_2 =$ cos$(\phi/3 + 120°)$ | $c_3 =$ cos$(\phi/3 + 240°)$ |
|---|---|---|---|---|---|
| .040 | 87.7 | 1.5308 | .8726 | −.8593 | −.0133 |
| .038 | 87.8 | 1.5328 | .8723 | −.8596 | −.0127 |
| .036 | 87.9 | 1.5348 | .8720 | −.8600 | −.0120 |
| .034 | 88.1 | 1.5368 | .8716 | −.8603 | −.0113 |
| .032 | 88.2 | 1.5388 | .8713 | −.8606 | −.0107 |
| .030 | 88.3 | 1.5408 | .8710 | −.8610 | −.0100 |
| .028 | 88.4 | 1.5428 | .8706 | −.8613 | −.0093 |
| .026 | 88.5 | 1.5448 | .8703 | −.8617 | −.0087 |
| .024 | 88.6 | 1.5468 | .8700 | −.8620 | −.0080 |
| .022 | 88.7 | 1.5488 | .8697 | −.8623 | −.0073 |
| .020 | 88.9 | 1.5508 | .8693 | −.8627 | −.0067 |
| .018 | 89.0 | 1.5528 | .8690 | −.8630 | −.0060 |
| .016 | 89.1 | 1.5548 | .8687 | −.8634 | −.0053 |
| .014 | 89.2 | 1.5568 | .8684 | −.8637 | −.0047 |
| .012 | 89.3 | 1.5588 | .8680 | −.8640 | −.0040 |
| .010 | 89.4 | 1.5608 | .8677 | −.8644 | −.0033 |
| .008 | 89.5 | 1.5628 | .674 | −.8647 | −.0027 |
| .006 | 89.7 | 1.5648 | .8670 | −.8650 | −.0020 |
| .004 | 89.8 | 1.5668 | .8667 | −.8654 | −.0013 |
| .002 | 89.9 | 1.5688 | .8664 | −.8657 | −.0007 |
| .000 | 90.0 | 1.5708 | .8660 | −.8660 | .0000 |

[a] When cos $\phi = -w$, Deg$(-w) = 180 -$ Deg$(w)$; Rad$(-w) = \pi -$ Rad$(w)$; $c_1(-w) = -c_2(w)$; $c_2(-w) = -c_1(w)$; and $c_3(-w) = -c_3(w)$.

# 4 | DIGRESSION
## Use of Canned Computer Programs

The present Primer could not have been written twenty or even ten years ago, before packages of statistical programs which could take advantage of the speed and storage capacity of "third generation" computers became readily available throughout the United States. Much more space would have had to be devoted to instructions on setting up work sheets and selecting arithmetic algorithms in such a way as to reduce somewhat the hours of labor and the almost inevitable computational errors involved in conducting a multivariate analysis on any problem of practical size if aided only by a desk calculator. However, with the aid of widely available programs, factor analyses or multivariate analyses of variance which once would have required hundreds of hours of desk-calculator-aided computation can now be completed (with much less chance of a numerical mistake) in a few *seconds* of machine time.

This is not to say that computation time or storage space no longer place limits on the kinds of analyses that can be performed, or the number of variables which can be considered. Few factor analysis or canonical correlation programs can handle more than 100 variables at a time, because of space limitations, and most factor analysis programs are restricted to principal factor analysis, not permitting the user the option of employing the more time-consuming minres or maximum-likelihood methods of deriving a factor pattern. However, improvement in computer speed and storage capacity is rapid, and additions to or extensions of the programs contained within various "packages" are frequent.

It is not to be assumed that the benefits derived from the use of computers in multivariate analysis are without cost. The very fact that computers and applications thereof are undergoing such rapid change forces the frequent user of multivariate statistics to devote considerable time to keeping up with what programs are available and (unfortunately) with changes in procedures for inputting data to the programs he is already using. For instance, when we came back to the task of writing this digression after a semester in which a heavy course load had kept the author away from UNM's computing center, we found that the format of the job card used for accounting purposes and required at the beginning of every program had been revamped because of a hardware change initiated during our "absence"; and that a new "release" (edition) of the job control language specifications for all IBM 360 series computers had been issued (the previous release having been in effect for about six months), thereby rendering obsolete (and a source of job failure and cryptic error messages) all instances in which we had used the previous procedure for specifying the maximum amount of computer time a particular job step might take. Other tasks which awaited the author were: becoming familiar with the OMNITAB statistical package, which had just recently been made available at the computing center; trying to determine why the canonical correlation program added in a recent update of the SPSS package either failed to run or gave incorrect answers for our data while performing perfectly on the update manual's sample problem; and impatiently awaiting the completion of the computing center staff's efforts to "implement" the latest addition to the BMD series of programs, the BMDP series. All of this can quickly add up to a considerable investment of time and a growing backlog of rapidly outdated minutiae which might have been devoted to more substantive learning. The tempting solution employed by many university departments of having available a computer-applications expert to keep track of all the changes carries with it the cost of removing the researcher farther and farther from the processing of his own data. This exacerbates a tendency resulting from the extreme speed of the computer as compared with desk calculator operations for the user to become more and more dependent on the computer, less and less willing or even capable of checking the validity of the results obtained from the computer. However, the user of all but the smallest sets of data has no practical alternative to the use of computer programs. Hopefully, familiarity with the logic of computer operations and some forewarning of the pitfalls awaiting the unwary will enable the user to remain in control of this indispensable tool.

The present digression will therefore

(a)   provide a brief description of the logic and the organization of a computer, and of the successive levels of computer languages which have been developed to mediate between the user's command of English and mathematics and the computer's reliance on patterns of on–off settings of binary devices,

(b)   discuss the relative merits of the most commonly available packages of statistical programs,

(c)   provide some practical hints as to the kinds of problems which arise in using computer programs for data analysis, and how to avoid and/or correct these problems, and

(d)   provide descriptions of the functions and limitations of some specific programs, including a few developed by the author to fill in "gaps" in available packages.

## D4.1   Computer Logic and Organization

Computers are fast, reliable, stupid, and rigid. The speed of computers needs no further comment, except to note that one's conception of what computational procedures are "practical" must be revised with each jump in the speed of the computer, to the point that iterative procedures involving repeating a complicated analysis with slightly different parameter values (for example, guesses as to factor loadings) until the successive guesses converge on an optimal set of values, are becoming quite feasible on recent (circa 1972) computer models. An aspect of computers' "behavior" which deserves more emphasis than it is popularly accorded is their high reliability, that is, their ability to (indeed, their inability not to) perform any set of instructions programmed into the computer hundreds of thousands of times without once departing from those specifications in the slightest detail. This is especially important in numerical computations, where a human operator employing a desk calculator will almost inevitably make at least one error in the course of entering numbers and recording results in any but the shortest string of computations. Elaborate precautions involving devising various internal checks on the calculations and having each analysis performed at least twice, must therefore be taken against these errors in any "hand" analysis of more than the most trivial size. These precautions add to the time differential between hand and computer calculations.

A less desirable aspect of the computer's reliability is its rigidity, a rigidity which in a human calculator would warrant the label, "stupidity." While the computer will not "misplace" or "forget" the location of the decimal point in a number once it has stored that number in one of its memory cells, neither will it recognize that the comma in an entry of "1,000" in a correlation matrix being input to it must "really" be a decimal point, nor will it note that the data on one card have been punched consistently one space too far to the left, requiring a simple departure from the prespecified format for reading that card. (Actually, checks for these kinds of goofs in data preparation, together with instructions as to how to handle them before proceeding with the analysis, can be built into the computer program, but it is virtually impossible to anticipate in advance all of the contingencies which a reasonably intelligent human calculator operator can recognize and handle "reasonably" as they arise.) The point is that a computer performs what it is specifically instructed to do by some human programmer, and nothing else. The computer is in many respects like a human calculator operator who is extremely dependable and extremely fast, but who has no "initiative" or "reasonableness" or "common sense" or "intelligence" which is not specifically incorporated into his instructions.

Indeed, at the most fundamental level (the "hardware" level) running a computer program involves changing the state of a large number of on–off "switches" (in the earliest models, sets of relay contacts and in current models, tiny metal chips whose direction of magnetization can be reversed 180° in a few billionths of a second, that is, a few nanoseconds) which have been organized into sets of fixed size each of which sets is considered a "cell" or a "word" having a specific location (address) within the computer. The contents (switch settings) of each such word can be interpreted, depending on the context in which that particular address is referenced, either as a datum or as one of a severely restricted set of instructions for altering the contents of other cells. These directly representable, "hard-wired" instructions constitute the computer's *machine language*, and usually include little more than instructions to transfer to a special register (often called the "accumulator") the contents of a particular memory cell or of a designated peripheral device (card reader, magnetic tape unit, etc.) and vice versa; to carry out basic arithmetic operations on the contents of the accumulator in combination with those of the memory cell whose address appears in the "address field" of the instruction; to treat the contents of the particular memory cell as the next instruction to be executed, rather than continuing with the basic procedure of reading successive instructions from adjacent cells; and to skip the next instruction (which ordinarily is a "jump" instruction of the sort just described) if the contents of the accumulator meet some criterion, such as being zero or negative. It is this last, "conditional jump" instruction which distinguishes a computer from an electronic calculator and which enables the computer user to build into his programs a degree of flexibility and responsiveness to particular circumstances of the input data which belie the basic rigidity and inflexibility of the computer's internal procedures. To execute a program, then, the contents of a block of cells in the computer's memory must be set to correspond to the desired machine language instructions, other cells must be set with the initial data values with which the program is to begin (numerical constants, counter settings, etc.), other blocks of cells must be set aside for storage of data to be read in from peripheral devices and output to these same devices after manipulation in various ways, and the computer must be instructed to begin execution of the program with the instruction stored in the first cell of the block of instructions.

A program written in machine language which accomplishes these things is known as an *object program*. The earliest applications of computers involved loading this object program as a set of binary numbers (the most natural representation of on–off switch settings) entered into the computer's memory through manipulation of switches on the main computer console or through patterns of holes punched into a paper tape. However, this is an extremely tedious process. To reduce the tedium, successive levels of programming languages have been developed. A program written in one of these higher-level programming languages is referred to as a *source program*. It is fed as input into a machine-language program whose sole function is to produce another machine-language program, the object program, which duplicates the logic and functions of the source program. The simplest type of programming language is an *assembler language.* The instructions in an assembler language bear a direct, one-to-one correspondence to the machine-language instructions they represent. However, the assembler instructions are alphabetic in character and are designed to bear a mnemonic resemblance to the function of the corresponding machine language instruction. This is a considerable aid, since it is much easier to remember what ADD and JMP do, say, than to recall that 01110111 and 111010111 are the machine-language codes for addition and an unconditional jump. The programmer is also much less likely to make a mistake in coding the mnemonic version than in writing (or keypunching) the binary version. Any assembler language will also provide for symbolic labeling of the addresses of the cells in which instructions and data are to be stored, and to which the flow of the program is to jump when a jump instruction is encountered. This is especially helpful in modifying a program, since adding one instruction to the machine-language program can require changing the address specifications of dozens or even hundreds of following instructions, while the addition of an assembler-language instruction requires no change in symbolic addresses, all adjustments being handled by the machine-language *assembler* program which translates the mnemonic instructions and symbolic addresses of the assembler-language source program into the binary instruction codes and specific addresses of the corresponding object program.

Assembler languages, being very closely tied to their corresponding machine languages, are highly machine-specific since they must take into account the particular hardware of that machine: the length (number of binary bits) of its words, the number and nature of various special registers, whether or not multiplication is handled by a single machine-language instruction, what peripheral devices are available, etc. At the next level, that of the *compiler languages* such as FORTRAN, COBOL, and PL/I, an attempt is made to develop a generally useful set of instructions expressed in a language which will seem as "natural" as possible to the user. The specifications as to what each of the instructions is to accomplish are then made available to computing facilities throughout the world in the form of manuals, and it is up to each computing center (or, more realistically and more commonly, each computer manufacturing company) to write an assembler-language or machine-language program called a *compiler* which will translate instructions (statements of a source program) written in the computer language into a machine-language object program which, when run on that particular computer, will meet the specifications of the compiler language. Thus, for instance, the FORTRAN statement $X = Y * Z + W$ would be translated by the compiler into assembler or machine-language statements which (i) transfer the contents of cell $Y$ to the accumulator, (ii) multiply the contents of the accumulator by the contents of cell Z and store the results in the accumulator, (iii) add the contents of cell W to the contents of the accumulator, and (iv) transfer the contents of the accumulator to cell X. A single input or output statement of a compiler language will often compile into (be translated into) a string of dozens of machine-language instructions. The person who learns the conventions of FORTRAN, ALGOL, or PL/I need never concern himself with learning the assembler language for any given machine, however, as the compiler language will be nearly machine-independent while the compiler itself will be very different from one computer to the next. In addition to making programming skills transferable from one computer installation to the next (or from one computer model to its inevitable replacement within the same computing center), the existence of compiler languages makes possible the exchange of programs between computing centers having quite different machines, and thereby cuts down on needless duplication of programming effort. This also facilitates yet another laudable goal, that of making the advantages of computers available to persons other than professional computer programmers, and has led to the development of canned computer programs.

A *canned program* (or library program) is a program written in compiler language to perform a very specific function or set of functions (for example, conduct multiple regression analyses), compiled into an object program, and then saved on a direct access storage device (such as magnetic tape or disk files) in such a way that it can be called into the computer for use by any person who wishes to use this program on his own data. The user need only submit a few systems control cards which satisfy local accounting requirements and which call the desired program from the storage device into the computer's "core," together with a few parameter cards which specify such things as the number of variables involved in this particular problem and whether data input is to be from cards or from magnetic tape, followed finally by the data for this particular "run." A canned program is thus essentially a programming language with very few commands (the systems control cards and the entries on the parameter cards) which sacrifices flexibility for ease of learning and ease of use. The range of problems which can be solved by a person who knows FORTRAN is (literally) incalculably large. A person who has learned the format of the one or two parameter cards and systems control cards required to run a particular program can perform only the one or two functions (usually a particular statistical analysis) specifically provided for in that program. On the other hand, the procedures for using a particular program can generally be learned in less than an hour, while reaching a level of facility in FORTRAN or PL/I programming adequate for the solution of multivariate statistical problems requires many hours of study and practice, while additional hours will be required to write and "debug" each program.

This tradeoff between flexibility and ease of use is a recurrent theme in computer applications. Thus, for instance, a program which permits the user to specify any input format he wishes for his data will be much more flexible than one which requires that all users prepare their data in the same way, often saving hours of effort that would otherwise be spent on converting already collected data to the "correct" format. On the other hand, leaving specification of the input format up to the user means that the user must know or learn how to write format statements—usually FORTRAN format statements, which are the most difficult (and most arbitrary) aspect of the FORTRAN language to learn. Moreover, providing for user-supplied data formats requires a more sophisticated compiler than having fixed formats and may require use of an operating system (the executive program which resides in a portion of the computer's core and handles such housekeeping chores as checking user IDs, sending the operator messages as to what tape or disk units have been requested, assigning and updating job priorities and switching from one job to another when in time-sharing mode, collecting together all the output which is to go to the printer and the card punch, etc.) which requires more numerous and more complex systems cards of the user.

The final level of programming "language" relevant to the present text is the *statistical package* of programs. This is a collection of subprograms, each performing a particular statistical procedure, and all sharing a common executive program for handling data input, transforming the data in various ways before analysis, etc. One advantage of the package over simply collecting separate programs is the high similarity in the conventions concerning preparation of the control cards for the use of the various programs in the package. Another is that by becoming familiar with just two or three of the most widely disseminated statistical packages one can be confident of finding familiar programs for most of his needs wherever he goes. Yet a third advantage is that the widespread use of these packages makes their subprograms the most thoroughly tested, most nearly "bug-free," of all available programs. For these reasons, the present text concentrates in its discussion of computer programs available for various procedures on those programs found within three nationally disseminated packages. Before getting into a procedure-oriented discussion of particular programs, it may prove helpful to discuss the general features of these three packages.

### D4.2   Three Nationally Distributed Packages of Statistical Programs

The oldest, most widely used, and most thoroughly debugged of the statistical packages available at the time of this writing is the BMD (for Biomedical) series. More recently developed but also widely available are SPSS (Statistical Package for the Social Sciences) and OMNITAB (which acronym has no official translation). These systems are described at length in their respective manuals (Nie, Bent, and Hull, 1970; Nie, Hull, Kim, and Steinbrenner, 1971; Dixon, 1968; Dixon, 1969; Chamberlain and Jowett, 1970), and specific examples of the control cards necessary to run programs contained in these packages are given in Section D4.4. The purpose of the present section is to compare the capabilities and liabilities of the three packages so as to aid the reader in deciding which of the three he should become acquainted with first. These advantages and disadvantages follow fairly directly from the basic organization of each of the packages.

The BMD series is essentially a collection of very powerful, thoroughly debugged programs, only one of which can be called up from the system's library in a given computer run (though some computer operating systems permit a user to chain together several computer runs with a single set of systems control cards, thereby controlling their order of execution) but most (not all) of which employ a common set of control card conventions and options, such as the option to attach identifying labels to the variables which will appear in the printed output from that program. The entries on each BMD control card must appear in prespecified card columns in order to be correctly interpreted by the program.

The SPSS system is also a collection of programs, the collection being somewhat less complete and the individual programs generally less powerful than in the BMD series. (The power of a statistical computer program is an ill defined but monotonically increasing function of the size of the data matrix it can handle, especially the number of variables; the range and complexity of the analyses it can perform; and the speed with which it performs these functions.) However, the various programs are controlled by a general "housekeeping" program which handles modification of the data file and to which control reverts upon completion of each analysis. Thus as many of the programs as the user wishes to apply to a given set of data may be employed in a single computer run, and any changes in the data file are available to all programs used in that run (though the user may specify that a given modification is to be in effect for only the next analysis performed) or even to programs used in subsequent runs if the modified data file is saved on disk or tape at the end of the run. Another attractive feature of SPSS is the very extensive labeling of output which is available to the user, who may specify both short (six character) and more extensive (up to 60 character) labels both for variables *and* for the specific values these variables take on. This labeling information becomes a permanent part of the data file when the file is saved at the end of the run and is automatically input along with the data themselves when the file is called for in a subsequent computer run. The control card conventions in SPSS are somewhat more "forgiving" than those of BMD. The only restriction to specific card columns is that columns 1–15 are reserved for the keyword which defines the function of that control card. The other parameters on the card can be placed anywhere in columns 16–80 of the card (and of continuation cards) so long as at least one blank space or comma is placed between successive parameter values.

OMNITAB is more similar in many respects to compiler languages like FORTRAN than it is to statistical packages like BMD and SPSS. Only two "fully canned" multivariate analyses are available: multiple regression and polynomial regression. However, OMNITAB provides for direct manipulation by the user of matrices which he sets up in an internal "worksheet" so that any of the analyses for which the user knows the matrix expressions can be performed by requesting the appropriate matrix operations. The OMNITAB system thus has great value as a teaching aid, since the learner can follow the various matrix operations involved in getting from the input data to the final results step by step. Even the two "canned" programs automatically print out many more of the intervening steps than is true of the programs available within BMD and SPSS. The OMNITAB system thus also has much greater flexibility than the other two—though on the other hand the user who wishes to perform an analysis other than multiple regression or polynomial regression must know much more about the computational procedures employed in that analysis than he would if he used one of the "fully canned" programs in the BMD series or the SPSS system. A further draw-back of the OMNITAB system is that it cannot operate on a matrix the product of whose dimensions exceeds 15,000, which makes it unsuitable for analysis "from scratch" of data involving large numbers of subjects and variables. This problem can be circumvented by feeding in the data for, say, 200 subjects at a time, computing sums, sums of squares, etc. for those Ss, clearing the worksheet, reading in the data for another 200 subjects, etc. and finally adding together all of the cumulated partial sums. However, most users will find it more practical to use the SPSS or BMD systems to construct the basic cross-product or correlation matrices needed in the analysis from raw data, and then use these matrices as input to the OMNITAB system. The control card conventions for OMNITAB are less rigid and more easily learned than for either of the other two systems. The program which reads these control cards examines the first word on the card and then looks for a few further numbers or symbols which provide parameters for that particular function, ignoring anything else typed on the card. Thus, for instance, the OMNITAB command for inverting a 4 × 4 matrix can be written either as

```
MINV (5,4) 4X4 (1,1)
```

or as

```
MINVERT COVARIANCE MATRIX EXTENDING FROM LOCATION (5,4) FOR 4 ROWS X 4 COLS AND
     STORE RESULT IN THE SAME SIZE MATRIX BEGINNING AT LOCATION (1,1)
```

The user can thus make his control cards very informative of what is being done.

Table D4.1 summarizes these and other properties of the three systems. It should be kept in mind that the three systems are constantly being improved by the addition of further programs, the correction of bugs in earlier editions, etc. so that some of the author's comments on the systems—particularly the negative comments—may no longer apply by the time you read this. As of this writing, for instance, the documentation of the SPSS system is definitely misleading in several ways. As an example, the description in the original (Nie, Bent, and Hull, 1970) manual of the highly useful RECODE feature states that "RECODE specification lists are cumulative so that it is possible to recode a recoded variable," and the figure immediately above (page 63 of the manual) certainly suggests that the CONVERT specification (which converts alphabetic representations of the digits 0 to 9 to the corresponding numeric values and at the same time converts blanks, minuses, pluses to the numeric values 11, 12, and 13) can come anywhere in the list. In practice, however, all recode requests following a CONVERT request are ignored, and the sequence of requests $\cdots (1, 2 = 3)(3 = 2) \cdots$ does not, as the manual description implies, lead to all previous values of 1, 2, or 3 being recoded as 2s. As another example, nowhere in the manual (including the section on JCL for the IBM 360 series computers) is it mentioned that a file *cannot* be read from a magnetic tape, modified, and then saved in its modified form on the same tape. Instead, the user must save the modified file onto an intermediate "scratch" tape in the first job step, then read it from the scratch tape and write it back onto the original tape in a second job step.

The SPSS system is also plagued by more bugs than the other two systems. For example, as of this writing the canonical correlation subprogram added in the Update Manual (Nie, Hull, Kim, and Steinbrenner, 1971) gives correct answers only for problems in which the two

*Table D4.1*

*A Comparison of the* BMD, SPSS, *and* OMNITAB *Statistical Packages*

| Property | BMD | SPSS | OMNITAB |
|---|---|---|---|
| Rigidity of control card conventions | High | Moderate | Low |
| Simplicity, ease of learning control card conventions | Low | Moderate | High |
| Accuracy of documentation | High | Low | High |
| Completeness of documentation | High | Low | Moderate |
| Ease of labeling output | Low | High | Moderate |
| Completeness of output labeling | Moderate to high, depending on program | High | Moderate |
| Ease of providing job control language for tape, disk input/output | Low | High | Moderate |
| Availability of output of one program for analysis by subsequent programs | Low | High | Moderate |
| Ease of modifying input data file | Moderate | High | High |
| How long are modifications in data file in effect? | Specific program which makes the modifications only | Can be specific to this program, in effect for all programs in this run, or until further modified | Same as SPSS |
| Completeness of coverage of multivariate statistics | High | Low | Low, unless user writes his own analysis using matrix manipulation commands |

sets of variables being related each contain the same number of variables *and* they are mentioned in the same order as they appear in the variables list—neither of which limitations is mentioned in the manual. However, the SPSS system is the newest of the three discussed in this section and can be expected to improve dramatically over the next year or so as these programming errors and inconsistencies in documentation are corrected.

Improvement is also to be expected in the other two systems. For instance, a new set of BMD programs, the BMDP series (Dixon, 1971), have recently been distributed to computing centers around the country. The programs in these series share a housekeeping master program much like that employed by SPSS which greatly facilitates interchange of data and results between programs. As of this writing, however, the staff at UNM's Computing Center have been unable to obtain a copy of the system which will compile in our computer. OMNITAB will also be improved, as illustrated by the fact that the manual's stated limit of 5000 "cells" in the worksheet has been increased to 15,000 cells in the version sent UNM's Computing Center. Despite these changes, the fundamental organization of each system can be expected to remain fixed, so that this description will hopefully continue to provide a basis for choosing among the systems the ones which should be given priority in the reader's process of becoming familiar with canned computer programs. There are a few suggestions which might aid them in avoiding the most dangerous pitfalls of this familiarization process. These are discussed in the next section.

### D4.3   Handy Hints and Warnings on the Use of Canned Computer Programs

This section begins with a brief discussion of how to become friends with your local computer, and concludes with a point by point list of principles to keep in mind in using canned programs.

Perhaps the most difficult step in becoming familiar with computer operations is the first one—that of learning the details of your local computing center's operating procedures, such as how to obtain computer time and a job number, where to turn programs in to be run and where to pick them up, etc. The best way to learn these things is to call the secretary of the computing center and ask. (This is also all it takes at UNM's Computing Center to obtain free computing time—at other installations, such as Stanford's, a formal application with many signatures may be required.) Most computing centers will have a brief handout in which such matters are detailed.

The next step is to learn how to use a keypunch to prepare IBM cards. There are manuals (for example, IBM A24332) which describe the operation of this equipment, but it is generally much easier to have someone else who is already familiar with keypunching show you the few details (mostly involved with feeding cards into the machine) which differentiate keypunching from use of a typewriter. If you have access to one of the increasingly common remote terminals which use a teletype or a keyboard-and-scope for inputting programs to a large central computer in time-sharing mode, the similarity to typing will be even greater. However, data input for practical problems is impractical solely from a keyboard, so that you will in any case need to learn keypunching procedures. After familiarizing yourself with and gaining some practice in basic keypunch operations, you will find it useful to learn to use the drum card to handle spacing of cards automatically (which turns out to be just as valuable as a means of avoiding errors as it is a means of increasing the speed of punching large amounts of data).

Next one should become familiar with the organization of the computing center's library of manuals, especially (a) those for the BMD, SPSS, OMNITAB, and other statistical packages of canned programs, (b) those which document the systems control cards for and special limitations on the use of these statistical packages, and (c) those which list the meanings of the various error messages produced by the computer's operating system.

Once you have run a few programs and seen the speed and power of the computer as a tool in statistical analyses, you should be sufficiently "hooked" for the next step, which is to pick up enough facility with a locally supported compiler language such as FORTRAN to be able to program simple data sorting and data rearrangement tasks. Most of these tasks can be handled by service routines within the canned programs of the statistical package, but it is often much simpler to write a short program specifically for that task. Moreover, the more flexible canned programs require that the user use what amounts to a subset of a compiler language (for example, FORTRAN format statements) in indicating data input formats, formulas for transforming the input data, etc. The introduction to a programming language can be made relatively painless by beginning with a simplified version of the full language, such as SKELETRAN (Bell, 1968). Most computing centers offer noncredit short courses, meeting one or two times a week for 2–6 weeks, on various programming languages. This can be an excellent way of beginning to learn a programming language *if* you take the "short course" with one or more very specific problems in mind for which you would like to be able to write programs.

As a step to insert anywhere along the line, if your local computer has a time-sharing system which provides for input from remote terminals, try to witness or preferably perform yourself the execution of one or two analyses from such a terminal. In the usual batch processing operations, where you submit a deck of cards to a dispatcher and get back your output an hour or more later, it is difficult to appreciate the fantastic speed of the computer. However, when you request from a remote terminal the inverse of a $5 \times 5$ matrix which took you an hour to do by hand and have the answer rattle off the teletype console or appear on the scope almost before you can release the "enter data" key, the indispensable virtues of computer operations really hit home. Throughout the process of acquainting yourself with computer operations, the following points should be kept in mind:

(1)   *No program is ever completely debugged.* As an example, we had been using BMDX69, a program for multivariate analysis of variance, for a couple of years without any problems, as had many users around the country, when a student submitted some data to the program which apparently caused it indigestion in that, after a certain point in the program, every line of output was followed by about 60 totally blank pages, leading the operator to cancel execution of the program after 300 pages or so had come flying naked out of the printer. It turns out that in FORTRAN the first character in each line of output is not printed, but is used instead as a carriage control character to tell the printer whether to double space, single space, or not space at all before printing the next line. One of the statements in the BMDX69 source program had failed to provide an explicit carriage control character, so that the label designated by the user for the first variable involved in a row of the summary table was treated as the control character on the line. Usually this caused no problem, since most alphabetic characters are uninterpretable as carriage control characters, leading the printer to "default" to single spacing. However, the student had labeled one of her factors as "A," which has a very special meaning (eject 50 or so pages) to the local high speed printer. As a corollary to this principle,

(2)   *Check the output of any program frequently, in as many ways as possible.*   Some of the kinds of checking procedures available to the user are to:

(a)   Compare this program's answers to the output of one or more other programs.

(b)   Work some simple problems involving only two or three variables by hand, and then submit these same problems to the program in question. Especially easy checks are usually provided by contrived data in which all variables are uncorrelated, or in which all intercorrelations between any pair of variables are identical. However, just because a series of sample problems with known answers are solved correctly does not guarantee that the program is bugfree. Remember, for instance, the example of the SPSS canonical correlation program which worked perfectly on the sample problem provided in the Update Manual, but either failed to run at all or gave nonsensical answers when problems for which the two sets of variables were of different size, were attempted.

(c)   Perform all the internal checks you can on the basis of your knowledge of the particular statistical technique being applied. Is the entry for $SS_{res}$ really equal to the sum of the squares of the residuals (errors of prediction)? Is the correlation between the canonical variates output by the program truly equal to the square root of the corresponding eigenvalue? etc.

(3)   Each month brings at least one change in local or nationwide operating procedures, so *keep in touch with your computing center*. The introduction to this digression provided a couple of examples of this problem of rapid changes in procedure. As yet another example: UNM's Computing Center recently installed a new "print train" on its high-speed printer which made available some special characters (such as ? and ※) which, while listed on the 029 keyboard, were printed as blanks on the old print train. Unfortunately, one consequence of the change was a loss of the former ability to mix cards punched on the 026 keypunch with cards punched on the newer, 029 keypunch (which has a different set of punch combinations for certain characters) so long as all of each card was punched on one or the other. Programs which we had been using for years, but whose source deck included a mixture of 026 and 029 punched cards, now produce rather peculiar table headings in some cases, with "(" being replaced in the output by "%," ")" by "@," etc. Since we have long since lost track of which statements were compiled from cards punched on the old machine, only trial and error use of these programs will enable us to get the labels back to their originally intended form. (The numerals and the letters of the alphabet were unaffected by this hardware change.) It may be a necessary cost of progress, but we must agree with Cooley and Lohnes's (1971) estimate that close to half of one's time in computer operations must be devoted to updating old programs so that they will continue to run despite hardware (such as the print train) and software (such as job control language specifications) changes.

(4)   *No one ever has a program—even a canned one—run correctly the first time he attempts to use it*, so do not get discouraged. On the other hand,

(5)   *It is never safe to assume that an error you locate* in your program or in the control cards you prepare for that program *is the only error you have made*, so always take the time to check over your *entire* deck after a failure. It is very tempting to make a change and resubmit the deck in the hopes that the problem has been corrected, especially since most failures in running a canned program come from errors which are detected very early in program execution and therefore waste only a few seconds of computer time. However, it will probably take an hour or more of clock time for you to obtain the results of a computer run, so that gradual elimination of errors can be quite wasteful of the user's time, if not the computer's. This caveat is less necessary if you are inputting the program from an on-line, time-sharing, remote terminal in which the feedback is almost immediate.

(6)   *The most common sources of program failure, the ones to check for first, are:*

(a)   exceeding the limits you set in some systems control card on the amount of time or number of lines of output your program would require;

(b)   inadvertently punching a special character such as a comma in place of a decimal point;

(c)   moving a data field over one or more columns so that, for example, the card columns which are supposed to contain the subject's scores on variable 5 have two decimal points, one "belonging" to the subject's score on variable 6;

(d)   getting your control cards out of order;

(e)   failure to provide on the first systems control card (which is usually of a distinctive color) all of the information needed by the operator as to what disk packs or other special storage devices (on which the desired statistical package, as well as, perhaps, your data, reside) will be required.

### D4.4   Specific Programs Useful in Multivariate Statistical Analyses

In this section we discuss functions (analyses performed) and limitations (maximum number of variables or subjects, for instance) of specific programs which can be used to conduct multivariate analyses. Most of these programs are available in one of the three statistical packages discussed in Section D4.2: BMD, SPSS, or OMNITAB. A few programs written by the author to cover "gaps" in these three statistical packages are also described here, with the listings of the actual programs being provided in Appendix B. If he has not already done so, the reader is urged to read the previous section's cautions and hints on the use of canned programs before attempting to use the programs listed in this section.

The descriptions which follow are not intended to replace the manuals for the OMNITAB, BMD, and SPSS systems, but rather to serve as a guide to which programs within each package are apt to be most useful for a particular multivariate problem. Since recurrent modifications of the packages may require changing the format of the control cards, control card requirements are not given in detail, except for the programs supplied by the author.

#### D4.4.1   Basic Statistics

The following listing is the deck which was used to set up an SPSS file for data from a questionnaire on drug use which was responded to by 3418 of UNM's undergraduates (with concentration on lowerclassmen) in the spring of 1970. The questionnaire was constructed and the data collected by Professor Douglas Ferraro. The data were initially punched on cards and then read from the cards onto the second file of a reel of magnetic tape (file DRUGF2 of tape volume PSY005) by a locally written assembler language program.

```
//RJHDRUG5  JOB  (PS482,7886,180,12),R.J.HARRIS,CLASS=F,TYPRUN=HOLD
//STEP1  EXEC  SPSS,TIME=(1,15),INFILE=DRUGF2,INVOL=PSY005,INLP=SL,
//       INBLK=1680,OUTFILE=DRUGSPSS,OUTVOL=PSY020,OUTLP=SL,OUTBLK=1680
//SPSS.SYSIN DD *
RUN NAME        ESTABLISHING SPSS FILE FOR FERRARO DRUG DATA
FILE NAME       DRUGFERR
VARIABLE LIST    SEX,AGE,CLASS,MAJOR,HISCHL,MARSTAT,ETHNIC,PANNUINC,PARRELIG,
                 YRRELIG,RELIGACT,PPOL,YRPOL,YRPOLACT,AGRPARVN,AGRPSEX,NCIG,
                 NCOFFTEA,FNDPZ,FNONPRES,FLIQUOR,FBEERWIN,NSNIFF,NTRANQNP,
                 FSTTRANQ,WLDTRANQ,NSEDBARB,FSTSED,WLDSED,NAMPHET,FSTAMPH,WLDAMPH,
                 NHALLUC,FSTHLLUC,WLDHLLUC,NNARCOT,FSTNRCOT,WLDNRCOT,NMARIJ,
                 FSTMARIJ,HJWFSTMJ,PARUSEMJ,FRNDSMJ,NOWUSEMJ,WLDUSEMJ,LEGALMJ,
                 RESTRMJ,ISMJADD,WHNUSEMJ,RSNFSTMJ,RSNNOTMJ
VAR LABELS      SEX,1.WHAT IS YOUR SEX?/AGE,2.WHAT IS YOUR AGE?/CLASS,3.WHAT
                IS YOUR CLASS STATUS?/MAJOR,4.WHAT IS YOUR ... MAJOR AREA OF
                STUDY?/HISCHL,5. WHAT HIGH SCHOOL DID YOU ATTEND?/MARSTAT,6.WHAT
                IS YOUR PRESENT MARITAL STATUS?/ETHNIC,7.WHAT IS YOUR ETHNIC BACK
                GROUND?/PANNUINC,8.WHAT IS YOUR PARENTS' ANNUAL INCOME?/
```

                (40 more cards providing variable labels)

```
                RSNNOTMJ,51.PRIMARY REASON YOUR NEV HVNG USED MJ?/
VALUE LABELS    SEX(0) MALE(1  FEMALE / AGE(0) 17 OR UNDER (1) 18 (2) 19
                (3) 20  (4) 21 (5) 22 (6) 23 (7) 24 (8) 25-29 (9) OVER 30 /
```

                (54 more cards labeling values of variables)

```
                RSNNOTMJ (0) NOT APPLICABLE (1) NEV HAD OPPTY (2) THREAT PUN,DISA
                P (3) MORAL REASONS (4)  PHYS,MENTAL DANGS (5)  NOT BEC DRUG DEP
                (6) HAVE NOT HAD DES TO /
MISSING VALUES SEX TO RSNNOTMJ (11,12,13)
# OF CASES     3418
INPUT MEDIUM   TAPE
INPUT FORMAT   FIXED(10X,51A1)
PRINT FORMATS  SEX TO RSNNOTMJ (0)
RECODE         SEX TO RSNNOTMJ (CONVERT)
CODEBOOK       ALL
SAVE FILE
FINISH
//
/*
```

In this SPSS run, the basic definitions of the data in the file, such as what each question was and what responses were available to each, were established. The data were then read from the tape as alphabetic variables (note the "51A1" in the input format statement) and converted to corresponding numerical values, with minus signs, plus signs, and blanks being coded as 11, 12, and 13, respectively, via the "RECODE" statement. This was necessary because some of the questions involved 10 responses coded as 0, 1, …, 9 so that it was decided to let a blank represent a nonresponse. Had the data been read in as numeric variables initially, each such blank would have been read as a value of 0, rather than as a missing observation. Tables of the frequency of each response to each question were constructed (via the "CODEBOOK" subprogram), and the data (now in numeric form), together with all of the variable labels, value labels, missing value specifications, etc., were then written onto a second tape (PYS020) via the SAVE FILE command, for future use. An example of one of the 51 tables included in the output of the run is given in Table D4.2. Note that, in this example, "5" and "9" were not among the responses made available to the subjects and must therefore represent response errors. Moreover, "11," "12," and "13" are standard values used by the RECODE command in converting minus signs, plus signs, and blanks to numerical values. Note that the process of setting up the labels for the variables and the possible values each can take on is a tedious one, accounting for 156 of the 173 cards in the input deck. We would have shortened the deck and its preparation time considerably by replacing the VARIABLE LIST cards by the command, "VARIABLE LIST VAR001 TO VAR051," and omitting the "VAR LABELS" and "VALUE LABELS" cards. The resulting tables would, however, have been much less informative about what was related to what, requiring constant reference back to a list of questions and question numbers. Moreover, once the file has been established the labeling cards may be dispensed with, since all of this information is stored along with the data and can be retrieved via a single "GET FILE" command.

The data are not yet ready for detailed analysis after this initial run. For instance, Table D4.2 shows that there were no minus or plus signs on the original data cards, so that there is

*Table D4.2*

| VARIABLE        AGRPSEX<br>VALUE LABEL | 16.AGR PARENTS PREMARITAL SEX,  ABORTION?<br>VALUE | ABSOLUTE<br>FREQUENCY | RELATIVE<br>FREQUENCY<br>(PERCENT) | ADJUSTED<br>FREQUENCY<br>(PERCENT) | CUMULATIVE<br>ADJ FREQ<br>(PERCENT) |
|---|---|---|---|---|---|
| STRNGLY AGREE | 0. | 322 | 9.4 | 9.5 | 9.5 |
| GEN'LLY AGREE | 1. | 981 | 28.7 | 28.8 | 38.3 |
| GEN'LLY DISAGREE | 2. | 872 | 25.5 | 25.6 | 63.9 |
| STRNGLY DISAGREE | 3. | 681 | 19.9 | 20.0 | 83.9 |
| DO NOT KNOW | 4. | 539 | 15.8 | 15.9 | 99.8 |
|  | 5. | 3 | 0.1 | 0.1 | 99.9 |
|  | 9. | 5 | 0.1 | 0.1 | 100.0 |
|  | 11. | 0 | 0.0 | MISSING | 100.0 |
|  | 12. | 0 | 0.0 | MISSING | 100.0 |
|  | 13. | 15 | 0.4 | MISSING | 100.0 |
|  | TOTAL | 3418 | 100.0 | 100.0 | 100.0 |

```
VALID   OBSERVATIONS -  3395
MISSING OBSERVATIONS -    23
```

no need to include 11 and 12 as missing values. (Recall that 13 represents a blank, that is, no response.) Moreover, there are 8 "wild" responses to that item which have no interpretation within the coding scheme and are probably punching errors. We would like to have these declared as missing values so that they may be excluded from most subsequent analyses. Finally, it could be argued that "I don't know" should be treated as a neutral response, falling between "Generally agree" and "Generally disagree." Certainly without this change, the AGRPSEX variable could be treated as a quantitative variable (ordinal, anyway) only if the "Do not know" response were treated as missing data. Similar changes were made in other variables to make them at least ordinal and to collect "wild" responses under the missing data category. The following second run was used to make these changes and to compute more extensive statistics on the modified data:

```
//RJHDRUG6 JOB (PS482,7886,180,12),R.J.HARRIS,CLASS=F,TYPRUN=HOLD
//STEP1 EXEC SPSS,TIME=(2,15),INFILE=DRUGSPSSS,DISP3='(OLD,PASS)',
//     INVOL=PSY020,INLP=SL,INBLK=1680,OUTFILE=INTMED,OUTBLK=1680
//SPSS.SYSIN DD *
RUN NAME        CONVERTING DRUG QUEST. RESPS TO ORDINAL MEAS*S,SVNG ON SCRTCH
GET FILE        DRUGFERR
RECODE          PPOL,YRPOL (5,6,7 = 1)(0,4=2)(8=3)(1,2=4)(3=5)(6 THRU 8 = 9)/
                AGRPARVN,AGRPSEX (0=4)(1=3)(4=2)(2=0)(3=0)(5 THRU 8 = 9)/
```

              (9 more cards)

```
                MARSTAT(4 THRU 8 = 9)/
                RSNNOTMJ(7,8=9)
MISSING VALUES  SEX,CLASS,MARSTAT TO RSNNOTMJ (9,13)/ AGE,MAJOR,HISCHL (13)
VALUE LABELS    PPOL,YRPOL (1) RAD,NEW LEF,MARXIST (2) DEMOCRAT,LIBERAL
                (3) NO PREF (4) REPUBL,CONSERV (5) VERY CONSERV /
                AGRPARVN,AGRPSEX (0) STRNGLY DISAGREE (1) GEN*LLY DISAGREE
                (2) DO NOT KNOW (3) GEN*LLY AGREE (4) STRNGLY AGREE /
```

              (6 more cards)

```
                FRNDSMJ,LEGALMJ,ISMJADD (0) NO (1) I DON*T KNOW (2) YES /
DOCUMENT        FOLLOWING CHANGES WERE MADE IN ORIGINAL DATA IN CREATING
                FINAL VERSION OF SPSS FILE OF DRUG DATA
                BLANKS AND OTHER SPECIAL SYMBOLS WERE CONVERTED TO MISSING
```

              (10 more cards)

```
                FRNDSMJ,LEGALMJ,ISMJADD (QS 43,46,48) WERE CHANGED TO MAKE
                DONT KNOW INTERMEDIATE BETWEEN YES AND NO
CODEBOOK        ALL
STATISTICS      ALL
PEARSON CORR    SEX,AGE,PANNUINC,RELIGACT TO FSTMARIJ,PARUSEMJ TO ISMJADD
OPTIONS         3
STATISTICS      2
CROSSTABS       CLASS TO ETHNIC,PARRELIG,YRRELIG BY NAMPHET TO NMARIJ,
                NOWUSEMJ,WLDUSEMJ
STATISTICS      1,2,3
SAVE FILE
FINISH
/*
//STEP2 EXEC SPSS,TIME=(1,15),INFILE=INTMED,DISP3='(OLD,PASS)',
//     INBLK=1680,INVOL=SCRTCH,OUTFILE=DRUGSPSS,OUTVOL=PSY020,OUTLP=SL,
//SPSS.SYSIN DD *
RUN NAME        TRANSFERRING MODIFIED DRUG DATA BACK TO PSY020
GET FILE        DRUGFERR
DUMP            VARLIST,VARINFO,LABELS,DOCUMENT
SAVE FILE
FINISH
//
/*
```

Note that, because of a restriction within the job control language for the IBM 360 series computers, it was not possible to read the SPSS file in from the PSY020 tape and then write the modified file back onto the same tape. Instead, the modified file had to be written onto a temporary (scratch) tape in one job step, then read in from the scratch tape and written onto the original tape (PSY020) in a separate job step. The output from the CODEBOOK subprogram is the same as it was for the initial run except for the changes made in value assignments, and the inclusion in this run of the "STATISTICS ALL" command, which causes the mean, standard error, median, mode, standard deviation, variance, kurtosis, skewness, range, maximum, and minimum of each variable to be printed at the bottom of its frequency table. The PEARSON CORR subprogram computes an intercorrelation matrix for the specified variables (all of those which have ordinal properties). The "OPTIONS 3" instruction specifies that two-tailed significance levels are to be used in specifying which of the correlations are statistically significant, and "STATISTICS 2" requests the covariance matrix as well as the intercorrelation matrix. The CROSSTABS subprogram constructs a two-way contingency table depicting the relationship between each of the variables to the left of the "BY" and each of the variables to the right of "BY." The "STATISTICS 1, 2, 3" instruction directs that a chi-square test for the independence of the two variables, a phi coefficient (in the case of a $2 \times 2$ table) or a Cramer's $V$ measure of the strength of the association between the two, and a contingency coefficient (which equals the square root of $\chi^2/N$) be computed for each of the 84 tables. In the second job step, the DUMP subprogram simply prints out all of the file-defining information for the entire SPSS file, including variable names, variable labels, value labels, missing values, and print formats. Note also that we could *not* have combined the recoding operations of this computer run with the CONVERT operation of the initial run since, despite what the manual says, all recode requests beyond a CONVERT request are ignored.

Several other subprograms for basic data description are available within the SPSS system. These include a program for constructing histograms, a faster version of CROSSTABS which omits the value labels, a similarly "stripped down" version of CODEBOOK, and a program which computes Spearman and/or Kendall rank-order correlation coefficients. These are described in the manual. An SPSS data file may have up to 500 variables, with no limit on number of subjects.

The OMNITAB system would not be able to handle a set of data as large as the DRUG-FERR file just mentioned, since the numbers of subjects multiplied by the number of variables cannot exceed 15,000. However, within these constraints the STATISTICAL ANALYSIS command can be useful in producing statistics descriptive of the frequency distributions of individual variables, and the matrix manipulation commands can be used to construct and print the intercorrelations among the variables. Following is an example of an OMNITAB program which generates descriptive statistics on 25 variables measured on each of 122 subjects:

```
//RJHOMNI  JOB (PS482,7886,60,23),R.J.HARRIS,CLASS=F,TYPRUN=HOLD
//S1 EXEC OMNITAB,SIZE=BIG
//OMNI.SYSIN DD *
OMNITAB (BASIC STATISTICS, 2-WAY PLOTS, AND INTERCORRELATION MATRIX)
READ INTO COLS 1 *** 25
```

> Data cards inserted here, in "free format," requiring only one or more blank spaces between successive values and a decimal point somewhere in each.

```
1/STATISTICAL ANALYSIS OF COL 1
2/ INCREMENT INST 1 BY (1)
EXECUTE INSTS 1 THRU 2 25 TIMES
3/ PLOT COLS 1 *** 5 VS COL 1
4/ INCREMENT 3 BY (5,5,0)
```

```
EXECUTE 3 THRU 4  5 TIMES
5/ SUM COL 1 STORE IN COL 26
6/ INCREMENT INS T 5 BY (1,1)
EXECUTE 5 THRU 6 25 TIMES
M(X'X) (1,1) 122 X 25, STORE IN (1,1)
M(X'X) (1,26) 1 X 25, STORE IN (1,26)
MSCALAR (1,26) 25 X 25 BY .0081967
MSUB (1,1) 25 X 25 MINUS (1,26) AND STORE IN (1,1)
MPRINT (1,1) 25 X 25
NOTE (MATRIX OF CROSSPRODUCTS OF DEVIATION SCORES)
MVECDIAG (1,1) 25 X 25 STORE IN COL 26
SQRT OF 26 AND STORE IN COL 26
DIVIDE 1.0 BY COL 26 AND STORE IN COL 26
MDIAG (1,27) 25 X 25 USING COL 26
M(X'AX) A = (1,1)  25 X 25 X =(1,27) 25 X 25 STORE IN (1,1)
MPRINT (1,1) 25 X 25
NOTE (CORRELATION MATRIX)
STCP
//
/*
```

Instructions 1, 2, and the following "EXECUTE" command produce for each variable the following statistics: number of observations, mean, median, midrange, 5% trimmed mean, standard deviation, standard deviation of the mean, range, mean deviation, 95% confidence interval about the mean, 95% confidence interval about the standard deviation, assorted tests of nonrandomness, a ten-cell frequency distribution, and the rank and deviation from the sample mean corresponding to each raw score. Instructions 3, 4, and the following "EXECUTE" command plot each variable (including itself) against variable 1, with a single graph containing five of the plots appearing on each of five pages. The OMNITAB system automatically scales the axes on the basis of the ranges of the variables involved. The sequence of instructions from ※5 to STOP print the cross-product and intercorrelation matrices after computing them on the basis of the matrix formulas, $\mathbf{x'x} = \mathbf{X'X} - \mathbf{TT'}/n$ and $\mathbf{R} = \mathbf{D}^{-1}\mathbf{x'xD}^{-1}$, where $\mathbf{T}$ is a column vector whose $i$th entry is the sum of the scores on variable $i$ and $\mathbf{D}$ is a diagonal matrix whose $i$th main diagonal entry is the square root of the sum of squared deviations for variable $i$.

Labels for various aspects of the output could have been provided via the "NOTE," "TITLEX," and "TITLEY" instructions, and more of the intermediate results could have been printed by inserting additional "PRINT" or "MPRINT" commands. It would also be possible to make the program more general by substituting "※V※" for "25" everywhere it occurs and ※W※ for 26, then using a "RESET V TO___" and a "RESET W TO___" command to specify the number of variables being handled in a particular run.

## D4.4.2  Matrix Operations

The basic statistics program illustrated in the preceding section gives some idea of the versatility of the OMNITAB system in handling matrix operations. A fuller description is provided by the following partial list of the matrix manipulation commands included within OMNITAB:

```
MADD(___,___)___X___TO (___,___) AND STORE  IN (___,___)
MSUB (___,___)___X___MINUS (___,___) AND STORE IN (___,___)'
MMULT(___,___)___X___TIMES (___,___) ___X___AND STORE IN (___,___)
MDEFINE (___,___)___X___TO BE ___
MDIAG (___,___)___X___USING COL ___
MZERO (___,___)___X___
MIDENT (___,___)___X___
MINVERT (___,___)___STORE INVERSE IN (___,___)
```

```
MSCALAR  (__,__)__ X __ BY_____ AND STORE IN (__,__)
MRAISE  (__,__)__X__ TO__ POWER , STORE IN (__,__)
MTRANSPOSE  (__,__)__X__ AND STORE IN (__,__)
MMOVE  (__,__)__ X__TO (__,__)
MVECDIAG (__,__)__X__ STORE IN (__,__)
M(X'X)  (__,__) STORE IN (__,__)
M(XX')  (__,__) STORE IN (__,__)
M(X'AX)  A=(__,__)__X__ X=(__,__)__ X__ STORE IN (__,__)
M(XAX')  A=(__,__)__X__ X=(__,__)__X__STORE IN (__,__)
M(AV)  A = (__,__)__ X__ VECTOR IN COL ___ STORE IN (__,__)
M(V'A)  A =(__,__)__ X__ VECTOR IN COL ___STORE IN (__,__)
M(AD)  A=(__,__)__ X__ D IN COL ___STORE IN (__,__)
M(DA)  A=(__,__)__X__ D IN COL ___ STORE IN (__,__)
MEIGEN  A(__,__)__X__STORE VALUES IN___, VECTORS IN (__,__)
```

The function of each command is self-explanatory, though full descriptions are provided in the manual. Each "(__, __)" gives the location in the OMNITAB worksheet of the upper left-hand entry of the matrix, with each "_ X _" giving the dimensions (number of rows × number of columns) of the matrix. Where the dimensionality of the matrix is implied by the operation being performed (as is always the case for the matrix in which the results of the operation are stored), the dimensions are omitted. One word of caution: the eigenvalue command outputs the values and vectors in an apparently random order.

A program which inverts matrices and which will run under the RAX operating system is RJH004, which is used as follows:

```
/INCLUDE RJH004
/INCLUDE MINV
/DATA
```

nv   (Use last 76 columns of card to punch identifying information.)

Data cards, punched in 10F8.0 format, that is, with each entry of the matrix occupying an eight-column field with the entry treated as an eight-digit integer unless a decimal point is punched. Begin each row of the matrix on a new card.

nv   (Parameter card for the next matrix you want inverted.)

Data cards for this matrix.

   etc.

nv   (Parameter card for the last matrix you want inverted.)

Data cards for the last matrix you want inverted.

99   (Or any other number greater than or equal to 20.)

```
/END
```

In the above listing, "nv" stands for the number of rows and columns in the matrix, which must be square. This program will handle up to a 19 × 19 matrix, and could easily be modified to accommodate larger matrices. If nv is less than 10, this entry must be punched right-justified, that is, with the single digit appearing in card column 2. The output of RJH004 includes a listing of the input matrix, the determinant of the matrix, and the inverse of the matrix.

### D4.4.3   Multiple Regression

The first multiple regression program to be described is a BMD program, BMDO2R. This program starts out with a univariate regression analysis employing only that single predictor variable which has the highest Pearson $r$ with the outcome variable. It then adds or subtracts one variable at a time to or from the regression equation, printing a summary table of the predictive efficacy of the resulting regression equation at each stage. The most important consideration in any run through the program is, of course, the manner in which the variable to be added or subtracted at each stage is selected. This is partially under the control of the program user and partially dependent on the results of the analysis at the present stage. More specifically, at each stage the matrix of sums of cross-products of deviation scores $\mathbf{A}$ is partitioned (by rearrangement of rows and columns, if necessary) into

$$\mathbf{A} = \begin{bmatrix} \mathbf{A}_{11} & \mathbf{A}_{12} \\ \mathbf{A}_{21} & \mathbf{A}_{22} \end{bmatrix},$$

where $\mathbf{A}_{11}$ involves only the $m$ variables presently in the regression equation, and $\mathbf{A}_{22}$ includes those variables (including the outcome variable) which are not included in the present form of the regression equation. A second matrix $\mathbf{B}$ is computed as follows:

$$\mathbf{B} = \begin{bmatrix} \mathbf{A}_{11}^{-1} & \mathbf{A}_{11}^{-1}\mathbf{A}_{12} \\ \mathbf{A}_{21}\mathbf{A}_{11}^{-1} & \mathbf{A}_{22} - \mathbf{A}_{21}\mathbf{A}_{11}^{-1}\mathbf{A}_{12} \end{bmatrix}.$$

Each column $j$ of $\mathbf{A}_{11}^{-1}\mathbf{A}_{12}$ (and thus each row $j$ of $\mathbf{A}_{21}\mathbf{A}_{11}^{-1}$) contains the regression coefficients which would be used in predicting variable $j$ from the $m$ variables presently in the regression equation. (Column d, of course, represents the coefficients for predicting the outcome variable, variable $d$, from the $m$ predictors.) Thus $\mathbf{A}_{22} - \mathbf{A}_{21}\mathbf{A}_{11}^{-1}\mathbf{A}_{12}$ is the *residual cross-product matrix* of the nonincluded variables, that is, the cross-product matrix which would be obtained if each column $j$ of $\mathbf{A}_{11}^{-1}\mathbf{A}_{12}$ were used to generate predicted scores on variable $j$, these scores were then subtracted from the observed scores on $X$, to obtain a set of *residual* scores (scores with the linear effect of the $m$ predictor variables removed) for each nonincluded variable, and the sums of squares and sums of cross-products of these residual scores were computed. The program then uses all this information, together with criterion values supplied by the user, as follows:

(1)   Each coefficient in column d of $\mathbf{B}$ (that is, the coefficient representing the contribution of each of the $m$ predictor variables to the regression equation relating these variables to the outcome variable) is tested for statistical significance. If one or more of these tests yield $F$-values below the "$F$-to-remove" value specified on the subproblem card, *and* if one or more of the variables producing these below-criterion $F$-values have been specified by the user as a free variable (by assigning it a *control value* of 2 on the control–delete card), that variable having the lowest $F$-value is deleted from the set of predictors, and the program proceeds to the next stage of the analysis (a multiple regression without the deleted variable).

(2)   If step (1) did not result in deletion of any variable, the *tolerance level* of each variable not yet included in the regression analysis is computed. The tolerance level $T_j = b_{ii}/a_{ii}$ and is the ratio of the residual variance of variable $i$ to the original variance of that variable. Thus if $b_{ii} = 0$ (and thus $T_i = 0$), variable $i$ is perfectly predictable from the variables already included in the analysis, and the addition would add nothing to our ability to predict the outcome variable. (In fact, a $T_i$ of exactly zero indicates that addition of that variable would produce an $A_{11}$ whose inverse does not exist, and cause the program to halt because of the persistent nuisance, a "floating point divide/check.") If there are one or more nonincluded variables whose $T_i$s exceed the "minimum tolerance level" specified on the subproblem card *and* whose control values are 3 or higher, that one of such variables whose control value is highest is added to the equation for the next stage of the analysis. If two or more of the qualifying variables have the same control value, then the one whose partial correlation with the outcome variable (that is, the correlation between residual scores on that variable and residual scores on the outcome variable), $R_i = b_{id}/(b_{ii}\,b_{dd})^{1/2}$, is highest (and which would therefore add most to our ability to predict scores on the outcome variable) is added.

(3)   If no variable is deleted by (1) or added by (2), then those nonincluded variables having a control value of 2 and a $T_i$ greater than the minimum tolerance level are tested against

the "$F$-to-add" value specified by the user. It can be shown that if variable $i$ were added to the regression equation, the regression analysis carried out, and the resulting regression coefficients computed, the $F$ test for the significance of the difference between the regression coefficient for the newly added variable and zero would produce an $F$-value equal to $F_i = (N - m - 1)R_i^2/(1 - R_i^2)$. That variable having the largest value of $F_i$ is therefore added to the regression equation *if $F_i$ exceeds the "$F$-to-add"* value.

(4)   If tests (1)–(3) do not result in addition or deletion of a variable, this stage of the analysis is taken as the final result, and the residual scores on the outcome variable for each subject are printed, as well as plots of these residuals against the scores on various original variables, if these have been requested by the user.

It should be pointed out that the standard significance tests for MRA discussed in Chapter 2 assume that the variables to be included in each equation—including those to be inserted at each stage of a multistage analysis—have been specified in advance of examining the data. Some researchers (for example, Cooley and Lohnes, 1971, p. 57) feel that the capitalization on chance which takes place in stepwise regression renders such approaches inappropriate for most behavioral research. Others (for example, Daniel and Wood, 1971) emphasize the descriptive and exploratory advantages of stepwise procedures.

BMDO2R includes transgeneration and labeling options. Transgeneration cards provide a means of generating new variables as a function of the original variables by, for example, squaring them or substituting arbitrary values for missing data, etc. Labels cards provide alphameric "tags" to be specified for the variables. These labels appear in every table employing the variable and thus improve interpretability of such tables. (Otherwise, the variables are labeled solely by the order in which they appear on the input cards or tape.) However, since transgeneration and labels cards use numerical order of appearance in identifying the variables to be transformed or labeled, the card setup for labels and transformations is usually not reusable in subsequent runs with these data.

The user of BMDO2R must provide a variable format card specifying the way in which data are to be read from cards, tape, or disk. This variable format card is essentially a FORTRAN format statement with the FORMAT omitted from the beginning. It begins with an open parenthesis in column one and closes with a close parenthesis. In between occur a series of expressions of the form kX (indicating skipping of $k$ columns of the card) or kFi.d (indicating $k$ successive fields of $i$ columns each, with the decimal point assumed to occur $d$ columns from the right unless overridden by an explicitly punched decimal point), separated by commas. In addition, one level of "group repetition specification" may be used by enclosing a series of expressions in parentheses and punching a number in front of the parentheses, k(...). Thus, for instance, a variable format card punched as follows:

```
( 1CX,1C(F5.C,1X),F5.1   /   10(F5.C,1X),1X, 2(2F1.0,1X), 3X, F5.2 )
```

indicates that the 26 variables per subject occur on a total of two cards per subject (the slash indicates the program is to skip to a new input card), with the first 10 variables occurring in five-column fields beginning with column 11 of the first card and with a space after each five-column field, variable 11 appearing in columns 61–65 of the first card, variables 12–21 appearing in five-column fields beginning in column one of the first card with a space inserted after each field; variables 22 and 23 appear in columns 65 and 66 of the second card; and variable 26 appears in columns 71–75 of card 2. All scores are assumed to be integers (the decimal point appearing to the right of the last column of that data field) unless overridden by a decimal point punched on the card in that field, *except* that variables 11 and 26 have implicit decimal points occurring immediately before the last and second-to-last columns of their respective data fields. If this seems confusing, the user may always employ a (30X,10F5.0) format—data values appearing in columns 31–35, 36–40, etc. of each card, with the first 30 columns left free for identifying information.

BMDO2R also provides for the plotting of residual scores, with each IDXPLT card permitting the user to plot the residual scores on the outcome variable against each of up to 30 of the predictor variables.

This program can handle up to 80 variables (including those added by transgeneration) and up to 9999 cases per problem. However, if the number of variables times the number of subproblems exceeds about 70, the program output will require more pages than are normally set aside for it, and more complicated JOB card and EXEC card entries will be required. (See your computing center's User's Guide for the BMD package for details.) The control card setup is quite rigid and a common source of errors. Moreover, no provision is made for inputting a correlation matrix and beginning the analysis from that point, so that the large amount of time involved in computing a correlation matrix for a large set of variables must be expended every time this program is run, no matter how much previous information about (for example, the intercorrelations among) the variables is already available from earlier analyses. It is, however, a very powerful, efficient program.

A multiple regression program which has less imposing control card requirements is the REGRESSION subprogram of the SPSS package. An example of the setup for an SPSS run in which a multiple regression analysis of the relationship between present marijuana usage and various background variables, together with a second analysis relating these variables to use of hard narcotics, is listed below. The data come from the Ferraro and Billings (in press, a,b) drug data which was used as an example of initialization of an SPSS file in the section on basic statistics programs.

```
//JOBNAME  JOB  (PS482,7886,120,2 ,I.I.LASTNAME,CLASS=F,TYPRUN=HOLD
//S1  EXEC  SPSS,INFILE=DRUGSPSS,INVOL=PSY020,INLP=SL,INBLK=1680
//SPSS.SYSIN DD *
GET FILE        DRUGFERR
REGRESSION      VARIABLES = SEX TO CLASS, PANNUINC, RELIGACT TO AGRPSEX,
                   FRNDSMJ, NOWUSEMJ,NNARCOT /
                REGRESSION = NOWUSEJM WITH SEX TO FRNDSMJ (1) /
                REGRESSION = NNARCOT WITH SEX TO FRNDSMJ (2)
OPTIONS         2
STATISTICS      ALL
FINISH
//
/*
```

The REGRESSION subprogram computes a stepwise regression analysis in the same way as BMDO2R, except that (i) no variable is ever deleted after it has been added to the regression equation, that is, there is no parameter comparable to BMDO2R's Fdelete; and (ii) REGRESSION provides 49 priority levels for forced variables and 50 for free variables versus BMDO2R's 7 and 1 levels, respectively. These priority levels are called *inclusion levels* in the SPSS documentation. An inclusion level of 0 prevents that variable from ever entering the regression equation. All variables having the same even, nonzero inclusion level are entered in a single step, with sets of variables having a high even inclusion level being entered before sets of variables having a low even inclusion level. Within a set of variables having the same odd inclusion level, the variables are entered one at a time, with the variable which would add most to the regression equation's predictive ability at each step being the one selected for inclusion. A variable within the set whose Ftoadd is too low or whose tolerance level is too low (cf. the discussion of these criteria within the BMDO2R description) does *not* enter the equation.

In the above example, the independent variables in the first analysis (of present frequency of use of marijuana) are all entered in stepwise mode, with Ftoadd and Tolvl being set at the default values of .01 and .001. These default values can be overridden quite simply by including substitute values within parentheses immediately following the name of the dependent variable and before the "WITH," using the form (M, F, T), where M is the maximum number of steps, F is the *F*-value required for a variable to be added if it is a free variable, and T is the tolerance level. The "(1)" following "FRNDSMJ" indicates that all variables are to have an inclusion level of 1; if different inclusion levels had been desired, they could have been indicated by placing the inclusion level in parentheses immediately after each group of independent variables that level applied to. The second analysis is performed using an inclusion level of 2 for all predictor variables, whence all variables are included in a single regression equation in a single step.

Had the correlation matrix for the variables we wished to analyze been available, it could have been used as the starting point for the analyses by way of the READ MATRIX card, as illustrated in the following setup for such a run:

```
//RJHSPSS JOB (PS482,7886,120,2 ,R.J.HARRIS,CLASS=F,TYPRUN=HOLD
//S1 EXEC SPSS
//SPSS.SYSIN DD *
RUN NAME            ILLUSTRATING MATRIX INPUT TO REGRESSION PROGRAM, DATA SET 1
VARIABLE LIST      VAR001 TO VAR004
# OF CASES         5
INPUT MEDIUM       CARD
REGRESSION         VARIABLES=VAR001 TO VAR004/
                   REGRESSION=VAR004 WITH VAR001 TO VAR003 (1) /
OPTIONS            5
STATISTICS         1
READ MATRIX
1.0         0.134       0.748       0.546
0.134       1.0         0.650       -.437
0.748       0.650       1.0         -.065
C.546       -.437       -.065       1.0
FINISH
//
/*
```

The options which can be selected on the OPTIONS card include options 2 (pairwise, rather than listwise deletion of cases having missing values), 4 (input of correlation matrix, preceded by means and standard deviations of variables), 5 (input of correlation matrix without means and standard deviations, in which case only the *z*-score regression equation is computed), and 8 (storage of the correlation matrix on punched cards, magnetic, or disk file). The two statistics which can be requested are statistics 1 (correlation matrix or matrices) and 2 (means, standard deviations, and number of valid cases for each variable). Analyses involving up to 199 predictors may be performed.

A fully canned subprogram for multiple regression is also available within the OMNITAB system. This subprogram has no stepwise features, all variables being entered into the regression equation in the order in which they are listed on the "FIT" instruction. An example of such an analysis follows:

```
//JOBNAME JOB (PS482,1234,30),M.R.COMP,CLASS=F,TYPRUN=HOLD
//STP1 EXEC OMNITAB,SIZE=BIG
//OMNI.SYSIN DD *
OMNITAB (MULTIPLE REGRESSION ANALYSIS OF DATA SET 1)
READ INTO COLS 1 *** 4
102.    17.      4.      96.
104.    19.      5.      87.
101.    21.      7.      62.
93.     18.      1.      68.
100.    15.      4.      78.
ADD 1.C TO 0.0 AND STORE IN COL 5
FIT Y IN 4,WTS IN 5, X IN 5,1,2,3  STORE COEF IN 6, RES IN COL 7
PLCT COL 4 VS COL 7
STOP
//
/*
```

The FIT subprogram permits weighting of the scores on the predicted variable, that is, multiplication of each $Y$-score by a (possibly different) constant, before the analysis begins. It also forces the regression line to pass through the origin, that is, analyzes the raw-score cross-product matrix, rather than the deviation-score matrix, unless a dummy variable on which each subject has a score of 1 is included in the analysis. Since an unweighted analysis including fit to the means was desired in the above problem (this will be the usual case), a column containing all ones had to be constructed (via the "ADD 1.0..." command) before the FIT request was made. The FIT command itself produces a very complete listing of the intermediate steps and final results of the analysis, including each regression coefficient and its standard deviation, $(X'X)^{-1}$, the variance–covariance matrix of regression coefficients, and the amount of variance in $Y$ remaining unaccounted for after each variable is added to the regression equation. The PLOT command plots $Y$ against its predicted value, and is a more certain test of the existence of nonlinearity in the relationship between $Y$ and the $X$s than would be provided by plotting $Y$ as a function of each separate predictor variable. Stepwise regression could of course be set up by the user in any way he wished with the help of the matrix manipulation commands in the OMNITAB system.

### D4.4.4  $T^2$ and Manova

The RJH004 program for obtaining determinants and inverses of square matrices, which was described in Section D4.4.2, can be useful, especially when the determinantal formulas for $T^2$ are employed. Another RAX-compatible program which can be useful, especially where greatest characteristic root tests are desired, is HEINV, the setup for which is described below:

```
/ID 467587886.PS482  HARIS, RICHARD
/INCLUDE HEINV
/INCLUDE NROOT
/INCLUDE EIGEN
/DATA
```

ɃNvɃ   (Fill in rest of card with up to 76 columns of identifying information.)
Cards for **H** matrix, punched in (10F8.0) format.
Cards for **E** matrix, punched in (10F8.0) format.
Parameter card, **H** matrix cards, and **E** matrix cards for next problem.
  $\vdots$
Parameter card, **H** matrix cards, and **E** matrix cards for next problem.
999

```
/END
```

"Nv" in the above listing is the number of dependent variables (that is, the size of the **H** and **E** matrices). The output includes the input **H** and **E** matrices and the eigenvalues and characteristic roots of $E^{-1}H$. The major drawback of this program is that it does not provide for computation of the **H** and **E** matrices, but instead requires them as input. It finds its most frequent use as a means of correctly computing significance tests for between-subjects effects when a within-subjects design has been analyzed using BMDX69.

BMDX69 is a very powerful and efficient program within the BMD series which takes raw data as input and conducts multivariate analyses of variance and covariance on these data for any factorial or completely nested design. One parameter on the PROBLM card and two of the other control cards are apt to be confusing to a first-time user of BMDX69: The "number of analysis of variance indices" refers to the number of factors in the analysis of variance of variance part of the design, *including* replications as a factor.

On the DESIGN card, the user must specify an "Index list" and a set of "nesting specifications." The index list is a list of one-character alphabetic names for the Anova indices. For a one-way MANOVA there will be only two indices, one standing for the between-groups factor and one for the within-groups (or subjects, or replications) factor. This list can be punched anywhere within columns 7–16, with no blank spaces or commas between indices.

The nesting specifications, which begin in column 17, indicate nesting of one factor within one or more other factors by enclosing within parentheses immediately following the letter representing a given factor, the letters of all factors within which that factor is nested (if any). These nesting specifications are separated by commas and ended with a period. Cell means for each variable and adjusted cell means for each univariate analysis of covariance will be printed for any term in the list preceded by a dollar sign.

On the INDEX card, the user must specify the number of levels of each of his factors. Trickily enough, however, the number of levels of the subjects factor must be the number of subjects *per cell*, not the total number of subjects.

On the SUPRO card(s), the user indicates which variables are to be included in this analysis, by specifying "control values." These are punched into columns 7–72, one control value per column, with no spaces between control values. A blank column indicates that that variable is not to be included in this subproblem. A value of 1 indicates that this is a dependent (outcome) variable. A control value of 2 indicates that this variable is a covariate.

BMDX69 can handle only analyses of variance where each cell has the same number of subjects, and where all effects are fixed effects. It will handle repeated-measures designs through the nesting specifications, but the $U$- and $F$-statistics will be incorrect in that the within-$S$ error term is used for testing *all* effects, including the between-subject effects. Similarly, if a mixture of fixed and random factors is employed, the error matrix used for all significance tests is the cross-product matrix corresponding to the most detailed interaction with the subjects or replications factor. In other words, BMDX69 can *not* be trusted to give correct significance tests for designs other than the completely factorial fixed-effects design. If the user can write the summary table and specify the relevant error terms for the univariate case, the summary tables for the univariate analyses performed by BMDX69 can be easily corrected. Moreover, if the cross-product **H** and **E** matrices have been printed, the appropriate pairings of hypothesis and error matrices can be punched onto cards and submitted to the HEINV program described above in order to obtain correct significance tests.

The "indices" for the analysis of variance correspond to factors of a $k$-way univariate Anova. For instance, for a one-way MANOVA involving five groups, with twelve subjects per group, the index and design cards would read as follows:

```
INDEX    5  12
DESIGN   AR        $A,R(A).
```

The index and design cards for a $2 \times 2 \times 3$ MANOVA with five subjects per group would be

```
INDEX    2  2  3  5
DESIGN   AHCR      A,B,C,$ABC,R(ABC).
```

Finally, the design and index cards for a $3 \times 2 \times 3$ MANOVA with ten subjects per group and with the third factor a repeated-measures (within-subjects) factor, would be

```
INDEX    3  2  3 10
DESIGN   UVWR      U,V,W,$UVW,R(UVW).
```

Incidentally, perhaps the most common and least easily detectable mistake in preparing a BMDX69 run is omitting the period at the end of the specification list on the DESIGN card. Note, too, that the LABELS and transgeneration options are *not* available with this BMD program.

A third program, MANOV1, has been written which computes both Wilks's lambda and greatest characteristic root (gcr) tests for one-way Manova, including unequal $n$ situations. Assignment of subjects to a particular level of a given independent variable is accomplished by the reading of an alphameric index value on each subject's data card, so that the data need not be read in any particular order. Each set of data can be analyzed on the basis of several independent variables, and several sets of data can be analyzed within a given computer run. It was initially planned to include a listing of MANOV1 in Appendix B. However, as the Primer was being written it was learned that Jeremy Finn's MULTIVARIANCE program, which has been made available by Finn on a private basis for a number of years, is now being distributed nationally by National Educational Resources. MULTIVARIANCE (Finn, 1973) incorporates all of the features of BMDX69 and of MANOV1, plus the generality of the multivariate general linear hypothesis (Sections 8.4 and D4.4.7) if needed. It has the further advantage of extensive use and consequent debugging at many installations. The reader is urged to employ MULTIVARIANCE if it is available to him. If not, and if unequal $n$ capability and/or gcr tests are desired, a program listing and description of MANOV1 can be obtained from the author.

### D4.4.5   Canonical Correlation

The BMD package includes a canonical correlation program, BMDX75. It, like the other programs in the BMD series, permits the user to specify his own format for inputting data. However, only one canonical analysis can be used for a given set of data, though use of tape input and the default option of rewinding the tape after each problem avoids having to punch duplicate decks when several analyses are to be run on the same set of data. Note, too, that the labeling and transgeneration options available with some BMD programs may not be used with this program. Further, since there is no option for the direct inputting of a correlation matrix, this program can be quite time consuming with large sets of data.

An SPSS program, CANCORR, is available which allows considerable flexibility in whether raw data or a correlation matrix is taken as the starting point of the analysis, and which also permits several canonical analyses to be performed on various pairs of subsets of the variables in a single set of data. An example of an SPSS run using the CANCORR subprogram follows:

```
//RJHSPSS  JOB  (PS482,7886,60,2),R.J.HARRIS,CLASS=F,TYPRUN=HOLD
//STEP1  EXEC  SPSS,INFILE=DRUGSPSS,INVOL=PSY020,INLP=SL,
//       INBLK=1680

//SPSS.SYSIN  DD  *
RUN  NAME          DEMONSTRATING  CANCORR  USE
GET  FILE          DRUGFERR
CANCORR            VARIABLES=SEX  TO  CLASS,PANNUINC,NMARIJ,NOWUSEMJ,WLDUSEMJ,
                     LEGALMJ,NNARCOT,NHALLUC,NAMPHET,NSEDBARB,RELIGACT,PPOL,
                     YRPOL,YRPOLACT /
                   RELATE=SEX  TO  PANNUINC  WITH  NMARIJ  TO  LEGALMJ /
                   RELATE=SEX  TO  PANNUINC  WITH  NNARCOT  TO  NSEDBARB /
                   RELATE=RELIGACT  TO  YRPOLACT  WITH  NMARIJ  TO  LEGALMJ
OPTIONS            2
STATISTICS         1,2,3
FINISH
//
/*
```

This deck setup would call the SPSS file, DRUGFERR from a tape labeled PSYO20, compute the intercorrelation matrix for the variables listed after " VARIABLES =," including AGE, which comes between SEX and CLASS in the variables list used in defining the file in the first place, and then conduct canonical analyses of the following relationships: sex, age, class in college, and parents' annual income as they relate to past marijuana use, present marijuana use, intention to use marijuana in the future, and opinion as to whether marijuana ought to be legalized; sex, age, class, and parents's income as "predictors" of reported use of narcotics, hallucinogens, and sedatives or barbiturates; and religious activity, parents's and own political positions, and own political activity as they relate to the marijuana items.

The options which can be specified on the "OPTIONS" card include such things as whether to use pairwise or listwise deletion of cases having missing values in constructing the intercorrelation matrix and whether to use a previously computed intercorrelation matrix as basic input to the subprogram. The statistics available include the intercorrelation matrix, the canonical coefficients, and the scores of each subject on the canonical variates.

### D4.4.6   Factor Analysis

The BMD program for factor analysis is BMDX72. This program performs principal component or principal factor analysis with a wide variety of rotation options. Rotations are performed so as to minimize the criterion,

$$G(l_{ij}) = \sum_{a=b} \left[ \sum_{i=1}^{m} l_{ia}^2 l_{ib}^2 - (\gamma/m)(\sum l_{ia}^2)(\sum l_{ib}^2) \right],$$

where $l_{ia}$ is the loading of original variable $i$ on factor $a$, and where $\gamma$ is specified by the user. Standard rotation criteria are:

| Type of rotation | $\gamma = 0$ | $\gamma = \frac{1}{2}$ | $\gamma = 1.0$ |
|---|---|---|---|
| Orthogonal | Quartimax | | Varimax |
| Oblimin | Quartimin | Biquartimin | Covarimin |
| Simple loadings | Direct quartimin | | |

HEINV can be used to perform principal component analysis and principal factor analysis, but without rotations. This application requires that an identity matrix be input as the **E** matrix.

The SPSS package includes a FACTOR subprogram. However, the author has not used this subprogram, and therefore cannot offer advice on its use.

Finally, Joreskög's program (Joreskög *et al.*, 1971) for carrying out confirmatory analysis (which permits the user to specify in advance that certain of the loadings will be zero or one and compares the fit to the original correlation matrix obtained under these restrictions to the unrestricted case) is available by writing to Marielle van Thillo, Educational Testing Service, Princeton, New Jersey 08540. (Professor Joreskög is now Professor of Statistics at the University of Uppsala, P.O. Box 2300, 750 02 Uppsala 2, Sweden. However, facilities at Uppsala do not make it feasible for him to handle requests for tapes and listings of his program.)

### D4.4.7   Multivariate General Linear Hypothesis

The BMD program for tests of the multivariate general linear hypothesis [mglh, Eq. (8.3)] is BMDX63. Actually, this program is somewhat more general than Eq. (8.3), providing for tests of the hypothesis that $\mathbf{C\beta M} = \mathbf{D}$, where **D** is a $g \times u$ matrix which need not be the null matrix. BMDX63 will handle any set of data for which the total number of variables, dependent or independent, is less than 55. However, it requires very extensive card preparation, since each hypothesis tested (for example, a main effect, interaction, or contrast in Manova) requires a new set of **C**, **M**, and **D** matrices, and the only default options which avoid punching these entire matrices occur when **C** and/or **M** = **I** and when **D** = **0**. This program is therefore not a convenient substitute for the more specialized programs described in preceding sections, but may nevertheless be the only canned program available for unequal *n* Manova and Mancova—unless your computing center has MULTIVARIANCE, to be described next. A more substantive shortcoming of BMDX63 is that all analyses are based on likelihood-ratio approaches, so that the appropriate characteristic roots and vectors are not computed.

Finn's (1973) MULTIVARIANCE program also provides for tests of the mglh stated in Eq. (8.3). However, rather than requiring the user to employ the most general form of the mglh for all analyses, MULTIVARIANCE automatically generates the appropriate matrices for most of the commonly used analyses, such as equal *n* Anova, Manova, Canona, and Mancova. Card preparation is therefore considerably easier than for BMDX63. Moreover, both likelihood-ratio and gcr tests are performed; the characteristic vectors (discriminant functions) are reported; and measures of strength of relationship based on $R_c^2$ are computed. The program provides so many options as to input and output formats and types of analysis to be performed that it should perhaps be considered a statistical package in and of itself. In fact, Bock (in press) states that MULTIVARIANCE was used in every one of the problems used as illustrations in his entire multivariate statistics text.

The author's direct knowledge of this program is limited to examination of a mimeographed description of its features. However, MULTIVARIANCE has been spontaneously mentioned by colleagues at several other universities and has always received enthusiastic endorsement when so mentioned.

# A | APPENDIX
## Statistical Tables

The tables listed in this appendix were generated by Fortran IV computer programs written by the author and run on the University of New Mexico's IBM 360/67 computer with the help of a grant of computing time from the UNM Computing Center. The programs were, of course, based on previous results and subroutines developed by other programmers.

The normal curve and chi-square programs employed subroutines written by Stoughton Bell to compute the basic cumulative probabilities, while the $t$, $F$, and greatest characteristic root (gcr) tables used IBM's SSP subroutines BDTR, NDTR, CDTR, and DLGAM in computing the incomplete beta distribution probabilities involved in those statistics. The table of natural logarithms used the FORTRAN IV built-in subroutine ALOG.

Initial approximations to the inverses of chi-square and $F$ were obtained from formulas 26.4.17 and 26.6.16 of Abramowitz and Stegun's (1964) *Handbook of Mathematical Functions*, before beginning zeroing-in procedures for computing a more precise estimate of the inverse.

The program for computing critical values of the gcr distribution used the iterative procedure derived by Pillai (1965) to compute the basic probabilities, with the exception of the entries for $n = 1$ and $n = 3$. Computation of these latter entries employed a subroutine written by Venables (personal communication) and based on the algorithm reported by Venables (1973).

293

## Table A1   Normal Distribution

The entries in this table represent $F(z) = \Pr(x < z)$, where $x$ has the unit normal distribution. Only the cumulative probabilities for positive values of $z$ are listed. However, due to the symmetry of the normal distribution,

$$F(-z) = \Pr(x < -z) = 1 - \Pr(x < z) = 1 - F(z).$$

The rows of the table represent $z$ to the nearest tenth, while the columns represent the hundredths position. Thus, for instance, $F(1.46)$ is found in the "1.4" row under the ".06" column, and equals .92785.

*Normal Probability Distribution $F(z)$*

| z | .00 | .01 | .02 | .03 | .04 | .05 | .06 | .07 | .08 | .09 |
|---|---|---|---|---|---|---|---|---|---|---|
| 0.0 | .50000 | .50399 | .50798 | .51197 | .51595 | .51994 | .52392 | .52790 | .53188 | .53586 |
| 0.1 | .53983 | .54380 | .54776 | .55172 | .55567 | .55962 | .56356 | .56749 | .57142 | .57535 |
| 0.2 | .57926 | .58317 | .58706 | .59095 | .59483 | .59871 | .60257 | .60642 | .61026 | .61409 |
| 0.3 | .61791 | .62172 | .62552 | .62930 | .63307 | .63683 | .64058 | .64431 | .64803 | .65173 |
| 0.4 | .65542 | .65910 | .66276 | .66640 | .67003 | .67364 | .67724 | .68082 | .68439 | .68793 |
| 0.5 | .69146 | .69497 | .69847 | .70194 | .70540 | .70884 | .71226 | .71566 | .71904 | .72240 |
| 0.6 | .72575 | .72907 | .73237 | .73565 | .73891 | .74215 | .74537 | .74857 | .75175 | .75490 |
| 0.7 | .75804 | .76115 | .76424 | .76730 | .77035 | .77337 | .77637 | .77935 | .78230 | .78524 |
| 0.8 | .78814 | .79103 | .79389 | .79673 | .79955 | .80234 | .80511 | .80785 | .81057 | .81327 |
| 0.9 | .81594 | .81859 | .82121 | .82381 | .82639 | .82894 | .83147 | .83398 | .83646 | .83891 |
| 1.0 | .84134 | .84375 | .84614 | .84849 | .85083 | .85314 | .85543 | .85769 | .85993 | .86214 |
| 1.1 | .86433 | .86650 | .86864 | .87076 | .87286 | .87493 | .87698 | .87900 | .88100 | .88298 |
| 1.2 | .88493 | .88686 | .88877 | .89065 | .89251 | .89435 | .89617 | .89796 | .89973 | .90147 |
| 1.3 | .90320 | .90490 | .90658 | .90824 | .90988 | .91149 | .91308 | .91466 | .91621 | .91774 |
| 1.4 | .91924 | .92073 | .92220 | .92364 | .92507 | .92647 | .92785 | .92922 | .93056 | .93189 |
| 1.5 | .93319 | .93448 | .93574 | .93699 | .93822 | .93943 | .94062 | .94179 | .94295 | .94408 |
| 1.6 | .94520 | .94630 | .94738 | .94845 | .94950 | .95053 | .95154 | .95254 | .95352 | .95449 |
| 1.7 | .95543 | .95637 | .95728 | .95818 | .95907 | .95994 | .96080 | .96164 | .96246 | .96327 |
| 1.8 | .96407 | .96485 | .96562 | .96637 | .96712 | .96784 | .96856 | .96926 | .96995 | .97062 |
| 1.9 | .97128 | .97193 | .97257 | .97320 | .97381 | .97441 | .97500 | .97558 | .97615 | .97670 |
| 2.0 | .97725 | .97778 | .97831 | .97882 | .97932 | .97982 | .98030 | .98077 | .98124 | .98169 |
| 2.1 | .98214 | .98257 | .98300 | .98341 | .98382 | .98422 | .98461 | .98500 | .98537 | .98574 |
| 2.2 | .98610 | .98645 | .98679 | .98713 | .98745 | .98778 | .98809 | .98840 | .98870 | .98899 |
| 2.3 | .98928 | .98956 | .98983 | .99010 | .99036 | .99061 | .99086 | .99111 | .99134 | .99158 |
| 2.4 | .99180 | .99202 | .99224 | .99245 | .99266 | .99286 | .99305 | .99324 | .99343 | .99361 |
| 2.5 | .99379 | .99396 | .99413 | .99430 | .99446 | .99461 | .99477 | .99492 | .99506 | .99520 |
| 2.6 | .99534 | .99547 | .99560 | .99573 | .99585 | .99598 | .99609 | .99621 | .99632 | .99643 |
| 2.7 | .99653 | .99664 | .99674 | .99683 | .99693 | .99702 | .99711 | .99720 | .99728 | .99736 |
| 2.8 | .99744 | .99752 | .99760 | .99767 | .99774 | .99781 | .99788 | .99795 | .99801 | .99807 |
| 2.9 | .99813 | .99819 | .99825 | .99831 | .99836 | .99841 | .99846 | .99851 | .99856 | .99861 |
| 3.0 | .99865 | .99869 | .99874 | .99878 | .99882 | .99886 | .99889 | .99893 | .99896 | .99900 |
| 3.1 | .99903 | .99906 | .99910 | .99913 | .99916 | .99918 | .99921 | .99924 | .99926 | .99929 |
| 3.2 | .99931 | .99934 | .99936 | .99938 | .99940 | .99942 | .99944 | .99946 | .99948 | .99950 |
| 3.3 | .99952 | .99953 | .99955 | .99957 | .99958 | .99960 | .99961 | .99962 | .99964 | .99965 |
| 3.4 | .99966 | .99968 | .99969 | .99970 | .99971 | .99972 | .99973 | .99974 | .99975 | .99976 |
| 3.5 | .99977 | .99978 | .99978 | .99979 | .99980 | .99981 | .99981 | .99982 | .99983 | .99983 |
| 3.6 | .99984 | .99985 | .99985 | .99986 | .99986 | .99987 | .99987 | .99988 | .99988 | .99989 |
| 3.7 | .99989 | .99990 | .99990 | .99990 | .99991 | .99991 | .99992 | .99992 | .99992 | .99992 |
| 3.8 | .99993 | .99993 | .99993 | .99994 | .99994 | .99994 | .99994 | .99995 | .99995 | .99995 |
| 3.9 | .99995 | .99995 | .99996 | .99996 | .99996 | .99996 | .99996 | .99996 | .99997 | .99997 |
| 4.0 | .99997 | .99997 | .99997 | .99997 | .99997 | .99997 | .99998 | .99998 | .99998 | .99998 |

## Table A2   *t*-Distribution

The entries in this table are $t_\alpha(\mathrm{df}) = c$ such that $\Pr(|t| > c) = \alpha$, where $t$ is distributed as Student's $t$-distribution, which is the case, for instance, when $t$ is the ratio between a unit normally distributed variable and the square root of a chi-square variate divided by its degrees of freedom, for example, $t = (X - \bar{X})/\sqrt{s_{\bar{X}}^2}$. In other words, $t_\alpha(\mathrm{df})$ is the $\alpha$-level critical value of the $t$-distribution with df degrees of freedom. Note that these are two-tailed critical values.

### Critical Values of t-Distribution

| DF | .40 | .25 | ALPHA .10 | .05 | .01 | .001 |
|---|---|---|---|---|---|---|
| 1 | 1.376 | 2.414 | 6.314 | 12.706 | 63.671 | 636.689 |
| 2 | 1.061 | 1.604 | 2.920 | 4.303 | 9.925 | 31.615 |
| 3 | .978 | 1.423 | 2.353 | 3.182 | 5.842 | 12.908 |
| 4 | .941 | 1.344 | 2.132 | 2.776 | 4.604 | 8.606 |
| 5 | .920 | 1.301 | 2.015 | 2.571 | 4.032 | 6.866 |
| | * | | * | | * | * |
| 6 | .906 | 1.273 | 1.943 | 2.447 | 3.707 | 5.962 |
| 7 | .896 | 1.254 | 1.895 | 2.365 | 3.499 | 5.411 |
| 8 | .889 | 1.240 | 1.860 | 2.306 | 3.355 | 5.039 |
| 9 | .883 | 1.230 | 1.833 | 2.262 | 3.250 | 4.780 |
| 10 | .879 | 1.221 | 1.812 | 2.228 | 3.169 | 4.586 |
| | * | | * | | * | * |
| 11 | .876 | 1.214 | 1.796 | 2.201 | 3.106 | 4.436 |
| 12 | .873 | 1.209 | 1.782 | 2.179 | 3.054 | 4.318 |
| 13 | .870 | 1.204 | 1.771 | 2.160 | 3.012 | 4.219 |
| 14 | .868 | 1.200 | 1.761 | 2.145 | 2.977 | 4.139 |
| 15 | .866 | 1.197 | 1.753 | 2.131 | 2.947 | 4.074 |
| | * | | * | | * | * |
| 16 | .865 | 1.194 | 1.746 | 2.120 | 2.921 | 4.017 |
| 17 | .863 | 1.191 | 1.740 | 2.110 | 2.898 | 3.964 |
| 18 | .862 | 1.189 | 1.734 | 2.101 | 2.878 | 3.922 |
| 19 | .861 | 1.187 | 1.729 | 2.093 | 2.861 | 3.882 |
| 20 | .860 | 1.185 | 1.725 | 2.086 | 2.845 | 3.848 |
| | * | | * | | * | * |
| 21 | .859 | 1.183 | 1.721 | 2.080 | 2.831 | 3.820 |
| 22 | .858 | 1.182 | 1.717 | 2.074 | 2.819 | 3.792 |
| 23 | .858 | 1.180 | 1.714 | 2.069 | 2.807 | 3.767 |
| 24 | .857 | 1.179 | 1.711 | 2.064 | 2.797 | 3.745 |
| 25 | .856 | 1.178 | 1.708 | 2.060 | 2.787 | 3.725 |
| | * | | * | | * | * |
| 26 | .856 | 1.177 | 1.706 | 2.056 | 2.779 | 3.707 |
| 27 | .855 | 1.176 | 1.703 | 2.052 | 2.771 | 3.688 |
| 28 | .855 | 1.175 | 1.701 | 2.048 | 2.763 | 3.675 |
| 29 | .854 | 1.174 | 1.699 | 2.045 | 2.756 | 3.661 |
| 30 | .854 | 1.173 | 1.697 | 2.042 | 2.750 | 3.646 |
| | * | | * | | * | * |
| 50 | .849 | 1.164 | 1.676 | 2.009 | 2.678 | 3.496 |
| 100 | .845 | 1.157 | 1.660 | 1.984 | 2.626 | 3.391 |
| 200 | .843 | 1.154 | 1.653 | 1.972 | 2.601 | 3.341 |
| INF | .842 | 1.150 | 1.645 | 1.960 | 2.576 | 3.291 |

## Table A3   Chi-Square Distribution

The entries in this table are the $\alpha$-level critical values of the chi-square distribution having df degrees of freedom. In other words, the entries are $\chi^2_\alpha(df) = c$ such that $\Pr(\chi^2 < c) = \alpha$. The sum of the squares of $p$ independent unit normal variates $\sum z_i^2$ has the chi-square distribution with $p$ degrees of freedom, and chi-square distributions arise as well in almost all applications of likelihood-ratio tests.

## Critical Values of Chi-Square and (Chi-Square/df)

| DF | .01 | .05 | .10 | .30 | .50 | .70 | .90 | .95 | .99 |
|---|---|---|---|---|---|---|---|---|---|
| 1 | 0.000 | 0.004 | 0.016 | 0.149 | 0.455 | 1.074 | 2.706 | 3.842 | 6.635 |
|  | (0.000) | (0.004) | (0.016) | (0.149) | (0.455) | (1.074) | (2.706) | (3.842) | (6.635) |
| 2 | 0.020 | 0.103 | 0.211 | 0.713 | 1.386 | 2.408 | 4.605 | 5.991 | 9.210 |
|  | (0.010) | (0.051) | (0.105) | (0.357) | (0.693) | (1.204) | (2.303) | (2.996) | (4.605) |
| 3 | 0.115 | 0.352 | 0.584 | 1.424 | 2.366 | 3.665 | 6.251 | 7.814 | 11.345 |
|  | (0.038) | (0.117) | (0.195) | (0.475) | (0.789) | (1.222) | (2.084) | (2.605) | (3.782) |
| 4 | 0.297 | 0.711 | 1.064 | 2.195 | 3.357 | 4.878 | 7.780 | 9.488 | 13.276 |
|  | (0.074) | (0.178) | (0.266) | (0.549) | (0.839) | (1.220) | (1.945) | (2.372) | (3.319) |
| 5 | 0.554 | 1.146 | 1.610 | 3.000 | 4.351 | 6.064 | 9.236 | 11.071 | 15.086 |
|  | (0.111) | (0.229) | (0.322) | (0.600) | (0.870) | (1.213) | (1.847) | (2.214) | (3.017) |
|  | * |  | * |  | * |  | * |  | * |
| 6 | 0.872 | 1.635 | 2.204 | 3.828 | 5.348 | 7.231 | 10.645 | 12.592 | 16.812 |
|  | (0.145) | (0.273) | (0.367) | (0.638) | (0.891) | (1.205) | (1.774) | (2.099) | (2.802) |
| 7 | 1.239 | 2.167 | 2.833 | 4.671 | 6.346 | 8.384 | 12.017 | 14.067 | 18.475 |
|  | (0.177) | (0.310) | (0.405) | (0.667) | (0.907) | (1.198) | (1.717) | (2.010) | (2.639) |
| 8 | 1.646 | 2.733 | 3.489 | 5.527 | 7.344 | 9.525 | 13.362 | 15.507 | 20.090 |
|  | (0.206) | (0.342) | (0.436) | (0.691) | (0.918) | (1.191) | (1.670) | (1.938) | (2.511) |
| 9 | 2.088 | 3.325 | 4.168 | 6.393 | 8.343 | 10.656 | 14.684 | 16.919 | 21.666 |
|  | (0.232) | (0.369) | (0.463) | (0.710) | (0.927) | (1.184) | (1.632) | (1.880) | (2.407) |
| 10 | 2.558 | 3.941 | 4.865 | 7.268 | 9.342 | 11.781 | 15.987 | 18.307 | 23.209 |
|  | (0.256) | (0.394) | (0.487) | (0.727) | (0.934) | (1.178) | (1.599) | (1.831) | (2.321) |
|  | * |  | * |  | * |  | * |  | * |
| 11 | 3.053 | 4.575 | 5.578 | 8.148 | 10.341 | 12.899 | 17.275 | 19.675 | 24.725 |
|  | (0.278) | (0.416) | (0.507) | (0.741) | (0.940) | (1.173) | (1.570) | (1.789) | (2.248) |
| 12 | 3.571 | 5.226 | 6.304 | 9.034 | 11.341 | 14.011 | 18.549 | 21.026 | 26.217 |
|  | (0.298) | (0.436) | (0.525) | (0.753) | (0.945) | (1.168) | (1.546) | (1.752) | (2.185) |
| 13 | 4.107 | 5.892 | 7.041 | 9.926 | 12.340 | 15.119 | 19.812 | 22.362 | 27.688 |
|  | (0.316) | (0.453) | (0.542) | (0.764) | (0.949) | (1.163) | (1.524) | (1.720) | (2.130) |
| 14 | 4.660 | 6.571 | 7.790 | 10.821 | 13.340 | 16.222 | 21.064 | 23.685 | 29.141 |
|  | (0.333) | (0.469) | (0.556) | (0.773) | (0.953) | (1.159) | (1.505) | (1.692) | (2.082) |
| 15 | 5.229 | 7.261 | 8.547 | 11.722 | 14.339 | 17.322 | 22.307 | 24.996 | 30.578 |
|  | (0.349) | (0.484) | (0.570) | (0.781) | (0.956) | (1.155) | (1.487) | (1.666) | (2.039) |
|  | * |  | * |  | * |  | * |  | * |
| 16 | 5.813 | 7.962 | 9.312 | 12.624 | 15.338 | 18.418 | 23.542 | 26.296 | 32.000 |
|  | (0.363) | (0.498) | (0.582) | (0.789) | (0.959) | (1.151) | (1.471) | (1.644) | (2.000) |
| 17 | 6.407 | 8.671 | 10.085 | 13.530 | 16.338 | 19.511 | 24.769 | 27.587 | 33.409 |
|  | (0.377) | (0.510) | (0.593) | (0.796) | (0.961) | (1.148) | (1.457) | (1.623) | (1.965) |
| 18 | 7.015 | 9.390 | 10.865 | 14.440 | 17.338 | 20.601 | 25.990 | 28.869 | 34.805 |
|  | (0.390) | (0.522) | (0.604) | (0.802) | (0.963) | (1.145) | (1.444) | (1.604) | (1.934) |
| 19 | 7.633 | 10.117 | 11.651 | 15.352 | 18.337 | 21.689 | 27.204 | 30.143 | 36.191 |
|  | (0.402) | (0.532) | (0.613) | (0.808) | (0.965) | (1.142) | (1.432) | (1.586) | (1.905) |
| 20 | 8.260 | 10.851 | 12.443 | 16.266 | 19.337 | 22.774 | 28.412 | 31.410 | 37.566 |
|  | (0.413) | (0.543) | (0.622) | (0.813) | (0.967) | (1.139) | (1.421) | (1.571) | (1.878) |
|  | * |  | * |  | * |  | * |  | * |
| 21 | 8.897 | 11.591 | 13.240 | 17.182 | 20.338 | 23.858 | 29.615 | 32.670 | 38.932 |
|  | (0.424) | (0.552) | (0.630) | (0.818) | (0.968) | (1.136) | (1.410) | (1.556) | (1.854) |
| 22 | 9.542 | 12.338 | 14.042 | 18.101 | 21.337 | 24.939 | 30.814 | 33.924 | 40.289 |
|  | (0.434) | (0.561) | (0.638) | (0.823) | (0.970) | (1.134) | (1.401) | (1.542) | (1.831) |
| 23 | 10.196 | 13.090 | 14.848 | 19.021 | 22.337 | 26.018 | 32.007 | 35.173 | 41.639 |
|  | (0.443) | (0.569) | (0.646) | (0.827) | (0.971) | (1.131) | (1.392) | (1.529) | (1.810) |
| 24 | 10.856 | 13.848 | 15.659 | 19.944 | 23.337 | 27.096 | 33.196 | 36.415 | 42.980 |
|  | (0.452) | (0.577) | (0.652) | (0.831) | (0.972) | (1.129) | (1.383) | (1.517) | (1.791) |
| 25 | 11.524 | 14.611 | 16.474 | 20.867 | 24.336 | 28.172 | 34.381 | 37.652 | 44.314 |
|  | (0.461) | (0.584) | (0.659) | (0.835) | (0.973) | (1.127) | (1.375) | (1.506) | (1.773) |
|  | * |  | * |  | * |  | * |  | * |
| 26 | 12.198 | 15.379 | 17.292 | 21.792 | 25.336 | 29.247 | 35.563 | 38.885 | 45.642 |
|  | (0.469) | (0.592) | (0.665) | (0.838) | (0.974) | (1.125) | (1.368) | (1.496) | (1.755) |
| 27 | 12.878 | 16.152 | 18.114 | 22.719 | 26.336 | 30.320 | 36.741 | 40.114 | 46.963 |
|  | (0.477) | (0.598) | (0.671) | (0.841) | (0.975) | (1.123) | (1.361) | (1.486) | (1.739) |
| 28 | 13.564 | 16.928 | 18.939 | 23.647 | 27.337 | 31.391 | 37.916 | 41.337 | 48.278 |
|  | (0.484) | (0.605) | (0.676) | (0.845) | (0.976) | (1.121) | (1.354) | (1.476) | (1.724) |
| 29 | 14.257 | 17.708 | 19.768 | 24.577 | 28.336 | 32.461 | 39.087 | 42.557 | 49.588 |
|  | (0.492) | (0.611) | (0.682) | (0.847) | (0.977) | (1.119) | (1.348) | (1.467) | (1.710) |
| 30 | 14.953 | 18.492 | 20.599 | 25.508 | 29.336 | 33.530 | 40.256 | 43.773 | 50.892 |
|  | (0.498) | (0.616) | (0.687) | (0.850) | (0.978) | (1.118) | (1.342) | (1.459) | (1.696) |
|  | * |  | * |  | * |  | * |  | * |
| 50 | 29.707 | 34.765 | 37.689 | 44.313 | 49.335 | 54.722 | 63.167 | 67.505 | 76.154 |
|  | (0.594) | (0.695) | (0.754) | (0.886) | (0.987) | (1.094) | (1.263) | (1.350) | (1.523) |
| 00 | 70.065 | 77.930 | 82.358 | 92.129 | 99.334 | 106.906 | 118.498 | 124.342 | 135.807 |
|  | (0.701) | (0.779) | (0.824) | (0.921) | (0.993) | (1.069) | (1.185) | (1.243) | (1.358) |
| 200 | 156.432 | 168.278 | 174.835 | 189.049 | 199.334 | 209.985 | 226.021 | 233.995 | 249.445 |
|  | (0.782) | (0.841) | (0.874) | (0.945) | (0.997) | (1.050) | (1.130) | (1.170) | (1.247) |
| 300 | 245.973 | 260.878 | 269.068 | 286.688 | 299.333 | 312.346 | 331.789 | 341.395 | 359.906 |
|  | (0.820) | (0.870) | (0.897) | (0.956) | (0.998) | (1.041) | (1.106) | (1.138) | (1.200) |
| NF | A | R | B | I | T | R | A | R | I L Y   L A R G E |
|  | (1.000) | (1.000) | (1.000) | (1.000) | (1.000) | (1.000) | (1.000) | (1.000) | (1.000) |
|  | * |  | * |  | * |  | * |  | * |

# Table A4 F-Distribution

The entries in this table are the α-level critical values of the F-distribution with DFH and DFE degrees of freedom, that is, $F_\alpha(\text{DFH}, \text{DFE}) = c$ such that $\Pr(F > c) = \alpha$, where $F$ is the

## Critical Values of F-Distribution for α = .05 and (.01)

| DFE/DFH | 1 | 2 | 3 | 4 | 5 | 6 | 7 | 8 | 9 | 10 | 12 |
|---|---|---|---|---|---|---|---|---|---|---|---|
| 1 | 161.4 | 199.5 | 215.7 | 224.6 | 230.2 | 234.0 | 236.8 | 238.9 | 240.5 | 241.9 | 243.9 |
|  | (4052.) | (5000.) | (5403.) | (5625.) | (5764.) | (5859.) | (5928.) | (5981.) | (6022.) | (6056.) | (6106. |
| 2 | 18.51 | 19.00 | 19.16 | 19.24 | 19.29 | 19.33 | 19.35 | 19.37 | 19.38 | 19.40 | 19.41 |
|  | (98.50) | (99.00) | (99.16) | (99.25) | (99.30) | (99.33) | (99.36) | (99.38) | (99.39) | (99.40) | (99.41 |
| 3 | 10.12 | 9.55 | 9.28 | 9.12 | 9.01 | 8.94 | 8.89 | 8.85 | 8.81 | 8.79 | 8.74 |
|  | (34.11) | (30.82) | (29.46) | (28.71) | (28.24) | (27.91) | (27.67) | (27.49) | (27.35) | (27.23) | (27.05 |
| 4 | 7.71 | 6.94 | 6.59 | 6.39 | 6.26 | 6.17 | 6.09 | 6.04 | 6.00 | 5.97 | 5.91 |
|  | (21.20) | (18.00) | (16.69) | (15.98) | (15.52) | (15.20) | (14.98) | (14.80) | (14.66) | (14.55) | (14.38 |
| 5 | 6.61 | 5.79 | 5.41 | 5.19 | 5.05 | 4.95 | 4.87 | 4.82 | 4.77 | 4.73 | 4.68 |
|  | (16.26) | (13.27) | (12.06) | (11.39) | (10.97) | (10.67) | (10.45) | (10.29) | (10.16) | (10.05) | ( 9.89 |
| 6 | 5.98 | 5.15 | 4.76 | 4.53 | 4.39 | 4.28 | 4.21 | 4.15 | 4.10 | 4.06 | 4.00 |
|  | (13.74) | (10.93) | ( 9.78) | ( 9.15) | ( 8.75) | ( 8.47) | ( 8.26) | ( 8.10) | ( 7.97) | ( 7.88) | ( 7.72 |
| 7 | 5.59 | 4.74 | 4.35 | 4.12 | 3.97 | 3.87 | 3.79 | 3.73 | 3.68 | 3.64 | 3.57 |
|  | (12.24) | ( 9.55) | ( 8.45) | ( 7.85) | ( 7.46) | ( 7.19) | ( 6.99) | ( 6.84) | ( 6.72) | ( 6.62) | ( 6.47 |
| 8 | 5.32 | 4.46 | 4.07 | 3.84 | 3.69 | 3.58 | 3.50 | 3.44 | 3.39 | 3.35 | 3.28 |
|  | (11.26) | ( 8.65) | ( 7.59) | ( 7.01) | ( 6.63) | ( 6.37) | ( 6.18) | ( 6.03) | ( 5.91) | ( 5.81) | ( 5.67 |
| 9 | 5.12 | 4.26 | 3.86 | 3.63 | 3.48 | 3.37 | 3.29 | 3.23 | 3.18 | 3.13 | 3.07 |
|  | (10.56) | ( 8.03) | ( 6.99) | ( 6.42) | ( 6.06) | ( 5.80) | ( 5.61) | ( 5.47) | ( 5.35) | ( 5.26) | ( 5.11 |
| 10 | 4.97 | 4.11 | 3.71 | 3.48 | 3.32 | 3.22 | 3.14 | 3.07 | 3.02 | 2.98 | 2.91 |
|  | (10.05) | ( 7.56) | ( 6.55) | ( 5.99) | ( 5.64) | ( 5.39) | ( 5.20) | ( 5.06) | ( 4.94) | ( 4.85) | ( 4.70 |
| 11 | 4.84 | 3.98 | 3.58 | 3.36 | 3.20 | 3.10 | 3.01 | 2.95 | 2.89 | 2.85 | 2.79 |
|  | ( 9.64) | ( 7.21) | ( 6.22) | ( 5.67) | ( 5.31) | ( 5.07) | ( 4.89) | ( 4.75) | ( 4.63) | ( 4.54) | ( 4.40 |
| 12 | 4.75 | 3.89 | 3.49 | 3.26 | 3.11 | 2.99 | 2.92 | 2.85 | 2.80 | 2.76 | 2.69 |
|  | ( 9.33) | ( 6.93) | ( 5.96) | ( 5.41) | ( 5.07) | ( 4.82) | ( 4.64) | ( 4.50) | ( 4.39) | ( 4.30) | ( 4.16 |
| 13 | 4.67 | 3.80 | 3.41 | 3.18 | 3.03 | 2.92 | 2.83 | 2.77 | 2.72 | 2.67 | 2.60 |
|  | ( 9.07) | ( 6.70) | ( 5.74) | ( 5.21) | ( 4.86) | ( 4.62) | ( 4.44) | ( 4.30) | ( 4.19) | ( 4.10) | ( 3.96 |
| 14 | 4.60 | 3.74 | 3.35 | 3.11 | 2.96 | 2.85 | 2.76 | 2.70 | 2.64 | 2.60 | 2.54 |
|  | ( 8.86) | ( 6.51) | ( 5.56) | ( 5.03) | ( 4.69) | ( 4.46) | ( 4.28) | ( 4.14) | ( 4.03) | ( 3.94) | ( 3.80 |
| 15 | 4.55 | 3.68 | 3.29 | 3.06 | 2.90 | 2.79 | 2.71 | 2.64 | 2.59 | 2.54 | 2.47 |
|  | ( 8.68) | ( 6.36) | ( 5.42) | ( 4.89) | ( 4.56) | ( 4.32) | ( 4.14) | ( 4.01) | ( 3.89) | ( 3.80) | ( 3.67 |
| 16 | 4.49 | 3.63 | 3.24 | 3.01 | 2.85 | 2.74 | 2.66 | 2.59 | 2.54 | 2.49 | 2.43 |
|  | ( 8.53) | ( 6.23) | ( 5.30) | ( 4.77) | ( 4.44) | ( 4.20) | ( 4.03) | ( 3.89) | ( 3.78) | ( 3.69) | ( 3.55 |
| 17 | 4.45 | 3.59 | 3.20 | 2.96 | 2.81 | 2.70 | 2.61 | 2.55 | 2.49 | 2.45 | 2.38 |
|  | ( 8.40) | ( 6.11) | ( 5.19) | ( 4.67) | ( 4.34) | ( 4.10) | ( 3.93) | ( 3.79) | ( 3.68) | ( 3.59) | ( 3.45 |
| 18 | 4.41 | 3.55 | 3.16 | 2.93 | 2.77 | 2.66 | 2.58 | 2.51 | 2.45 | 2.41 | 2.34 |
|  | ( 8.29) | ( 6.01) | ( 5.09) | ( 4.58) | ( 4.25) | ( 4.02) | ( 3.84) | ( 3.71) | ( 3.60) | ( 3.51) | ( 3.37 |
| 19 | 4.38 | 3.52 | 3.13 | 2.89 | 2.74 | 2.63 | 2.54 | 2.48 | 2.42 | 2.38 | 2.31 |
|  | ( 8.19) | ( 5.92) | ( 5.01) | ( 4.50) | ( 4.17) | ( 3.94) | ( 3.76) | ( 3.63) | ( 3.52) | ( 3.43) | ( 3.30 |
| 20 | 4.35 | 3.49 | 3.10 | 2.86 | 2.71 | 2.60 | 2.52 | 2.45 | 2.39 | 2.35 | 2.28 |
|  | ( 8.10) | ( 5.85) | ( 4.94) | ( 4.43) | ( 4.10) | ( 3.87) | ( 3.70) | ( 3.56) | ( 3.45) | ( 3.37) | ( 3.23 |
| 21 | 4.33 | 3.47 | 3.08 | 2.84 | 2.68 | 2.57 | 2.49 | 2.42 | 2.37 | 2.32 | 2.25 |
|  | ( 8.02) | ( 5.78) | ( 4.87) | ( 4.37) | ( 4.04) | ( 3.81) | ( 3.64) | ( 3.51) | ( 3.40) | ( 3.31) | ( 3.17 |
| 22 | 4.30 | 3.44 | 3.05 | 2.82 | 2.66 | 2.55 | 2.47 | 2.40 | 2.34 | 2.29 | 2.23 |
|  | ( 7.94) | ( 5.72) | ( 4.81) | ( 4.32) | ( 3.99) | ( 3.76) | ( 3.59) | ( 3.46) | ( 3.35) | ( 3.26) | ( 3.12 |
| 23 | 4.28 | 3.42 | 3.03 | 2.79 | 2.64 | 2.53 | 2.44 | 2.37 | 2.32 | 2.28 | 2.20 |
|  | ( 7.88) | ( 5.66) | ( 4.76) | ( 4.26) | ( 3.94) | ( 3.71) | ( 3.54) | ( 3.41) | ( 3.30) | ( 3.21) | ( 3.07 |
| 24 | 4.26 | 3.40 | 3.01 | 2.77 | 2.62 | 2.51 | 2.42 | 2.35 | 2.30 | 2.26 | 2.18 |
|  | ( 7.82) | ( 5.62) | ( 4.72) | ( 4.22) | ( 3.90) | ( 3.67) | ( 3.50) | ( 3.37) | ( 3.26) | ( 3.17) | ( 3.03 |
| 25 | 4.24 | 3.39 | 2.99 | 2.76 | 2.60 | 2.49 | 2.41 | 2.34 | 2.28 | 2.24 | 2.16 |
|  | ( 7.77) | ( 5.57) | ( 4.68) | ( 4.18) | ( 3.85) | ( 3.63) | ( 3.46) | ( 3.33) | ( 3.22) | ( 3.13) | ( 2.99 |
| 26 | 4.23 | 3.37 | 2.97 | 2.74 | 2.59 | 2.47 | 2.39 | 2.32 | 2.26 | 2.22 | 2.15 |
|  | ( 7.72) | ( 5.53) | ( 4.63) | ( 4.14) | ( 3.82) | ( 3.59) | ( 3.42) | ( 3.29) | ( 3.18) | ( 3.09) | ( 2.96 |
| 27 | 4.21 | 3.35 | 2.96 | 2.73 | 2.57 | 2.46 | 2.37 | 2.31 | 2.25 | 2.20 | 2.13 |
|  | ( 7.68) | ( 5.49) | ( 4.60) | ( 4.11) | ( 3.78) | ( 3.56) | ( 3.39) | ( 3.26) | ( 3.15) | ( 3.06) | ( 2.93 |
| 28 | 4.19 | 3.34 | 2.94 | 2.71 | 2.56 | 2.44 | 2.36 | 2.29 | 2.23 | 2.19 | 2.12 |
|  | ( 7.64) | ( 5.45) | ( 4.57) | ( 4.07) | ( 3.76) | ( 3.53) | ( 3.35) | ( 3.22) | ( 3.12) | ( 3.03) | ( 2.89 |
| 29 | 4.19 | 3.33 | 2.93 | 2.70 | 2.54 | 2.43 | 2.35 | 2.28 | 2.22 | 2.18 | 2.10 |
|  | ( 7.60) | ( 5.42) | ( 4.54) | ( 4.04) | ( 3.72) | ( 3.50) | ( 3.33) | ( 3.20) | ( 3.09) | ( 3.00) | ( 2.87 |
| 30 | 4.17 | 3.32 | 2.92 | 2.69 | 2.53 | 2.42 | 2.33 | 2.27 | 2.21 | 2.16 | 2.09 |
|  | ( 7.56) | ( 5.39) | ( 4.51) | ( 4.02) | ( 3.70) | ( 3.48) | ( 3.30) | ( 3.17) | ( 3.07) | ( 2.98) | ( 2.84 |
| 50 | 4.03 | 3.18 | 2.79 | 2.56 | 2.40 | 2.29 | 2.20 | 2.13 | 2.07 | 2.03 | 1.95 |
|  | ( 7.17) | ( 5.06) | ( 4.20) | ( 3.72) | ( 3.41) | ( 3.19) | ( 3.02) | ( 2.89) | ( 2.79) | ( 2.70) | ( 2.56 |
| 100 | 3.94 | 3.08 | 2.69 | 2.46 | 2.30 | 2.19 | 2.10 | 2.03 | 1.97 | 1.93 | 1.85 |
|  | ( 6.90) | ( 4.82) | ( 3.98) | ( 3.51) | ( 3.20) | ( 2.99) | ( 2.82) | ( 2.69) | ( 2.59) | ( 2.50) | ( 2.37 |
| INF | 3.84 | 3.00 | 2.60 | 2.37 | 2.21 | 2.10 | 2.01 | 1.94 | 1.88 | 1.83 | 1.75 |
|  | ( 6.63) | ( 4.61) | ( 3.78) | ( 3.32) | ( 3.02) | ( 2.80) | ( 2.64) | ( 2.51) | ( 2.41) | ( 2.32) | ( 2.18 |

ratio of two chi-square variates, each divided by its degrees of freedom. For instance, $MS_H/MS_E$ is distributed under the null hypothesis as $F$ with DFH and DFE degrees of freedom.

*Critical Values of F-Distribution for $\alpha = .05$ and $(.01)$* (continued)

| DFE/ DFH | 14 | 16 | 18 | 20 | 25 | 30 | 40 | 50 | 75 | 100 | INF |
|---|---|---|---|---|---|---|---|---|---|---|---|
| 1 | 245.4 | 246.5 | 247.3 | 248.0 | 249.3 | 250.1 | 251.1 | 251.8 | 252.6 | 253.0 | 252.2 |
| | (6143.) | (6170.) | (6192.) | (6209.) | (6240.) | (6261.) | (6287.) | (6303.) | (6324.) | (6334.) | (6392.) |
| 2 | 19.42 | 19.43 | 19.44 | 19.44 | 19.46 | 19.46 | 19.47 | 19.48 | 19.48 | 19.49 | 19.44 |
| | (99.43) | (99.44) | (99.44) | (99.45) | (99.46) | (99.47) | (99.47) | (99.48) | (99.49) | (99.49) | (101.4) |
| 3 | 8.71 | 8.69 | 8.67 | 8.66 | 8.63 | 8.62 | 8.59 | 8.58 | 8.56 | 8.55 | 8.52 |
| | (26.92) | (26.83) | (26.75) | (26.69) | (26.58) | (26.50) | (26.41) | (26.35) | (26.28) | (26.24) | (26.15) |
| 4 | 5.87 | 5.85 | 5.82 | 5.80 | 5.77 | 5.75 | 5.72 | 5.70 | 5.68 | 5.67 | 5.63 |
| | (14.25) | (14.16) | (14.08) | (14.02) | (13.91) | (13.84) | (13.75) | (13.69) | (13.61) | (13.58) | (13.45) |
| 5 | 4.64 | 4.61 | 4.58 | 4.56 | 4.52 | 4.50 | 4.46 | 4.45 | 4.42 | 4.40 | 4.36 |
| | ( 9.77) | ( 9.68) | ( 9.61) | ( 9.56) | ( 9.45) | ( 9.38) | ( 9.29) | ( 9.24) | ( 9.16) | ( 9.13) | ( 9.02) |
| 6 | 3.96 | 3.92 | 3.89 | 3.88 | 3.84 | 3.81 | 3.78 | 3.75 | 3.73 | 3.71 | 3.67 |
| | ( 7.60) | ( 7.52) | ( 7.45) | ( 7.40) | ( 7.30) | ( 7.23) | ( 7.14) | ( 7.09) | ( 7.02) | ( 6.99) | ( 6.88) |
| 7 | 3.53 | 3.50 | 3.47 | 3.44 | 3.40 | 3.38 | 3.34 | 3.32 | 3.29 | 3.27 | 3.23 |
| | ( 6.36) | ( 6.28) | ( 6.21) | ( 6.16) | ( 6.06) | ( 5.99) | ( 5.91) | ( 5.86) | ( 5.79) | ( 5.75) | ( 5.65) |
| 8 | 3.23 | 3.20 | 3.17 | 3.15 | 3.11 | 3.08 | 3.04 | 3.02 | 2.99 | 2.98 | 2.93 |
| | ( 5.56) | ( 5.48) | ( 5.41) | ( 5.36) | ( 5.26) | ( 5.20) | ( 5.12) | ( 5.07) | ( 5.00) | ( 4.96) | ( 4.86) |
| 9 | 3.02 | 2.99 | 2.96 | 2.94 | 2.89 | 2.86 | 2.83 | 2.80 | 2.77 | 2.76 | 2.71 |
| | ( 5.00) | ( 4.92) | ( 4.86) | ( 4.81) | ( 4.71) | ( 4.65) | ( 4.57) | ( 4.52) | ( 4.45) | ( 4.41) | ( 4.31) |
| 10 | 2.86 | 2.83 | 2.80 | 2.77 | 2.73 | 2.70 | 2.66 | 2.64 | 2.60 | 2.59 | 2.54 |
| | ( 4.60) | ( 4.52) | ( 4.46) | ( 4.40) | ( 4.31) | ( 4.25) | ( 4.17) | ( 4.11) | ( 4.05) | ( 4.01) | ( 3.91) |
| 11 | 2.74 | 2.70 | 2.67 | 2.65 | 2.60 | 2.57 | 2.53 | 2.51 | 2.48 | 2.46 | 2.40 |
| | ( 4.29) | ( 4.21) | ( 4.15) | ( 4.10) | ( 4.01) | ( 3.94) | ( 3.86) | ( 3.81) | ( 3.74) | ( 3.71) | ( 3.60) |
| 12 | 2.64 | 2.60 | 2.57 | 2.54 | 2.50 | 2.47 | 2.43 | 2.40 | 2.37 | 2.35 | 2.30 |
| | ( 4.05) | ( 3.97) | ( 3.91) | ( 3.86) | ( 3.77) | ( 3.70) | ( 3.62) | ( 3.57) | ( 3.50) | ( 3.47) | ( 3.36) |
| 13 | 2.56 | 2.51 | 2.49 | 2.46 | 2.41 | 2.38 | 2.34 | 2.31 | 2.28 | 2.26 | 2.21 |
| | ( 3.85) | ( 3.78) | ( 3.72) | ( 3.67) | ( 3.57) | ( 3.51) | ( 3.42) | ( 3.37) | ( 3.30) | ( 3.27) | ( 3.17) |
| 14 | 2.48 | 2.45 | 2.41 | 2.39 | 2.34 | 2.31 | 2.27 | 2.24 | 2.20 | 2.19 | 2.13 |
| | ( 3.69) | ( 3.62) | ( 3.56) | ( 3.51) | ( 3.41) | ( 3.35) | ( 3.27) | ( 3.21) | ( 3.15) | ( 3.11) | ( 3.00) |
| 15 | 2.43 | 2.38 | 2.35 | 2.33 | 2.28 | 2.25 | 2.21 | 2.18 | 2.14 | 2.12 | 2.07 |
| | ( 3.57) | ( 3.48) | ( 3.43) | ( 3.37) | ( 3.28) | ( 3.21) | ( 3.13) | ( 3.08) | ( 3.01) | ( 2.98) | ( 2.87) |
| 16 | 2.37 | 2.33 | 2.30 | 2.28 | 2.23 | 2.20 | 2.15 | 2.13 | 2.09 | 2.07 | 2.01 |
| | ( 3.45) | ( 3.37) | ( 3.31) | ( 3.26) | ( 3.17) | ( 3.10) | ( 3.02) | ( 2.97) | ( 2.90) | ( 2.86) | ( 2.75) |
| 17 | 2.33 | 2.29 | 2.25 | 2.23 | 2.18 | 2.15 | 2.10 | 2.08 | 2.04 | 2.02 | 1.96 |
| | ( 3.36) | ( 3.28) | ( 3.21) | ( 3.16) | ( 3.07) | ( 3.00) | ( 2.92) | ( 2.87) | ( 2.80) | ( 2.76) | ( 2.65) |
| 18 | 2.29 | 2.25 | 2.22 | 2.19 | 2.14 | 2.11 | 2.06 | 2.04 | 2.00 | 1.98 | 1.92 |
| | ( 3.27) | ( 3.19) | ( 3.13) | ( 3.08) | ( 2.98) | ( 2.92) | ( 2.83) | ( 2.78) | ( 2.72) | ( 2.68) | ( 2.57) |
| 19 | 2.26 | 2.21 | 2.18 | 2.16 | 2.11 | 2.07 | 2.02 | 2.00 | 1.96 | 1.94 | 1.88 |
| | ( 3.19) | ( 3.12) | ( 3.05) | ( 3.00) | ( 2.91) | ( 2.84) | ( 2.76) | ( 2.71) | ( 2.64) | ( 2.60) | ( 2.49) |
| 20 | 2.23 | 2.18 | 2.15 | 2.12 | 2.07 | 2.04 | 1.99 | 1.96 | 1.93 | 1.91 | 1.84 |
| | ( 3.13) | ( 3.05) | ( 2.99) | ( 2.94) | ( 2.84) | ( 2.78) | ( 2.70) | ( 2.64) | ( 2.57) | ( 2.54) | ( 2.42) |
| 21 | 2.20 | 2.16 | 2.12 | 2.10 | 2.04 | 2.01 | 1.97 | 1.93 | 1.90 | 1.88 | 1.81 |
| | ( 3.07) | ( 2.99) | ( 2.93) | ( 2.88) | ( 2.78) | ( 2.72) | ( 2.63) | ( 2.58) | ( 2.51) | ( 2.48) | ( 2.36) |
| 22 | 2.17 | 2.13 | 2.10 | 2.07 | 2.02 | 1.99 | 1.94 | 1.91 | 1.87 | 1.85 | 1.78 |
| | ( 3.02) | ( 2.94) | ( 2.88) | ( 2.83) | ( 2.73) | ( 2.67) | ( 2.58) | ( 2.53) | ( 2.46) | ( 2.42) | ( 2.31) |
| 23 | 2.15 | 2.11 | 2.08 | 2.05 | 2.00 | 1.96 | 1.92 | 1.89 | 1.84 | 1.82 | 1.76 |
| | ( 2.97) | ( 2.89) | ( 2.83) | ( 2.78) | ( 2.69) | ( 2.62) | ( 2.53) | ( 2.48) | ( 2.41) | ( 2.37) | ( 2.26) |
| 24 | 2.13 | 2.09 | 2.06 | 2.03 | 1.97 | 1.94 | 1.89 | 1.86 | 1.82 | 1.80 | 1.73 |
| | ( 2.93) | ( 2.85) | ( 2.79) | ( 2.74) | ( 2.64) | ( 2.58) | ( 2.49) | ( 2.44) | ( 2.37) | ( 2.33) | ( 2.21) |
| 25 | 2.11 | 2.07 | 2.04 | 2.01 | 1.95 | 1.92 | 1.87 | 1.84 | 1.80 | 1.78 | 1.71 |
| | ( 2.89) | ( 2.81) | ( 2.75) | ( 2.70) | ( 2.60) | ( 2.54) | ( 2.45) | ( 2.40) | ( 2.33) | ( 2.29) | ( 2.17) |
| 26 | 2.10 | 2.05 | 2.02 | 1.99 | 1.94 | 1.90 | 1.85 | 1.83 | 1.78 | 1.76 | 1.69 |
| | ( 2.86) | ( 2.78) | ( 2.72) | ( 2.66) | ( 2.57) | ( 2.50) | ( 2.42) | ( 2.36) | ( 2.29) | ( 2.25) | ( 2.13) |
| 27 | 2.08 | 2.04 | 2.00 | 1.98 | 1.92 | 1.88 | 1.84 | 1.81 | 1.76 | 1.74 | 1.67 |
| | ( 2.82) | ( 2.75) | ( 2.68) | ( 2.63) | ( 2.54) | ( 2.47) | ( 2.38) | ( 2.33) | ( 2.25) | ( 2.22) | ( 2.10) |
| 28 | 2.06 | 2.02 | 1.99 | 1.96 | 1.91 | 1.87 | 1.82 | 1.79 | 1.75 | 1.73 | 1.65 |
| | ( 2.80) | ( 2.72) | ( 2.65) | ( 2.60) | ( 2.51) | ( 2.44) | ( 2.36) | ( 2.30) | ( 2.23) | ( 2.19) | ( 2.06) |
| 29 | 2.05 | 2.01 | 1.97 | 1.95 | 1.89 | 1.85 | 1.81 | 1.78 | 1.73 | 1.71 | 1.64 |
| | ( 2.77) | ( 2.69) | ( 2.63) | ( 2.57) | ( 2.48) | ( 2.41) | ( 2.33) | ( 2.27) | ( 2.20) | ( 2.16) | ( 2.03) |
| 30 | 2.04 | 1.99 | 1.96 | 1.93 | 1.88 | 1.84 | 1.79 | 1.76 | 1.72 | 1.70 | 1.62 |
| | ( 2.74) | ( 2.66) | ( 2.60) | ( 2.55) | ( 2.45) | ( 2.38) | ( 2.30) | ( 2.24) | ( 2.17) | ( 2.13) | ( 2.01) |
| 50 | 1.90 | 1.85 | 1.81 | 1.78 | 1.73 | 1.69 | 1.63 | 1.60 | 1.55 | 1.52 | 1.44 |
| | ( 2.46) | ( 2.38) | ( 2.32) | ( 2.27) | ( 2.16) | ( 2.10) | ( 2.00) | ( 1.95) | ( 1.86) | ( 1.82) | ( 1.68) |
| 100 | 1.79 | 1.75 | 1.71 | 1.67 | 1.62 | 1.57 | 1.52 | 1.48 | 1.42 | 1.39 | 1.28 |
| | ( 2.27) | ( 2.18) | ( 2.12) | ( 2.07) | ( 1.96) | ( 1.89) | ( 1.80) | ( 1.74) | ( 1.65) | ( 1.60) | ( 1.43) |
| INF | 1.69 | 1.64 | 1.60 | 1.57 | 1.51 | 1.46 | 1.39 | 1.35 | 1.28 | 1.24 | 1.00 |
| | ( 2.08) | ( 2.00) | ( 1.93) | ( 1.88) | ( 1.77) | ( 1.70) | ( 1.59) | ( 1.52) | ( 1.42) | ( 1.36) | ( 1.00) |

## Table A5   Greatest Characteristic Root Distribution

The entries in this table are the $\alpha$-level critical values of the greatest characteristic root (gcr) distribution, that is, $\theta_\alpha(s, m, n) = c$ such that $\Pr(\theta_{max} > c) = \alpha$, where $\theta_{max}$ is the largest of the $s$ characteristic roots of $(\mathbf{H} + \mathbf{E})^{-1}\mathbf{H}$ or of $\mathbf{S}_{11}^{-1}\mathbf{S}_{12}\mathbf{S}_{22}^{-1}\mathbf{S}_{21}$; $s$ is the rank of the matrix and equals $\min(r_1, r_2)$, where $r_1$ and $r_2$ are the ranks of $\mathbf{H}$ and of $\mathbf{E}$ or of $\mathbf{S}_{11}$ and $\mathbf{S}_{22}$; $m = (|r_1 - r_2| - 1)/2$; and $n = (\mathrm{df}_e - r_2 - 1)/2$, where $\mathrm{df}_e$ is the rank of the data matrix, that is, the number of degrees of freedom going into each variance or covariance estimate in the covariance matrix.

For values of $s$, $m$, and $n$ not included in these tables, the user may either interpolate within the tables or use the GCRCLC program listed in Appendix B to compute the probability (under the null hypothesis) of his observed value of $\theta_{max}$. If $n < 5$, more accurate values can be obtained from a subroutine written by Venables (in press) on the basis of the algorithm derived by Venables (1973).

*Critical Values of Distribution of* gcr

$$s = 2, \quad \alpha = .05(.01)$$

| N / M | -.5 | 0 | 1 | 2 | 3 | 4 | 5 | 6 | 7 | 8 | 9 | 10 | 15 |
|---|---|---|---|---|---|---|---|---|---|---|---|---|---|
| 1 | .858 | .894 | .930 | .947 | .958 | .965 | .970 | .973 | .976 | .979 | .981 | .982 | .987 |
|  | (.937) | (.954) | (.970) | (.977) | (.982) | (.985) | (.987) | (.989) | (.990) | (.991) | (.992) | (.992) | (.995) |
| 3 | .638 | .702 | .776 | .820 | .849 | .870 | .885 | .898 | .908 | .916 | .922 | .928 | .948 |
|  | (.764) | (.808) | (.858) | (.886) | (.905) | (.919) | (.929) | (.936) | (.943) | (.948) | (.952) | (.956) | (.968) |
| 5 | .498 | .565 | .651 | .706 | .746 | .776 | .799 | .818 | .834 | .847 | .858 | .868 | .901 |
|  | (.623) | (.677) | (.745) | (.787) | (.817) | (.839) | (.857) | (.871) | (.882) | (.892) | (.900) | (.907) | (.931) |
| 10 | .318 | .374 | .455 | .514 | .561 | .598 | .629 | .656 | .679 | .698 | .716 | .732 | .789 |
|  | (.418) | (.470) | (.544) | (.597) | (.638) | (.670) | (.697) | (.720) | (.739) | (.756) | (.771) | (.783) | (.831) |
| 15 | .232 | .278 | .348 | .402 | .446 | .483 | .515 | .542 | .567 | .589 | .609 | .627 | .696 |
|  | (.312) | (.357) | (.425) | (.476) | (.517) | (.551) | (.580) | (.606) | (.628) | (.648) | (.665) | (.681) | (.742) |
| 20 | .183 | .221 | .281 | .329 | .369 | .404 | .434 | .461 | .486 | .508 | .528 | .546 | .620 |
|  | (.249) | (.288) | (.347) | (.394) | (.433) | (.466) | (.495) | (.521) | (.543) | (.564) | (.583) | (.600) | (.667) |
| 25 | .151 | .184 | .236 | .278 | .314 | .346 | .375 | .401 | .424 | .446 | .465 | .484 | .558 |
|  | (.207) | (.240) | (.293) | (.336) | (.372) | (.403) | (.431) | (.456) | (.478) | (.499) | (.517) | (.535) | (.604) |
| 30 | .129 | .157 | .203 | .241 | .274 | .303 | .330 | .354 | .376 | .396 | .416 | .433 | .507 |
|  | (.177) | (.207) | (.254) | (.293) | (.326) | (.355) | (.381) | (.405) | (.427) | (.447) | (.465) | (.482) | (.552) |
| 40 | .099 | .122 | .159 | .190 | .218 | .243 | .266 | .287 | .306 | .325 | .342 | .359 | .428 |
|  | (.137) | (.161) | (.200) | (.232) | (.261) | (.286) | (.309) | (.331) | (.350) | (.369) | (.386) | (.402) | (.470) |
| 50 | .081 | .099 | .130 | .157 | .180 | .202 | .222 | .241 | .259 | .275 | .291 | .306 | .370 |
|  | (.112) | (.132) | (.165) | (.193) | (.217) | (.240) | (.260) | (.279) | (.297) | (.313) | (.329) | (.344) | (.408) |
| 100 | .042 | .052 | .069 | .084 | .097 | .110 | .122 | .134 | .145 | .155 | .166 | .176 | .220 |
|  | (.058) | (.069) | (.088) | (.104) | (.118) | (.132) | (.145) | (.157) | (.168) | (.179) | (.190) | (.200) | (.246) |
| 500 | .009 | .011 | .014 | .018 | .021 | .024 | .027 | .029 | .032 | .035 | .037 | .040 | .052 |
|  | (.012) | (.015) | (.019) | (.022) | (.026) | (.029) | (.032) | (.035) | (.038) | (.041) | (.043) | (.046) | (.059) |
| 1000 | .004 | .005 | .007 | .009 | .010 | .012 | .013 | .015 | .016 | .018 | .019 | .020 | .027 |
|  | (.006) | (.007) | (.010) | (.011) | (.013) | (.015) | (.016) | (.018) | (.019) | (.021) | (.022) | (.023) | (.030) |

## $s = 3,\quad \alpha = .05(.01)$

| N / M | -.5 | 0 | 1 | 2 | 3 | 4 | 5 | 6 | 7 | 8 | 9 | 10 | 15 |
|---|---|---|---|---|---|---|---|---|---|---|---|---|---|
| 1 | .922 | .938 | .956 | .965 | .972 | .976 | .979 | .982 | .984 | .985 | .986 | .987 | .991 |
|  | (.966) | (.973) | (.981) | (.985) | (.988) | (.990) | (.991) | (.992) | (.993) | (.994) | (.994) | (.995) | (.996) |
| 3 | .756 | .792 | .839 | .868 | .887 | .902 | .913 | .922 | .930 | .936 | .941 | .945 | .960 |
|  | (.844) | (.868) | (.898) | (.917) | (.930) | (.939) | (.946) | (.952) | (.956) | (.960) | (.963) | (.966) | (.975) |
| 5 | .625 | .669 | .729 | .770 | .800 | .822 | .840 | .855 | .867 | .877 | .886 | .894 | .920 |
|  | (.725) | (.758) | (.804) | (.834) | (.856) | (.873) | (.886) | (.897) | (.906) | (.913) | (.919) | (.925) | (.944) |
| 10 | .429 | .472 | .537 | .586 | .625 | .656 | .683 | .705 | .725 | .741 | .756 | .770 | .819 |
|  | (.520) | (.559) | (.616) | (.659) | (.692) | (.719) | (.742) | (.760) | (.777) | (.791) | (.803) | (.814) | (.854) |
| 15 | .324 | .362 | .422 | .469 | .508 | .541 | .569 | .594 | .616 | .635 | .653 | .669 | .730 |
|  | (.402) | (.438) | (.494) | (.537) | (.573) | (.603) | (.628) | (.651) | (.670) | (.688) | (.703) | (.717) | (.771) |
| 20 | .260 | .293 | .346 | .390 | .427 | .458 | .486 | .511 | .533 | .554 | .572 | .589 | .656 |
|  | (.327) | (.359) | (.410) | (.452) | (.487) | (.517) | (.543) | (.566) | (.587) | (.605) | (.622) | (.638) | (.698) |
| 25 | .218 | .246 | .294 | .333 | .367 | .397 | .424 | .448 | .470 | .490 | .508 | .525 | .594 |
|  | (.275) | (.303) | (.351) | (.389) | (.422) | (.451) | (.477) | (.500) | (.520) | (.539) | (.557) | (.573) | (.637) |
| 30 | .187 | .212 | .255 | .291 | .322 | .350 | .375 | .398 | .419 | .439 | .457 | .473 | .543 |
|  | (.237) | (.263) | (.306) | (.342) | (.373) | (.400) | (.425) | (.447) | (.467) | (.486) | (.503) | (.519) | (.585) |
| 40 | .146 | .166 | .201 | .232 | .259 | .283 | .305 | .326 | .345 | .363 | .379 | .395 | .462 |
|  | (.186) | (.207) | (.243) | (.274) | (.301) | (.326) | (.348) | (.368) | (.387) | (.405) | (.421) | (.436) | (.501) |
| 50 | .119 | .136 | .167 | .192 | .216 | .237 | .257 | .275 | .292 | .309 | .324 | .339 | .402 |
|  | (.153) | (.171) | (.202) | (.229) | (.253) | (.274) | (.294) | (.313) | (.330) | (.346) | (.361) | (.376) | (.438) |
| 100 | .063 | .072 | .089 | .104 | .118 | .131 | .143 | .155 | .166 | .177 | .187 | .197 | .242 |
|  | (.081) | (.091) | (.109) | (.125) | (.140) | (.153) | (.166) | (.178) | (.189) | (.200) | (.211) | (.221) | (.267) |
| 500 | .013 | .015 | .019 | .022 | .026 | .029 | .031 | .034 | .037 | .040 | .043 | .045 | .058 |
|  | (.017) | (.019) | (.023) | (.027) | (.031) | (.034) | (.037) | (.040) | (.043) | (.046) | (.049) | (.052) | (.064) |
| 1000 | .007 | .008 | .010 | .011 | .013 | .015 | .016 | .018 | .019 | .020 | .022 | .023 | .030 |
|  | (.009) | (.010) | (.012) | (.014) | (.015) | (.017) | (.019) | (.020) | (.022) | (.023) | (.025) | (.026) | (.033) |

## $s = 4,\quad \alpha = .05(.01)$

| N / M | -.5 | 0 | 1 | 2 | 3 | 4 | 5 | 6 | 7 | 8 | 9 | 10 | 15 |
|---|---|---|---|---|---|---|---|---|---|---|---|---|---|
| 1 | .950 | .959 | .969 | .975 | .979 | .982 | .985 | .986 | .988 | .989 | .990 | .991 | .993 |
|  | (.979) | (.982) | (.987) | (.990) | (.991) | (.993) | (.993) | (.994) | (.995) | (.995) | (.996) | (.996) | (.997) |
| 3 | .824 | .846 | .877 | .898 | .912 | .923 | .931 | .938 | .944 | .948 | .952 | .956 | .967 |
|  | (.888) | (.903) | (.923) | (.936) | (.945) | (.952) | (.957) | (.962) | (.965) | (.968) | (.971) | (.973) | (.980) |
| 5 | .708 | .739 | .782 | .813 | .836 | .854 | .868 | .880 | .889 | .898 | .905 | .911 | .933 |
|  | (.788) | (.811) | (.844) | (.866) | (.883) | (.896) | (.906) | (.915) | (.922) | (.927) | (.933) | (.937) | (.953) |
| 10 | .513 | .547 | .601 | .641 | .674 | .700 | .723 | .742 | .759 | .773 | .786 | .798 | .840 |
|  | (.595) | (.625) | (.671) | (.706) | (.733) | (.756) | (.775) | (.791) | (.805) | (.817) | (.827) | (.837) | (.872) |
| 15 | .399 | .431 | .482 | .523 | .558 | .587 | .612 | .634 | .654 | .671 | .687 | .701 | .756 |
|  | (.472) | (.501) | (.549) | (.587) | (.618) | (.644) | (.667) | (.686) | (.703) | (.719) | (.733) | (.745) | (.793) |
| 20 | .326 | .354 | .402 | .441 | .474 | .503 | .529 | .552 | .572 | .591 | .607 | .623 | .684 |
|  | (.390) | (.417) | (.463) | (.500) | (.531) | (.558) | (.582) | (.603) | (.621) | (.638) | (.654) | (.668) | (.723) |
| 25 | .275 | .301 | .344 | .380 | .412 | .440 | .464 | .487 | .507 | .526 | .543 | .559 | .624 |
|  | (.332) | (.357) | (.399) | (.434) | (.465) | (.491) | (.515) | (.536) | (.555) | (.573) | (.589) | (.603) | (.663) |
| 30 | .238 | .261 | .301 | .334 | .364 | .390 | .414 | .435 | .455 | .474 | .491 | .507 | .572 |
|  | (.289) | (.312) | (.351) | (.384) | (.412) | (.438) | (.461) | (.482) | (.501) | (.518) | (.534) | (.549) | (.611) |
| 40 | .188 | .207 | .240 | .269 | .294 | .318 | .339 | .359 | .377 | .395 | .411 | .426 | .490 |
|  | (.229) | (.248) | (.282) | (.311) | (.336) | (.360) | (.381) | (.400) | (.418) | (.435) | (.450) | (.465) | (.527) |
| 50 | .155 | .171 | .199 | .224 | .247 | .268 | .287 | .305 | .322 | .338 | .353 | .367 | .428 |
|  | (.190) | (.206) | (.236) | (.261) | (.284) | (.305) | (.324) | (.342) | (.358) | (.374) | (.389) | (.403) | (.463) |
| 100 | .082 | .091 | .108 | .123 | .137 | .150 | .162 | .174 | .185 | .196 | .206 | .216 | .261 |
|  | (.102) | (.112) | (.129) | (.145) | (.159) | (.172) | (.185) | (.197) | (.208) | (.219) | (.230) | (.240) | (.285) |
| 500 | .017 | .020 | .023 | .027 | .030 | .033 | .036 | .039 | .042 | .045 | .048 | .050 | .063 |
|  | (.022) | (.024) | (.028) | (.032) | (.035) | (.039) | (.042) | (.045) | (.048) | (.051) | (.054) | (.056) | (.070) |
| 1000 | .009 | .010 | .012 | .014 | .015 | .017 | .018 | .020 | .021 | .023 | .024 | .026 | .032 |
|  | (.011) | (.012) | (.014) | (.016) | (.018) | (.020) | (.021) | (.023) | (.024) | (.026) | (.027) | (.029) | (.036) |

## $s = 5, \quad \alpha = .05(.01)$

| N / M | -.5 | 0 | 1 | 2 | 3 | 4 | 5 | 6 | 7 | 8 | 9 | 10 | 15 |
|---|---|---|---|---|---|---|---|---|---|---|---|---|---|
| 1 | .966 | .971 | .977 | .981 | .984 | .986 | .988 | .989 | .990 | .991 | .992 | .992 | .994 |
|  | (.986) | (.988) | (.990) | (.992) | (.993) | (.994) | (.995) | (.995) | (.996) | (.996) | (.997) | (.997) | (.998) |
| 3 | .866 | .881 | .903 | .918 | .929 | .937 | .944 | .949 | .953 | .957 | .960 | .963 | .972 |
|  | (.916) | (.926) | (.939) | (.949) | (.956) | (.961) | (.965) | (.968) | (.971) | (.973) | (.975) | (.977) | (.983) |
| 5 | .766 | .788 | .821 | .845 | .863 | .877 | .888 | .898 | .906 | .913 | .918 | .924 | .942 |
|  | (.832) | (.848) | (.872) | (.889) | (.902) | (.913) | (.921) | (.928) | (.933) | (.938) | (.943) | (.946) | (.959) |
| 10 | .580 | .607 | .651 | .685 | .713 | .735 | .755 | .771 | .786 | .799 | .810 | .820 | .857 |
|  | (.653) | (.676) | (.714) | (.743) | (.766) | (.785) | (.801) | (.815) | (.827) | (.837) | (.847) | (.855) | (.885) |
| 15 | .462 | .483 | .533 | .569 | .599 | .625 | .648 | .667 | .685 | .701 | .715 | .728 | .777 |
|  | (.530) | (.554) | (.595) | (.627) | (.655) | (.678) | (.698) | (.715) | (.731) | (.744) | (.757) | (.768) | (.811) |
| 20 | .383 | .407 | .449 | .485 | .515 | .542 | .565 | .586 | .604 | .621 | .637 | .651 | .708 |
|  | (.444) | (.468) | (.507) | (.540) | (.568) | (.593) | (.614) | (.633) | (.651) | (.666) | (.680) | (.693) | (.744) |
| 25 | .326 | .349 | .388 | .422 | .451 | .477 | .500 | .521 | .540 | .557 | .573 | .588 | .648 |
|  | (.382) | (.404) | (.442) | (.473) | (.501) | (.526) | (.547) | (.567) | (.585) | (.601) | (.616) | (.630) | (.685) |
| 30 | .284 | .305 | .341 | .373 | .400 | .425 | .448 | .468 | .487 | .504 | .520 | .535 | .597 |
|  | (.334) | (.355) | (.390) | (.421) | (.448) | (.471) | (.493) | (.512) | (.530) | (.546) | (.562) | (.576) | (.634) |
| 40 | .226 | .243 | .275 | .302 | .327 | .349 | .370 | .389 | .406 | .423 | .438 | .453 | .514 |
|  | (.268) | (.285) | (.317) | (.344) | (.368) | (.390) | (.410) | (.429) | (.446) | (.462) | (.477) | (.491) | (.550) |
| 50 | .187 | .202 | .230 | .254 | .276 | .296 | .315 | .332 | .348 | .364 | .378 | .392 | .451 |
|  | (.223) | (.239) | (.266) | (.290) | (.312) | (.332) | (.351) | (.368) | (.384) | (.399) | (.413) | (.427) | (.484) |
| 100 | .101 | .110 | .126 | .141 | .155 | .168 | .180 | .191 | .203 | .213 | .224 | .233 | .278 |
|  | (.122) | (.131) | (.148) | (.163) | (.177) | (.190) | (.203) | (.215) | (.226) | (.237) | (.247) | (.257) | (.302) |
| 500 | .022 | .024 | .027 | .031 | .034 | .038 | .041 | .044 | .047 | .049 | .052 | .055 | .068 |
|  | (.026) | (.028) | (.032) | (.036) | (.040) | (.043) | (.046) | (.049) | (.052) | (.055) | (.058) | (.061) | (.075) |
| 1000 | .011 | .012 | .014 | .016 | .017 | .019 | .021 | .022 | .024 | .025 | .027 | .028 | .035 |
|  | (.013) | (.014) | (.016) | (.018) | (.020) | (.022) | (.024) | (.025) | (.027) | (.028) | (.030) | (.031) | (.039) |

## $s = 6, \quad \alpha = .05(.01)$

| N / M | -.5 | 0 | 1 | 2 | 3 | 4 | 5 | 6 | 7 | 8 | 9 | 10 | 15 |
|---|---|---|---|---|---|---|---|---|---|---|---|---|---|
| 1 | .975 | .978 | .983 | .985 | .988 | .989 | .990 | .991 | .992 | .993 | .993 | .994 | .995 |
|  | (.989) | (.991) | (.993) | (.994) | (.995) | (.995) | (.996) | (.996) | (.997) | (.997) | (.997) | (.997) | (.998) |
| 3 | .895 | .906 | .922 | .933 | .941 | .948 | .953 | .957 | .961 | .964 | .966 | .968 | .976 |
|  | (.934) | (.941) | (.951) | (.958) | (.963) | (.968) | (.971) | (.973) | (.976) | (.978) | (.979) | (.980) | (.985) |
| 5 | .808 | .825 | .850 | .869 | .883 | .895 | .904 | .912 | .918 | .924 | .929 | .934 | .949 |
|  | (.863) | (.875) | (.893) | (.906) | (.917) | (.925) | (.932) | (.938) | (.942) | (.947) | (.950) | (.953) | (.964) |
| 10 | .633 | .655 | .692 | .721 | .744 | .764 | .781 | .795 | .808 | .819 | .829 | .838 | .871 |
|  | (.698) | (.717) | (.748) | (.772) | (.792) | (.809) | (.823) | (.834) | (.845) | (.854) | (.862) | (.869) | (.896) |
| 15 | .514 | .537 | .576 | .608 | .635 | .658 | .678 | .696 | .711 | .726 | .739 | .750 | .795 |
|  | (.578) | (.599) | (.633) | (.662) | (.686) | (.706) | (.724) | (.740) | (.754) | (.766) | (.777) | (.787) | (.826) |
| 20 | .432 | .454 | .491 | .523 | .551 | .575 | .596 | .615 | .632 | .648 | .662 | .676 | .728 |
|  | (.491) | (.511) | (.546) | (.576) | (.601) | (.623) | (.643) | (.660) | (.676) | (.690) | (.703) | (.715) | (.762) |
| 25 | .372 | .392 | .428 | .458 | .485 | .509 | .531 | .550 | .568 | .584 | .599 | .613 | .669 |
|  | (.426) | (.445) | (.479) | (.508) | (.533) | (.556) | (.576) | (.594) | (.611) | (.626) | (.640) | (.652) | (.704) |
| 30 | .326 | .345 | .378 | .407 | .433 | .457 | .478 | .497 | .514 | .531 | .546 | .560 | .618 |
|  | (.375) | (.394) | (.426) | (.454) | (.479) | (.501) | (.521) | (.539) | (.556) | (.572) | (.586) | (.599) | (.654) |
| 40 | .261 | .278 | .307 | .333 | .356 | .378 | .397 | .416 | .432 | .448 | .463 | .477 | .536 |
|  | (.303) | (.319) | (.348) | (.374) | (.397) | (.418) | (.437) | (.454) | (.471) | (.486) | (.500) | (.513) | (.570) |
| 50 | .218 | .232 | .258 | .281 | .302 | .322 | .340 | .357 | .372 | .387 | .401 | .414 | .472 |
|  | (.254) | (.269) | (.294) | (.318) | (.338) | (.358) | (.375) | (.392) | (.407) | (.422) | (.436) | (.448) | (.504) |
| 100 | .119 | .128 | .144 | .158 | .172 | .184 | .197 | .208 | .219 | .230 | .240 | .250 | .294 |
|  | (.140) | (.149) | (.166) | (.180) | (.194) | (.207) | (.220) | (.231) | (.242) | (.253) | (.263) | (.273) | (.318) |
| 500 | .026 | .028 | .031 | .035 | .039 | .042 | .045 | .048 | .051 | .054 | .057 | .059 | .073 |
|  | (.031) | (.033) | (.037) | (.041) | (.044) | (.047) | (.051) | (.054) | (.057) | (.060) | (.063) | (.066) | (.080) |
| 1000 | .013 | .014 | .016 | .018 | .020 | .021 | .023 | .024 | .026 | .028 | .029 | .031 | .037 |
|  | (.016) | (.017) | (.019) | (.021) | (.022) | (.024) | (.026) | (.028) | (.029) | (.031) | (.032) | (.034) | (.041) |

## Critical Values of Distribution of gcr (continued)

### s = 7,   α = .05(.01)

| N / M | -.5 | 0 | 1 | 2 | 3 | 4 | 5 | 6 | 7 | 8 | 9 | 10 | 15 |
|---|---|---|---|---|---|---|---|---|---|---|---|---|---|
| 1 | .981 | .983 | .986 | .988 | .990 | .991 | .992 | .993 | .993 | .994 | .994 | .995 | .996 |
|  | (.992) | (.993) | (.994) | (.995) | (.996) | (.996) | (.997) | (.997) | (.997) | (.997) | (.998) | (.998) | (.998) |
| 3 | .915 | .923 | .935 | .944 | .950 | .956 | .960 | .963 | .966 | .969 | .971 | .973 | .979 |
|  | (.947) | (.952) | (.960) | (.965) | (.969) | (.972) | (.975) | (.977) | (.979) | (.981) | (.982) | (.983) | (.987) |
| 5 | .840 | .852 | .872 | .887 | .899 | .908 | .917 | .923 | .929 | .934 | .938 | .941 | .955 |
|  | (.885) | (.895) | (.909) | (.920) | (.928) | (.935) | (.941) | (.946) | (.950) | (.953) | (.956) | (.959) | (.968) |
| 10 | .677 | .695 | .726 | .750 | .771 | .788 | .802 | .815 | .826 | .836 | .845 | .853 | .882 |
|  | (.735) | (.751) | (.777) | (.797) | (.814) | (.828) | (.840) | (.851) | (.860) | (.868) | (.875) | (.882) | (.906) |
| 15 | .560 | .579 | .613 | .641 | .665 | .686 | .704 | .720 | .734 | .747 | .759 | .769 | .810 |
|  | (.619) | (.636) | (.667) | (.691) | (.713) | (.731) | (.747) | (.761) | (.773) | (.784) | (.794) | (.804) | (.839) |
| 20 | .475 | .494 | .528 | .557 | .582 | .604 | .624 | .641 | .657 | .671 | .685 | .697 | .745 |
|  | (.531) | (.549) | (.580) | (.607) | (.630) | (.650) | (.667) | (.684) | (.698) | (.711) | (.723) | (.734) | (.777) |
| 25 | .412 | .431 | .463 | .491 | .516 | .538 | .558 | .576 | .593 | .608 | .622 | .635 | .688 |
|  | (.464) | (.482) | (.512) | (.539) | (.563) | (.583) | (.602) | (.618) | (.634) | (.648) | (.661) | (.673) | (.721) |
| 30 | .364 | .381 | .412 | .439 | .463 | .485 | .505 | .523 | .540 | .555 | .569 | .583 | .638 |
|  | (.412) | (.429) | (.458) | (.484) | (.507) | (.528) | (.547) | (.564) | (.580) | (.594) | (.607) | (.620) | (.671) |
| 40 | .294 | .309 | .337 | .362 | .384 | .404 | .423 | .440 | .456 | .471 | .485 | .499 | .555 |
|  | (.336) | (.351) | (.378) | (.402) | (.423) | (.443) | (.461) | (.478) | (.493) | (.508) | (.521) | (.534) | (.588) |
| 50 | .247 | .260 | .285 | .307 | .327 | .346 | .363 | .379 | .395 | .409 | .422 | .435 | .491 |
|  | (.283) | (.296) | (.321) | (.343) | (.363) | (.381) | (.398) | (.414) | (.429) | (.443) | (.456) | (.469) | (.522) |
| 100 | .136 | .145 | .160 | .175 | .188 | .200 | .212 | .224 | .235 | .245 | .255 | .265 | .309 |
|  | (.158) | (.167) | (.182) | (.197) | (.211) | (.223) | (.236) | (.247) | (.258) | (.269) | (.279) | (.288) | (.332) |
| 500 | .030 | .032 | .036 | .039 | .042 | .046 | .049 | .052 | .055 | .058 | .061 | .064 | .077 |
|  | (.035) | (.037) | (.041) | (.045) | (.048) | (.052) | (.055) | (.058) | (.061) | (.064) | (.067) | (.070) | (.084) |
| 1000 | .015 | .016 | .018 | .020 | .022 | .023 | .025 | .027 | .028 | .030 | .031 | .033 | .040 |
|  | (.018) | (.019) | (.021) | (.023) | (.025) | (.026) | (.028) | (.030) | (.031) | (.033) | (.034) | (.036) | (.043) |

### s = 8,   α = .05(.01)

| N / M | -.5 | 0 | 1 | 2 | 3 | 4 | 5 | 6 | 7 | 8 | 9 | 10 | 15 |
|---|---|---|---|---|---|---|---|---|---|---|---|---|---|
| 1 | .985 | .986 | .989 | .990 | .992 | .992 | .993 | .994 | .994 | .995 | .995 | .995 | .997 |
|  | (.994) | (.994) | (.995) | (.996) | (.996) | (.997) | (.997) | (.997) | (.998) | (.998) | (.998) | (.998) | (.999) |
| 3 | .930 | .936 | .945 | .952 | .958 | .962 | .965 | .968 | .971 | .973 | .974 | .976 | .982 |
|  | (.957) | (.960) | (.966) | (.970) | (.974) | (.976) | (.979) | (.980) | (.982) | (.983) | (.984) | (.985) | (.989) |
| 5 | .864 | .874 | .890 | .902 | .912 | .920 | .927 | .932 | .937 | .941 | .945 | .948 | .959 |
|  | (.903) | (.910) | (.922) | (.931) | (.938) | (.943) | (.948) | (.952) | (.956) | (.958) | (.961) | (.963) | (.972) |
| 10 | .713 | .728 | .754 | .775 | .793 | .808 | .821 | .832 | .842 | .851 | .859 | .865 | .892 |
|  | (.766) | (.779) | (.800) | (.818) | (.833) | (.845) | (.855) | (.865) | (.873) | (.880) | (.886) | (.892) | (.914) |
| 15 | .598 | .615 | .645 | .670 | .692 | .710 | .727 | .741 | .754 | .766 | .776 | .786 | .824 |
|  | (.653) | (.669) | (.695) | (.717) | (.736) | (.752) | (.767) | (.779) | (.791) | (.801) | (.810) | (.818) | (.851) |
| 20 | .513 | .531 | .561 | .587 | .610 | .630 | .648 | .664 | .679 | .692 | .705 | .716 | .761 |
|  | (.567) | (.583) | (.610) | (.634) | (.655) | (.673) | (.690) | (.704) | (.718) | (.729) | (.740) | (.750) | (.791) |
| 25 | .449 | .466 | .495 | .521 | .544 | .565 | .583 | .600 | .616 | .630 | .643 | .655 | .705 |
|  | (.499) | (.515) | (.543) | (.567) | (.588) | (.607) | (.625) | (.640) | (.655) | (.668) | (.680) | (.691) | (.736) |
| 30 | .398 | .414 | .443 | .468 | .491 | .511 | .530 | .547 | .563 | .577 | .590 | .603 | .655 |
|  | (.445) | (.460) | (.488) | (.512) | (.533) | (.552) | (.570) | (.586) | (.601) | (.614) | (.627) | (.639) | (.687) |
| 40 | .325 | .339 | .365 | .388 | .409 | .428 | .446 | .463 | .478 | .493 | .506 | .519 | .573 |
|  | (.365) | (.379) | (.405) | (.427) | (.448) | (.467) | (.484) | (.500) | (.514) | (.528) | (.541) | (.553) | (.605) |
| 50 | .274 | .286 | .310 | .331 | .351 | .368 | .385 | .401 | .416 | .429 | .442 | .455 | .508 |
|  | (.310) | (.323) | (.345) | (.366) | (.385) | (.403) | (.419) | (.435) | (.449) | (.462) | (.475) | (.487) | (.539) |
| 100 | .153 | .161 | .176 | .190 | .203 | .216 | .228 | .239 | .250 | .260 | .270 | .279 | .323 |
|  | (.175) | (.184) | (.199) | (.213) | (.226) | (.239) | (.251) | (.262) | (.273) | (.283) | (.293) | (.302) | (.346) |
| 500 | .034 | .036 | .040 | .043 | .047 | .050 | .053 | .056 | .059 | .062 | .065 | .068 | .081 |
|  | (.039) | (.041) | (.045) | (.049) | (.052) | (.056) | (.059) | (.062) | (.065) | (.069) | (.072) | (.074) | (.088) |
| 1000 | .017 | .018 | .020 | .022 | .024 | .025 | .027 | .029 | .030 | .032 | .033 | .035 | .042 |
|  | (.020) | (.021) | (.023) | (.025) | (.027) | (.029) | (.030) | (.032) | (.034) | (.035) | (.037) | (.038) | (.046) |

## $s = 9, \quad \alpha = .05(.01)$

| N / M | -.5 | 0 | 1 | 2 | 3 | 4 | 5 | 6 | 7 | 8 | 9 | 10 | 15 |
|---|---|---|---|---|---|---|---|---|---|---|---|---|---|
| 1 | .988 | .989 | .991 | .992 | .993 | .994 | .994 | .995 | .995 | .995 | .996 | .996 | .997 |
|  | (.995) | (.995) | (.996) | (.997) | (.997) | (.997) | (.998) | (.998) | (.998) | (.998) | (.998) | (.998) | (.999) |
| 3 | .942 | .946 | .953 | .959 | .963 | .967 | .970 | .972 | .974 | .976 | .977 | .979 | .984 |
|  | (.964) | (.967) | (.971) | (.975) | (.977) | (.979) | (.981) | (.983) | (.984) | (.985) | (.986) | (.987) | (.990) |
| 5 | .883 | .891 | .904 | .914 | .922 | .929 | .935 | .939 | .944 | .947 | .950 | .953 | .963 |
|  | (.917) | (.923) | (.932) | (.939) | (.945) | (.950) | (.954) | (.957) | (.960) | (.963) | (.965) | (.967) | (.974) |
| 10 | .743 | .756 | .778 | .797 | .812 | .825 | .837 | .847 | .855 | .863 | .870 | .876 | .901 |
|  | (.791) | (.802) | (.820) | (.835) | (.848) | (.859) | (.868) | (.876) | (.884) | (.890) | (.896) | (.901) | (.920) |
| 15 | .632 | .647 | .674 | .696 | .715 | .732 | .747 | .760 | .771 | .782 | .792 | .801 | .835 |
|  | (.683) | (.697) | (.720) | (.740) | (.756) | (.771) | (.784) | (.795) | (.805) | (.815) | (.823) | (.831) | (.861) |
| 20 | .548 | .563 | .591 | .614 | .635 | .654 | .670 | .685 | .698 | .711 | .722 | .733 | .775 |
|  | (.598) | (.612) | (.637) | (.659) | (.678) | (.695) | (.709) | (.723) | (.735) | (.746) | (.756) | (.766) | (.803) |
| 25 | .482 | .497 | .525 | .549 | .570 | .589 | .606 | .622 | .636 | .650 | .662 | .673 | .720 |
|  | (.530) | (.544) | (.570) | (.592) | (.612) | (.630) | (.646) | (.660) | (.673) | (.686) | (.697) | (.707) | (.750) |
| 30 | .430 | .445 | .471 | .495 | .516 | .535 | .552 | .569 | .583 | .597 | .610 | .622 | .671 |
|  | (.475) | (.490) | (.515) | (.537) | (.557) | (.575) | (.591) | (.606) | (.620) | (.633) | (.645) | (.656) | (.702) |
| 40 | .353 | .366 | .391 | .413 | .433 | .451 | .468 | .484 | .499 | .512 | .525 | .538 | .590 |
|  | (.393) | (.406) | (.430) | (.451) | (.471) | (.488) | (.505) | (.520) | (.534) | (.547) | (.559) | (.571) | (.620) |
| 50 | .299 | .311 | .333 | .354 | .373 | .390 | .406 | .421 | .435 | .448 | .461 | .473 | .524 |
|  | (.335) | (.347) | (.369) | (.389) | (.407) | (.424) | (.439) | (.454) | (.468) | (.481) | (.493) | (.504) | (.554) |
| 100 | .169 | .177 | .192 | .206 | .219 | .231 | .242 | .253 | .264 | .274 | .284 | .293 | .336 |
|  | (.192) | (.200) | (.215) | (.229) | (.241) | (.254) | (.265) | (.276) | (.287) | (.297) | (.307) | (.316) | (.359) |
| 500 | .038 | .040 | .043 | .047 | .051 | .054 | .057 | .060 | .063 | .066 | .069 | .072 | .086 |
|  | (.043) | (.045) | (.049) | (.053) | (.056) | (.060) | (.063) | (.066) | (.070) | (.073) | (.076) | (.079) | (.093) |
| 1000 | .019 | .020 | .022 | .024 | .026 | .028 | .029 | .031 | .032 | .034 | .036 | .037 | .044 |
|  | (.022) | (.023) | (.025) | (.027) | (.029) | (.031) | (.032) | (.034) | (.036) | (.037) | (.039) | (.041) | (.048) |

## $s = 10, \quad \alpha = .05(.01)$

| N / M | -.5 | 0 | 1 | 2 | 3 | 4 | 5 | 6 | 7 | 8 | 9 | 10 | 15 |
|---|---|---|---|---|---|---|---|---|---|---|---|---|---|
| 1 | .990 | .991 | .992 | .993 | .994 | .994 | .995 | .995 | .996 | .996 | .996 | .997 | .997 |
|  | (.996) | (.996) | (.997) | (.997) | (.997) | (.998) | (.998) | (.998) | (.998) | (.998) | (.998) | (.999) | (.999) |
| 3 | .950 | .954 | .960 | .964 | .968 | .971 | .973 | .975 | .977 | .978 | .980 | .981 | .985 |
|  | (.969) | (.971) | (.975) | (.978) | (.980) | (.982) | (.983) | (.985) | (.986) | (.987) | (.988) | (.988) | (.991) |
| 5 | .898 | .905 | .916 | .924 | .931 | .937 | .941 | .946 | .949 | .952 | .955 | .958 | .967 |
|  | (.928) | (.933) | (.940) | (.946) | (.951) | (.955) | (.959) | (.962) | (.964) | (.967) | (.969) | (.970) | (.977) |
| 10 | .769 | .780 | .799 | .815 | .829 | .840 | .851 | .859 | .867 | .874 | .880 | .886 | .908 |
|  | (.812) | (.822) | (.837) | (.851) | (.862) | (.871) | (.880) | (.887) | (.893) | (.899) | (.904) | (.909) | (.926) |
| 15 | .662 | .675 | .699 | .719 | .736 | .751 | .764 | .776 | .787 | .797 | .806 | .814 | .846 |
|  | (.710) | (.721) | (.742) | (.759) | (.774) | (.788) | (.799) | (.810) | (.819) | (.827) | (.835) | (.842) | (.869) |
| 20 | .578 | .592 | .617 | .639 | .658 | .675 | .690 | .704 | .716 | .728 | .738 | .748 | .787 |
|  | (.626) | (.639) | (.661) | (.681) | (.698) | (.713) | (.727) | (.740) | (.751) | (.761) | (.771) | (.779) | (.814) |
| 25 | .512 | .526 | .551 | .573 | .593 | .611 | .627 | .642 | .655 | .667 | .679 | .690 | .734 |
|  | (.558) | (.571) | (.595) | (.615) | (.634) | (.650) | (.665) | (.678) | (.691) | (.702) | (.712) | (.722) | (.762) |
| 30 | .459 | .473 | .497 | .519 | .539 | .557 | .573 | .589 | .603 | .615 | .627 | .639 | .686 |
|  | (.503) | (.516) | (.539) | (.560) | (.579) | (.595) | (.611) | (.625) | (.638) | (.650) | (.661) | (.672) | (.715) |
| 40 | .380 | .392 | .415 | .436 | .455 | .473 | .489 | .504 | .518 | .531 | .543 | .555 | .605 |
|  | (.419) | (.431) | (.454) | (.474) | (.492) | (.509) | (.524) | (.539) | (.552) | (.564) | (.576) | (.587) | (.635) |
| 50 | .323 | .335 | .356 | .375 | .393 | .410 | .425 | .440 | .453 | .466 | .478 | .490 | .540 |
|  | (.359) | (.370) | (.391) | (.410) | (.427) | (.443) | (.458) | (.472) | (.485) | (.498) | (.510) | (.521) | (.569) |
| 100 | .185 | .193 | .207 | .220 | .233 | .245 | .256 | .267 | .278 | .287 | .297 | .306 | .348 |
|  | (.208) | (.215) | (.230) | (.243) | (.256) | (.268) | (.279) | (.290) | (.301) | (.311) | (.320) | (.329) | (.371) |
| 500 | .042 | .044 | .047 | .051 | .054 | .058 | .061 | .064 | .067 | .070 | .073 | .076 | .090 |
|  | (.047) | (.049) | (.053) | (.057) | (.061) | (.064) | (.067) | (.071) | (.074) | (.077) | (.080) | (.083) | (.097) |
| 1000 | .021 | .022 | .024 | .026 | .028 | .030 | .031 | .033 | .034 | .036 | .038 | .039 | .047 |
|  | (.024) | (.025) | (.027) | (.029) | (.031) | (.033) | (.034) | (.036) | (.038) | (.040) | (.041) | (.043) | (.050) |

## $s = 11, \quad \alpha = .05(.01)$

| N / M | −.5 | 0 | 1 | 2 | 3 | 4 | 5 | 6 | 7 | 8 | 9 | 10 | 15 |
|---|---|---|---|---|---|---|---|---|---|---|---|---|---|
| 1 | .992 | .992 | .993 | .994 | .995 | .995 | .996 | .996 | .996 | .997 | .997 | .997 | .998 |
| | (.996) | (.997) | (.997) | (.997) | (.998) | (.998) | (.998) | (.998) | (.998) | (.999) | (.999) | (.999) | (.999) |
| 3 | .957 | .960 | .965 | .969 | .972 | .974 | .976 | .978 | .979 | .981 | .982 | .983 | .987 |
| | (.974) | (.975) | (.978) | (.981) | (.982) | (.984) | (.985) | (.986) | (.987) | (.988) | (.989) | (.989) | (.992) |
| 5 | .911 | .916 | .925 | .932 | .938 | .943 | .947 | .951 | .954 | .957 | .959 | .961 | .970 |
| | (.937) | (.941) | (.947) | (.952) | (.956) | (.960) | (.963) | (.966) | (.968) | (.970) | (.971) | (.973) | (.979) |
| 10 | .791 | .800 | .817 | .831 | .843 | .854 | .863 | .870 | .877 | .884 | .889 | .895 | .915 |
| | (.830) | (.838) | (.852) | (.864) | (.874) | (.882) | (.889) | (.896) | (.902) | (.907) | (.911) | (.916) | (.931) |
| 15 | .688 | .700 | .721 | .739 | .754 | .768 | .780 | .791 | .801 | .810 | .818 | .825 | .855 |
| | (.733) | (.743) | (.761) | (.777) | (.790) | (.802) | (.813) | (.822) | (.831) | (.839) | (.845) | (.852) | (.877) |
| 20 | .606 | .618 | .641 | .661 | .678 | .694 | .708 | .721 | .732 | .743 | .753 | .762 | .798 |
| | (.651) | (.663) | (.683) | (.701) | (.717) | (.731) | (.743) | (.755) | (.765) | (.775) | (.783) | (.791) | (.824) |
| 25 | .540 | .552 | .576 | .596 | .615 | .631 | .646 | .660 | .672 | .684 | .695 | .705 | .746 |
| | (.584) | (.596) | (.617) | (.636) | (.653) | (.668) | (.682) | (.695) | (.706) | (.717) | (.727) | (.736) | (.774) |
| 30 | .486 | .499 | .521 | .542 | .561 | .577 | .593 | .607 | .620 | .633 | .644 | .655 | .699 |
| | (.529) | (.541) | (.562) | (.581) | (.599) | (.615) | (.629) | (.642) | (.655) | (.666) | (.677) | (.686) | (.728) |
| 40 | .405 | .417 | .438 | .458 | .476 | .492 | .508 | .522 | .536 | .548 | .560 | .571 | .619 |
| | (.443) | (.455) | (.476) | (.495) | (.512) | (.528) | (.542) | (.556) | (.569) | (.581) | (.592) | (.603) | (.648) |
| 50 | .346 | .357 | .377 | .396 | .413 | .429 | .444 | .458 | .470 | .483 | .495 | .506 | .554 |
| | (.381) | (.392) | (.411) | (.429) | (.446) | (.461) | (.476) | (.489) | (.502) | (.514) | (.525) | (.536) | (.582) |
| 100 | .200 | .208 | .222 | .235 | .247 | .259 | .270 | .281 | .291 | .301 | .310 | .319 | .360 |
| | (.223) | (.230) | (.244) | (.258) | (.270) | (.282) | (.293) | (.303) | (.314) | (.323) | (.333) | (.342) | (.383) |
| 500 | .046 | .048 | .051 | .055 | .058 | .062 | .065 | .068 | .071 | .074 | .077 | .080 | .094 |
| | (.052) | (.053) | (.057) | (.061) | (.064) | (.068) | (.071) | (.074) | (.078) | (.081) | (.084) | (.087) | (.101) |
| 1000 | .023 | .024 | .026 | .028 | .030 | .032 | .033 | .035 | .037 | .038 | .040 | .041 | .049 |
| | (.026) | (.027) | (.029) | (.031) | (.033) | (.035) | (.037) | (.038) | (.040) | (.042) | (.043) | (.045) | (.052) |

## $s = 12, \quad \alpha = .05(.01)$

| N / M | −.5 | 0 | 1 | 2 | 3 | 4 | 5 | 6 | 7 | 8 | 9 | 10 | 15 |
|---|---|---|---|---|---|---|---|---|---|---|---|---|---|
| 1 | .993 | .993 | .994 | .995 | .995 | .996 | .996 | .996 | .997 | .997 | .997 | .997 | .998 |
| | (.997) | (.997) | (.998) | (.998) | (.998) | (.998) | (.998) | (.998) | (.999) | (.999) | (.999) | (.999) | (.999) |
| 3 | .963 | .965 | .969 | .972 | .975 | .977 | .979 | .980 | .981 | .983 | .984 | .984 | .988 |
| | (.977) | (.979) | (.981) | (.983) | (.984) | (.986) | (.987) | (.988) | (.989) | (.989) | (.990) | (.990) | (.993) |
| 5 | .921 | .926 | .933 | .939 | .944 | .949 | .952 | .956 | .958 | .961 | .963 | .965 | .972 |
| | (.944) | (.948) | (.953) | (.957) | (.961) | (.964) | (.966) | (.969) | (.971) | (.972) | (.974) | (.975) | (.980) |
| 10 | .810 | .818 | .833 | .845 | .856 | .865 | .873 | .880 | .886 | .892 | .897 | .902 | .920 |
| | (.846) | (.853) | (.865) | (.875) | (.884) | (.891) | (.898) | (.904) | (.909) | (.914) | (.918) | (.921) | (.936) |
| 15 | .711 | .722 | .740 | .757 | .771 | .783 | .794 | .804 | .813 | .822 | .829 | .836 | .864 |
| | (.753) | (.762) | (.778) | (.792) | (.805) | (.815) | (.825) | (.834) | (.842) | (.849) | (.855) | (.861) | (.885) |
| 20 | .631 | .642 | .663 | .681 | .697 | .711 | .724 | .736 | .747 | .757 | .766 | .774 | .809 |
| | (.674) | (.684) | (.703) | (.719) | (.733) | (.746) | (.758) | (.769) | (.778) | (.787) | (.795) | (.802) | (.833) |
| 25 | .565 | .577 | .598 | .617 | .634 | .649 | .664 | .677 | .688 | .699 | .709 | .719 | .758 |
| | (.607) | (.618) | (.638) | (.656) | (.671) | (.685) | (.698) | (.710) | (.721) | (.731) | (.740) | (.749) | (.784) |
| 30 | .511 | .523 | .544 | .563 | .581 | .596 | .611 | .625 | .637 | .648 | .659 | .669 | .712 |
| | (.552) | (.563) | (.583) | (.601) | (.617) | (.632) | (.646) | (.658) | (.670) | (.680) | (.690) | (.700) | (.739) |
| 40 | .428 | .439 | .460 | .478 | .496 | .511 | .526 | .540 | .552 | .564 | .576 | .587 | .632 |
| | (.466) | (.477) | (.496) | (.514) | (.531) | (.546) | (.560) | (.573) | (.585) | (.596) | (.607) | (.617) | (.660) |
| 50 | .368 | .378 | .397 | .415 | .431 | .447 | .461 | .474 | .487 | .499 | .510 | .521 | .567 |
| | (.402) | (.412) | (.431) | (.448) | (.464) | (.479) | (.493) | (.506) | (.518) | (.529) | (.540) | (.550) | (.595) |
| 100 | .215 | .222 | .236 | .249 | .261 | .272 | .283 | .293 | .303 | .313 | .323 | .331 | .372 |
| | (.238) | (.245) | (.259) | (.271) | (.283) | (.295) | (.306) | (.316) | (.326) | (.336) | (.345) | (.354) | (.394) |
| 500 | .050 | .052 | .055 | .059 | .062 | .065 | .069 | .072 | .075 | .078 | .081 | .084 | .098 |
| | (.055) | (.057) | (.061) | (.065) | (.068) | (.072) | (.075) | (.078) | (.082) | (.085) | (.088) | (.091) | (.105) |
| 1000 | .025 | .026 | .028 | .030 | .032 | .034 | .035 | .037 | .039 | .040 | .042 | .043 | .051 |
| | (.028) | (.029) | (.031) | (.033) | (.035) | (.037) | (.039) | (.041) | (.042) | (.044) | (.045) | (.047) | (.055) |

## s = 13,  α = .05(.01)

| N / M | -.5 | 0 | 1 | 2 | 3 | 4 | 5 | 6 | 7 | 8 | 9 | 10 | 15 |
|---|---|---|---|---|---|---|---|---|---|---|---|---|---|
| 1 | .994 | .994 | .995 | .995 | .996 | .996 | .997 | .997 | .997 | .997 | .997 | .998 | .998 |
|  | (.997) | (.998) | (.998) | (.998) | (.998) | (.998) | (.999) | (.999) | (.999) | (.999) | (.999) | (.999) | (.999) |
| 3 | .967 | .969 | .973 | .975 | .977 | .979 | .981 | .982 | .983 | .984 | .985 | .986 | .989 |
|  | (.980) | (.981) | (.983) | (.985) | (.986) | (.987) | (.988) | (.989) | (.990) | (.990) | (.991) | (.991) | (.993) |
| 5 | .930 | .934 | .940 | .945 | .950 | .953 | .957 | .959 | .962 | .964 | .966 | .968 | .974 |
|  | (.950) | (.953) | (.958) | (.961) | (.965) | (.967) | (.969) | (.971) | (.973) | (.975) | (.976) | (.977) | (.982) |
| 10 | .826 | .833 | .846 | .857 | .867 | .875 | .882 | .889 | .895 | .900 | .905 | .909 | .925 |
|  | (.860) | (.865) | (.876) | (.885) | (.893) | (.899) | (.906) | (.911) | (.915) | (.919) | (.923) | (.927) | (.940) |
| 15 | .732 | .741 | .758 | .773 | .786 | .797 | .807 | .816 | .825 | .833 | .839 | .846 | .871 |
|  | (.771) | (.779) | (.794) | (.806) | (.818) | (.827) | (.836) | (.844) | (.851) | (.858) | (.864) | (.869) | (.891) |
| 20 | .653 | .664 | .683 | .699 | .714 | .727 | .740 | .750 | .760 | .770 | .778 | .786 | .818 |
|  | (.694) | (.703) | (.720) | (.735) | (.749) | (.760) | (.771) | (.781) | (.790) | (.798) | (.806) | (.813) | (.841) |
| 25 | .588 | .599 | .619 | .636 | .652 | .667 | .680 | .692 | .703 | .713 | .723 | .732 | .769 |
|  | (.629) | (.639) | (.657) | (.673) | (.688) | (.701) | (.713) | (.724) | (.734) | (.743) | (.752) | (.760) | (.794) |
| 30 | .534 | .545 | .565 | .583 | .599 | .614 | .628 | .641 | .652 | .663 | .674 | .683 | .723 |
|  | (.574) | (.584) | (.603) | (.619) | (.635) | (.648) | (.661) | (.673) | (.684) | (.694) | (.704) | (.712) | (.750) |
| 40 | .450 | .461 | .480 | .498 | .514 | .529 | .543 | .556 | .568 | .580 | .591 | .601 | .645 |
|  | (.487) | (.497) | (.516) | (.532) | (.548) | (.562) | (.576) | (.588) | (.600) | (.611) | (.621) | (.631) | (.672) |
| 50 | .388 | .398 | .417 | .433 | .449 | .464 | .478 | .490 | .502 | .514 | .525 | .535 | .580 |
|  | (.422) | (.432) | (.450) | (.466) | (.481) | (.495) | (.509) | (.521) | (.533) | (.544) | (.554) | (.564) | (.607) |
| 100 | .229 | .236 | .250 | .262 | .274 | .285 | .296 | .306 | .316 | .325 | .334 | .343 | .383 |
|  | (.252) | (.259) | (.272) | (.285) | (.297) | (.308) | (.319) | (.329) | (.339) | (.348) | (.357) | (.366) | (.405) |
| 500 | .053 | .055 | .059 | .063 | .066 | .069 | .073 | .076 | .079 | .082 | .085 | .088 | .102 |
|  | (.059) | (.061) | (.065) | (.069) | (.072) | (.076) | (.079) | (.082) | (.085) | (.088) | (.092) | (.094) | (.109) |
| 1000 | .027 | .028 | .030 | .032 | .034 | .036 | .037 | .039 | .041 | .042 | .044 | .045 | .053 |
|  | (.031) | (.031) | (.033) | (.035) | (.037) | (.039) | (.041) | (.042) | (.044) | (.046) | (.047) | (.049) | (.057) |

## s = 14,  α = .05(.01)

| N / M | -.5 | 0 | 1 | 2 | 3 | 4 | 5 | 6 | 7 | 8 | 9 | 10 | 15 |
|---|---|---|---|---|---|---|---|---|---|---|---|---|---|
| 1 | .995 | .995 | .995 | .996 | .996 | .997 | .997 | .997 | .997 | .997 | .998 | .998 | .998 |
|  | (.998) | (.998) | (.998) | (.998) | (.998) | (.999) | (.999) | (.999) | (.999) | (.999) | (.999) | (.999) | (.999) |
| 3 | .971 | .973 | .975 | .978 | .980 | .981 | .982 | .984 | .985 | .986 | .986 | .987 | .990 |
|  | (.982) | (.983) | (.985) | (.986) | (.987) | (.988) | (.989) | (.990) | (.991) | (.991) | (.992) | (.992) | (.994) |
| 5 | .937 | .940 | .946 | .950 | .954 | .958 | .960 | .963 | .965 | .967 | .969 | .970 | .976 |
|  | (.956) | (.958) | (.962) | (.965) | (.968) | (.970) | (.972) | (.974) | (.975) | (.977) | (.978) | (.979) | (.983) |
| 10 | .841 | .847 | .858 | .868 | .877 | .884 | .891 | .896 | .902 | .906 | .911 | .915 | .930 |
|  | (.871) | (.877) | (.886) | (.894) | (.901) | (.907) | (.912) | (.917) | (.921) | (.925) | (.928) | (.932) | (.944) |
| 15 | .751 | .759 | .774 | .788 | .799 | .810 | .819 | .827 | .835 | .842 | .848 | .854 | .878 |
|  | (.787) | (.794) | (.808) | (.819) | (.829) | (.838) | (.846) | (.854) | (.860) | (.866) | (.872) | (.877) | (.897) |
| 20 | .674 | .683 | .701 | .716 | .730 | .742 | .753 | .763 | .773 | .781 | .789 | .797 | .827 |
|  | (.713) | (.721) | (.737) | (.750) | (.762) | (.773) | (.783) | (.792) | (.801) | (.809) | (.816) | (.822) | (.849) |
| 25 | .610 | .620 | .638 | .654 | .669 | .682 | .695 | .706 | .717 | .726 | .735 | .744 | .779 |
|  | (.649) | (.658) | (.675) | (.689) | (.703) | (.715) | (.727) | (.737) | (.746) | (.755) | (.763) | (.771) | (.802) |
| 30 | .556 | .566 | .584 | .601 | .616 | .631 | .644 | .656 | .667 | .677 | .687 | .696 | .734 |
|  | (.594) | (.604) | (.621) | (.636) | (.651) | (.664) | (.676) | (.687) | (.697) | (.707) | (.716) | (.724) | (.759) |
| 40 | .471 | .481 | .499 | .516 | .531 | .546 | .559 | .572 | .583 | .594 | .605 | .615 | .656 |
|  | (.507) | (.516) | (.534) | (.550) | (.564) | (.578) | (.591) | (.603) | (.614) | (.624) | (.634) | (.643) | (.683) |
| 50 | .408 | .417 | .435 | .451 | .466 | .480 | .493 | .506 | .517 | .528 | .539 | .549 | .592 |
|  | (.441) | (.450) | (.467) | (.483) | (.497) | (.511) | (.524) | (.535) | (.547) | (.557) | (.567) | (.577) | (.618) |
| 100 | .243 | .250 | .263 | .275 | .287 | .298 | .308 | .318 | .328 | .337 | .346 | .355 | .394 |
|  | (.266) | (.273) | (.286) | (.298) | (.309) | (.320) | (.331) | (.341) | (.350) | (.360) | (.368) | (.377) | (.416) |
| 500 | .057 | .059 | .063 | .066 | .070 | .073 | .076 | .079 | .083 | .085 | .088 | .091 | .105 |
|  | (.063) | (.065) | (.069) | (.073) | (.076) | (.080) | (.083) | (.086) | (.089) | (.092) | (.095) | (.098) | (.113) |
| 1000 | .029 | .030 | .032 | .034 | .036 | .038 | .039 | .041 | .043 | .044 | .046 | .047 | .055 |
|  | (.032) | (.033) | (.035) | (.037) | (.039) | (.041) | (.043) | (.044) | (.046) | (.048) | (.050) | (.051) | (.059) |

## $s = 15, \quad \alpha = .05(.01)$

| N / M | -.5 | 0 | 1 | 2 | 3 | 4 | 5 | 6 | 7 | 8 | 9 | 10 | 15 |
|---|---|---|---|---|---|---|---|---|---|---|---|---|---|
| 1 | .995 | .996 | .996 | .996 | .997 | .997 | .997 | .997 | .998 | .998 | .998 | .998 | .998 |
|  | (.998) | (.998) | (.998) | (.998) | (.999) | (.999) | (.999) | (.999) | (.999) | (.999) | (.999) | (.999) | (.999) |
| 3 | .974 | .976 | .978 | .980 | .981 | .983 | .984 | .985 | .986 | .987 | .987 | .988 | .991 |
|  | (.984) | (.985) | (.986) | (.988) | (.989) | (.989) | (.990) | (.991) | (.991) | (.992) | (.992) | (.993) | (.994) |
| 5 | .943 | .946 | .951 | .955 | .958 | .961 | .964 | .966 | .968 | .969 | .971 | .972 | .978 |
|  | (.960) | (.962) | (.965) | (.968) | (.970) | (.973) | (.974) | (.976) | (.977) | (.979) | (.980) | (.981) | (.984) |
| 10 | .853 | .859 | .869 | .878 | .885 | .892 | .898 | .903 | .908 | .913 | .917 | .920 | .934 |
|  | (.882) | (.886) | (.895) | (.902) | (.908) | (.913) | (.918) | (.923) | (.926) | (.930) | (.933) | (.936) | (.947) |
| 15 | .767 | .775 | .789 | .801 | .812 | .821 | .830 | .837 | .844 | .851 | .857 | .862 | .884 |
|  | (.802) | (.808) | (.820) | (.831) | (.840) | (.848) | (.855) | (.862) | (.868) | (.874) | (.879) | (.883) | (.902) |
| 20 | .693 | .701 | .717 | .731 | .744 | .755 | .766 | .775 | .784 | .792 | .800 | .806 | .835 |
|  | (.729) | (.737) | (.751) | (.764) | (.775) | (.785) | (.795) | (.803) | (.811) | (.818) | (.825) | (.831) | (.856) |
| 25 | .629 | .639 | .656 | .671 | .685 | .697 | .709 | .719 | .729 | .738 | .747 | .755 | .788 |
|  | (.667) | (.675) | (.691) | (.705) | (.717) | (.729) | (.739) | (.749) | (.758) | (.766) | (.774) | (.781) | (.811) |
| 30 | .576 | .585 | .603 | .618 | .633 | .646 | .658 | .669 | .680 | .690 | .699 | .708 | .744 |
|  | (.613) | (.622) | (.638) | (.652) | (.666) | (.678) | (.689) | (.700) | (.710) | (.719) | (.727) | (.735) | (.769) |
| 40 | .490 | .500 | .517 | .533 | .548 | .562 | .574 | .586 | .597 | .608 | .618 | .627 | .668 |
|  | (.526) | (.534) | (.551) | (.566) | (.580) | (.593) | (.605) | (.616) | (.627) | (.637) | (.646) | (.655) | (.693) |
| 50 | .427 | .436 | .452 | .468 | .482 | .496 | .508 | .520 | .531 | .542 | .552 | .562 | .604 |
|  | (.459) | (.468) | (.484) | (.499) | (.513) | (.526) | (.538) | (.549) | (.560) | (.570) | (.580) | (.589) | (.629) |
| 100 | .257 | .263 | .276 | .288 | .299 | .310 | .320 | .330 | .339 | .349 | .357 | .366 | .404 |
|  | (.280) | (.286) | (.299) | (.311) | (.322) | (.333) | (.343) | (.353) | (.362) | (.371) | (.379) | (.388) | (.426) |
| 500 | .061 | .063 | .067 | .070 | .073 | .077 | .080 | .083 | .086 | .089 | .092 | .095 | .109 |
|  | (.067) | (.069) | (.073) | (.076) | (.080) | (.083) | (.087) | (.090) | (.093) | (.096) | (.099) | (.102) | (.116) |
| 1000 | .031 | .032 | .034 | .036 | .038 | .040 | .041 | .043 | .045 | .046 | .048 | .049 | .057 |
|  | (.034) | (.036) | (.038) | (.039) | (.041) | (.043) | (.045) | (.047) | (.048) | (.050) | (.052) | (.053) | (.061) |

## $s = 16, \quad \alpha = .05(.01)$

| N / M | -.5 | 0 | 1 | 2 | 3 | 4 | 5 | 6 | 7 | 8 | 9 | 10 | 15 |
|---|---|---|---|---|---|---|---|---|---|---|---|---|---|
| 1 | .996 | .996 | .996 | .997 | .997 | .997 | .997 | .998 | .998 | .998 | .998 | .998 | .999 |
|  | (.998) | (.998) | (.998) | (.999) | (.999) | (.999) | (.999) | (.999) | (.999) | (.999) | (.999) | (.999) | (.999) |
| 3 | .977 | .978 | .980 | .982 | .983 | .984 | .985 | .986 | .987 | .988 | .988 | .989 | .991 |
|  | (.986) | (.986) | (.988) | (.989) | (.990) | (.990) | (.991) | (.992) | (.992) | (.992) | (.993) | (.993) | (.995) |
| 5 | .949 | .951 | .955 | .958 | .962 | .964 | .966 | .968 | .970 | .972 | .973 | .974 | .979 |
|  | (.964) | (.966) | (.969) | (.971) | (.973) | (.975) | (.976) | (.978) | (.979) | (.980) | (.981) | (.982) | (.986) |
| 10 | .865 | .870 | .879 | .886 | .893 | .899 | .905 | .910 | .914 | .918 | .922 | .925 | .938 |
|  | (.891) | (.895) | (.902) | (.909) | (.914) | (.919) | (.924) | (.927) | (.931) | (.934) | (.937) | (.940) | (.950) |
| 15 | .782 | .789 | .802 | .813 | .823 | .831 | .839 | .846 | .853 | .859 | .865 | .870 | .890 |
|  | (.815) | (.821) | (.831) | (.841) | (.849) | (.857) | (.864) | (.870) | (.875) | (.881) | (.885) | (.890) | (.907) |
| 20 | .710 | .718 | .732 | .745 | .757 | .768 | .778 | .786 | .794 | .802 | .809 | .815 | .842 |
|  | (.745) | (.752) | (.765) | (.777) | (.787) | (.796) | (.805) | (.813) | (.820) | (.827) | (.833) | (.839) | (.862) |
| 25 | .648 | .656 | .672 | .686 | .699 | .711 | .722 | .732 | .741 | .750 | .758 | .765 | .796 |
|  | (.683) | (.691) | (.706) | (.719) | (.730) | (.741) | (.751) | (.760) | (.769) | (.776) | (.783) | (.790) | (.819) |
| 30 | .594 | .603 | .619 | .634 | .648 | .660 | .672 | .683 | .693 | .702 | .711 | .719 | .754 |
|  | (.630) | (.638) | (.653) | (.667) | (.680) | (.691) | (.702) | (.712) | (.721) | (.730) | (.738) | (.745) | (.777) |
| 40 | .509 | .518 | .534 | .549 | .563 | .576 | .589 | .600 | .611 | .621 | .630 | .639 | .678 |
|  | (.543) | (.552) | (.567) | (.582) | (.595) | (.607) | (.619) | (.629) | (.640) | (.649) | (.658) | (.667) | (.703) |
| 50 | .444 | .453 | .469 | .483 | .497 | .510 | .522 | .534 | .545 | .555 | .565 | .574 | .615 |
|  | (.476) | (.485) | (.500) | (.514) | (.528) | (.540) | (.552) | (.563) | (.573) | (.583) | (.592) | (.601) | (.640) |
| 100 | .270 | .276 | .289 | .300 | .311 | .322 | .332 | .342 | .351 | .360 | .368 | .377 | .414 |
|  | (.293) | (.299) | (.311) | (.323) | (.334) | (.344) | (.354) | (.364) | (.373) | (.382) | (.390) | (.398) | (.436) |
| 500 | .065 | .067 | .070 | .074 | .077 | .081 | .084 | .087 | .090 | .093 | .096 | .099 | .113 |
|  | (.071) | (.073) | (.077) | (.080) | (.084) | (.087) | (.090) | (.094) | (.097) | (.100) | (.103) | (.106) | (.120) |
| 1000 | .033 | .034 | .036 | .038 | .040 | .042 | .043 | .045 | .047 | .048 | .050 | .052 | .059 |
|  | (.037) | (.038) | (.040) | (.042) | (.043) | (.045) | (.047) | (.049) | (.050) | (.052) | (.053) | (.055) | (.063) |

## $s = 17, \quad \alpha = .05(.01)$

| N / M | -.5 | 0 | 1 | 2 | 3 | 4 | 5 | 6 | 7 | 8 | 9 | 10 | 15 |
|---|---|---|---|---|---|---|---|---|---|---|---|---|---|
| 1 | .996 | .996 | .997 | .997 | .997 | .998 | .998 | .998 | .998 | .998 | .998 | .998 | .999 |
|  | (.998) | (.998) | (.999) | (.999) | (.999) | (.999) | (.999) | (.999) | (.999) | (.999) | (.999) | (.999) | (.999) |
| 3 | .979 | .980 | .982 | .983 | .985 | .986 | .987 | .987 | .988 | .989 | .989 | .990 | .992 |
|  | (.987) | (.988) | (.989) | (.990) | (.990) | (.991) | (.992) | (.992) | (.993) | (.993) | (.993) | (.994) | (.995) |
| 5 | .953 | .955 | .959 | .962 | .965 | .967 | .969 | .971 | .972 | .974 | .975 | .976 | .981 |
|  | (.967) | (.969) | (.971) | (.973) | (.975) | (.977) | (.978) | (.979) | (.980) | (.981) | (.982) | (.983) | (.986) |
| 10 | .875 | .879 | .887 | .894 | .900 | .906 | .911 | .915 | .919 | .923 | .926 | .929 | .941 |
|  | (.899) | (.903) | (.909) | (.915) | (.920) | (.925) | (.928) | (.932) | (.935) | (.938) | (.941) | (.943) | (.953) |
| 15 | .796 | .802 | .813 | .824 | .833 | .841 | .848 | .855 | .861 | .866 | .872 | .876 | .896 |
|  | (.826) | (.832) | (.842) | (.850) | (.858) | (.865) | (.871) | (.877) | (.882) | (.887) | (.891) | (.895) | (.912) |
| 20 | .726 | .733 | .746 | .759 | .769 | .779 | .788 | .797 | .804 | .811 | .818 | .824 | .849 |
|  | (.759) | (.765) | (.777) | (.788) | (.798) | (.807) | (.815) | (.822) | (.829) | (.835) | (.841) | (.846) | (.868) |
| 25 | .665 | .672 | .687 | .700 | .712 | .724 | .734 | .743 | .752 | .760 | .768 | .775 | .804 |
|  | (.699) | (.706) | (.719) | (.732) | (.743) | (.753) | (.762) | (.771) | (.778) | (.786) | (.793) | (.799) | (.826) |
| 30 | .612 | .620 | .635 | .649 | .662 | .674 | .685 | .695 | .705 | .713 | .722 | .729 | .762 |
|  | (.646) | (.654) | (.668) | (.681) | (.693) | (.704) | (.714) | (.723) | (.732) | (.740) | (.748) | (.755) | (.785) |
| 40 | .526 | .535 | .550 | .565 | .578 | .591 | .602 | .613 | .623 | .633 | .642 | .651 | .688 |
|  | (.560) | (.568) | (.583) | (.596) | (.609) | (.621) | (.632) | (.642) | (.652) | (.661) | (.669) | (.677) | (.712) |
| 50 | .461 | .469 | .484 | .499 | .512 | .524 | .536 | .547 | .558 | .568 | .577 | .586 | .625 |
|  | (.493) | (.500) | (.515) | (.529) | (.542) | (.553) | (.565) | (.575) | (.585) | (.595) | (.604) | (.612) | (.650) |
| 100 | .283 | .289 | .301 | .312 | .323 | .333 | .343 | .353 | .362 | .370 | .379 | .387 | .424 |
|  | (.306) | (.312) | (.323) | (.335) | (.345) | (.356) | (.365) | (.375) | (.384) | (.392) | (.401) | (.409) | (.445) |
| 500 | .069 | .070 | .074 | .077 | .081 | .084 | .087 | .091 | .094 | .097 | .100 | .103 | .116 |
|  | (.075) | (.077) | (.081) | (.084) | (.087) | (.091) | (.094) | (.097) | (.101) | (.104) | (.107) | (.110) | (.124) |
| 1000 | .035 | .036 | .038 | .040 | .042 | .043 | .045 | .047 | .049 | .050 | .052 | .053 | .061 |
|  | (.039) | (.040) | (.042) | (.043) | (.045) | (.047) | (.049) | (.051) | (.052) | (.054) | (.055) | (.057) | (.065) |

## $s = 18, \quad \alpha = .05(.01)$

| N / M | -.5 | 0 | 1 | 2 | 3 | 4 | 5 | 6 | 7 | 8 | 9 | 10 | 15 |
|---|---|---|---|---|---|---|---|---|---|---|---|---|---|
| 1 | .997 | .997 | .997 | .997 | .998 | .998 | .998 | .998 | .998 | .998 | .998 | .998 | .999 |
|  | (.999) | (.999) | (.999) | (.999) | (.999) | (.999) | (.999) | (.999) | (.999) | (.999) | (.999) | (.999) | (.999) |
| 3 | .981 | .982 | .983 | .985 | .986 | .987 | .988 | .988 | .989 | .990 | .990 | .991 | .992 |
|  | (.988) | (.989) | (.990) | (.991) | (.991) | (.992) | (.992) | (.993) | (.993) | (.994) | (.994) | (.994) | (.995) |
| 5 | .957 | .959 | .962 | .965 | .967 | .969 | .971 | .973 | .974 | .976 | .977 | .978 | .982 |
|  | (.970) | (.971) | (.973) | (.975) | (.977) | (.979) | (.980) | (.981) | (.982) | (.983) | (.984) | (.984) | (.987) |
| 10 | .883 | .887 | .895 | .901 | .907 | .912 | .916 | .920 | .924 | .927 | .930 | .933 | .945 |
|  | (.906) | (.909) | (.916) | (.921) | (.925) | (.929) | (.933) | (.936) | (.939) | (.942) | (.944) | (.947) | (.956) |
| 15 | .808 | .814 | .824 | .834 | .842 | .849 | .856 | .862 | .868 | .873 | .878 | .882 | .900 |
|  | (.837) | (.842) | (.851) | (.859) | (.866) | (.872) | (.878) | (.883) | (.888) | (.893) | (.897) | (.901) | (.916) |
| 20 | .740 | .747 | .759 | .771 | .781 | .790 | .798 | .806 | .813 | .820 | .826 | .832 | .855 |
|  | (.772) | (.778) | (.789) | (.799) | (.808) | (.816) | (.823) | (.830) | (.837) | (.843) | (.848) | (.853) | (.874) |
| 25 | .680 | .688 | .701 | .714 | .725 | .736 | .745 | .754 | .762 | .770 | .777 | .784 | .812 |
|  | (.713) | (.720) | (.732) | (.744) | (.754) | (.763) | (.772) | (.780) | (.788) | (.795) | (.801) | (.807) | (.833) |
| 30 | .628 | .636 | .650 | .663 | .676 | .687 | .697 | .707 | .716 | .724 | .732 | .740 | .771 |
|  | (.661) | (.669) | (.682) | (.694) | (.705) | (.716) | (.725) | (.734) | (.742) | (.750) | (.757) | (.764) | (.793) |
| 40 | .543 | .551 | .566 | .579 | .592 | .604 | .615 | .625 | .635 | .645 | .654 | .662 | .698 |
|  | (.575) | (.583) | (.597) | (.610) | (.622) | (.633) | (.644) | (.654) | (.663) | (.671) | (.680) | (.688) | (.721) |
| 50 | .477 | .485 | .500 | .513 | .526 | .538 | .549 | .560 | .570 | .579 | .589 | .597 | .635 |
|  | (.508) | (.516) | (.530) | (.543) | (.555) | (.566) | (.577) | (.587) | (.597) | (.606) | (.615) | (.623) | (.659) |
| 100 | .295 | .301 | .313 | .324 | .334 | .344 | .354 | .364 | .372 | .381 | .389 | .397 | .434 |
|  | (.318) | (.324) | (.335) | (.346) | (.357) | (.367) | (.376) | (.385) | (.394) | (.403) | (.411) | (.419) | (.455) |
| 500 | .072 | .074 | .078 | .081 | .084 | .088 | .091 | .094 | .097 | .100 | .103 | .106 | .120 |
|  | (.079) | (.081) | (.084) | (.088) | (.091) | (.094) | (.098) | (.101) | (.104) | (.107) | (.110) | (.113) | (.127) |
| 1000 | .037 | .038 | .040 | .042 | .044 | .045 | .047 | .049 | .051 | .052 | .054 | .055 | .063 |
|  | (.041) | (.042) | (.043) | (.045) | (.047) | (.049) | (.051) | (.052) | (.054) | (.056) | (.058) | (.059) | (.067) |

$$s = 19, \quad \alpha = .05(.01)$$

| N / M | -.5 | 0 | 1 | 2 | 3 | 4 | 5 | 6 | 7 | 8 | 9 | 10 | 15 |
|---|---|---|---|---|---|---|---|---|---|---|---|---|---|
| 1 | .997 | .997 | .997 | .998 | .998 | .998 | .998 | .998 | .998 | .998 | .998 | .999 | .999 |
|  | (.999) | (.999) | (.999) | (.999) | (.999) | (.999) | (.999) | (.999) | (.999) | (.999) | (.999) | (.999) | (.999) |
| 3 | .983 | .984 | .985 | .986 | .987 | .988 | .989 | .989 | .990 | .990 | .991 | .991 | .993 |
|  | (.989) | (.990) | (.991) | (.991) | (.992) | (.992) | (.993) | (.993) | (.994) | (.994) | (.994) | (.995) | (.996) |
| 5 | .961 | .962 | .965 | .968 | .970 | .971 | .973 | .975 | .976 | .613 | .978 | .464 | .530 |
|  | (.972) | (.974) | (.976) | (.977) | (.979) | (.980) | (.981) | (.982) | (.983) | (.618) | (.985) | (.474) | (.534) |
| 10 | .891 | .895 | .902 | .907 | .913 | .917 | .921 | .925 | .928 | .931 | .934 | .937 | .947 |
|  | (.913) | (.916) | (.921) | (.926) | (.930) | (.934) | (.937) | (.940) | (.943) | (.945) | (.948) | (.949) | (.958) |
| 15 | .820 | .825 | .834 | .843 | .851 | .857 | .864 | .869 | .875 | .879 | .884 | .888 | .905 |
|  | (.847) | (.851) | (.859) | (.867) | (.873) | (.879) | (.885) | (.889) | (.894) | (.898) | (.902) | (.905) | (.920) |
| 20 | .753 | .760 | .771 | .782 | .791 | .800 | .808 | .815 | .822 | .828 | .833 | .839 | .861 |
|  | (.784) | (.789) | (.800) | (.809) | (.817) | (.825) | (.832) | (.838) | (.844) | (.850) | (.855) | (.859) | (.879) |
| 25 | .695 | .702 | .715 | .726 | .737 | .747 | .756 | .764 | .772 | .779 | .786 | .792 | .819 |
|  | (.727) | (.733) | (.744) | (.755) | (.765) | (.774) | (.782) | (.789) | (.796) | (.803) | (.809) | (.815) | (.839) |
| 30 | .644 | .651 | .664 | .677 | .688 | .699 | .708 | .718 | .726 | .734 | .742 | .749 | .779 |
|  | (.676) | (.682) | (.695) | (.706) | (.717) | (.727) | (.736) | (.744) | (.752) | (.759) | (.766) | (.773) | (.800) |
| 40 | .559 | .566 | .580 | .593 | .605 | .617 | .627 | .637 | .647 | .656 | .664 | .672 | .707 |
|  | (.590) | (.597) | (.611) | (.623) | (.635) | (.645) | (.655) | (.665) | (.674) | (.682) | (.690) | (.697) | (.729) |
| 50 | .493 | .500 | .514 | .527 | .539 | .551 | .562 | .572 | .582 | .591 | .600 | .608 | .645 |
|  | (.523) | (.530) | (.543) | (.556) | (.568) | (.579) | (.589) | (.599) | (.608) | (.617) | (.625) | (.633) | (.668) |
| 100 | .307 | .313 | .324 | .335 | .346 | .355 | .365 | .374 | .383 | .391 | .399 | .407 | .443 |
|  | (.330) | (.336) | (.347) | (.358) | (.368) | (.377) | (.387) | (.396) | (.404) | (.413) | (.421) | (.428) | (.464) |
| 500 | .076 | .078 | .081 | .085 | .088 | .091 | .094 | .098 | .101 | .104 | .107 | .110 | .124 |
|  | (.083) | (.084) | (.088) | (.092) | (.095) | (.098) | (.102) | (.105) | (.108) | (.111) | (.114) | (.117) | (.131) |
| 1000 | .039 | .040 | .042 | .044 | .046 | .047 | .049 | .051 | .052 | .054 | .056 | .057 | .065 |
|  | (.042) | (.043) | (.045) | (.047) | (.049) | (.051) | (.053) | (.054) | (.056) | (.058) | (.060) | (.061) | (.069) |

$$s = 20, \quad \alpha = .05(.01)$$

| N / M | -.5 | 0 | 1 | 2 | 3 | 4 | 5 | 6 | 7 | 8 | 9 | 10 | 15 |
|---|---|---|---|---|---|---|---|---|---|---|---|---|---|
| 1 | .997 | .997 | .998 | .998 | .998 | .998 | .998 | .998 | .998 | .998 | .999 | .999 | .999 |
|  | (.999) | (.999) | (.999) | (.999) | (.999) | (.999) | (.999) | (.999) | (.999) | (.999) | (.999) | (.999) | (1.00) |
| 3 | .984 | .985 | .986 | .987 | .988 | .989 | .989 | .990 | .990 | .991 | .991 | .992 | .993 |
|  | (.990) | (.991) | (.991) | (.992) | (.993) | (.993) | (.993) | (.994) | (.994) | (.994) | (.995) | (.995) | (.996) |
| 5 | .964 | .965 | .968 | .970 | .972 | .974 | .975 | .976 | .488 | .979 | .396 | .980 | .984 |
|  | (.975) | (.976) | (.977) | (.979) | (.980) | (.981) | (.982) | (.983) | (.494) | (.985) | (.405) | (.986) | (.989) |
| 10 | .899 | .902 | .908 | .913 | .918 | .922 | .926 | .929 | .932 | .935 | .938 | .940 | .614 |
|  | (.918) | (.921) | (.926) | (.930) | (.934) | (.938) | (.941) | (.943) | (.946) | (.948) | (.950) | (.952) | (.619) |
| 15 | .830 | .835 | .844 | .851 | .858 | .865 | .871 | .876 | .881 | .885 | .889 | .893 | .909 |
|  | (.855) | (.860) | (.867) | (.874) | (.880) | (.885) | (.890) | (.895) | (.899) | (.903) | (.906) | (.910) | (.923) |
| 20 | .766 | .772 | .782 | .792 | .801 | .809 | .816 | .823 | .829 | .835 | .841 | .846 | .867 |
|  | (.795) | (.800) | (.809) | (.818) | (.826) | (.833) | (.839) | (.845) | (.851) | (.856) | (.861) | (.865) | (.884) |
| 25 | .708 | .715 | .727 | .738 | .748 | .757 | .766 | .773 | .781 | .788 | .794 | .800 | .326 |
|  | (.739) | (.745) | (.756) | (.766) | (.775) | (.783) | (.791) | (.798) | (.804) | (.811) | (.817) | (.822) | (.329) |
| 30 | .658 | .665 | .677 | .689 | .700 | .710 | .719 | .728 | .736 | .744 | .751 | .758 | .305 |
|  | (.689) | (.695) | (.707) | (.718) | (.728) | (.737) | (.746) | (.753) | (.761) | (.768) | (.774) | (.781) | (.807) |
| 40 | .573 | .581 | .594 | .606 | .618 | .629 | .639 | .649 | .658 | .666 | .674 | .682 | .715 |
|  | (.604) | (.611) | (.624) | (.635) | (.646) | (.657) | (.666) | (.675) | (.684) | (.692) | (.699) | (.706) | (.737) |
| 50 | .507 | .514 | .528 | .540 | .552 | .563 | .573 | .583 | .593 | .602 | .610 | .618 | .654 |
|  | (.537) | (.544) | (.557) | (.569) | (.580) | (.590) | (.600) | (.610) | (.619) | (.627) | (.635) | (.643) | (.677) |
| 100 | .319 | .325 | .336 | .346 | .356 | .366 | .375 | .384 | .393 | .401 | .409 | .417 | .452 |
|  | (.342) | (.347) | (.358) | (.369) | (.378) | (.388) | (.397) | (.406) | (.414) | (.422) | (.430) | (.438) | (.472) |
| 500 | .080 | .081 | .085 | .088 | .092 | .095 | .098 | .101 | .104 | .107 | .110 | .113 | .127 |
|  | (.086) | (.088) | (.092) | (.095) | (.099) | (.102) | (.105) | (.108) | (.111) | (.115) | (.117) | (.120) | (.135) |
| 1000 | .041 | .042 | .044 | .046 | .048 | .049 | .051 | .053 | .054 | .056 | .058 | .059 | .067 |
|  | (.044) | (.046) | (.047) | (.049) | (.051) | (.053) | (.055) | (.056) | (.058) | (.060) | (.062) | (.063) | (.071) |

## Table A6   Natural Logarithms

The entries in this table are $\ln(x) = c$ such that $e^c = x$, where $e = 2.71828 \cdots$ is the base of the natural logarithms. Only values of $\ln(x)$ for $0 < x < 1$ are provided, since this is by far the most common case in statistical applications.

The natural logarithm of 0 is $-\infty$. Values of $\ln(x)$ for $x > 1.0$, and for three-place values of $x$, are available in almost any handbook of mathematical tables, or can be computed from tables of common logarithms (logarithms to the base 10) from the relationship $\ln(x) = 2.30258 \cdots \log_{10}(x)$.

*Natural Logarithms of Fractions*

| X | .00 | .01 | .02 | .03 | .04 | .05 | .06 | .07 | .08 | .09 |
|---|---|---|---|---|---|---|---|---|---|---|
| | | | | | LN(X) | | | | | |
| 0.0 | ******** | -4.6052 | -3.9120 | -3.5066 | -3.2189 | -2.9957 | -2.8134 | -2.6593 | -2.5257 | -2.4079 |
| 0.1 | -2.3026 | -2.2073 | -2.1203 | -2.0402 | -1.9661 | -1.8971 | -1.8326 | -1.7720 | -1.7148 | -1.6607 |
| 0.2 | -1.6094 | -1.5606 | -1.5141 | -1.4697 | -1.4271 | -1.3863 | -1.3471 | -1.3093 | -1.2730 | -1.2379 |
| 0.3 | -1.2040 | -1.1712 | -1.1394 | -1.1087 | -1.0788 | -1.0498 | -1.0217 | -0.9943 | -0.9676 | -0.9416 |
| 0.4 | -0.9163 | -0.8916 | -0.8675 | -0.8440 | -0.8210 | -0.7985 | -0.7765 | -0.7550 | -0.7340 | -0.7133 |
| 0.5 | -0.6931 | -0.6733 | -0.6539 | -0.6349 | -0.6162 | -0.5978 | -0.5798 | -0.5621 | -0.5447 | -0.5276 |
| 0.6 | -0.5108 | -0.4943 | -0.4780 | -0.4620 | -0.4463 | -0.4308 | -0.4155 | -0.4005 | -0.3857 | -0.3711 |
| 0.7 | -0.3567 | -0.3425 | -0.3285 | -0.3147 | -0.3011 | -0.2877 | -0.2744 | -0.2614 | -0.2485 | -0.2357 |
| 0.8 | -0.2231 | -0.2107 | -0.1985 | -0.1863 | -0.1744 | -0.1625 | -0.1508 | -0.1393 | -0.1278 | -0.1165 |
| 0.9 | -0.1054 | -0.0943 | -0.0834 | -0.0726 | -0.0619 | -0.0513 | -0.0408 | -0.0305 | -0.0202 | -0.0101 |

NATURAL LOGS OF SMALL FRACTIONS

| X | .001 | .002 | .003 | .004 | .005 | .006 | .007 | .008 | .009 |
|---|---|---|---|---|---|---|---|---|---|
| LN(X) | -6.9078 | -6.2146 | -5.8091 | -5.5215 | -5.2983 | -5.1160 | -4.9618 | -4.8283 | -4.7105 |

NATURAL LOGS FOR X NEAR 1.0

| X | .991 | .992 | .993 | .994 | .995 | .996 | .997 | .998 | .999 |
|---|---|---|---|---|---|---|---|---|---|
| LN(X) | -0.0090 | -0.0080 | -0.0070 | -0.0060 | -0.0050 | -0.0040 | -0.0030 | -0.0020 | -0.0010 |

# APPENDIX
## FORTRAN IV Program Listings

This Primer has stressed, where possible, the use of already available (and already debugged) computer programs such as those contained in the BMD, SPSS, and OMNITAB packages. The four programs whose listings are provided in this appendix, however, seemed to the author to perform functions which are not readily available through the most widely distributed packages. The first two—RJH004 and HEINV—allow the user to obtain inverses and determinants of $20 \times 20$ or smaller matrices, and to compute the characteristic roots and vectors of a matrix product of the form $E^{-1}H$, where $E$ and $H$ are symmetric matrices of size $50 \times 50$ or smaller. The third program, GCRCLC, permits computation of the significance level of a greatest characteristic root statistic for degree-of-freedom parameters not included in Table A.5. The fourth program, TINV, permits computation of critical values of Student's $t$-distribution for unusual levels of significance such as arise in applications of the Bonferroni approach to multiple comparisons.

All four programs make use of IBM Scientific Subroutine Package (SSP) subroutines. Most computing centers will either have these subroutines resident in the system "library" for direct call by any programmer using the computer or will have tape or card deck listings of the subroutines which can be easily copied by the user for insertion into his program. If not available in one of these two forms, the printed listings in the SSP manual can be used to punch card decks.

## Listing B1    RJH004 (Matrix Inversion)

The arrangement of cards for input to this program was described in Section D4.4.2.

```
      DIMENSION A(3600),L(60),M(60),TITLE(20)
    2 READ(5,501) N,(TITLE(I),I=1,19)
      IF(N-99) 5,105,105
    5 WRITE(6,503) (TITLE(I),I=1,19)
      WRITE(6,505)
      DO 10 I =1,N
      J1=N*(I-1)+1
      J2=J1+N-1
      READ(5,507) (A(J),J=J1,J2)
      WRITE(6,509) I, (A(J),J=J1,J2)
   10 CONTINUE
      CALL MINV(A,N,D,L,M)
      WRITE(6,511) D
      DO 20 I =1,N
      J1=N*(I-1)+1
      J2=J1+N-1
      WRITE(6,509) I, (A(J),J=J1,J2)
   20 CONTINUE
      GO TO 2
  105 STOP
  501 FORMAT(I2,2X,19A4)
  503 FORMAT(1H1,19A4)
  505 FORMAT(1H0,'INPUT MATRIX')
  507 FORMAT(10F8.0)
  509 FORMAT(1H ,I2,2X, 5E16.5/(1H ,4X,5E16.5))
  511 FORMAT(1H0,'DETERMINANT =',E16.5/1H0,'INVERSE OF INPUT MATRIX' )
      END
          INSERT IBM SSP SUBROUTINE MINV HERE
```

## Listing B2   HEINV (Characteristic Roots and Vectors of $E^{-1}H$)

Use of this program was described in Section D4.4.4.

```
   DIMENSION H(400), E(400), XL(20), X(400), TITLE(19), S(20)
 2 READ(5,501) M, (TITLE(I),I=1,19)
   IF(M-20) 3,3,50
 3 WRITE(6,601) M, (TITLE(I),I=1,19)
   WRITE(6,602)
   DO 5 I = 1,M
   K = M*(I-1) + 1
   KP = K + M - 1
   READ(5,503) (H(J),J=K,KP)
 5 WRITE(6,603) I, (H(J),J= K,KP)
   WRITE(6,604)
   DO 10 I = 1,M
   K = M*(I-1) + 1
   KP = K + M - 1
   READ(5,503) (E(J),J=K,KP)
10 WRITE(6,603) I, (E(J),J=K,KP)
   CALL NROOT(M,H,E,XL,X)
   WRITE(6,605) (XL(I),I=1,M)
   WRITE(6,609)
   DO 20 J = 1,M
   K = M*(J-1) + 1
   KP = K +M - 1
   WRITE(6,611) XL(J), (X(I), I = K,KP)
20 CONTINUE
   GO TO 2
50 WRITE(6,607)
   STOP
501 FORMAT(I3,1X,19A4)
503 FORMAT(10F8.0)
601 FORMAT(1H1,I3,1X,19A4)
602 FORMAT(1H0,'INPUT H, OR BETWEEN-GROUPS COVARIANCE MATRIX:')
603 FORMAT(1H ,I2,2X,5F20.5/(1H ,5F20.5) )
604 FORMAT(1H0,'INPUT E, OR WITHIN-GROUPS COVARIANCE MATRIX:')
605 FORMAT(1H0,'THE CHARACTERISTIC ROOTS OF HE-1 ARE'/(5E20.5) )
607 FORMAT(1H0,'M READ IS > 20. ALL DONE.' )
609 FORMAT(1H0,'CHARACTERISTIC VECTORS OF HE-1')
611 FORMAT(1H ,'FOR THE ROOT,',E20.5,' THE CHARACTERISTIC VECTOR IS'/
   1(5E20.5))
   END
        INSERT IBM SSP SUBROUTINE NROOT HERE
        INSERT IBM SSP SUBROUTINE EIGEN HERE
```

## Listing B3   GCRCLC (Percentiles of the Greatest Characteristic Root Distribution)

The input format for this program consists of one card for each test, with degree-of-freedom parameter $s$ punched into columns 1–5, right-justified; $m$ and $n$ into columns 6–10 and 11–15, respectively, with a decimal punched somewhere in the field; and the observed value of $\theta$ in columns 16–25.

```
  2 READ(9,901) IS, AM, AN, X
901 FORMAT(I5,2F5.0,F10.5)
    S = IS
    P = GCR(X,S,AM,AN)
    WRITE(6,601) X,S,AM,AN,P
601 FORMAT(1H ,'PR(THETA < ',F8.3,') FOR S,M,N = ',
   1 3F8.1,' = ',F10.5   )
    GO TO 2
    END
    FUNCTION GCR(THETA, SS, AM, AN)
    DOUBLE PRECISION S, M, N, PI, SGN
    DOUBLE PRECISION LOGR,R,COMB,PROD,BINCO,X,SUM,COEFK,TERMR,TERML
    DOUBLE PRECISION A1,A2,A3,A4,AI
C THIS SUBPROGRAM COMPUTES THE P-VALUE ASSOCIATED WITH A GCR PARAMETER
C OF THETA
    S = SS
    M = AM
    N = AN
    PI = 3.1415926535898
    IF(ABS(THETA) - 1.E-6) 201,201,205
201 WRITE(6,605) THETA
605 FORMAT(1H ,'THETA = ',F12.8,', WHICH IS SO CLOSE TO ZERO THAT GCR
   1HAS BEEN SET TO 0.0' )
    GCR = 0.0
    RETURN
205 IF(ABS(1.0 - THETA) - 1.E-6) 206,206,210
206 WRITE(6,607) THETA
607 FORMAT(1H ,'THETA = ',F12.8,', WHICH IS SO CLOSE TO ONE   THAT GCR
   1HAS BEEN SET TO 1.0' )
    GCR = 1.0
    RETURN
210 CONTINUE
    IM = M
    TM = 2*IM
    ADD = 0.0
    A = M
    B = N
    X = THETA
    IS = S
    IF(IS-1)  50,100,2
  2 LOGR = ALGAMI(M+N+S+1.0) - ALGAMI(M+S/2.0 + 0.5)
   1-ALGAMI(N+ S/2.0 + 0.5) + ALGAMI(M+N+S+0.5)
   2-ALGAMI(M+N+S/2.0 + 1.0) - ALGAMI(S/2.0) + 0.5*ALOG(PI)
    COEFK = 0.0
    BINCO = (X**(M+S))*((1.-X)**(N+1.))
    R = DEXP(LOGR)
    SUM = 0.0
    SGN = 1.0
    PROD = 1.0
    COMB = 1.0
```

```
      A1 = M + S + 1.0
      A2 = M + N + S + 1.0
      A3 = M + 0.5 + S/2.
      A4 = M + 0.5 + N + S
      ISM1 = IS - 1
      DO 20 I = 1,ISM1
      AI = I
      TERML = R*PROD*COMB/(A2-AI)
   15 TERMR = COEFK*(A1-AI)/(A2-AI)
      COEFK = TERML - TERMR
      BINCO = BINCO/X
      SGN = -SGN
      SUM = SUM + COEFK*BINCO*SGN
      PROD = PROD*(A3 - AI/2.)/(A4-AI/2.)
      COMB = COMB*(S-AI)/AI
   20 CONTINUE
      IF(2*(IS/2) - IS)    30,25,30
   25 GCR = SUM + 1.0
      RETURN
   30 CALL BDTR(X,A+1.0,N+1.0,P,D,IER)
      GCR = SUM + P
      RETURN
  100 CALL BDTR(X,A+1.0,N+1.0,P,D,IER)
      GCR = P
      RETURN
   50 WRITE(6,603) S
  603 FORMAT(' S = ',F8.3,',WHENCE NO FURTHER COMPUTATION OF P(GCR) IS P
     1OSSIBLE.'    )
      GCR = -9.0
      RETURN
      END
      DOUBLE PRECISION FUNCTION ALGAMI(X)
      DOUBLE PRECISION X, ALG
      CALL DLGAM(X,ALG,IER)
      ALGAMI = ALG
      RETURN
      END
           INSERT SSP SUBROUTINE NDTR HERE
           INSERT SSP SUBROUTINE CDTR HERE
           INSERT SSP SUBROUTINE BDTR HERE
           INSERT SSP SUBROUTINE DLGAM HERE
```

## Listing B4   TINV (Critical Values of Student's *t*-Distribution)

The input format for this program consists of one card for each test, with the degrees of freedom for the test punched into columns 1–5, right-justified, and the desired significance level punched into columns 6–15, with a decimal punched somewhere in the field. The critical value printed by the program is for a two-tailed test.

```
  2 READ(5,501) NDF, ALPHA
501 FORMAT(I5,F10.0)
    CALL TINV(NDF,1.-ALPHA,T)
    WRITE(6,601) ALPHA, NDF,T
601 FORMAT(1H0,1CRITICAL VALUE FOR ',F10.7,'-LEVEL SIGNIFICANT',
   1' STUDENT''S T WITH ',I5, ' DF = ',F10.3)
    GO TO 2
    END
    SUBROUTINE TINV(NDF, ALPHA, T)
    DF = NDF
    X = .00001
    STEP = 4.0*DF/(1.-ALPHA)
  2 R = DF/(DF+X*X)
    A = DF/2.
    B = .5
    CALL BDTR(R,A,B,P,D,IER)
    Z = 1. - P
    IF(ABS(Z-ALPHA) - .000004) 30,30,5
  5 IF(Z - ALPHA) 10,30,20
 10 X = X + STEP
    GO TO 2
 20 STEP = STEP/2.
    X = X - STEP
    GO TO 2
 30 T = X
    RETURN
    END
          INSERT SSP SUBROUTINE NDTR HERE
          INSERT SSP SUBROUTINE CDTR HERE
          INSERT SSP SUBROUTINE BDTR HERE
          INSERT SSP SUBROUTINE DLGAM HERE
```

# REFERENCES

ABRAMOWITZ, M. and STEGUN, I. A.  *Handbook of Mathematical Functions*. National Bureau of Standards, Applied Mathematics Series ✳ 55. Washington, D. C., 1964.

ANDERSON, N. H.  Scales and statistics: Parametric and non-parametric. *Psychological Bulletin* **58** (1961): 305–316.

ANDERSON, T. W.  The asymptotic distribution of certain characteristic roots and vectors. *Proceedings of the Second Berkeley Symposium on Mathematical Statistics and Probability*, pp. 103–130. Berkeley: Univ. of Calif. Press, 1951.

ANDERSON, T. W.  *Introduction to Multivariate Statistical Analysis*. New York: Wiley, 1958.

ANDERSON, T. W.  Asymptotic theory for principal component analysis. *Annals of Mathematical Statistics* **34** (1963): 122–148.

ANSCOMBE, F. J.  Graphs in statistical analysis. *American Statistician* **27** (1973): 17–21.

ATKINSON, R. C., BOWER, G., and CROTHERS, E. J.  *Introduction to Mathematical Learning Theory*. New York: Wiley, 1965.

BARTLETT, M. S.  Multivariate analysis. *Journal of the Royal Statistical Society, Series B* **9** (1947): 176–197.

BARTLETT, M. S.  A note on the multiplying factors for various chi-square approximations. *Journal of the Royal Statistical Society, Series B* **16** (1954): 296–298.

BELL, STOUGHTON.  *SKELETRAN*. Computing Center, Univ. of New Mexico, Albuquerque, 1968. (Mimeo)

BOCK, R. D.  *Multivariate Statistical Methods in Behavioral Research*. Chicago: R. D. Bock, in press.

BONEAU, C. A.  The effects of violations of assumptions underlying the *t* test. *Psychological Bulletin* **57** (1960): 49–64.

BOX, G. E. P.  A general distribution theory for a class of likelihood criteria. *Biometrika* **36** (1949): 317–346.

BOX, G. E. P.  Problems in the analysis of growth and wear curves. *Biometrics* **6** (1950): 362–389.

BROWN, R.  *Social Psychology*. Glencoe, Illinois: Free Press, 1965.

CARROLL, J. B.  An analytical solution for approximating simple structure in factor analysis. *Psychometrika* **18** (1953): 23–38.

CARROLL, J. D.  Generalization of canonical correlation analysis to three or more sets of variables. *Proceedings of the 76th Annual Convention, American Psychological Association*, pp. 227–228. American Psychological Association, 1968.

CHAMBERLAIN, R. I. and JOWETT, D.  *The Omnitab Programming System: A Guide for Users*. Ames, Iowa: Iowa State University Bookstore, 1970.

COCHRAN, W. G.  Some consequences when the assumptions for the analysis of variance are not satisfied. *Biometrics* **3** (1947): 22–38.

COHEN, B. P.   *Conflict and Conformity.* Cambridge, Mass.: M.I.T. Press, 1963.

COMREY, A. L.   *A First Course in Factor Analysis.* New York: Academic Press, 1973.

COOLEY, W. W. and LOHNES, P. R.   *Multivariate Data Analysis.* New York: Wiley, 1971.

COOPER, J. G. and LUFT, M.   *Statistical Programs for Behavioral Scientists.* College of Education, University of New Mexico, 1969. (mimeo)

COTTON, J. W.   *Elementary Statistical Theory for Behavior Scientists.* Reading, Massachusetts: Addison-Wesley, 1967.

CRAMÉR, H.   *Mathematical Methods of Statistics.* Princeton: Princeton Univ. Press, 1946.

DANIEL, C. and WOOD, F. S.   *Fitting Equations to Data.* New York: Wiley, 1971.

DARLINGTON, R. B.   Multiple regression in psychological research and practice. *Psychological Bulletin* **79** (1968): 161–182.

DAWES, ROBYN M., and CORRIGAN, B.   Linear models in decision making. *Psychological Bulletin* **81** (1974): 95–106.

DEMPSTER, A. P.   *Elements of Continuous Multivariate Analysis.* Menlo Park, Calif.: Addison-Wesley, 1969.

DEUTSCH, M.   The effect of motivational orientation upon threat and suspicion. *Human Relations* **13** (1960): 122–139.

DIXON, W. J. (ed.)   *BMD Biomedical Computer Programs.* Los Angeles, Calif.: UCLA, 1968.

DIXON, W. J. (ed.)   *BMD Biomedical Computer Programs, X-Series Supplement.* Los Angeles, Calif.: UCLA, 1969.

DIXON, W. J. (ed.)   *BMD Biomedical Computer Programs, BMDP-Series Supplement* (xerox). Los Angeles, Calif.: UCLA, 1971.

DONALDSON, T. S.   Robustness of the *F*-test to errors of both kinds and the correlation between the numerator and denominator of the *F*-ratio. *American Statistical Association Journal* **63** (1968): 660–676.

DRAPER, N. R. and SMITH, H.   *Applied Regression Analysis.* New York: Wiley, 1966.

DWYER, P. S. and MACPHAIL, M. S.   Symbolic matrix derivatives. *Annals of Mathematical Statistics* **19** (1948): 517–534.

ESTES, W. K.   Of models and men. *American Psychologist* **12** (1957): 609–617.

FERRARO, D. P. and BILLINGS, D. K.   Illicit drug use by college students. Three year trends, 1970–1972. *International Journal of the Addictions,* in press (a).

FERRARO, D. P. and BILLINGS, D. K.   Marihuana use by college students. Three year trends, 1970–1972. *International Journal of the Addictions,* in press (b).

FERGUSON, G. A.   The concept of parsimony in factor analysis. *Psychometrika* **19** (1954): 281–290.

FERGUSON, G. A.   *Statistical Analysis in Psychology and Education* (3rd ed.). New York: McGraw-Hill, 1971.

FINN, J. D.   *Multivariance: Univariate and Multivariate Analysis of Variance, Covariance, and Regression* (Version V). Ann Arbor, Mich.: National Educational Resources, Inc., 1973.

FLINT, Ronald A.   *The Relative Importance of Structure and Individual Differences in Determining Behavior in Two Person Games.* Unpublished doctoral dissertation, University of New Mexico, Albuquerque, 1970.

FRANKEL, M. R.   *Inference from Survey Samples: An Empirical Investigation.* Ann Arbor, Mich.: Univ. of Michigan Press, 1971.

GABRIEL, K. R.   A comparison of some methods of simultaneous inference in Manova. *In* Krishnaiah, P. R. (ed.), *Multivariate Analysis—II,* pp. 67–86. New York: Academic Press, 1969.

GABRIEL, K. R.   On the relation between union intersection and likelihood ratio tests. *In* Bose, R. C., Chakravarti, I. M., Mahalanobis, P. C., Rao, C. R., and Smith, K. J. C. (eds.), *Essays in Probability and Statistics,* pp. 251–266. Chapel Hill, N. C.: Univ. of N. Carolina Press, 1970.

GAMES, P.   Multiple comparisons of means. *American Educational Research Journal* **8** (1971): 531–565.

GLASS, G. V. and STANLEY, J. C.   *Statistical Methods in Education and Psychology.* Englewood Cliffs, New Jersey: Prentice-Hall, 1970.

GOODMAN, L. A., and KRUSKAL, W. H.   Measures of association for cross-classifications. *Journal of the American Statistical Association* **49** (1954): 732–764.

GREENHOUSE, S. W. and GEISSER, S.   On methods in the analysis of profile data. *Psychometrika* **24** (1959): 95–112.

GRICE, G. R.   Dependence of empirical laws upon the source of experimental variation. *Psychological Bulletin* **66** (1966): 488–498.

GUTTMAN, L.   Image theory for the structure of quantitative variates. *Psychometrika* **18** (1953): 277–296.

GUTTMAN, L.   The determining of factor score matrices with implications for five other basic problems of common-factor analysis. *British Journal of Statistical Psychology* **8** (1955): 65–81.

GUTTMAN, L.   "Best possible" systematic estimates of communalities. *Psychometrika* **21** (1956): 273–285.

HARMAN, H. H.   *Modern Factor Analysis* (2nd ed.). Chicago: Univ. of Chicago Press, 1967.

HARRIS, C. W.   Some Rao–Guttman relationships. *Psychometrika* **27** (1962): 247–263.

HARRIS, C. W.   *Problems in Measuring Change.* Madison: Univ. of Wisconsin Press, 1963.

HARRIS, R. J.   Deterministic nature of probabilistic choices among identifiable stimuli. *Journal of Experimental Psychology* **79** (1969): 552–560.

HARRIS, R. J.   Repeated-battery canonical correlation analysis. Unpublished manuscript, University of New Mexico, Albuquerque 1973a.

HARRIS, R. J.   *The Useless Power of Likelihood-Ratio Tests in Multivariate Analysis of Variance and Canonical Correlation.* Unpublished manuscript, University of New Mexico, Albuquerque, 1973b.

HARRIS, R. J.   Multivariate analysis of variance. Paper presented at meetings of American Educational Research Association. Chicago, Illinois, April 19, 1974a.

HARRIS, R. J.   *The Invalidity of Partitioned-U Tests of Individual Roots and Residuals therefrom in Multivariate Analysis of Variance and Canonical Correlation.* Unpublished manuscript, University of New Mexico, Albuquerque, 1974b.

HARRIS, R. J., FLINT, R. A., and EVERITT, G.   Directive versus non-directive instructions in the prisoner's dilemma. Paper presented at Rocky Mountain Psychological Association, Denver, Colorado, 1970.

HARRIS, R. J., WITTNER, W., KOPPELL, B., and HILF, F. D.   MMPI scales versus interviewer ratings of paranoia. *Psychological Reports* **27** (1970): 447–450.

HAYS, W. L. and WINKLER, R. L.   *Statistics: Probability, Inference, and Decision.* San Francisco: Holt, Rinehart and Winston, 1971.

HECK, D. L.   Charts of some upper percentage points of the distribution of the largest characteristic root. *Annals of Mathematical Statistics* **31** (1960): 625–642.

HORST, P.   Generalized canonical correlations and their applications to experimental data. *Journal of Clinical Psychology* (monograph supplement), **14** (1961a): 331–347.

HORST, P.   Relations among $m$ sets of measures. *Psychometrika* **27** (1961b): 129–149.

HORST, P.   *Matrix Algebra for Social Scientists.* New York: Holt, 1963.

HUYNH, H., and FELDT, L. S.   Conditions under which mean square ratios in repeated measurements designs have exact $F$-distributions. *Journal of the American Statistical Association* **65** (1970): 1582–1589.

INTERNATIONAL BUSINESS MACHINES COMPANY.   *IBM 29 Card Punch.* Manual ✳ A2433332, 1970.

ITO, K.   On the effect of heteroscedasticity and nonnormality upon some multivariate test procedures. *In* Krishnaiah, P. R. (ed.), *Multivariate Analysis—II*, pp. 87–120. New York: Academic Press, 1969.

ITO, K. and SCHULL, W. J.   On the robustness of the $T_0^2$ test in multivariate analysis of variance when variance–covariance matrices are not equal. *Biometrika* **51** (1964): 71–82.

JONES, L. V.   Analysis of variance in its multivariate developments. *In* Cattell, R. B. (ed.), *Handbook of Multivariate Experimental Psychology.* Chicago: Rand McNally, 1966.

JORESKÖG, K. G.   On the statistical treatment of residuals in factor analysis. *Psychometrika* **27** (1962): 335–354.

JORESKÖG, K. G.   Some contributions to maximum likelihood factor analysis. *Psychometrika* **32** (1967): 443–482.

JORESKÖG, K. G.   A general approach to confirmatory maximum likelihood factor analysis. *Psychometrika* **34** (1969): 183–202.

JORESKÖG, K. G.   A general method for analysis of covariance structures. *Biometrika* **57** (1970): 239–251.

JORESKÖG, K. G.   Analysis of covariance structures. *In* Krishnaiah, P. R. (ed.), *Multivariate Analysis—III*, pp. 263–285. New York: Academic Press, 1973.

JORESKÖG, K. G., VAN THILLO, MARIELLE, and GRUVAEUS, G. T. *ACOVSM: A General Computer Program for Analysis of Covariance Structures Including Generalized Manova.* Research Bulletin 71-1. Educational Testing Service, January, 1971.

KAISER, H. F.   *The Varimax Method of Factor Analysis.* Unpublished Ph.D. thesis, Univ. of Calif. at Berkeley, 1956.

KAISER, H. F.   The varimax criterion for analytic rotation in factor analysis. *Psychometrika* **23** (1958): 187–200.

KAISER, H. F.   The application of electronic computers to factor analysis. *Educational and Psychological Measurement* **20** (1960): 141–151.

KENDALL, M. G.   On the future of statistics: A second look. *Journal of the Royal Statistical Society, Series A* **131** (1968): 182–192.

KENDALL, M. G., and STUART, A.   *The Advanced Theory of Statistics, Vol. 2, Inference and Relationship.* London: Hafner, 1961. (rev. ed., 1968.)

KIRK, R. E.   *Experimental Design: Procedures for the Behavioral Sciences.* Belmont, Calif.: Brooks/Cole, 1968.

KIRK, R. E. (ed.)   *Statistical Issues: A Reader for the Behavioral Sciences.* Belmont, Calif.: Wadsworth, 1972.

KORIN, B. P.   Some comments on the homoscedasticity criterion, $M$, and the multivariate analysis of variance tests $T^2$, $W$ and $R$. *Biometrika* **59** (1972): 215–217.

KRISHNAIAH, P. R. (ed.)   *Multivariate Analysis—II.* New York: Academic Press, 1969.

KSHIRSAGAR, A. M.   *Multivariate Analysis.* New York: Dekker, 1972.

LANCASTER, H. O.   Canonical correlations and partitions of $\chi^2$. *Quarterly Journal of Mathematics* **14** (1963): 220–224.

LAWLEY, D. N.   The estimation of factor loadings by the method of maximum likelihood. *Proceedings of the Royal Society of Edinburgh* **60** (1940): 64–82.

LAWLEY, D. N.   On testing a set of correlation coefficients for equality. *Annals of Mathematical Statistics* **34** (1963): 149–151.

LAWLEY, D. N. and MAXWELL, A. E.   *Factor Analysis as a Statistical Method.* London: Butterworth & Co., 1963.

LINDQUIST, E. F.   *Design and Analysis of Experiments in Psychology and Education.* Boston, Mass.: Houghton-Mifflin, 1953.

LORD, F. M.   On the statistical treatment of football numbers. *American Psychologist* **8** (1953): 750–751.

MARKS, M. R.   Two kinds of regression weights that are better than betas in crossed samples. Paper read to Convention of the American Psychological Association. September, 1966.

McGUIRE, W. J.   The Yin and Yang of progress in social psychology: Sevan koan. *Journal of Personality and Social Psychology* **26** (1973): 446–456.

MEREDITH, W.   Canonical correlations with fallible data. *Psychometrika* **29** (1964): 55–66.

MORRISON, D. F.   *Multivariate Statistical Methods.* San Francisco, Calif.: McGraw-Hill, 1967.

MOSTELLER, F. and TUKEY, J. W.   Data analysis, including statistics, Chapter 10, pp. 80–203. *In* Lindzey, G. and Aronson, E. (eds.), *Handbook of Social Psychology* (2nd ed.), Vol. Two. Reading, Massachusetts: Addison-Wesley, 1968.

MULAIK, S. A.   *The Foundations of Factor Analysis.* San Francisco, Calif.: McGraw-Hill, 1972.

MYERS, J. L.   *Fundamentals of Experimental Design* (2nd ed.). Boston, Mass.: Allyn and Bacon, 1972.

NEWHAUS, J. and WRIGLEY, C.   The quartimax method: An analytical approach to orthogonal simple structure. *British Journal of Statistical Psychology* **7** (1954): 81–91.

NIE, Norman H., BENT, D. H., and HULL, C. H.   *SPSS: Statistical Package for the Social Sciences.* New York: McGraw-Hill, 1970.

NIE, N. H., HULL, C. H., KIM, J., and STEINBRENNER, K.   *SPSS Update Manual.* National Opinion Research Center, Chicago, Ill.: Univ. of Chicago, July, 1971. (mimeo)

NORRIS, R. C. and HJELM, H. F.   Non-normality and product moment correlation. *Journal of Experimental Education* **29** (1961): 261–270.

NORTON, D. W.   *An Empirical Investigation of Some Effects of Non-Normality and Heterogeneity on the F-Distribution.* Unpublished doctoral dissertation, State University of Iowa, Iowa City, 1952. As described *in* Lindquist, E. F. (1953, pp. 78–86).

OLKIN, I. and PRATT, J. W.   Unbiased estimation of certain correlation coefficients. *Annals of Mathematical Statistics* **29** (1958): 201–211.

OVERALL, J. and KLETT, C. J.   *Applied Multivariate Analysis.* San Francisco, Calif.: McGraw-Hill, 1972.

PILLAI, K. C. S.   On the distribution of the largest characteristic root of a matrix in multivariate analysis. *Biometrika* **52** (1965): 405–414.

PILLAI, K. C. S.   Upper percentage points of the largest root of a matrix in multivariate analysis. *Biometrika* **54** (1967): 189–194.

POTHOFF, R. F. and ROY, S. N.   A generalized multivariate analysis of variance model useful especially for growth curve problems. *Biometrika* **51** (1964): 313–326.

PRESS, S. J.   *Applied Multivariate Analysis.* San Francisco, Calif.: Holt, 1972.

PRUZEK, R. M.   Methods and problems in the analysis of multivariate data. *Review of Educational Research* **41** (1971): 163–190.

PRUZEK, R. M.   Small sample generalizability of factor analytic results. Paper presented at the Annual Research Meetings of American Educational Research Association. New Orleans, Louisiana, February, 1973.

RAO, C. R.   Some problems involving linear hypotheses in multivariate analysis. *Biometrika* **46** (1959): 49–58.

ROSENBERG, S.   Mathematical models of social behavior. *In* Lindzey, G., and Aronson, E. (eds.), *Handbook of Social Psychology* (2nd ed.), Vol. I, Chap. 3, pp. 245–319. Menlo Park, Calif.: Addison-Wesley, 1968.

ROY, S. N.   *Some Aspects of Multivariate Analysis.* New York: Wiley, 1957.

ROY, S. N. and BOSE, R. C.   On a heuristic method of test construction and its use in multivariate analysis. *Annals of Mathematical Statistics* **24** (1953): 220–238.

ROY, S. N. and BOSE, R. C.   Simultaneous confidence interval estimation. *Annals of Mathematical Statistics* **24** (1953): 513–536.

ROZEBOOM, W. W.   *Foundations of the Theory of Prediction.* Homewood, Ill.: Dorsey Press, 1966.

SAUNDERS, D. R.   An analytic method for rotation to orthogonal simple structure. Research Bulletin RB 53-10. Princeton, N. J.: Educational Testing Service, 1953.

SCHATZOFF, M.   Exact distributions of Wilk's likelihood ratio criterion. *Biometrika* **53** (1966a): 347–358.

SCHATZOFF, M.   Sensitivity comparisons among tests of the general linear hypothesis. *Journal of the American Statistical Association* **61** (1966b): 415–435.

SCHMIDT, F. L.   The relative efficiency of regression in simple unit predictor weights in applied differential psychology. *Educational and Psychological Measurement* **31** (1971): 699–714.

SEARLE, S. R.   *Matrix Algebra for the Biological Sciences.* New York: Wiley, 1966.

SELLTIZ, C., JAHODA, M., DEUTSCH, M., and COOK, S. W.   *Research Methods in Social Relations* (rev. ed.). New York: Holt, 1959.

SENDERS, V. L.   *Measurement and Statistics.* New York: Oxford, 1958.

SIEGEL, S.   *Nonparametric Statistics for the Behavioral Sciences.* New York: McGraw-Hill, 1956.

SKELLUM, J. G.   Models, inference, and strategy. *Biometrics* **25** (1969): 457–475.

SPEARMAN, C. and HOLZINGER, K. J.   Note on the sampling error of tetrad differences. *British Journal of Psychology* **16** (1925): 86–89.

STEELE, M. W. and TEDESCHI, J.   Matrix indices and strategy choices in mixed-motive games. *Journal of Conflict Resolution* **11** (1967): 198–205.

STEGER, J. A. (ed.)   *Readings in Statistics for the Behavioral Scientist.* Dallas, Texas: Holt, 1971.

STEVENS, S. S.   Mathematics, measurement, and psychophysics. *In* Stevens, S. S. (ed.). *Handbook of Experimental Psychology.* New York: Wiley, 1951.

STEVENS, S. S.   Measurement, statistics and the schemapiric view. *Science* **161** (1968): 849–856.

TATSUOKA, M. M.   *Validation Studies: The Use of Multiple Regression Equations.* Booklet ※ LB 322. Champaign, Ill. Institute for Personality and Ability Testing, 1971a.

TATSUOKA, M. M.   *Discriminant Analysis: The Study of Group Differences.* Booklet ※ LB 328, Champaign, Ill. Institute for Personality and Ability Testing, 1971b.

TATSUOKA, M. M.   *Multivariate Analysis: Techniques for Educational and Psychological Research.* New York: Wiley, 1971c.

TATSUOKA, M. M.   *Significance Tests: Univariate and Multivariate.* Booklet ※ LB 332. Champaign, Ill. Institute for Personality and Ability Testing, 1972.

THURSTONE, L. L.   *Multiple Factor Analysis.* Chicago: Univ. of Chicago Press, 1947.

TIMM, N.   *Multivariate Analysis with Applications in Education and Psychology.* New York: Brooks Cole, 1974.

TRYON, R. C. and BAILEY, D. E.   *Cluster Analysis.* New York: McGraw-Hill, 1970.

VAN DE GEER, J. P.   *Introduction to Multivariate Analysis for the Social Sciences.* San Francisco, Calif.: Freeman, 1971.

VENABLES, W. N.   Computation of the null distribution of the largest or smallest latent roots of a beta matrix. *Journal of Multivariate Analysis* **3** (1973): 125–131.

VENABLES, W. N.   Computer subroutines for computation of the null distribution of the largest or smallest latent roots of a beta matrix. *Journal of the Royal Statistical Society, Series C: Applied Statistics.* In press.

WALL, F. J.   *The generalized variance ratio of U-statistics.* Albuquerque, N. M.: The Dikewood Corporation, 1968.

WILKS, S. S.   Certain generalizations in the analysis of variance. *Biometrika* **24** (1932): 471–494.

WILKS, S. S.   On the independence of $k$ sets of normally distributed statistical variables. *Econometrica* **3** (1935): 309–326.

WINER, B. J.   *Statistical Principles in Experimental Design* (2nd ed.). San Francisco, Calif.: McGraw-Hill, 1971.

WISHART, J.   The mean and second moment coefficient of the multiple correlation coefficient in samples from a normal population. *Biometrika* **22** (1931): 353–361.

WISHNER, J.   Reanalysis of "Impressions of personality." *Psychological Review* **67** (1960): 96–112.

WRIGLEY, C.   Objectivity in factor analysis. *Educational and Psychological Measurement* **18** (1958): 463–476.

WRIGLEY, CHARLES.   An empirical question of various methods for the estimation of communalities. Contract report, No. 1. Berkeley, Calif.: Univ. of Calif., 1956.

# INDEX